Food biopolymers: Structural, functional and nutraceutical properties

Adil Gani • Bilal Ahmad Ashwar
Editors

Food biopolymers: Structural, functional and nutraceutical properties

Springer

Editors
Adil Gani
Department of Food Science and Technology
University of Kashmir
Srinagar, India

Bilal Ahmad Ashwar
Department of Food Science and Technology
University of Kashmir
Srinagar, India

ISBN 978-3-030-27060-5 ISBN 978-3-030-27061-2 (eBook)
https://doi.org/10.1007/978-3-030-27061-2

This Springer imprint is published by the registered company Springer Nature Switzerland AG
The registered company address is: Gewerbestrasse 11, 6330 Cham, Switzerland

Contents

Part I
Starch: Structure, Functions, Bioactivity and Applications

Starch: An Overview

Khalid Gul, Nisar Ahmad Mir, Basharat Yousuf, Farhana Mehraj Allai, and Savita Sharma

Introduction

Starch is one of the most abundant biopolymers and serves as energy reserve in many plants including cereals, tubers, roots, fruits and seeds. Starch, the second largest biomass on earth, is a natural, abundant, cheap, available, renewable, and biodegradable polymer (Doi et al. 2002; Chandanasree et al. 2016; Le Corre et al. 2010). It is the primary source of stored energy in cereal grains. Although the amount of starch contained in grains varies, it is generally between 60 and 75% of the weight of the grain and provides 70–80% of the calories consumed by humans worldwide.

Commercial starches are obtained from cereal grains, particularly from corn, wheat, and rice, and from tubers and roots, particularly potato, sweet potato, and tapioca (cassava). Starch obtained from corn, potatoes, cassava, and wheat in the

K. Gul (✉)
Department of Food Process Engineering, National Institute of Technology
Odisha, India

Department of Food Science and Engineering, School of Agriculture and Biology,
Shanghai Jiao Tong University, Shanghai, China

N. A. Mir
Department of Food Engineering and Technology, Sant Longowal Institute of Engineering and Technology, Longowal, India

B. Yousuf
Department of Post-Harvest Engineering and Technology, Faculty of Agricultural Sciences,
Aligarh Muslim University, Aligarh, India

F. M. Allai
Department of Food Technology, Islamic University of Science and Technology,
Awantipora, India

S. Sharma
Department of Food Science and Technology, Punjab Agricultural University, Ludhiana, India

A. Gani, B. A. Ashwar (eds.), *Food biopolymers: Structural, functional and nutraceutical properties*, https://doi.org/10.1007/978-3-030-27061-2_1

native and modified form accounts for 99% of the world production. The importance of starch, both biologically and technologically, is well known, as is its central role in the human diet. Starches have an enormous number of food uses, including adhesive, binding, clouding, dusting, film forming, foam strengthening, anti-staling, gelling, glazing, moisture retaining, stabilizing, texturizing, and thickening applications. The demand of starch has increased enormously in recent years as starch is being widely used in production of ethanol and biodegradable plastics (Wani et al. 2012) apart from being used in food processing industries (Gul et al. 2014). Native starches when cooked can easily retrograde and there is a gelling tendency of pastes besides easily undergoing syneresis. Therefore, starches are modified primarily to overcome the shortcomings of native starches and increase their usefulness for industrial applications. To meet specific requirements or demands, different modification methods such as physical, chemical, and enzymatic have been used to enhance or inhibit their inherent properties or to endow specific properties of starch.

Starch Chemistry

Starch is organized in discrete particles called granules whose size, shape, morphology, composition, and supramolecular structure depend on the botanical source (Fig. 1). Depending on the origin of starch, the granules can vary in shape, size, structure and chemical composition (Smith 2001). Starch granules are relatively dense and insoluble and hydrate only slightly in cold water. They can be dispersed in water, producing low-viscous slurries that can be easily mixed and pumped. Starch granules occur in all shapes and sizes (spheres, ellipsoids, polygons, platelets, irregular tubules). Their dimensions range from 0.1 to over 200 μm, depending

Fig. 1 Scanning electron micrograph of rice starch granules

on the botanical source. Differences in external granule morphology are generally sufficient to provide unambiguous characterization of the botanical source, via optical microscopy.

Regardless of the botanical source, starch is a polymer of the six-carbon sugar D-glucose, often referred to as the "building block" of starch. Starch granules are partially crystalline particles composed mainly of two homopolymers of glucopyranose with different structures: amylose, which is composed of units of D-glucose linked through α–D–(1–4) linkages and amylopectin,the branching polymer of starch, composed of α–D–(1–4)-linked glucose segments containing glucose units in α–D (1–6) branches (Fig. 2).

Although amylose and amylopectin are both composed of D-glucopyranose molecules, dissimilarities between these two polymers result in major differences in functional properties. Some important characteristics of amylose and amylopectin are presented in Table 1.

Amylose is a linear polymer, although evidence has suggested some branches in its structure. Amylose can also be present as a hydrophobic helix, allowing the formation of a complex with free fatty acids, fatty acid components of glycerides, some alcohols, and iodine (Thomas and Atwell 2005). Amylopectin is larger than amylose and their chains are classified as small chains, with an average degree of polymerization (DP) of about 15, and large chains, in which the DP is larger than 45. This unique configuration ordered in the packing arrangement contributes to the crystalline nature of the starch granule. The crystallinity reflects the organization of amylopectin molecules within the starch granules, whereas amylose makes up most of the amorphous materials that are randomly distributed among the amylopectin clusters (Blanshard 1987). Native starch granules observed microscopically under plane-polarized light exhibit the characteristic polarization of "Maltese Cross", which is due to birefringence properties of the crystalline portion of the granule (Fig. 3).

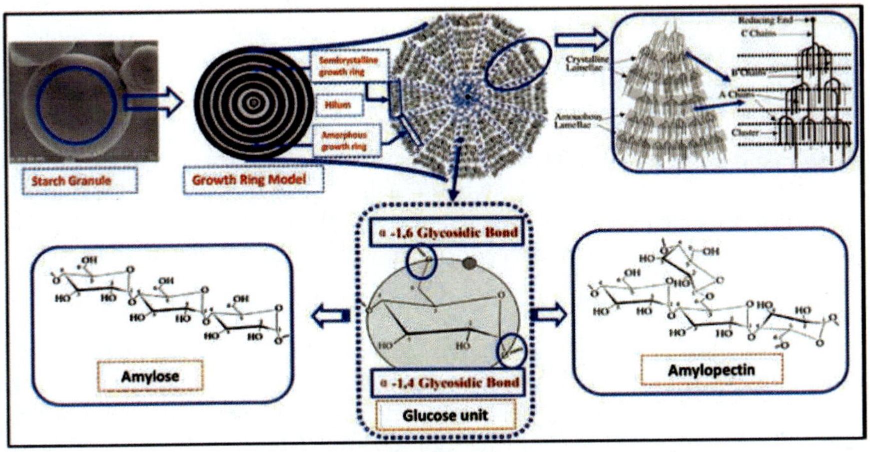

Fig. 2 Schematic diagram of the starch structure. Adapted from Liu et al. (2017)

Table 1 Characteristic features of amylose and amylopectin

Characteristic	Amylose	Amylopectin
Structure	Linear	Branched
Linkages	α-1,4	α-1,4 and α-1,6
Molecular weight	<0.5 million	50–500 million
Gel formation	Firm	Soft
Color with iodine	Blue	Reddish brown

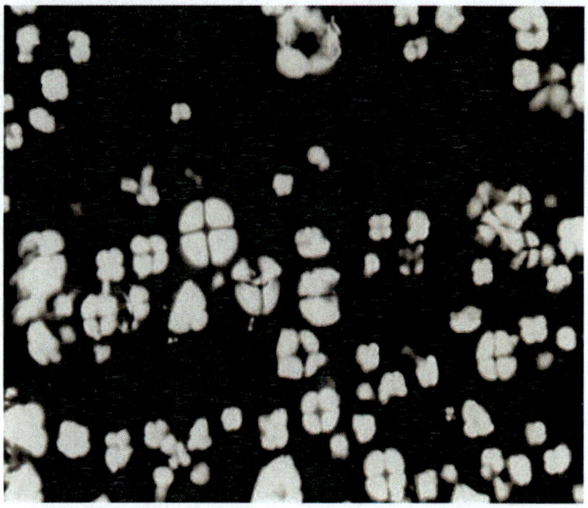

Fig. 3 Maltese cross in starch granules observed under polarized light

Amylose and amylopectin molecules are packaged into semi-crystalline aggregates and these aggregates are systematically packaged into starch granules. In these tightly packaged starch granules, the starch molecules have little affinity for water and are not functional (Blanshard 1987). The amylose content of starch varies: waxy, normal, and high-amylose starch consist of 0–8%, 20–30%, and more than 40% amylose, respectively (Li et al. 2008a, b). Branch chains of amylopectin form double helices and contribute to the crystalline structure of the granules, whereas amylose is amorphous and interspersed among amylopectin molecules (Jane et al. 1992).

Amylose and amylopectin do not exist free in nature, but as components of discrete, semicrystalline aggregates called granules. The size, shape, and structure of granules varies among botanical sources (Table 2). The diameters of the granules generally range from less than 1 μm to more than 100 μm, regular (spherical, ovoid, or angular) or quite irregular shapes. Wheat, barley, and rye granules exhibit two separate distributions of granule sizes and distinctly different shapes. Some granules, such as those in oats and rice, have a higher level of structure in which many small, individual granules are cohesively bound together in an organized manner. These are called compound starch granules. Although the major components of all types of starch granules are amylose and amylopectin polymers, there is great diversity in the structure and characteristics of native starch granules.

Table 2 Granule size and shape of starches of different botanical sources

Property	Dent corn	Waxy corn	High amylose corn	Wheat	Rice	Potato	Tapioca
Diameter (μm)	5–30	5–30	5–30	1–45	1–3	5–100	4–35
Shape	Polygonal, round	Polygonal, round	Polygonal, round, irregular	Round lenticular	Polygonal, spherical compound granules	Oval, spherical	Oval, truncated

Adapted and modified from Alexander (1995)

Physicochemical and Functional Characteristics of Starch

Starch when heated in excess of water results in the disruption of crystalline structure, due to breakage of hydrogen bonds, and water molecules form hydrogen bonds to the exposed hydroxyl groups of amylose and amylopectin. This results in an increase in granule swelling and solubility. Swelling power and solubility provide evidence of the magnitude of interaction between starch chains within the amorphous and crystalline domains. The extent of this interaction is influenced by the amylose/amylopectin ratio, and by the characteristics of amylose and amylopectin in terms of molecular weight distribution, degree of branching, length of branches and conformation of the molecules. Amylose-lipid complexes have been shown to restrict swelling and solubilization (Hoover and Hadziyev 1981). Swelling power provides an evidence of non-covalent bonding between starch molecules. Factors that determine the degree of swelling and solubility include amylose-amylopectin ratio, chain length and molecular weight distribution, degree/length of branching and conformation.

Starch Gelatinization and Retrogradation

Gelatinization is an endothermic process resulting in the disruption of molecular order of the starch granules. Double helical and crystalline structures are disrupted in starches during gelatinization. The granules are disrupted with absorption of water, losing their organized molecular structure, to facilitate swelling (García-Alonso et al. 1999). Native starches are insoluble in cold water and gelatinization is essential to facilitate water absorption and enhance their chemical and physical properties, to suit food processing industries. An increase in temperature causes the starch crystallites to break apart and undergo hydration. This results in several changes such as granular swelling, native crystalline melting, loss of birefringence, and starch solubilization (Atwell et al. 1988). Gelatinization is a two-step process, first the starch granules swell due to breakage of hydrogen bonds in the amorphous portions and in the next event, water acts as a plasticizer, resulting in hydration and swelling of the amorphous regions (Wani et al. 2012).Starch gelatinization is normally achieved by heating starch with water, but can also be achieved by heating starch in the presence of plasticizers like glycerol, ethylene glycol, and 1, 4-butanediol, alkaline solutions like NaOH and KOH, in neutral salt solutions like $CaCl_2$ and LiCl, and solvents like dimethyl sulfoxide (DMSO) (Ai and Jane 2015). After gelatinization is complete, the amorphous starch readily absorbs water and forms a paste. Upon cooling, some starch pastes can develop gels.

Gelatinization, an important functional property of starches, varies with respect to the starch composition (amylose-to-amylopectin ratio, and the percentage of phosphorus, lipids, proteins, and enzymes), the molecular structure of amylopectin, granule architecture, granule morphology, and granule size distribution (Tester 1997; Hoover et al. 2010).The temperature of initial gelatinization and the range over which gelatinization occurs depend on the method of measurement, on the starch: water ratio and granule type. To determine the gelatinization temperature of starch, many instruments have been used. These include differential scanning calorimetry (DSC), polarized light microscopy, thermomechanical analysis, and nuclear magnetic resonance spectroscopy, and other methods for the degree of starch gelatinization, such as X-ray scattering and Fourier transform infrared spectroscopy. However, DSC has emerged as the preferred method of choice for the measurement of starch gelatinization and the properties that are reported using DSC include gelatinization onset (To), peak (Tp), and conclusion (Tc) temperatures, peak height index (PHI), and enthalpy (ΔHg). The gelatinization properties of different starches determined by differential scanning calorimetry are presented.

Cooling of gelatinized starch results in the re-association of the leached amylose from gelatinized granules. This process is called retrogradation. Retrogradation causes starch gels to become less soluble during cooling due to recrystallization of starch molecules (BeMiller and Whistler 1996) and is of great interest to food technologists and industries since it profoundly affects quality, acceptability and shelf-life of starch-containing foods (Karim et al. 2000).

Retrogradation, also referred to as setback, occurs with re-crystallization of amylose which is more susceptible to retrogradation than amylopectin. Amylopectin is only minimally involved in starch retrogradation (Mua and Jackson 1998). The re-association and re-crystallization of amylose causes release of the water absorbed and bound during gelatinization, leading to the phenomenon known as syneresis. Retrogradation of gelatinized starch involves formation and subsequent aggregation of double helices of amylose and amylopectin chains, thus governing elasticity, firmness, and textural staling of all starch-containing systems (Atwell et al. 1988). Therefore, changes leading to retrogradation can restrict the starch functional properties making it less desirable for food industries. Cereal starches in general retrograde more slowly, and to a lesser extent, than tuber and root starches (Roulet et al. 1990). Differences in retrogradation behavior might be related to amylopectin fine structure, lipid content, amylose/amylopectin ratio and molecular weights. Avarami equation is frequently used to analyze retrogradation kinetics, however, this analysis implies a single step process. Since retrogradation is more complex, this analysis has therefore limited applicability.

Retrogradation in starches is commonly measured using a DSC, texture analyzer or a rheometer. While the DSC analysis is sensitive to chain ordering of the amylopectin fraction, the mechanical tests are sensitive to chain entanglements with/without the formation of crystallites (Shevkani et al. 2017). Rheological techniques are suited to monitor gel firmness (rigidity) on ageing and are usually carried by measuring the storage modulus (G`). These techniques do not evaluate the role of one component or process only, but the combined effect of all the components of the gel. Karim et al. (2000) have reviewed the methods to study retrogradation of starches. DSC measures retrogradation enthalpy (ΔHret) and transition temperatures for the melting of recrystallized starch. ΔHret is the amount of energy required for dissociation of the re-associated amylopectin. Hence, it provides a quantitative measure of retrogradation, higher ΔHret means higher retrogradation tendencies.

Pasting Properties

Heating of starch continuously in excess of water causes starch granules to swell and burst. Then the amylose leaches out and the granules disintegrate resulting in the formation of a viscous material called paste (BeMiller 2007). Pasting occurs after or simultaneously with gelatinization. Pasting properties of starch are important indicators of how the starch will behave during processing and, starch is generally regarded as the most important constituent of rice in terms of cooking quality and functionality. The pasting properties of starch are used in determining the suitability of starch in different foods and other allied products. The most important pasting characteristic of granular starch dispersion is its viscosity. High paste viscosity suggests suitability as a thickening agent in foods and as a finishing agent in the textile and paper industries. The pasting characteristics of starch are determined either using a Brabender Visco Amylograph or a Rapid Visco Analyzer (RVA) (Wickramasinghe and Noda 2008; Tukomane and Varavinit 2008; Lin et al. 2009). Rheometers (Park et al. 2007; Li et al. 2008a, b) or viscometers, which record viscosity profiles continuously with changes in temperature are also used to measure the pasting properties. RVA is used to determine pasting properties such as peak, final, setback, and breakdown viscosity, pasting temperature, and peak time. In this test, starch is mixed with water and allowed to hydrate rapidly on heating, then held constant for a specific time, and finally cooled to measure the pasting properties. As heating continues, an increase in viscosity can be observed by swelling of starch granules, which reflects the process of pasting. The temperature at the onset of viscosity increase is termed pasting temperature. Viscosity increases with continued heating until the rate of granule swelling equals the rate of granule collapse, which is referred to as the peak viscosity (PV). PV reflects the swelling extent or water-binding capacity of starch and often correlates with final product quality since the swollen and collapsed granules relate to the texture of cooked starch. Once PV is achieved, a drop in viscosity, or breakdown, is observed as a result of disintegration of granules. Breakdown is a measure of the ease of disrupting swollen starch granules and suggests the degree of stability during cooking (Adebowale and Lawal 2003). Minimum viscosity, also called hot paste viscosity, holding strength, or trough, marks the end of the holding stage at the maximum temperature of the RVA test. The cooling stage begins and viscosity again rises (setback) which is caused by retrogradation of starch, particularly amylose. Setback is an indicator of final product texture and is linked to syneresis or weeping during freeze–thaw cycles. Viscosity normally stabilizes at a final viscosity or cold paste viscosity, which is related to the capacity of starch to form a viscous paste or gel after cooking and cooling (Newport Scientific 1998). Other components naturally present in the starchy material or additives interact with starch and influence pasting behavior (Newport Scientific 1998).

Modification of Starch

Native starch, extracted from plants, cannot always withstand the extreme processing conditions, e.g., high temperature, freezethaw cycles, strong acid and alkali treatments, and high shear rates (Hermansson and Svegmark 1996; Wang and Copeland 2015). In addition, retrogradation occurs after loss of ordered structure on starch gelatinization, which results in syneresis or water separation in starchy food systems. Therefore, its use is limited and it unacceptable in many industrial applications. However, these shortcomings of native starch could be overcome, by modifying the starches. Modification of starch alters the properties of starch, including solution viscosity, association behavior, and shelf life stability in final products. Different techniques are used to modify native starch to enhance or inhibit its inherent properties, or to endue its specific properties to meet the requirements of industrial applications. Common modification methods for starches include physical (e.g., high pressure autoclave, osmotic pressure treatment, extrusion, irradiation, etc.), chemical (e.g., oxidation, esterification, etherification, hydroxypropylation), and enzymatic modifications (e.g., dextrin) (Liu et al. 2017). The commonly used methods for starch modification are summarized in Table 4.

Chemical modification is the most widely explored modification method due to the non-destructive nature of a select few of the processes and potential increases in the functionality of the modified starch. Chemical modification is generally achieved through derivatization, such as acetylation, cationization, oxidation, acidhydrolysis, and cross-linking. These methods are however limited due to environmental concerns and consumer safety (Ashogbon and Akintayo 2014). Starches can also be modified physically to improve their solubility and change particle size. The physical modification involves the treatment of native starch granules under different temperature/moisture combinations, pressure, shear, and irradiation and this modification has been gaining wider acceptance because of the absence of chemical reagents in the modified starch. Pre-gelatinization is the simplest of all starch modifications. It is effected by the cooking of aqueous starch slurry and subsequent drum drying (Tharanathan 2005).

Table 4 Common methods of starch modification

Modification	Types	Reference
Physical	Thermally inhibited treatment (dry heating)	Chiu et al. (1998)
	Deep freezing	Szymonska et al. (2000)
	Multiple deep freezing and thawing	Szymonska et al. (2003)
	Iterated syneresis	Lewandowicz and Soral-Smietana (2004)
	Corona electrical discharges	Nemtanu and Minea (2006)
	Mechanical activation-with stirringball mill	Huang et al. (2007)
	Micronization in vacuum ball mill	Che et al. (2007)
	Osmotic-pressure treatment	Pukkahuta et al. (2007)
	Instantaneous controlled pressure drop (DIC) process	Zarguili et al. (2006); Maache-Rezzoug et al. (2009)
	Pulsed electric fields treatment	Han et al. (2009)
	Superheated starch	Steeneken and Woortman (2009)
	Ultrasound	Jambrak et al. (2010); Sujka and Jamroz (2013)
	Dry heating with ionic gums	Gul et al. (2014); Chandanasree et al. (2016); Lim et al. (2002)
	Gamma irradiation	Wani et al. (2014); Gul et al. (2016)
Chemical	Acid hydrolysis	Sanguanpong et al. (2003); Akubor (2007); Rolland-Sabate et al. (2012)
	Acid hydrolysis in alcohol	Ferrini et al. (2008); Cavallini and Franco (2010)
	Alkali treatment	Raja (1992)
	Acetylation	Tran et al. (2007); Osundahunsi and Mueller (2011); Mbougueng et al. (2012)
	Cross-linking	Nabeshima and Grossmann (2001); Jyothi et al. (2006), Tran et al. (2007); Varavinit et al. (2007)
	Cationization	Han and Sosulski (1998)
	Oxidation	Tran et al. (2007); Sangseethong et al. (2009, 2010); Amorim et al. (2011); Klein et al. (2014); Beninca et al. (2013)
	Ozone-oxidised starch	Kesselmans and Bleeker (1997);An and King (2009);Chan et al. (2009)
	Hydroxypropylation	Schmitz et al. (2006); Jyothi et al. (2007)

Recent developments in biotechnology have provided means for the modification of starch during the growth of the plant. Different amylose levels, amylopectin structure, and phosphorus contents from various plant sources can be produced using antisense reduction of enzyme activity of single or multiple enzymes (Jobling et al. 1999). Modifications can be achieved through transgenic means by introducing genes for new enzymes into a desired plant from other plants or novel mutations for enhanced carbohydrates.

Applications of Modified Starches

Starches have found enormous uses in food and other industries. Modification of starch is an ever evolving industry with numerous possibilities to generate novel starches which includes new functional and value added properties as demanded by the industry. Modified starches have wide applications as compared to native starches. Starch modification is primarily carried to overcome the shortcomings of native starch and increase its usefulness for industrial applications (Table 5). Starch can be modified by hydrolysis to serve as a fat replacer. Well-known fat replacer from starch base is maltodextrin beside some other starch hydrolysis products. It is widely used for low-fat butter spread/margarine, low fat mayonnaise, low fat milk type products and low-fat ice cream (Sajilata and Singhal 2005).

Several studies have been reported on the potential use of modified starches as texture improver in the food industry. It can provide crispness of crackers and biscuits, desired chewiness for certain foods, viscosity breakdown resistant for canned foods, and improvement in the quality of extruded products. Modified starches have

Table 5 Starch modifications and their applications

Modification	Outcome	Applications
Cross-linking	Increased cooked granule rigidity and stability, tolerance to high shear and temperature and low pH	Viscosity builders in high temperature and high shear processing (e.g. in canning and retort processing, in applications that require viscosity control during pumping operations
Genetic modification	Waxy, high amylose	High viscosity and potential for RS3 resistant starch respectively
Partial hydrolysis	Depolymerized, reduced viscosity, increased retrogradation	Confectionary, batters, coatings
Pregelatinized and cold-water swelling	Dispersible in cold water	Convenience foods (e.g. Instant puddings)
Oxidation	Depolymerized, reduced viscosity, improved clarity, reduced syneresis, improved cling	Batters and breadings, other coatings
Stabilization/ starch esters	Reduced gelatinization temperature, decreased retrogradation, and a reduced tendency to form gels, improved clarity	Refrigerated and frozen foods, emulsion stabilizers
Stabilization/ starch ethers	Reduced gelatinization temperature, high viscosity, decreased retrogradation and good freeze-thaw stability, improved clarity	Many applications: gravies, dips, sauces, fruit pie fillings, and puddings

Source: Adapted from Singh et al. (2007)

been used to prepare resistant starches which are further used to prepare slow digestible food products for example cookies, muffins and breads. In recent years, modified starches are being used for high shear and temperature stability, and for the encapsulation and controlled release of different pigments, flavors and oils. Other related benefits of modified starches are: stabilizer, emulsifier, thickening agent, dusting agent, drying aids, binder, clouding agent, suspending agent and for freeze-thaw stability (Abbas et al. 2010). Modified starches have also been used to produce biopolymer films (biodegradable films) to partially or entirely replace plastic polymers because of low cost and biodegradability and good mechanical properties (Xie et al. 2005).

Conclusions

Starch is the reserve carbohydrate in plant and plant materials and undoubtedly has been widely researched. The shortcomings in native starches, such as insolubility in cold water, high viscosity, and thickening after cooking, syneresis, tendency to easily retrograde and the loss of ordered structure after gelatinization in starchy food systems can be overcome. Modification of starches is being carried out since long and allows starches to be used in a variety of food applications. The physical, enzymatic and genetic modification of starch has been most promising with a large number of newer methods in starch modification being continuously explored. Though the efforts are large in chemical modification, the health of the consumer and environmental concerns should also be taken into consideration. Starch may be added to foods in order to provide thickening or product stability or potentially to carry flavors. There is still much to learn about starch granules and its modification, although considerable progress has been made in recent years.

References

Abbas, K. A., Khalil, S. K., & Hussin, A. S. M. (2010). Modified starches and their usages in selected food products: A review study. *Journal of Agricultural Science, 2*(2), 90.

Adebowale, K. O., & Lawal, O. S. (2003). Microstructure, physicochemical properties and retrogradation behaviour of Mucuna bean (*Mucuna pruriens*) starch on heat moisture treatments. *Food Hydrocolloids, 17*(3), 265–272.

Ai, Y., & Jane, J. L. (2015). Gelatinization and rheological properties of starch. *Starch-Stärke, 67*(3–4), 213–224.

Akubor, P. I. (2007). Chemical and functional properties of modified and unmodified cassava and sweet potato starches. *Journal of Food Science and Technology, 44*, 260–263.

Alexander, R. J. (1995). Potato starch: New prospects for an old product. *Cereal Foods World, 40*, 763–764.

Amorim, E. O. C., Doval, V. C., & Cristianini, M. (2011). Effect of ozonation on the sensory characteristics and pasting properties of cassava starch. *Procedia Food Science, 1*, 914–919.

An, H. J., & King, J. M. (2009). Using ozonation and amino acids to change pasting properties of rice starch. *Journal of Food Science, 74*, 278–283.
Ashogbon, A. O., & Akintayo, E. T. (2014). Recent trend in the physical and chemical modification of starches from different botanical sources: A review. *Starch-Stärke, 66*(1–2), 41–57.
Atwell, W. A., Hood, L. F., Lineback, D. R., Varriano-Marston, E., & Zobel, H. F. (1988). The terminology and methodology associated with basic starch phenomena. *Cereal Foods World, 33*, 306–311.
BeMiller, J. N. (2007). *Carbohydrate chemistry for food scientists* (2nd ed.). St. Paul, MN: AACC International.
BeMiller, J. N., & Whistler, R. L. (1996). Carbohydrates. In O. R. Fennema (Ed.), *Food chemistry* (pp. 157–223). Boca Raton, FL: CRC Press Taylor and Francis Group.
Beninca, C., Colman, T. A. D., Lacerda, L. G., da Silva Carvalho Filho, M. A., Demiate, I. M., Bannach, G., & Schnitzler, E. (2013). Thermal, rheological, and structural behaviors of natural and modified cassava starch granules, with sodium hypochlorite solutions. *Journal of thermal analysis and calorimetry, 111*(3), 2217–2222.
Blanshard, J. M. V. (1987). Starch granule structure and function: A physicochemical approach. In T. Galliard (Ed.), *Starch: Properties and potential* (2nd ed., pp. 16–54). New York, NY: Wiley.
Cavallini, C. M., & Franco, C. M. L. (2010). Effect of acid-ethanol treatment followed by ball milling on structural and physicochemical characteristics of cassava starch. *Starch-Stärke, 62*, 236–245.
Chan, H. T., Bhat, R., & Karim, A. A. (2009). Physicochemical and functional properties of ozone-oxidized starch. *Journal of Agricultural and Food Chemistry, 57*, 5965–5970.
Chandanasree, D., Gul, K., & Riar, C. S. (2016). Effect of hydrocolloids and dry heat modification on physicochemical, thermal, pasting and morphological characteristics of cassava (Manihot esculenta) starch. *Food Hydrocolloids, 52*, 175–182.
Che, L.-M., Li, D., Wang, L.-J., Chen, X. D., & Mao, Z.-H. (2007). Microniation and hydrophobic modification of cassava starch. *International Journal of Food Properties, 10*, 527–536.
Chiu, C.-W., Schiermeyer, E., Thomas, D. J., & Shah, M. B. (1998). *Thermally inhibited starches and flours and process for their production.* U.S. Patent 5,725,676.
Doi, S., Clark, J. H., Macquarrie, D. J., & Milkowski, K. (2002). New materials based on renewable resources: Chemically modified expanded corn starches as catalysts for liquid phase organic reactions. *Chemical Communications, 22*, 2632–2633.
Ferrini, L. M. K., Rocha, T. S., Demiate, I. M., & Franco, C. M. L. (2008). Effect of acid-methanol treatment on the physicochemical and structural characteristics of cassava and maize starches. *Starch-Stärke, 60*, 417–425.
García-Alonso, A., Jiménez-Escrig, A., Martín-Carrón, N., Bravo, L., & Saura-Calixto, F. (1999). Assessment of some parameters involved in the gelatinization and retrogration of starch. *Food Chemistry, 66*(2), 181–187.
Gul, K., Riar, C. S., Bala, A., & Sibian, M. S. (2014). Effect of ionic gums and dry heating on physicochemical, morphological, thermal and pasting properties of water chestnut starch. *LWT - Food Science and Technology, 59*(1), 348–355.
Gul, K., Singh, A. K., & Sonkawade, R. G. (2016). Physicochemical, thermal and pasting characteristics of gamma irradiated rice starches. *International Journal of Biological Macromolecules, 85*, 460–466.
Han, H. L., & Sosulski, F. W. (1998). Cationization of potato and tapioca starches using an aqueous alcoholic-alkaline process. *Starch-Stärke, 50*, 487–492.
Han, Z., Zeng, X., Zhang, B., & Yu, S. (2009). Effect of pulsed electric fields (PEF) treatment on the properties of corn starch. *Journal of Food Engineering, 93*, 318–323.
Hermansson, A. M., & Svegmark, K. (1996). Developments in the understanding of starch functionality. *Trends in Food Science & Technology, 7*, 345–353.
Hoover, R., & Hadziyev, D. (1981). The effect of monoglycerides on amylose complexing during a potato granule process. *Starch-Stärke, 33*, 346–355.

Hoover, R., Hughes, T., Chung, H. J., & Liu, Q. (2010). Composition, molecular structure, properties, and modification of pulse starches: A review. *Food Research International, 43*(2), 399–413.

Huang, Z.-Q., Lu, J.-P., Li, X.-0H., & Tong, Z. F. (2007). Effect of mechanical activation on physico-chemical properties and structure of cassava starch. *Carbohydrate Polymers, 68*, 128–135.

Jambrak, A. R., Herceg, Z., Šubarić, D., Babić, J., Brnčić, M., Brnčić, S. R., … Gelo, J. (2010). Ultrasound effect on physical properties of corn starch. *Carbohydrate Polymers, 79*(1), 91–100.

Jane, J., Xu, A., Radosavljevic, M., & Seib, P. A. (1992). Location of amylose in normal starch granules. I. Susceptibility of amylose and amylopectin to cross-linking reagents. *Cereal Chemistry, 69*, 405–409.

Jobling, S. A., Schwall, G. P., Westcott, R. J., Sidebottom, C. M., Debet, M., Gidley, M. J., … Safford, R. (1999). A minor form of starch branching enzyme in potato (*Solanum tuberosum* L.) tubers has a major effect on starch structure: Cloning and characterisation of multiple forms of SBE A. *Plant Journal, 18*, 163–171.

Jyothi, A. N., Moorthy, S. N., & Rajasekharan, K. N. (2007). Studies on the synthesis and properties of hydroxypropyl derivatives of cassava (*Manihot esculenta* Crantz) starch. *Journal of the Science of Food and Agriculture, 87*, 1964–1972.

Jyothi, A. N., Moorthy, S. N., & Rajasekharan, V. (2006). Effects of cross linking with epichlorohydrin on the properties of cassava (*Manihot esculenta* Crantz) starch. *Starch-Stärke, 58*, 292–299.

Karim, A. A., Norziah, M. H., & Seow, C. C. (2000). Methods for the study of starch retrogradation. *Food Chemistry, 71*(1), 9–36.

Kesselmans, R. P. W., & Bleeker, I. P. (1997). *Method for oxidizing dry starch using ozone*. World Intellectual Property Organization. WO/1997/032902.

Klein, B., Vanier, N. L., Moomand, K., Pinto, V. Z., Colussi, R., Zavareze, E. D. R., et al. (2014). Ozone oxidation of cassava starch in aqueous solution at different pH. *Food Chemistry, 155*, 167–173.

Le Corre, D., Bras, J., & Dufresne, A. (2010). Starch nanoparticles: A review. *Biomacromolecules, 11*, 1139–1153.

Lewandowicz, G., & Soral-Smietana, M. (2004). Starch modification by iterated syneresis. *Carbohydrate Polymers, 56*, 403–413.

Li, L., Jiang, H., Campbell, M., Blanco, M., & Jane, J.-l. (2008a). Characterization of maize amylose-extender (ae) mutant starches. Part I: Relationship between resistant starch contents and molecular structures. *Carbohydrate Polymers, 74*, 396–404.

Li, Y., Shoemaker, C. F., Shen, X., Ma, J., Ibáñez-Carranza, A. M., & Zhong, F. (2008b). The isolation of rice starch with food grade proteases combined with other treatments. *Food Science and Technology International, 14*(3), 215–224.

Lim, S.-T., Han, J.-A., Lim, H. S., & BeMiller, J. N. (2002). Modification of starch by dry heating with ionic gums. *Cereal Chemistry, 79*, 601–606.

Lin, Q., Xiao, H., Zhao, J., Li, L., & Yu, F. (2009). Characterization of the pasting, flow and rheological properties of native and phosphorylated rice starches. *Starch-Stärke, 61*(12), 709–715.

Liu, G., Gu, Z., Hong, Y., Cheng, L., & Li, C. (2017). Structure, functionality and applications of debranched starch: A review. *Trends in Food Science & Technology, 63*, 70–79.

Maache-Rezzoug, Z., Maugard, T., Zarguili, I., Bezzine, E., El Marzouki, M.-N., et al. (2009). Effect of instantaneous controlled pressure drop (DIC) on physicochemical properties of wheat, waxy and standard maize starches. *Journal of Cereal Science, 49*, 346–353.

Mbougueng, P. D., Tenin, D., Scher, J., & Tchiegang, C. (2012). Influence of acetylation on physicochemical, functional and thermal properties of potato and cassava starches. *Journal of Food Engineering, 108*, 320–326.

Mua, J. P., & Jackson, D. S. (1998). Retrogradation and gel textural attributes of corn starch amylose and amylopectin fractions. *Journal of Cereal Science, 27*(2), 157–166.

Nabeshima, E. H., & Grossmann, M. V. E. (2001). Functional properties of pregelatinized and cross-linked cassava starch obtained by extrusion with sodium trimetaphosphate. *Carbohydrate Polymers, 45*, 347–353.

Nemtanu, M. R., & Minea, R. (2006). Functional properties of corn starch treated with corona electrical discharges. *Macromolecular Symposia, 245-246*, 525–528.

Newport Scientific. (1998). *Applications manual for the Rapid Visco Analyzer*. Warriewood, NSW: Author.

Osundahunsi, O. F., & Mueller, R. (2011). Functional and dynamic rheological properties of acetylated starches from two cultivars of cassava. *Starch-Stärke, 63*, 3–10.

Park, I. M., Ibáñez, A. M., Zhong, F., & Shoemaker, C. F. (2007). Gelatinization and pasting properties of waxy and non-waxy rice starches. *Starch-Stärke, 59*(8), 388–396.

Pukkahuta, C., Shobsngob, S., & Varavinit, S. (2007). Effect of osmotic pressure on starch: New method of physical modification of starch. *Starch-Stärke, 58*, 78–90.

Raja, K. C. M. (1992). Studies on alkali-modified cassava starch-changes of structural and enzyme (α-amylase) susceptibility properties. *Starch-Stärke, 44*, 133–136.

Rolland-Sabate, A., Sanchez, T., Buléon, A., Colonna, P., Jaillais, B., Ceballos, H., et al. (2012). Structural characterization of novel cassava starches with low and high-amylose contents in comparison with other commercial sources. *Food Hydrocolloids, 27*, 161–174.

Roulet, P., Macinnes, W. M., Gumy, D., & Würsch, P. (1990). Retrogradation kinetics of eight starches. *Starch-Stärke, 42*(3), 99–101.

Sajilata, M. G., & Singhal, R. S. (2005). Specially starches for snack foods. *Carbohydrate Polymers, 59*, 131–151.

Sangseethong, K., Lertphanich, S., & Sriroth, K. (2009). Physicochemical properties of oxidized cassava starch prepared under various alkalinity levels. *Starch-Stärke, 61*, 92–100.

Sangseethong, K., Termvejsayanon, N., & Sriroth, K. (2010). Characterization of physicochemical properties of hypochlorite- and peroxide-oxidized cassava starches. *Carbohydrate Polymers, 82*, 446–453.

Sanguanpong, V., Chotineeranat, S., Piyachomkwan, K., Oates, C. G., Chinachoti, P., & Sriroth, K. (2003). Hydration and physicochemical properties of small-particle cassava starch. *Journal of the Science of Food and Agriculture, 83*, 123–132.

Schmitz, C. S., De Simas, K. N., Santos, K., Joao, J. J., Amboni, R. D. D. M. C., & Amante, E. R. (2006). Cassava starch functional properties by etherification-hydroxypropylation. *International Journal of Food Science & Technology, 41*, 681–687.

Shevkani, K., Singh, N., Bajaj, R., & Kaur, A. (2017). Wheat starch production, structure, functionality and applications—A review. *International Journal of Food Science & Technology, 52*(1), 38–58.

Singh, J., Kaur, L., & McCarthy, O. J. (2007). Factors influencing the physico-chemical, morphological, thermal and rheological properties of some chemically modified starches for food applications—A review. *Food Hydrocolloids, 21*(1), 1–22.

Smith, A. M. (2001). The biosynthesis of the starch granule. *Biomacromolecules, 2*, 335–341.

Steeneken, P. A. M., & Woortman, A. J. J. (2009). Superheated starch: A novel approach towards spreadable particle gels. *Food Hydrocolloids, 23*, 394–405.

Sujka, M., & Jamroz, J. (2013). Ultrasound-treated starch: SEM and TEM imaging, and functional behaviour. *Food Hydrocolloids, 31*(2), 413–419.

Szymonska, J., Krok, F., & Tomasik, P. (2000). Deep-freezing of potato starch. *International Journal of Biological Macromolecules, 27*, 307–314.

Szymonska, J., Krok, F., Komorowska-Czepirska, E., & Rebilas, K. (2003). Modification of granular potato starch by multiple deep-freezing and thawing. *Carbohydrate Polymers, 52*, 1–10.

Tester, R. F. (1997). Starch: The polysaccharide fractions. In P. J. Richmond & A. M. Donald (Eds.), *Structure starch: Structure and functionality* (pp. 163–171). London, UK: Royal Society of Chemistry.

Tharanathan, R. N. (2005). Starch—Value addition by modification. *Critical Reviews in Food Science and Nutrition, 45*(5), 371–384.

Thomas, D. J., & Atwell, W. A. (2005). *Starches*. St. Paul, MN: Eagan Press.
Tran, T., Piyachomkwan, K., & Sriroth, K. (2007). Gelatinization and thermal properties of modified cassava starches. *Starch-Stärke, 59*, 46–55.
Tukomane, T., & Varavinit, S. (2008). Classification of rice starch amylose content from rheological changes of starch paste after cold recrystallization. *Starch-Stärke, 60*(6), 292–297.
Varavinit, S., Paisanjit, W., Tukomane, T., & Pukkahuta, C. (2007). Effect of osmotic pressure on the cross-linking reaction of tapioca starch. *Starch-Stärke, 59*, 290–296.
Wang, S., & Copeland, L. (2015). Effect of acid hydrolysis on starch structure and functionality: A review. *Critical Reviews in Food Science and Nutrition, 55*, 1081–1097.
Wani, A. A., Singh, P., Shah, M. A., Schweiggert-Weisz, U., Gul, K., & Wani, I. A. (2012). Rice starch diversity: Effects on structural, morphological, thermal and physicochemical properties: A review. *Comprehensive Reviews in Food Science and Food Safety, 11*, 417–436.
Wani, I. A., Jabeen, M., Geelani, H., Masoodi, F. A., Saba, I., & Muzaffar, S. (2014). Effect of gamma irradiation on physicochemical properties of Indian Horse Chestnut (*Aesculus indica* Colebr.) starch. *Food Hydrocolloids, 35*, 253–263.
Wickramasinghe, H. A. M., & Noda, T. (2008). Physicochemical properties of starches from Sri Lankan rice varieties. *Food Science and Technology Research, 14*(1), 49–54.
Xie, S. X., Lui, Q., & Cui, S. W. (2005). Starch modification and applications. In S. W. Cui (Ed.), *Food carbohydrates: Chemistry, physical properties, and applications* (pp. 357–405). Boca Raton, FL: CRC Press.
Zarguili, I., Maache-Rezzoug, Z., Loisel, C., & Doublier, J.-L. (2006). Influence of DIL hydrothermal process conditions on the gelatinization properties of standard maize starch. *Journal of Food Engineering, 77*, 454–461.

Resistant Starch and Slowly Digestible Starch

Check for updates

Bilal Ahmad Ashwar, Adil Gani, Asima Shah, Mudasir Ahmad, Asir Gani, Faiza Jhan, and Nairah Noor

Introduction

In human nutrition, starch plays an important role in supplying metabolic energy, which enables the body to perform its functions. Based on the rate and extent of digestion, starch has been classified into different fractions viz. Rapidly Digestible Starch (RDS), Slowly Digestible Starch (SDS), and Resistant Starch (RS) (Englyst et al. 1992). Starch has been quantified into these fractions using the in vitro Englyst assay: the starch fraction digested within 20 min of incubation has been classified as RDS, the starch fraction digested between 20 and 120 min corresponds to SDS, and the remaining fraction that was not further digested has been classified as RS. RDS induces a rapid increase in the blood glucose and insulin levels, which may cause a series of health complications, such as diabetes and cardiovascular diseases (CVD). SDS is slowly digested in the small intestine, thereby resulting in a slow and prolonged release of glucose into the blood, coupled to the low glycemic response. Thus SDS can be helpful in controlling and preventing hyperglycemia-related diseases. Resistant Starch (RS) is the fraction of starch that is resistant to hydrolysis by α-amylase and pullulanase enzymes in vitro and may be fermented in the colon (Englyst et al. 1982). It is that fraction of starch, which escapes digestion in the GI tract but may be fermented in the colon (Englyst et al. 1996). The end-products of fermentation are carbon dioxide, hydrogen, methane and short chain fatty acids (SCFAs).

RS is measured as the difference between total starch (TS) and the sum of rapidly digestible starch (RDS) and slowly digestible starch (SDS) (Sajilata et al. 2006).

$$RS = TS - (RDS + SDS)$$

B. A. Ashwar · A. Gani (✉) · A. Shah · M. Ahmad · A. Gani · F. Jhan · N. Noor
Department of Food Science and Technology, University of Kashmir, Srinagar, India

A. Gani, B. A. Ashwar (eds.), *Food biopolymers: Structural, functional and nutraceutical properties*, https://doi.org/10.1007/978-3-030-27061-2_2

RS may escape digestion due to various reasons. The compact molecular structure of starch may limit the accessibility of hydrolytic enzymes (Haralampu 2000). The starch may be physically inaccessible to the hydrolytic enzymes as in grains, seeds and tubers. The starch granules may also be configured in such a manner which prevents their digestion e.g., unripe bananas, raw potatoes, and high amylose maize starch (Nugent 2005). When gelatinized starch is cooled retrograded starch (starch crystals) is formed which is resistant to digestive enzymes. This form of 'retrograded' starch is found in foods like corn flakes, and cooked and cooled potatoes (approximately 5%) (Haralampu 2000). Chemical modifications like esterification, etherification and cross linking of starch also makes it resistant to enzymatic hydrolysis (Lunn and Buttriss 2007). RS has been further classified into five types: RS1, RS2, RS3, RS4 and RS5.

RS1

This form of RS is physically inaccessible to hydrolytic enzymes because it is entrapped within the food matrix such as partly milled grains and seeds. It can be used in a variety of conventional foods as it is stable to most cooking operations (Sajilata et al. 2006).

RS2

The RS2 comprises of native, uncooked granules like raw potato or banana starches, whose crystallinity makes them resistant to enzymatic hydrolysis (Hernandez et al. 2008). These ungelatinized starches are resistant to digestion because of their compact structure (Sajilata et al. 2006). A particular type of RS2 called high amylose maize starch (HAM) is unique because it is stable to most cooking operations (Wepner et al. 1999).

RS3

RS3 is generally called retrograded starch (Wepner et al. 1999). It is formed when starch is first gelatinized and then cooled for retrogradation. During retrogradation the polymer chains reassociate by the formation of inter chain hydrogen bonds to form double helices. The double helices are left handed, parallel stranded and one turn of double helix is 20.8 Å. Retrograded starch has type A crystalline structure (Eerlingen et al. 1993a, b). RS3 content is also affected by degree of polymerization (DP) of amylose, with increase of DP, RS3 content increases, reaching maximum at

100 DP and then remains constant (Eerlingen et al. 1993a, b). DP level of 10–100 is necessary for the formation of double helix (Gidley et al. 1995).

RS4

RS4 is chemically modified starch and include modifications like esterification, acetylation, etherification, phosphorylation or crosslinking. RS4 is further grouped into subcategories based on their solubility in water and the experimental protocols used for analysis (Nugent 2005). Chemical modifications of starch prevent its digestion by blocking the access to hydrolytic enzymes and by the formation of atypical links (Sajilata et al. 2006; Kim et al. 2008).

RS5

Besides structural characteristics, other factors intrinsic to starch rich foods can affect enzyme activity and hence starch hydrolysis. These factors include the amylose-lipid complexes, the presence of native α-amylase inhibitors in starchy foods and non-starch polysaccharides. RS5 is formed by the complexation of amylose with lipids (Fuentes-Zaragoza et al. 2011).

Sources

Naturally cereals, seeds and other starch rich foods are excellent sources of SDS and RS (Charalampopoulos et al. 2002; Gani et al. 2020). Among non processed foods, unripe bananas are the richest source of RS (47–57%). Unripe banana flour was prepared with 17.5% RS, 73.4% total starch and 14.5% dietary fibre (Rodriguez et al. 2008). Tuber starch like potato shows B-type crystallinity which is highly resistant to enzymatic hydrolysis. Raw potato starch contains almost 75% RS (Bednar et al. 2001). Whole grains are rich sources of RS, dietary fibre and oligosaccharides while as flour group contains low concentrations of RS (Slavin 2004). Flour contains two principal components viz. protein and starch, where as whole cereal grain contains the pericarp, aleurone layers and germ that provides the lipids and fibre. Processing of cereal grains alters their chemical composition. The RS content of whole cereal grains was found to be five times higher than their respective flours (Bednar et al. 2001).

Pulse grains contain high contents of RS (Rochfort and Panozzo 2007). RS and total dietary fibre contents of 24.7% and 36.5%, respectively were found in legumes. Various factors are responsible for higher contents of RS in legumes. C-type pattern of crystallinity found in leguminous starches makes them more resistant to hydrolysis as compared to cereals which have A-type crystallinity (Mir et al. 2013). Leguminous starch contains higher levels of amylose as compared to cereal and pseudocereal starches which reflects their high RS content (Mikulíkova et al. 2008). Quick retrogradation of cooked leguminous starch makes it exceptionally resistant to hydrolysis (Tharanathan and Mahadevamma 2003).

RS content varies from a few percent to 80% in the legumes. It may further increase or decrease due to hydrothermal processing, depending on the variety of legumes and parameters of processing (Giczewska and Borowska 2003).

Processing had a great impact on the RS and SDS content of foods, and generally RS contents are reduced by severe and longer periods of processing. RS content of cooked rice reduced from 12% to 5% during grinding where as RS content of oats reduced from 16% to 3% during cooking (Muir and O'Dea 1992). Autoclaving, baking, flaking, and parboiling are known to influence starch digestibility and the yield of SDS (Brand et al. 1985; Holm et al. 1985; Casiraghi et al. 1993; Kingman and Englyst 1994). During pullulanase debranching and cooling treatment of the cooked waxy maize starch, short-term retrogradation occurs as a result of the crystallization of the amylose fraction, leading to maximum SDS formation (Miao et al. 2009).

Formation of Resistant Starch and Slowly Digestible Starch

Different modifications of starch like physical, chemical, enzymatic treatments, irradiation and genetic modifications have been employed for the formation of RS and SDS. Some of these modifications are described as follows:

Physical Modifications

Physical treatments for preparation of SDS and RS include hydrothermal treatments, recrystallization, polymer-entrapment, and extrusion. When starch is heated to various levels, it leads to the formation of RS and SDS. RS was obtained by cooking the starch above its gelatinization temperature and drying simultaneously on heated rolls like drum driers and extruders (Holm et al. 1988). Gelatinization of starch at 120 °C for 20 min, followed by cooling to room temperature also provides good yields of RS (Garcia-Alonso et al. 1999). High yields of SDS (39.3–56.7%) were obtained by dual retrogradation treatment (gelatinization-retrogradation-gelatinization-retrogradation) in rice starch (Tian et al. 2013). Good yields of RS3 were also obtained by various combinations of time and temperature treatments to various sources of native starch. The temperature treatments were autoclaving the starch at 110 °C (Berry 1986), at 121 °C (Sievert and Wursch 1993), at 127 °C (Berry 1986), at 134 °C (Berry 1986), or at 148 °C Sievert and Pomeranz 1989) for periods ranging from 30 min to 1 h. Recently effects of autoclaving temperature

(140–145 °C) and storing time (24, 48 and 72 h) on resistant starch (RS) formation from high amylose corn starch were investigated. High autoclaving temperature (145 °C) and long storage time (72 h) increased the yield of RS (Dundar and Gocmen 2013). Partial acid hydrolysis (PAH) of the high-amylose corn starch can be used to produce granular RS, which is stable to further heat treatment at atmospheric pressure (Brumovsky and Thompson 2001; Ozturk et al. 2011). PAH followed by heat moisture treatment increased the yield of boiling-stable granular RS to the maximum of 63.2%. Pyrodextrinization has been identified as a way of producing RS which is water-soluble and has non-starch linkages (Laurentin and Edwards 2004). Modification of dry starch through heat treatments, with or without addition of acids is referred to as Pyroconversion. The acids include hydrochloric acid at 0.15% (based on starch dry weight) and orthophosphoric or sulfuric acids at 0.17% (Wurzburg 1995). Pyrodextrins are commercially produced by heating dry, acidified starch in a reactor with agitation. During pyroconversion hydrolysis and transglycosidation of starch occurs which can be facilitated by spraying acid on the starch. A wide range of products that vary in available starch, digestibility, cold-water solubility, swelling power, viscosity, color, and stability were produced during pyroconversion (Ohkuma and Wakabayashi 2001).

Shin et al. (2005) reported that when granular sweet potato starch (50% moisture) is heated to 55 °C, the amount of SDS increases by 200%. It has been reported that hydrothermal treatment of granular sweet potato starch alters its structure from C_b type to A-type as a result of the melting of starch crystallites and subsequent recrystallization. This structure change converts a fraction of amorphous amylose molecules into the crystalline form, thereby decreasing enzyme susceptibility. Miao et al. (2009) showed that controlled retrogradation of partially debranched waxy maize starch can be used to make SDS and RS, which occurs due to the formation of imperfect, low-density B-type crystallites (Miao et al. 2009). Controlled debranching of waxy starch results in the formation of great number of short chains of amylose that are available for chain re-alignment, cross-linking and double helix formation, which leads to the formation of more SDS and RS contents. Other studies have shown that retrogradation correlates with the SDS and RS content of mutant maize; this maize has a higher proportion of long amylopectin chains and linear branch chains of amylopectin with *DP* 9–30. This type of amylopectin probably acts as an anchor point to slow the digestion of branched-chain fractions with *DP* > 30, which as physical entities are the primary constituents of SDS and RS (Zhang et al. 2008). Entrapment or encapsulation of the starch in the structured protein network can be used as a novel method for development of RS and SDS. Starch-encapsulated spheres with 44% SDS were prepared by dropping a homogeneous mixture of 1% sodium alginate (w/w) and 5 g of starch into a 2% $CaCl_2$ solution (w/v) (Hamaker et al. (2007). An SDS product has been generated by using partially gelatinized or plasticized materials to form a low-swelling network of mixed crystallites that consisted of short-chain amylose (*DP* < 300) and basic starch. This network has been formed through cooking or mixing processes, especially extrusion (Innereber and Mueller 2005). In addition SDS has been generated in feed by adding a reducing carbohydrate to comminuted cereal grain, heating the mixture followed by drying (Winowiski et al. 2005). In other words, physical modifications of the starch that affects enzyme binding and the rate of digestion can be used to modulate starch digestibility for formation of SDS and RS.

Enzyme Treatment

Controlled enzymatic treatment of starch with α-amylase, β-amylase, isoamylase, pullulanase and transglucosidase is an alternative approach to change the chain-length of starch supramolecular structure in order to achieve appropriate digestibility and glycemic response (Shah et al. 2018). RS has been prepared from high amylose starch by gelatinization followed by treatment of slurry with debranching enzymes like pullulanase and isolating the starch product by drying/extrusion (Haralampu and Gross 1998). RS products having at least 50% RS content were manufactured by forming a water-starch suspension, heating the suspension in an autoclave at 100 °C so that full starch gelatinization takes place and then cooling to allow retrogradation of amylose. The best results were obtained at a temperature of 134 °C, with four heating-cooling cycles and a starch: water ratio of 1:3.5 (Pomeranz and Sievert 1990). RS was also prepared by gelatinizing the starch (common corn starch and waxy maize starch), followed by treatment with a debranching enzyme, isoamylase or pullulanase and precipitation of the debranched starch. For precipitation, the suspension was allowed to cool at room temperature, which reduced the solubility of the starch and then the precipitate was heated at 70 °C to dissolve a small portion of the precipitate. Reprecipitation was then employed by cooling of the suspension. This repetition of the dissolving and precipitation processes improved the temperature stability of the resulting aqueous dispersion (Harris and Little 1995). Increased yields of RS were obtained by subjecting the starch to enzymatic hydrolysis (pullulanase, 40 U/g/10 h), autoclaving (121 °C/30 min), storing under refrigeration (4 °C/24 h), and lyophilizing (Reddy et al. 2013).

SDS has been prepared by debranching starch using pullulanase or isoamylase (Shi et al. 2003). In the case of waxy starches, shorter debranching time and high concentrations of debranching enzymes are more suitable for debranching starch to form SDS (Guraya et al. 2001b). A low GI maize starch with some branched structure has been developed by partial α-amylase treatment and retrogradation, and the slow digestibility was retained even after cooking (Han et al. 2006). Shorter chains of amylopectin and noncrystalline amylose molecules were rapidly digested, while DP_n 121 chains showed the greatest resistance to digestion, followed by DP_n 46 chains. A similar trend was reported in the formation of SDS from commercial starch by controlling the hydrolysis of gelatinized starch with α-amylase (Hamaker and Han et al. 2006). A novel slowly digestible storage carbohydrate comprising of more than 90% amylopectin was produced by treating a native root or tuber starch with a branching enzyme derived from a microorganism with a branching degree of at least 8.5–9% (Vander-Maarel et al. 2008). Moreover, it was reported that both the increase in branch density and the crystalline structure of starch enhances its slow digestion property through the partial shortening of amylopectin A and B1 exterior chains, as well as linear chains of amylose, through the action of β-amylase and maltogenic α-amylase (Ao et al. 2007). This correlated closely with an increase in the number of α-1, 6 linkages and a simultaneous decrease in the number of α-1, 4 linkages. The enzyme-treated starch consists of B- and V-type crystalline structures, which increases the resistance of starch to digestion. These studies suggested that enzymatic debranching of the exterior chains of amylopectin molecule can change its structure and form higher proportions of SDS and RS.

Chemical Modifications

In many processes, starch is being modified by chemical reagents to improve functionality and create commercially valuable, starch based products. The most common chemical treatments are acid treatment, oxidation, cross-linking and substitution including esterification and etherification. Recently studies have focused on such treatments in SDS and RS production (Zhao et al. 2012; Ashwar et al. 2016; Ashwar et al. 2018). The enzymatic resistance in RS4 is done by cross linking with chemical agents (Haynes et al. 2000). Cross linked starches are obtained by reaction of starch with bi- or polyfunctional reagents like phosphorus oxychloride, sodium trimetaphosphate, or mixed anhydrides of acetic acid and dicarboxylic acids like adipic acid. Cross-linking of rice, wheat, corn, potato, tapioca, oat and mung bean starches using sodium trimetaphosphate (STMP), sodium tripolyphosphate (STPP), epichlorohydrin or phosphoryl chloride ($POCl_3$), produced type 4 resistant starch (Seib and Woo 1999; Zhao et al. 2012). These authors explained that the levels of RS in wheat starch cross-linked with 2% $POCl_3$, 12% STMP/STPP, and 2% epichlorohydrin were 85.6, 75.6 and 75.8 g/100 g starch, respectively. Sang et al. (2010) prepared phosphorylated wheat starch with high levels of the RS (68.7%) and SDS (24.4%). Cross-linking when carried out by sulphonate and phosphate groups between starch molecules through their hydroxyl groups brings resistance to enzymatic hydrolysis (Hamilton and Paschall 1967). Cross-linking of starch with mixtures of STMP and STPP under alkaline conditions restricts swelling of starch and imparts increasing resistance to digestive enzymes (Woo and Seib 2002). Simsek et al. (2012) prepared acetylated bean starch with high levels of the RS (44%). Acetylation of starch increased the RS content in bean, as a result of acetyl groups which blocks the action of digestive enzymes (Chung et al. 2008). Modification of starch with octenyl succinic anhydride is known to increase levels of SDS and RS more than other modifications such as acetylation, hydroxypropylation, or crosslinking (Han and BeMiller 2007; Juansang et al. 2012). Esterification with octenyl succinic anhydride (OSA) has been shown to be the most potent method of modifying waxy starch to form SDS, followed by combined modifications like crosslinking-hydroxypropylation, acetylation and crosslinking (Han and BeMiller 2007). The modified starch with attached OSA molecules may act as uncompetitive inhibitors to reduce the enzyme activity and thereby cause slow digestion of starch. As these studies showed, chemical modifications of starch can be used to prepare SDS and RS, but clinical and toxicological trials needs to be performed in order to evaluate the safety and efficacy of SDS and RS consumption.

Genetic Modification

Genetic modification of starch biosynthesis has been used to develop a strategy for generating new cultivars with desired functionality through extensive breeding and characterization of the resulting cultivars. Genetically controlled factors that affect the starch functionality include structure of starch, content of starch, interaction of cell components, and starch granule architecture. Waxy starches may be more suitable for developing SDS, since their fine amylopectin structure (i.e., the distribution of branches and chain length) is more critical for SDS formation (Guraya et al. 2001b). In one study, SDS (long-chain amylopectin starch) was developed from maize by over expressing a particular enzyme involved in starch biosynthesis (Moallic et al. 2006). This starch, with high granule crystallinity, has few short chains of amylopectin and more intermediate and long chains. Other study showed that genetic mutants containing amylopectin molecules with either a high proportion of short chains with $DP < 13$ (particularly A chains with DP 5–9) or a high proportion of long chains with $DP \geq 13$ (particularly intermediate to long B chains with $DP > 30$) contain greater proportions of SDS than wild type (Zhang et al. 2008). According to a study by Benmoussa et al. (2007), development of SDS and RS is positively correlated with the presence of both long and intermediate/short chains, respectively, while it is negatively correlated with the lowest proportion of extremely short chains. They also found that the channels in starch granules can regulate starch digestibility, since starch granules with channels are digested from the interior, and more extensive channelization of starch gives more access to the hydrolytic enzymes (Benmoussa et al. 2006). Therefore, genetic modification has the potential to produce ideal starch with high contents of SDS and RS.

Factors Influencing the Formation of RS and SDS

Various factors have been reported to influence the formation of RS and SDS. These are described as:

Starch Structure

Starch is semicrystalline in nature synthesized mostly as spherical granules in plant tissues. These granules are composed of alternating concentric layers of ordered crystalline and less-ordered amorphous lamellae extending from hilum to the surface of granules. The crystalline lamellae are formed from the amylopectin short branch chains arranged in clusters; these crystalline lamellae are interspersed with amorphous lamellae that consist of branching points of amylopectin and amylose molecules. The sub-chains of the amylopectin has been classified into three types. A chains (outer chains) are the shortest among the three chains (CL 6–15) and are α-(1, 6)-linked to B chains. The B chains are linked in the same way and bear one or more A chains and/or B chains. Depending on their respective length and the number of clusters they span, B chains are further classified into B1, B2, B3 and B4 chains (with one to four clusters). B1 and B2 chains have CL of 15–25 and 40–50, respectively, with B3 and B4 chains being much longer. The single C chain per amylopectin molecule contains the sole terminal reducing group and carries other chains (Donald 2004). Within this structure, the linear chains lie in the region of high molecular order, and the branch-points lie in the region of low molecular order. These linear chains can form double helices to make up the crystalline structure. On the basis of wide-angle X-ray diffraction scattering studies, native starch has been classified into four types viz. A, B, C and V. The A type is characteristic of most cereal starches, while the B type is characteristic of potato starch, other root starches, amylomaize starch, and retrograded starch. The C type is a combination of A and B types, and is found in smooth pea and various bean starches. The V type can be found only in starch after gelatinization and the formation of amylose helical complexes with lipids or related compounds.

Studies of X-ray diffraction of RS showed that chain fragments were packed in a B type crystalline structure with enlarged crystal lattice which contributes to the formation of RS. Any treatment that eliminates starch crystallinity (e.g., gelatinization) or damages the integrity of the plant cell or tissue structure (e.g., milling) increases access to enzymes and reduces the RS content, whereas recrystallization and chemical modifications increases the RS content (Englyst and Cummings 1986; Adebowale et al. 2009; Kim and White 2013). Further high amylose content of starch is known to lower starch digestibility (Chung et al. 2009). High amylose maize starches with very long chains might be perfectly ordered into double helices to form resistant starch (Ozturk et al. 2011). Higher contents of resistant starch were found in Hylon VII than in Hylon V (high-amylose genetically modified corn starches) which might be because of higher amylose content in Hylon VII (Dimantov et al. 2004). Margareta Leeman et al. (2006) claimed that high amylose starch resists enzymatic digestion due to its internal structure and B-type crystallinity. Native cereal starch has been classified as an ideal SDS, since its structure makes it to be digested slowly (Zhang et al. 2006a, b). They found that the A-type semicrystalline structure of native cereal starch, including the distribution of perfect crystalline regions in both crystalline and amorphous lamellae explains this slow digestion property. The high proportion of SDS in cereal starch was also correlated with higher proportion of short A chains with *DP* 5–10. The mechanism of slow digestion property of native cereal starch involves enzymatic digestion from inside out and layer-by-layer. Enzymatic digestion begins in interior channels and at surface pores, and then side-by-side digestion gradually enlarges the channel by simultaneously hydrolyzing crystalline and amorphous regions. Native starch is hydrolysed more slowly than gelatinized starch; since gelatinization has lost the crystalline structure of starch, allowing greater access to enzymes without the obstructions caused by α-glucan associations, such as double helices or by amylose-lipid complexes in cereal starches (Tester et al. 2002). Other studies claimed that the dispersed, amylopectin fine structures with high branch density, either long or short internal chains as well as short terminal non reducing ends, leads to the slow digestion property, because of the inherent structure of amylopectin molecules (Hamaker et al. 2007; Zhang et al. 2008). The plasma glucose response after consuming raw maize starch was slow and sustained, which is characteristic of SDS (Seal et al. 2003). The structure of SDS is composed of imperfect crystallites and amylopectin with a high branching pattern and density, and this is most likely the cause of slow digestion property.

Heat and Moisture

Heat and moisture content are important factors for development of SDS and RS. When native starch is heated in excess water, the starch granules undergo gelatinization. The extent of gelatinization depends on the temperature, time, water content and degree of shear during the process. As described previously, native starch (A-type) is an ideal SDS and the slow digestibility changes during cooking or processing. Incomplete gelatinization can be achieved by lowering the temperature, decreasing the moisture content, or shortening the heating time. In this way, low GI benefits of SDS and RS may be retained. When partially gelatinized waxy rice starch was heated at different temperatures (60, 65, or 70 °C) for 5 min, they showed different digestibility rates after retrogradation (Chung et al. 2006). The amounts of SDS and RS positively correlated with the relative enthalpy of the partially gelatinized starches. In cereal products, such as parboiled rice, barley porridges, biscuits and pasta, the degree of gelatinization or limited swelling of starch, which is determined mainly by the cooking time and temperature, moisture level, largely influences the formation of SDS and RS (Wolever et al. 1986a, b; Holm et al. 1992; Granfeldt et al. 1994; Garsetti et al. 2005). Heat-moisture treatment usually refers to the incubation of starch at low moisture content (<35% w/w) for a certain period of time at a temperature below the gelatinization temperature, but above the glass transition temperature, while as annealing is performed in excess water or at an intermediate water level (≥40% w/w) (Jacobs and Delcour 1998; Tester and Debon 2000). Heat–moisture treatment does not destroy structure of starch granules, but it alters its crystalline packing; for example, the B type of starch can be converted to the A or C type, whereas the annealing technique can modify the binding forces between the crystalline and the amorphous matrix (Stute 1992). Therefore, hydrothermal treatment can be used as a method to form SDS and RS. Anderson et al. (2002) adjusted both nonwaxy and waxy rice starches to 20% moisture, after that heated them to their melting temperature (*T*m) in a differential scanning calorimeter (DSC), and held them there for 60 min. They observed that these starches were more slowly digested than unheated samples. Severijnen et al. (2007) used these principles to study the production of a sterilized liquid product with a low GI. When the modified high amylose starch was heated above 120 °C for 4–5 min., the SDS content increased and reached a maximum, where it remains stable for several months when stored at 4 °C. According to Woortman and Steeneken (2004), high SDS content was achieved when a starch product with an amylose content of below 50% was heated to at least 170 °C under mild acidic conditions, followed by rapid cooling. High amylose starch is rich source of RS2 (Berry 1986), which after heating and cooling gives RS3 in high yields (Sievert and Pomeranz 1989) or retrograded starch (Englyst et al. 1992). Retrograded amylose in wheat, maize, peas and potatoes was found to be highly resistant to digestion (Ring et al. 1988). Park et al. (2009) reported that temperature cycled storage increased the formation of resistant starch and reduced the GI (glycemic index) of waxy corn starch. Borczak et al. (2014) claimed that prolonged frozen storage of wheat-flour rolls significantly increased RS formation. Dual-retrogradation treatment was more efficient as compared to single retrogradation (Tian et al. 2013). Repeated autoclaving of wheat starch increased the RS upto 10%. Retrogradation of amylose was recognized as the main factor for the formation of RS and higher amounts were obtained with repeated autoclaving (Dundar and Gocmen 2013; Bjorck et al. 1990). On storage, gelatinized starch pastes undergo retrogradation to semicrystalline structure that resists enzymatic digestion. Wheat bread and corn flakes are rich sources of this type of RS where as cooked and cooled potatoes have only 25% of RS3 (retrograded starch) (Englyst and Cummings 1985).

Interactions of Starch with Other Components

Interactions of starch with other food components are known to influence the formation of SDS or RS. Two important types of starch interaction with other components involve formation of starch-lipid complexes and starch-protein interactions. Interaction of starch with the protein is thought to reduce the rate of α-amylolysis of starch in cereal and legume products (Wursch et al. 1986; Jenkins et al. 1987; Colonna et al. 1990; Biliaderis 1991). According to Granfeldt and Bjorck (1991), a dense and viscoelastic gluten network surrounds the starch granules in pasta products which restricts the swelling and leaching of starch molecules during boiling and also reduces the access of enzymes to the starch. The interaction of starch with protein also limited the glycemic response of starch in white bread made from regular flour, while gluten-free bread showed a higher glycemic and insulinemic response (Jenkins et al. 1987). In another study, mixture of potato starch and albumin protein was autoclaved and then cooled to −20 °C, and effect of albumin on digestibility of potato starch was studied. It was found that added albumin reduced the content of resistant starch (Escarpa et al. 1997). In a study by Holm et al. (1983), it was claimed that amylose molecules formed complexes with lysolecithin, and these complexes were degraded slowly and were completely absorbed in the GI tract of rats within

120 min. As a result, the amylose complex produces lower plasma glucose and liver glycogen levels than does free amylose. Murray et al. (1998) evaluated apparent digestibility of starch in ileal-cannulated dogs that were fed enteral diets containing debranched amylopectin-lipid V-complex or RS. They found that the ileal and total GI tract digestibilities of the control, V-complex, and RS diets were 89%, 76%, and 43%, respectively, which indicated that the diet containing V-complex starch lowered the carbohydrate digestibility and hence the serum glucose and insulin responses. Differential scanning calorimetry (DSC) was used to study the effect of sodium stearoyl lactylate (SSL), lysophosphatidyl choline (LPC), and hydroxylated lecithin (OHL) on autoclaved amylomaize starch (Czuchajowska et al. 1991). Differential scanning calorimetry (DSC) peaks at around 95–110 °C indicated the formation of complex compounds between amylose chains and lipid, and the peak at about 155 °C indicated the presence of resistant starch (RS). However lower yields of RS were observed from lipid complexed samples as compared to autoclaved and cooled control when subjected to amylolysis by thermostable bacterial α-amylase and amyloglucosidase. Amylose recrystallization which is important in resistant starch formation is adversely affected by complexation of amylose with LPC and SSL. In another study, influence of endogenous lipids on wheat starch showed that defatting of the starch samples resulted in decrease of the RS content. On addition of SDS to defatted wheat or amylomaize starch, resistant starch yields decreased significantly. X-ray diffraction and DSC techniques confirmed formation of amylose-lipid complexes in the presence of both endogenous lipids as well as added lipids (SDS) (Eerlingen et al. 1994).

In addition to the interaction of starch with lipids and proteins, it has also been found that starch can interact with soluble fibers (β-glucans, guar gum, psyllium, or pectin), antinutrients (enzyme inhibitors, phytates, tannins, lectins or saponins), organic acids and sugars (Biliaderis 1991; Bjorck et al. 2000; Pi-Sunyer 2002). Enzyme inhibitors like phytic acid, polyphenols, and lectins present in leguminous seeds, have been found to inhibit in vitro digestion and hence the glycemic index of starch (Thompson and Yoon 1984). Both amylases and intestinal maltase activity are inhibited by tannic acid (Bjorck et al. 1987). Since phytic acid inhibits the amylolysis, an increase in phytate content decreases starch digestibility (Thompson and Yoon 1984). Brennan et al. (1996) reported that the rate of starch hydrolysis slowed down significantly when the starch granules and surrounding bread matrix were coated with a layer of galactomannan mucilage, which acted as a physical barrier to enzyme-starch interactions and hence the release of hydrolyzed products. Guar gum has been found to increase the viscosity of digesta and reduce the rise in postprandial glycemic response that occurs due to the reduction in rate of gastric emptying. Starch blockers (α- amylase inhibitors) inhibits in vitro α-amylase activity or binds to starch substrate, indicating that these have the potential to interfere with the digestion of starch in vivo and hence modulate the glycemic effect of SDS and RS (Giri and Kachole 1998; Obiro et al. 2008). Potato starch gels showed decreased yield of resistant starch in the presence of ions like calcium and potassium (Escarpa et al. 1997) which may be reflected to the prevention of hydrogen bond formation between amylopectin and amylose chains.

Processing Conditions

Processing/cooking, post processing and storage conditions (retrogradation) influence the formation of SDS and RS in food. This fact is of great concern for the food industry, since it offers the possibility of increasing the SDS and RS contents of processed foods and foodstuffs. Autoclaving, pressure-cooking, baking, flaking and parboiling, among other methods, are known to influence the starch digestibility and the yield of RS and SDS (Brand et al. 1985; Holm et al. 1985; Casiraghi et al. 1993; Kingman and Englyst 1994; Ashwar et al. 2016; Ashwar et al. 2017). Holm et al. (1985) demonstrated that starch in flaked whole grained wheat was more resistant to digestion than that in boiled, steam-cooked and popped wheat during in vitro assay. In a study of Casiraghi et al. (1993), both parboiled and quick cooking parboiled rice were digested more slowly with a lower GI than polished rice, which was related to the availability of starch for α-amylase. Autoclaving of red kidney beans has been shown to increase the blood glucose and insulin response as compared to boiling at atmospheric pressure, and this may be due to thermal or mechanical alteration of the structure of seeds and may also be related to the release of physically inaccessible starch due to mechanical disruption of cell walls (Tovar et al. 1992). Granfeldt et al. (2000) reported that thick rolled oats cause lower glycemic responses than thin flakes or reference bread. Boiling and pressure-cooking significantly decreased the SDS content in three *Doongara*, *Inga* and *Japonica* varieties of rice and the amylose content affected starch digestibility, which has been attributed to the retrogradation process (Sagum and Arcot 2000). According to Guraya et al. (2001a), when 10% waxy and nonwaxy starch suspensions were debranched with pullulanase, followed by heating and cooling treatments to allow the crystallization or gelling to occur, the digestibility decreased because of the formation of crystalline structures or double helices. During pullulanase debranching and retrogradation treatments of the cooked waxy maize starch suspensions, short-term retrogradation have occurred as a result of gelation and crystallization of amylose molecules, leading to the formation of SDS; in contrast, long-term retrogradation occurs during storage of starch gels due to the amylopectin molecules (Miao et al. 2009).

Determination of Resistant Starch and Slowly Digestible Starch

Different methods have been developed for the determination of RS with significant differences in sample preparation, enzymes used, and the establishment of experimental conditions that mimic gastrointestinal environment. Most of the methods are employed for the determination of total RS, but some specific methods have been developed for the quantification of RS1, RS2 and RS3.

Basic method for the determination of RS was proposed by Englyst et al. (1982). Briefly, in this method 100–200 mg sample is mixed with sodium acetate buffer at pH 5.4 and heated for 1 h at 100 °C to gelatinize the starch. An enzyme mixture (α-amylase, pullulanase and amyloglucosidase) is added and hydrolysis is carried out at 40 °C for 16 h. Absolute ethanol is added to precipitate nonhydrolyzed starch and to terminate enzymatic activity. The pellet is collected by centrifugation and washed with 80% ethanol twice. The residue is dried with acetone and then treated with 2 M KOH for 30 min at room temperature to solubilize the starch. An aliquot of alkali digest is mixed with 2 M acetic acid and amyloglucosidase and then incubated at 65 °C for 1 h. After cooling and centrifugation, neutral sugars (glucose, mannose, galactose, arabinose and xylose) in the supernatant are analyzed by GLC (gas–liquid chromatography). The amount of glucose detected by GLC represents the amount of resistant starch (RS3) in the sample.

The above procedure was later modified by Englyst et al. (1992) to better mimic human gastrointestinal conditions. The authors proposed that enzymatic hydrolysis can be carried out at 37 °C instead of at 40 °C. Also significant reduction in the duration of enzymatic digestion is used. Instead of 16 h of sample digestion with enzymes, the protocol uses sequential removal of aliquots of enzyme digest at 20 min and at 120 min from the beginning of the digestion. The rationale to terminate enzymatic hydrolysis at 120 min is that the release of glucose reached a plateau at this time interval. The starch fraction digested within 20 min of incubation is classified as RDS, the starch fraction digested between 20 and 120 min is classified as SDS, and the remaining fraction that is not further digested is RS.

Goni et al. (1996) further modified the procedure by decreasing the sample pH to 1.5 with HCl–KCl buffer to simulate gastric pH. Hydrolysis with pepsin at 40 °C for 1 h is carried out. To simulate conditions in the small intestine sample pH is adjusted to 6.9 with tris–maleate buffer. Instead of using a group of enzymes, hydrolysis is carried out with α-amylase only for 16 h at 37 °C. Glucose content is determined by the colorimetric method.

The Megazyme assay kit is widely used in analytical laboratories for the determination of RS and is the basis of both AACC Method 32–40 and AOAC method 2002.02 (Megazyme 2008). In this method samples are ground to a coarse meal which can pass a 1-mm sieve. Initial boiling of samples in acetate buffer and the use of pullulanase is eliminated. Instead a mixture of α-amylase and amyloglucosidase is employed to hydrolyze starch in raw or processed food samples. Hydrolysis is carried out at 37 °C for 16 h. The incubation time with amyloglucosidase is 30 min at 50 °C. This method uses the glucose oxidase-peroxidase colorimetric assay (GOPOD) to determine glucose concentration in the final hydrolysate. The Megazyme protocol for RS determination is not applicable to the determination of SDS and RDS.

Other modifications have been made to the procedure of RS determination particularly in the sample preparation step with the intention of simulating the in vivo digestion. However it is obvious that significantly different levels of RS would be detected in similar foods because of wide variations in analytical protocals, including differences in the enzymes used and in their activity, concentration and sequence of application, and dissimilarities in the conditions of experimental protocals.

Conclusion

Epidemiological studies suggest that low-GI foods have beneficial metabolic effects like the potential to reduce insulin resistance and improve certain metabolic conditions. However, there are very few commercially available low-GI foods in the market, which fails to meet the growing needs of patients with diabetes, obesity and related disorders. SDS and RS as the novel functional components in products deliver a slow and prolonged release of glucose when ingested, resulting in a lower GI. RS has recently been recognized as a dietary starch that escapes digestion in the small intestine and is fermented in the colon producing short chain fatty acids that provides various health benefits including prevention of colon cancer, reduced risk of diabetes, reduction in total cholesterol, promotion of growth of beneficial microflora of colon, improved laxation etc. Interaction of RS and SDS with certain nutrients like wheat bran, cellulose and protein promotes its health benefits. Technically, it is possible to prepare RS and SDS by different physical, enzymatic and chemical modifications. The unique properties of RS and SDS, that is, bland flavor, fine particle size, high gelatinization temperature, good extrusion and film forming qualities and lower water holding properties make the formulation of wide range of food products possible with improved organoleptic qualities as compared with traditional high-fibre products.

Acknowledgements Dr. Bilal Ahmad Ashwar is highly thankful to Indian Council of Medical Research (ICMR), Government of India for providing Research Associate Fellowship [3/1/2/117/2019-(Nutri)].

References

Adebowale, K. O., Henle, T., Schwarzenbolz, U., & Doert, T. (2009). Modification and properties of African yam bean (*Sphenostylis stenocarpa* Hochst. Ex A. Rich.) Harms starch I: Heat moisture treatments and annealing. *Food Hydrocolloids, 23*, 1947–1957.

Anderson, A. K., Guraya, H. S., James, C., & Salvaggio, L. (2002). Digestibility and pasting properties of rice starch heat-moisture treated at the melting temperature (Tm). *Starch-Starke, 54*, 401–409.

Ao, Z., Simsek, S., Zhang, G., Venkatachalam, M., Reuhs, B. L., & Hamaker, B. R. (2007). Starch with a slow digestion property produced by altering its chain length, branch density, and crystalline structure. *Journal of Agricultural and Food Chemistry, 55*, 4540–4547.

Ashwar, B. A., Gani, A., Shah, A., & Masoodi, F. A. (2017). Production of RS4 from rice by acetylation: Physico-chemical, thermal, and structural characterization. *Starch-Starke, 69*, 1–10.

Ashwar, B. A., Gani, A., Wani, I. A., Shah, A., Masoodi, F. A., & Saxena, D. C. (2016). Production of resistant starch from rice by dual autoclaving- retrogradation treatment: Invitro digestibility, thermal and structural characterization. *Food Hydrocolloids, 56*, 108–117.

Ashwar, B. A., Gani, A., Gani, A., Shah, A., & Masoodi, F. A. (2018). Production of RS4 from rice starch and its utilization as an encapsulating agent for targeted delivery of probiotics. *Food Chemistry, 239*, 287–294.

Bednar, G. E., Patil, A. R., Murray, S. M., Grieshop, C. M., Merchen, N. R., & Fahey, G. C. (2001). Starch and fibre fractions in selected food and feed ingredients affect their small intestinal

digestibility and fermentability and their large bowel fermentability in vitro in a canine model. *Journal of Nutrition, 131*, 276–286.
Benmoussa, M., Moldenhauer, K. A. K., & Hamaker, B. R. (2007). Rice amylopectin fine structure variability affects starch digestion properties. *Journal of Agricultural and Food Chemistry, 55*, 1475–1479.
Benmoussa, M., Suhendra, B., Aboubacar, A., & Hamaker, B. R. (2006). Distinctive sorghum starch granule morphologies appear to improve raw starch digestibility. *Starch-Starke, 58*, 92–99.
Berry, C. S. (1986). Resistant starch: Formation and measurement of starch that survives exhaustive digestion with amylolytic enzymes during the determination of dietary fibre. *Journal of Cereal Science, 4*, 301–314.
Biliaderis, C. G. (1991). The structure and interactions of starch with food constituents. *Canadian Journal of Physiology and Pharmacology, 69*, 60–78.
Bjorck, I., Eliasson, A. C., Drews, A., Gudmundsson, M., & Karlson, R. (1990). Some nutritional properties of starch and dietary fiber in barley genotypes containing different levels of amylose. *Cereal Chemistry, 67*, 367–333.
Bjorck, I., Liljeberg, H., & Ostman, E. (2000). Low glycaemic-index food. *British Journal of Nutrition, 83*, 149S–155S.
Bjorck, I., Nyman, M., Pedersen, P., Siljestrom, M., Asp, N. G., & Eggum, B. O. (1987). Formation of enzyme resistant starch during autoclaving of wheat starch: Studies in vitro and in vivo. *Journal of Cereal Science, 6*, 159–172.
Borczak, B., Sikora, E., Sikora, M., & Kapusta-Duch, J. (2014). The influence of prolonged frozen storage of wheat-flour rolls on resistant starch development. *Starch-Starke, 66*, 533–538.
Brand, J. C., Nicholson, P. L., Thorburn, A. W., & Truswell, A. S. (1985). Food processing and the glycemic index. *American Journal of Clinical Nutrition, 42*, 1192–1196.
Brennan, C. S., Blake, D. E., Ellis, P. R., & Schofield, J. D. (1996). Effects of guar galactomannan on wheat bread microstructure and on the in vitro and in vivo digestibility of starch in bread. *Journal of Cereal Science, 24*, 151–160.
Brumovsky, J. O., & Thompson, D. B. (2001). Production of boiling-stable granular resistant starch by partial acid hydrolysis and hydrothermal treatments of high-amylose maize starch. *Cereal Chemistry, 78*(6), 680–689.
Casiraghi, M. C., Brighenti, F., Pellegrini, N., Leopardi, E., & Testolin, G. (1993). Effect of processing on rice starch digestibility evaluated by in vivo and in vitro methods. *Journal of Cereal Science, 17*, 147–156.
Charalampopoulos, D., Wang, R., Pandiella, S. S., & Webb, C. (2002). Application of cereals and cereal components in functional foods: A review. *International Journal of Food Microbiology, 79*, 131–141.
Chung, H., Liu, Q., Donner, E., Hoover, R., Warkentin, T. D., & Vandenberg, B. (2008). Composition, molecular structure, properties and in vitro digestibility of starches from newly released Canadian pulse cultivars. *Cereal Chemistry, 85*(4), 471–479.
Chung, H., Liu, Q., & Hoover, R. (2009). Impact of annealing and heat-moisture treatment on rapidly digestible, slowly digestible and resistant starch levels in native and gelatinized corn pea and lentil starches. *Carbohydrate Polymers, 75*, 436–447.
Chung, H. J., Lim, H. S., & Lim, S. T. (2006). Effect of partial gelatinization and retrogradation on the enzymatic digestion of waxy rice starch. *Journal of Cereal Science, 43*, 353–359.
Colonna, P., Barry, J. L., Cloarec, D., Bornet, F., Gouilloud, S., & Galmiche, J. P. (1990). Enzymic susceptibility of starch from pasta. *Journal of Cereal Science, 11*, 59–70.
Czuchajowska, Z., Sievert, D., & Pomeranz, Y. (1991). Enzyme-resistant starch. IV. Effects of complexing lipids. *Cereal Chemistry, 68*(5), 537–542.
Dimantov, A., Kesselman, E., & Shimoni, E. (2004). Surface characterization and dissolution properties of high amylose corn starch-pectin coatings. *Food Hydrocolloids, 18*(9), 29–37.
Donald, A. M. (2004). Understanding starch structure and functionality. In A. C. Eliasson (Ed.), *Starch in food: Structure, function and applications* (pp. 156–184). Boca Raton, FL: CRC Press Inc..

Dundar, A. N., & Gocmen, D. (2013). Effects of autoclaving temperature and storing time on resistant starch formation, and its functional and physicochemical properties. *Carbohydrate Polymers, 97*(2), 764–771.

Eerlingen, R. C., Crombez, M., & Delcour, J. A. (1993a). Enzyme-resistant starch. I. Quantitative and qualitative influence of incubation time and temperature of autoclaved starch on resistant starch formation. *Cereal Chemistry, 70*(3), 339–344.

Eerlingen, R. C., Deceuninck, M., & Delcour, J. A. (1993b). Enzyme-resistant starch. II. Influence of amylose chain length on resistant starch formation. *Cereal Chemistry, 70*(3), 345–350.

Eerlingen, R. C., Vanden-Broeck, I., Delcour, J. A., & Levine, H. (1994). Enzyme resistant starch. VI. Influence of sugars on resistant starch formation. *Cereal Chemistry, 70*, 345.

Englyst, H. N., & Cummings, J. H. (1985). Digestion of the polysaccharides of some cereal foods in the human small intestine. *American Journal of Clinical Nutrition, 42*, 778–787.

Englyst, H. N., & Cummings, J. H. (1986). Digestion of the carbohydrates of banana (*Musa paradisiaca sapientum*) in the human small intestine. *American Journal of Clinical Nutrition, 44*, 42–50.

Englyst, H. N., Kingman, S. M., & Cummings, J. H. (1992). Classification and measurement of nutritionally important starch fractions. *European Journal of Clinical Nutrition, 46*, S33–S50.

Englyst, H. N., Kingman, S. M., & Judson, G. J. (1996). Measurement of resistant starch in vitro and in vivo. *British Journal of Nutrition, 75*, 749–755.

Englyst, H. N., Wiggins, H. S., & Cummings, J. H. (1982). Determination of the non-starch polysaccharides in plant foods by gas-liquid chromatography of constituent sugars as alditol acetates. *Analyst, 107*, 307–318.

Escarpa, A., Gonzalez, M. C., Morales, M. D., & Saura-Calixto, F. (1997). An approach to the influence of nutrients and other food constituents on resistant starch formation. *Food Chemistry, 60*(4), 527–536.

Fuentes-Zaragoza, E., Sanchez-Zapata, E., Sendra, E., Sayas, E., Navarro, C., & Fernandez-Lopez, J. (2011). Resistant starch as prebiotic: A review. *Starch-Starke, 63*(7), 406–415.

Gani, A., Ashwar, B. A., Akhter, G., Gani, A., Shah, A., Masoodi, F. A., & Wani, I. A. (2020). Resistant starch from five Himalayan rice cultivars and Horse chestnut: Extraction method optimization and characterization. *Scientific Reports, 10*(1), 1–9.

Garcia-Alonso, A., Jimenez-Escrig, A., Martin-Carron, N., Bravo, L., & Saura-Calixto, F. (1999). Assessment of some parameters involved in the gelatinization and retrogradation of starch. *Food Chemistry, 66*, 181–187.

Garsetti, M., Vinoy, S., Lang, V., Holt, S., Loyer, S., & Brand-Miller, J. C. (2005). The glycemic and insulinemic index of plain sweet biscuits: Relationships to invitro starch digestibility. *Journal of the American College of Nutrition, 24*, 441–447.

Giczewska, A., & Borowska, J. (2003). Nutritional value of broad bean seeds. Part 1: Starch and fibre. *Food, 47*(2), 95–97.

Gidley, M. J., Cooke, D., Drake, A. H., Hoffman, R. A., Russell, A. L., & Greenwell, P. (1995). Molecular order and structure in enzyme-resistant retrograded starch. *Carbohydrate Polymers, 28*, 23–31.

Giri, A. P., & Kachole, M. S. (1998). Amylase inhibitors of pigeon pea (*Cajanus cajan*) seeds. *Phytochemistry, 47*, 197–202.

Goni, I., Garcia-Diza, L., Manasb, E., & Saura-Calixto, F. (1996). Analysis of resistant starch: A method for foods and food product. *Food Chemistry, 56*(4), 445–459.

Granfeldt, Y., & Bjorck, I. (1991). Glycemic response to starch in pasta: A study of mechanisms of limited enzyme availability. *Journal of Cereal Science, 14*, 47–61.

Granfeldt, Y., Eliasson, A. C., & Bjorck, I. (2000). An examination of the possibility of lowering the glycemic index of oat and barley flakes by minimal processing. *Journal of Nutrition, 130*, 2207–2214.

Granfeldt, Y., Liljeberg, H., Drews, A., Newman, R., & Bjorck, I. (1994). Glucose and insulin responses to barley products: Influence of food structure and amylose-amylopectin ratio. *American Journal of Clinical Nutrition, 59*, 1075–1082.

Guraya, H. S., James, C., & Champagne, E. T. (2001a). Effect of cooling and freezing on the digestibility of debranched rice starch and physical properties of the resulting material. *Starch-Starke, 53*, 64–74.

Guraya, H. S., James, C., & Champagne, E. T. (2001b). Effect of enzyme concentration and storage temperature on the formation of slowly digestible starch from cooked debranched rice starch. *Starch-Starke, 53*, 131–139.

Hamaker, B. R., Venkatachalam, M., Zhang, G., Keshavarzin, A., & Rose, D. J. (2007). *Slowly digesting starch and fermentable fiber*. US Patent 2007196437.

Hamilton, R. M., & Paschall, E. F. (1967). Production and uses of starch phosphates. In R. L. Whistler & E. F. Paschall (Eds.), *Starch: Chemistry & technology* (Vol. II, pp. 351–365). New York: Academic Press.

Han, J. A., & BeMiller, J. N. (2007). Preparation and physical characteristics of slowly digesting modified food starches. *Carbohydrate Polymers, 67*, 366–374.

Han, X. Z., Ao, Z., Janaswamy, S., Jane, J. L., Chandrasekaran, R., & Hamaker, B. R. (2006). Development of a low glycemic maize starch: Preparation and characterization. *Biomacromolecules, 7*, 1162–1168.

Haralampu, S. G. (2000). Resistant starch: A review of the physical properties and biological impact of RS3. *Carbohydrate Polymers, 41*, 285–292.

Haralampu, S.G., & Gross, A. (1998). *Granular RS and method of making*. U.S. Patent 5849 090. 15 Dec 1998. Bedford MA: Opta Food Ingredients Inc.

Harris, D. W., & Little, J. A. (1995). *Method of preparing reduced fat foods*. U.S. Patent 5395640. 7 Mar 1995. Decatur, IL: Stanley Manufacturing Co.

Haynes, L., Gimmler, N., Locke, J. P., Mee-Ra, K., Slade, L., & Levine, H. (2000) *Process for making enzyme-resistant starch for reduced-calorie flour replacer*. U.S. Patent 6013299.

Hernandez, O., Emaldi, U., & Tovar, J. (2008). In vitro digestibility of edible films from various starch sources. *Carbohydrate Polymers, 71*, 648–655.

Holm, J., Bjorck, I., Asp, N. G., Sjoberg, L. B., & Lundquist, I. (1985). Starch availability in vitro and in vivo after flaking, steam-cooking and popping of wheat. *Journal of Cereal Science, 3*, 193–206.

Holm, J., Bjorck, I., Ostrowska, S., Eliasson, A. C., Asp, N. G., Larsson, K., & Lundquist, I. (1983). Digestibility of amylose-lipid complexes in-vitro and in-vivo. *Starch-Starke, 35*, 294–297.

Holm, J., Koellreutter, B., & Wursch, P. (1992). Influence of sterilization, drying and oat bran enrichment of pasta on glucose and insulin responses in healthy subjects and on the rate and extent of in vitro starch digestion. *European Journal of Clinical Nutrition, 46*, 629–640.

Holm, J., Lundquist, I., Bjorck, I., Eliasson, A. C., & Asp, N. G. (1988). Relationship between degree of gelatinization, digestion rate in vitro and metabolic response in rats. *American Journal of Clinical Nutrition, 47*, 1010–1016.

Innereber, F., & Mueller, R. (2005). *Slowly digestible starch product*. WO Patent 2005058974.

Jacobs, H., & Delcour, J. A. (1998). Hydrothermal modifications of granular starch with retention of the granular structure: A review. *Journal of Agricultural and Food Chemistry, 46*, 2895–2905.

Jenkins, D. J. A., Thorne, M. J., Wolever, T. M. S., Jenkins, A. L., Rao, A. V., & Thompson, L. U. (1987). The effect of starch-protein interaction in wheat on the glycemic response and rate of in vitro digestion. *American Journal of Clinical Nutrition, 45*, 946–951.

Juansang, J., Puttanlek, C., Rungsardthong, V., Punchaarnon, S., & Uttapap, D. (2012). Effect of gelatinisation on slowly digestible starch and resistant starch of heat moisture treated and chemically modified canna starches. *Food Chemistry, 131*, 500–507.

Kim, H. J., & White, P. J. (2013). Impact of the molecular weight, viscosity and solubility of β-glucan on in vitro oat starch digestibility. *Journal of Agricultural and Food Chemistry, 61*, 3270–3277.

Kim, M. J., Choi, S. J., Shin, S. I., Sohn, M. R., Lee, C. J., & Kim, Y. (2008). Resistant glutarate starch from adlay: Preparation and properties. *Carbohydrate Polymers, 74*, 787–796.

Kingman, S. M., & Englyst, H. N. (1994). The influence of food preparation methods on the in-vitro digestibility of starch in potato. *Food Chemistry, 49*, 181–186.

Laurentin, A., & Edwards, C. A. (2004). Differential fermentation of glucose-based carbohydrates in vitro by human faecal bacteria. A study of pyrodextrinised starches from different sources. *European Journal of Nutrition, 43*(3), 183–189.

Lunn, J., & Buttriss, J. L. (2007). Carbohydrates and dietary fibre. *Nutrition Bulletin, 36*, 21–64.

Margareta Leeman, A., Karlsson, M. E., Eliasson, A. C., & Bjorck, I. M. E. (2006). Resistant starch formation in temperature treated potato starches varying in amylose/amylopectin ratio. *Carbohydrate Polymers, 65*(3), 306–313.

Megazyme. (2008). *Resistant starch assay procedure*. http://secure.megazyme.com/downloads/en/data/K-RSTAR.pdf. Accessed 25 July 2009.

Miao, M., Jiang, B., & Zhang, T. (2009). Effect of pullulanase debranching and recrystallization on structure and digestibility of waxy maize starch. *Carbohydrate Polymers, 76*, 214–221.

Mikulíkova, D., Masar, S., & Kraic, J. (2008). Biodiversity of legume health-promoting starch. *Starch-Starke, 60*, 426–436.

Mir, J. A., Srikaeo, K., & Garcia, J. (2013). Effects of amylose and resistant starch on starch digestibility of rice flours and starches. *International Food Research Journal, 20*(3), 1329–1335.

Moallic, C., Myers, A. M., & James, M. G. (2006). Development of novel slowly digestible starches from maize. In: *World Grains Summit: Foods and Beverages San Francisco AACC International*, San Francisco, CA, USA, 17–20 September 2006.

Muir, J. G., & O'Dea, K. (1992). Measurement of resistant starch. Factors affecting the amount of starch escaping digestion in vitro. *American Journal of Clinical Nutrition, 56*, 123–127.

Murray, S. M., Patil, A. R., Fahey, G. C., Jr., Merchen, N. R., Wolf, B. W., Lai, C. S., & Garleb, K. A. (1998). Apparent digestibility of a debranched amylopectin-lipid complex and resistant starch incorporated into enteral formulas fed to ileal- cannulated dogs. *Journal of Nutrition, 128*, 2032–2035.

Nugent, A. P. (2005). Health properties of resistant starch. *Nutrition Bulletin, 30*, 27–54.

Obiro, W. C., Zhang, T., & Jiang, B. (2008). The nutraceutical role of the *Phaseolus vulgaris* alpha-amylase inhibitor. *British Journal of Nutrition, 100*, 1–12.

Ohkuma, K., & Wakabayashi, S. (2001). Fibersol-2: Soluble, non-digestible starch-derived dietary fiber. In B. V. McCleary & L. Prosky (Eds.), *Advanced dietary fibre technology* (pp. 509–526). Oxford, UK: Blackwell Science.

Ozturk, S., Koksel, H., & Perry, K. W. N. (2011). Characterization of resistant starch samples prepared from two high-amylose maize starches through debranching and heat treatments. *Cereal Chemistry, 86*(5), 503–510.

Park, E. Y., Baik, B. K., & Lim, S. T. (2009). Influences of temperature-cycled storage on retrogradation and in vitro digestibility of waxy maize starch gel. *Journal of Cereal Science, 50*(1), 43–48.

Pi-Sunyer, F. X. (2002). Glycemic index and disease. *American Journal of Clinical Nutrition, 76*, 290S–298S.

Pomeranz, Y., & Sievert, D. (1990). *Purified resistant starch products and their preparation*. WO 9015147. 13 Dec 1990. University of Washington.

Reddy, C. K., Suriya, M., & Haripriya, S. (2013). Physico-chemical and functional properties of Resistant starch prepared from red kidney beans (*Phaseolus vulgaris*.L) starch by enzymatic method. *Carbohydrate Polymers, 95*(1), 220–226.

Ring, S. G., Gee, J. M., Whittam, M., Orford, P., & Johnson, I. (1988). Resistant starch. Its chemical form in foodstuffs and effect on digestibility in vitro. *Food Chemistry, 28*, 97–109.

Rochfort, S., & Panozzo, J. (2007). Phytochemicals for health, the role of pulses. *Journal of Agricultural and Food Chemistry, 55*, 7981–7994.

Rodriguez, S. L., Islas, J. J., Agama, E., Tovar, J., & Bello, L. A. (2008). Characterization of a fibre-rich powder prepared by liquefaction of unripe banana flour. *Food Chemistry, 107*, 1515–1521.

Sagum, R., & Arcot, J. (2000). Effect of domestic processing methods on the starch, non-starch polysaccharides and in vitro starch and protein digestibility of three varieties of rice with varying levels of amylose. *Food Chemistry, 70*, 107–111.

Sajilata, M. G., Singhal, R. S., & Kulkarni, P. R. (2006). Resistant starch—A review. *Comprehensive Reviews in Food Science and Food Safety, 5*, 1–17.

Sang, Y., Seib, P. A., Herrera, A. I., Prakash, O., & Shi, Y. (2010). Effects of alkaline treatment on the structure of phosphorylated wheat starch and its digestibility. *Food Chemistry, 118*, 323–327.

Seal, C. J., Daly, M. E., Thomas, L. C., Bal, W., Birkett, A. M., Jeffcoat, R., & Mathers, J. C. (2003). Postprandial carbohydrate metabolism in healthy subjects and those with type 2 diabetes fed starches with slow and rapid hydrolysis rates determined in vitro. *British Journal of Nutrition, 90*, 853–864.

Seib, P. A., & Woo, K. (1999). *Food grade starch resistant to α-amylase and the method of preparing the same*. US Patent Number 5855946.

Severijnen, C., Abrahamse, E., van der Beek, E. M., Buco, A., van de Heijning, B. J. M., van Laere, K., & Bouritius, H. (2007). Sterilization in a liquid of a specific starch makes it slowly digestible in vitro and low glycemic in rats. *Journal of Nutrition, 137*, 2202–2207.

Shah, A., Masoodi, F. A., Gani, A., & Ashwar, B. A. (2018). Dual enzyme modified oat starch: Structural characterisation, rheological properties, and digestibility in simulated GI tract. *International Journal of Biological Macromolecules, 106*, 140–147.

Shi, Y. G., Cui, X., Birkett, A. M., & Thatcher, M. G. (2003). *Slowly digestible starch product*. US Patent 20030215562.

Shin, S. I., Kim, H. J., Ha, H. J., Lee, S. H., & Moon, T. W. (2005). Effect of hydrothermal treatment on formation and structural characteristics of slowly digestible, non-pasted, granular sweet potato starch. *Starch-Starke, 57*, 421–430.

Sievert, D., & Pomeranz, Y. I. (1989). Enzyme-resistant starch. Characterization and evaluation by enzymatic, thermoanalytical and microscopic methods. *Cereal Chemistry, 66*(4), 342–347.

Sievert, D., & Wursch, P. (1993). Thermal behavior of potato amylose and enzyme-resistant starch from maize. *Cereal Chemistry, 70*, 333–338.

Simsek, S., Ovando-Martinez, M., Whitney, K., & Bello-Perez, L. A. (2012). Effect of acetylation, oxidation and annealing on physicochemical properties of bean starch. *Food Chemistry, 134*, 1796–1803.

Slavin, J. (2004). Whole grains and human health. *Nutrition Research Reviews, 17*, 99–110.

Stute, R. (1992). Hydrothermal modification of starches: The difference between annealing and heat/moisture-treatment. *Starch-Starke, 44*, 205–214.

Tester, R. F., & Debon, S. J. J. (2000). Annealing of starch—a review. *International Journal of Biological Macromolecules, 27*, 1–12.

Tester, R. F., Karkalas, J., & Qi, X. (2002). Starch structure and digestibility, enzyme- substrate relationship. *World's Poultry Science Journal, 60*, 186–195.

Tharanathan, R. N., & Mahadevamma, S. (2003). Grain legumes: A boon to human nutrition. *Trends in Food Science and Technology, 14*, 507–518.

Thompson, L. U., & Yoon, J. H. (1984). Starch digestibility as affected by polyphenol and phytic acid. *Journal of Food Science, 49*, 1268–1269.

Tian, Y., Zhan, J., Zhao, J., Xie, Z., Xu, X., & Jin, Z. (2013). Preparation of products rich in slowly digestible starch (SDS) from rice starch by a dual-retrogradation treatment. *Food Hydrocolloids, 31*, 1–4.

Tovar, J., Granfeldt, Y., & Bjorck, I. (1992). Effect of processing on blood glucose and insulin responses to starch in legumes. *Journal of Agricultural and Food Chemistry, 40*, 1848–1851.

Vander-Maarel, M. J. E. C., Bennema, D. J., Semeijn, C., & Buwalda, P. L. (2008). *Novel slowly digestible storage carbohydrate*. EP Patent 1943908.

Wepner, B., Berghofer, E., Miesenberger, E., & Tiefenbacher, K. (1999). Citrate starch: Application as resistant starch in different food systems. *Starch-Starke, 51*(10), 354–361.

Winowiski, T. S., Schade, O., & Sudekum, K. H. (2005). *Ruminant's feed containing slowly digestible starch*. WO Patent 2005025323.

Wolever, T. M. S., Jenkins, D. J. A., Kalmusky, J., Giordano, C., Giudici, S., Jenkins, A. L., … Josse, R. G. (1986a). Glycemic response to pasta: Effect of surface area, degree of cooking and protein enrichment. *Diabetes Care, 9*, 401–404.

Wolever, T. M. S., Jenkins, D. J. A., Kalmusky, J., Jenkins, A. L., Giordano, C., Giudici, S., … Wong, G. S. (1986b). Comparison of regular and parboiled rices: Explanation of discrepancies between reported glycemic responses to rice. *Nutrition Research, 6*, 349–357.

Woo, K. S., & Seib, P. A. (2002). Cross-linked resistant starch: Preparation & properties. *Cereal Chemistry, 79*(6), 819–825.

Woortman, A. J. J., & Steeneken, P. A. M. (2004). *Method for preparing a gellable starch product.* WO Patent 2004069877.

Wursch, P., Vedevo, S. D., & Koellreutter, B. (1986). Cell structure and starch nature as key determinants of the digestion rate of starch in legume. *American Journal of Clinical Nutrition, 43*, 25–29.

Wurzburg, O. B. (1995). Modified starches. In A. M. Stephen (Ed.), *Food polysaccharides and their applications* (pp. 67–97). New York: Marcel Dekker Inc..

Zhang, G., Ao, Z., & Hamaker, B. R. (2006a). Slow digestion property of native cereal starches. *Biomacromolecules, 7*, 3252–3258.

Zhang, G., Sofyan, M., & Hamaker, B. R. (2008). Slowly digestible state of starch: Mechanism of slow digestion property of gelatinized maize starch. *Journal of Agricultural and Food Chemistry, 56*, 4695–4702.

Zhang, G., Venkatachalam, M., & Hamaker, B. R. (2006b). Structural basis for the slow digestion property of native cereal starches. *Biomacromolecules, 7*, 3259–3266.

Zhao, J., Schols, H. A., Chen, Z., Jin, Z., Buwalda, P., & Gruppen, H. (2012). Substituent distribution within cross-linked and hydroxypropylated sweet potato starch and potato starch. *Food Chemistry, 133*, 1333–1340.

Neutraceutical Properties of Resistant Starch

Gazalla Akhtar, Saqib Farooq, Tariq Ahmad Ganaie, Sajad Ahmad Mir, and F. A. Masoodi

Overview

As Hippocrates, father of modern medicine rightly said, "Let food be thy medicine and medicine be thy food". Consumers nowadays mostly consider healthier foods at par with medicines. Awareness towards nutritious diet and its relation to health and well being has lead to an intense increase for high quality food products that in turn brought revolution in the growth and use of new technologies. The paradigm shift towards healthier foods has necessitated the need of such foods that can alleviate major health concerns (cholesterol, cancer, obesity, etc.) as well as provide convenience in distribution and price. In this case functional foods serve as main players that either contain components with positive health benefits or eliminate components with negative health effects from the system. Dietary fiber mostly consist of an important functional component that is mostly considered 'roughage' material which remains indigestible in the small intestines due to its resistant property towards most human digestive enzymes hence conferring various important health benefits. The inclusion of dietary fiber in the regular diet has been responded well by consumers that lead to the establishment of many companies launching fiber fortified foods (Pérez and Alvarez 2008). Various recommendations have been made in this regard for general carbohydrate consumption by FAO and WHO: the Acceptable Macronutrient Distribution Range (AMDR) for carbohydrates is 55–75% of energy intake, and the Adequate Intake (AI) for total fiber is 38 g for men and 25 g for women while added sugars should be limited to 10% of the total energy intake (FAO/WHO Expert Consultation 2002). Dietary fiber comprises of

G. Akhtar · S. A. Mir · F. A. Masoodi (✉)
Department of Food Science and Technology, University of Kashmir, Srinagar, Jammu and Kashmir, India

S. Farooq · T. A. Ganaie
Department of Food Science and Technology, Islamic University of Science and Technology, Awantipora, Jammu and Kashmir, India

A. Gani, B. A. Ashwar (eds.), *Food biopolymers: Structural, functional and nutraceutical properties*, https://doi.org/10.1007/978-3-030-27061-2_3

important components, one such healthier component being Resistant starch (RS) (Mikulíková, Masár and Kraic 2008), that is widely used as a functional ingredient in a range of foods containing high dietary fiber levels and slowly digestible starch (SDS). SDS also forms an important component of dietary fiber due to its incomplete digestion and absorption properties that ultimately leads to its partial fermentation in the large intestines (Marteau et al. 2001).

Definition

The growing importance of functional foods has lead to an increased consumer demand for nutritious, healthy and beneficial foods. Starch and starch derivatives being the major food components contribute an energy reserve roughly about 60–70%. Resistant starch (RS) refers to the fraction of starch and starch products that resist digestion as they pass through the gastrointestinal tract (Nugent 2005). Resistant starch on a whole is that part of dietary fiber that is not fully hydrolyzed to D- glucose or sugar after an hour or 120 min after being consumed but gets fermented in the distal part of the large intestine releasing some important fatty acids including acetate, propionate and butyrate (Fergusan et al. 2007). The resulting short chain fatty acids get absorbed in distal part of large intestines and account for almost 5–10% of energy (Nordgaard and Mortensen 1995). On an average 80–90% of the RS of the total food components ingested that passes into the human colon is fermented by microbial fauna rest is excreted via feacal elimination. As such RS significantly shifts the positive balance of microbial colonies that decreases the fermentation of non starchy components hence increasing the feacal bulk (Cummings et al. 1996). Many studies have shown that RS is a linear molecule of α-1,4-D-glucan, essentially derived from the retrograded amylose fraction, and has a relatively low molecular weight (1.2×10^5 Da) (Tharanathan 2002). RS has a small particle size, white appearance and bland flavor. RS also has a low water-holding capacity. It has desirable physicochemical properties such as swelling, viscosity increase, gel formation, and water-binding capacity, making it useful in a variety of foods (Fausto et al. 1997). RS is an extremely broad and diverse range of materials and a number of different types exist (RS1–RS4) (Nugent 2005). The early classification involved only three forms of starch (Englyst and Cummings 1987; Englyst et al. 1992): RS 1—physically inaccessible starch; RS 2—starch of raw (non-gelatinised) granules of some plant species; RS 3—retrograded starch. In successive years, those forms were supplemented with another one (Eerlingen and Delcour 1995; Haralampu 2000), namely—RS 4—chemically or physically modified starch. Presently, all the four types are categorized according to their physicochemical characteristics (Nugent 2005). RS1 type of starches are resistant to GI tract conditions due to physical inaccessibility offered by seed coats and outer walls of partly processed grains and dense starchy foods making it stable to almost all cooking operations (Sajilata et al. 2006). RS2 represents that fraction of starch that is packed tightly inside the starch granule in such a way rendering it difficult for enzymes to penetrate, break the chains and release the constituent units (Sajilata et al. 2006a). Some versatile types of RS2 are shown to retain their intact conformation and resistant nature even after undergoing rigorous food processing operations referred to as high amylose maize starch (Wepener et al. 1999). RS3 refers to the retrograded or crystalline type of starch (Eerlingen et al. 1993) possessing a vast thermo stability due to a higher content of retrograded amylose conferring on it a strong tendency to reassociate strongly making amylose content a governing factor in the formation of RS3 (Luckett and Wang 2012). Various enzymatic treatments of starch e.g. pullulanases, can lead to RS3 formation by enzymatic debranching of amylopectin as in case of waxy maize (Lin and Chang 2006). It facilitates the formation of short chain starch fractions by cleaving the α-1-6 glycosidic linkages of amylopectin which promotes formation of resistant starch. To aggravate the formation of RS3, debranched starch fractions can also be stored at lower temperatures that favor the formation of strong crystalline structures enhancing resistant starch formation (Shi et al. 2013). RS4 refers to chemically modified starch wherein starch contains bonds other than normally occurring 1-4 and 1-6 glycosidic linkages thereby limiting normal course of enzyme action in small intestines thereby reducing its digestibility and thus leading to the formation of RS4 (Mudgil et al. 2013). RS4 includes a group of starch and related compounds that have been chemically modified by the process of etherification in such a way that tends to decrease its digestibility (Nugent 2005). RS4 can be produced by chemical modifications, such as conversion, substitution or cross-linking which can prevent its digestion. The compact molecular structure of resistant starch in raw starch granules limits its accessibility to most of the digestive enzymes e. g amylases that explains its resistant nature (Haralampu 2000) while several researches also concluded other mechanisms involved ranging from restructuring of starch granules (Nugent et al. 2005) to gelatinization of starch after cooking as well as chemical modification by etherisation, esterificaion and addition of various chemical groups which cannot be broken down by digestive enzymes (Lunn and Buttriss 2007).

Naturally RS is found in cereal grains, seeds and in heated starch or starch-containing foods (Charalampopoulos et al. 2002). As a food ingredient, RS has a lower calorific (8 kJ/g) value compared with fully digestible starch (15 kJ/g) (Rochfort and Panozzo 2007). Resistant starch has received much attention for both its potential health benefits (similar to soluble fibre) and functional properties. Resistant starch positively influences the functioning of the digestive tract, microbial flora, the blood cholesterol level, the glycemic index and assists in the control of diabetes. Apart from the potential health benefits of resistant starch, another positive advantage is its lower impact on the sensory properties of food compared with traditional sources of fibre, as whole grains, fruits or bran (Zaragoza et al 2010). Nutritionists and food industries are paying increasing attention to RS mainly due to fiber fortification and the potential physiological benefits as well as unique technological properties. They have high gelatinization temperature, good extrusion and film-forming qualities, and lower water-holding properties than traditional fiber products. They allow the formation of low-bulk high-fiber products with improved texture, appearance, and mouth feel compared with traditional high-fiber products; they increase coating crispness of products and the bowl life of breakfast cereals. They are functional food ingredients lowering the calorific value of foods and useful in products for celiacs, as bulk laxatives and in products for oral rehydration therapy (Nugent 2005) (Table 1).

Sources of Resistant Starch

Resistant Starch (RS) is one of nature's most interesting bioactive compounds. There is a wide variety of starchy food plants in world that are good sources of RS such as cereal grains, root and tuber crops, legume seeds and unripe bananas. Furthermore, it was reported that green bananas contained highest amount of resistant starch ranging between 35.14 and 45.87%, whereas the RS content of legumes ranging from 2.33 to 10.63% and 1.16 to 4.85% in cereal grains. Most of the starchy food plants contained moderate to high level of amylose (11.45–34.85%), except the waxy rice (2.72%) (Moongngarm 2013). On the basis of its botanical source, physical or chemical processing, RS can be divided into four types. Type 1 resistant starch (RS 1) includes starch present in cells of plants with undamaged cell walls, occurring e.g. in not completely ground cereal grain. Type 2 resistant starch (RS 2) includes granules of raw starch of some plant species, e.g. potato or banana. Type 3 resistant starch (RS 3) is a substance precipitated from paste or starch gel in the retrogradation process. Type 4 resistant starch (RS 4) includes starch modified chemically or physically (mainly by thermal treatment), or with both those treatments (Leszczyñski 2004). However the natural types of RS are frequently destroyed when processed. The manufacture of resistant starch at industrial level usually involves partial acid hydrolysis and hydrothermal treatments, heating, retrogradation, extrusion cooking, chemical modification, and repolymerisation (Charalampopoulos et al. 2002).

Table 1 Types of resistant starch and physiological rate of digestion

RS type	Description	Rate of digestion
RS1	Physically inaccessible to digestion by entrapment in non digestible matrix	Slow rate: partial degree of digestion
RS2	Ungelatinised resistant granules with type B crystallinity, slowly hydrolysed by α- amylase	Very slow rate: little degree of digestion
RS3	Retrograded starch formed when starch containing foods are cooked and cooled	Slow rate : partial degree of digestion
RS4	Selected chemically modified resistant starches and industrially processed food ingredients	As a result of chemical modification, can resist hydrolysis

Source: Nugent et al. (2005)

Slowly Digestible Starch

For the purpose of definition starch can be quantified into different fractions by the application of in vitro techniques. Depending upon the rate of starch digestion, the starch can be classified as rapidly digestible starch (RDS) digested within 20 min, slowly digestible starch (SDS) digested between 20 and 120 min corresponds to SDS and the remaining fractions not further digested comprises resistant starch (Mio et al. 2015). Most of processed starchy foods as such are a good source of RDS, therefore rapidly digested by the small intestines leading to elevated glucose levels and evident occurrence of hyperglycemia (Zhang et al. 2009). SDS as such undergoes slow digestion from 20 to 120 min after being ingested providing for slow and sustained glucose release throughout the time period. The digestibility of SDS depends on a variety of factors most important being molecular structure which includes amorphous phase packing, amylose/amylopectin ratio and crystallite structure that actually manipulates enzyme action (Colonna et al. 1992). The practical value of SDS is negated while measuring initial glycemic responses since under these assays only the glycemic response from 0 to 120 min are taken into consideration sidelining the actual components of foods. Foods containing an appropriate amount of SDS serve as an efficient source of functional components in lowering blood glucose as it results in slow and prolonged release of glucose in blood stream thereby helpful in controlling and preventing hyperglycemia related diseases.

Nutraceutical Potential of Resistant Starch

Prebiotic Potential of Resistant Starch

Prebiotics are non-digestible carbohydrates that escape digestion in small intestine and play a vital role in the survival of probiotics by providing fermentable carbohydrates to colonic bacteria (Coşkun 2006). Since resistant starch as such escapes normal route of digestion from the small intestines but undergoes fermentation in distal part of large intestines producing varied amounts of short chain fatty acids (SCFA) mainly acetic acid, propionic acid and butyric acid (Yoa et al. 2009) thereby increasing faecal bulk and lowering down colonic pH (Demigné and Remésy 1995). RS seems to work as prebiotic by its ability to resist digestion in the small intestine by possessing an inherent property to be dead set against enzymatic cleavage while passing through the gastrointestinal tract that enables it to act as protective covering for the bacteria in vivo (Wang et al. 1999). RS promotes a higher amount of butyric acid in comparison to other non-digestible carbohydrates. Butyrate provides colonic microbiota with a major energy substrate and plays a vital role in maintaining the colonic health (Leeman et al. 2006; Scourboutakos 2010), by affecting cell metabolism, differentiation, cell growth and prevents the formation of colon tumors (Champ et al. 2003). Zhu and Zhao (2013) reported that the concentration of SCFA increased due to the fermentation of retrograded maize starch. In another study, an increasing trend was observed in propionate, acetate and butyrate when rats were fed with high amylose maize (RS3) (Charrier et al. 2013). Various in vitro studies confirmed the diminished hydrolysis of starch or starch ingredients that had undergone chemical or physical modification (b-cyclodextrinisation, acetylation, gelatinization etc.) to small intestine hydrolysis depending on the rate of modification or the degree of substitution. A study conducted by (Raben et al.) showed an immense enhancement in large bowel SCFA production by feeding a group of rats with acetylated starches. The probable mechanism involved production of esterified fatty acids and fermentation of the available starches. Furthermore, Kalmokoff et al. (2013) reported an increase of propionate in rats fed with high amylose maize starch (RS2). Jenkins et al. (1998) reported that 12 women and 12 men were fed diet with or without the supplementation of 30 g high amylose starch (RS2) or retrograded high amylase starch (RS3) daily for 2 weeks, followed by a 2 week washout intervals. It was reported that butyrate concentration in the fecal water was increased in those persons who were fed with RS supplemented diet (mean of RS2 and RS3) (22.7 mmol/L) in comparison to low fiber diet (19.2 mol/L). Martínez et al. (2010) studied the effects of RS (RS2 and RS4) on the fecal microbiota in humans. In this study 10 human participants were fed with crackers supplemented with granular form of high-amylose corn starch (RS2) or chemically modified by phosphate cross-linking (RS4) for three weeks, which were compared with control starch in the form of crackers. It was reported that both RS2 and RS4 significantly increased the *Actinobacteria* and *Bacteroidetesphyla* and decreased *Firmicutes*. However both forms of resistant starch differed in their ability to change species. It was further reported that RS2 increased the number of *Ruminococcusbromii* and *Eubacteriumrectale*, which is similar to previous results from in vitro studies of starch fermentation in the large intestine (Kovatcheva-Datchary et al. 2009; Leitch et al. 2007). In contrast, RS4 was linked with increased *Bifidobacterium adolescentis* and *Parabacteroid esdistasonis* (Martínez et al. 2010) (Fig. 1).

Recently, a number of studies have been conducted on the synergistic effects of RS. In one such study it has been reported that the prebiotic effect of RS can be increased by combining it with short chain fructooligosaccharides (FOS). This combination has reportedly caused changes in the microbial flora of intestines by increasing *lactobacilli* and *bifidobacteria* in caecum and colonic contents. These observations have been confirmed in vivo once the different prebiotics reach the large intestine: FOS are rapidly fermented, whereas RS is slowly degraded. In consequence, the particular kinetics would determine the region of the intestine where the effects will be clearer. Thus, FOS would be more active in the first parts of the large bowel whereas RS would reach the distal part of the colon (Rodríguez-Cabezas et al. 2010; Fuentes-Zaragoza et al. 2010). Various studies conducted by different researchers over different periods of time conclusively accomplished that RS serves as a feeding substrate for *Bifidobacterium* under in vitro conditions and simultaneously can act as an efficient shield for protection of bacteria under in vivo conditions while surpassing the gastrointestinal conditions (Wang et al 1999). Other in vitro

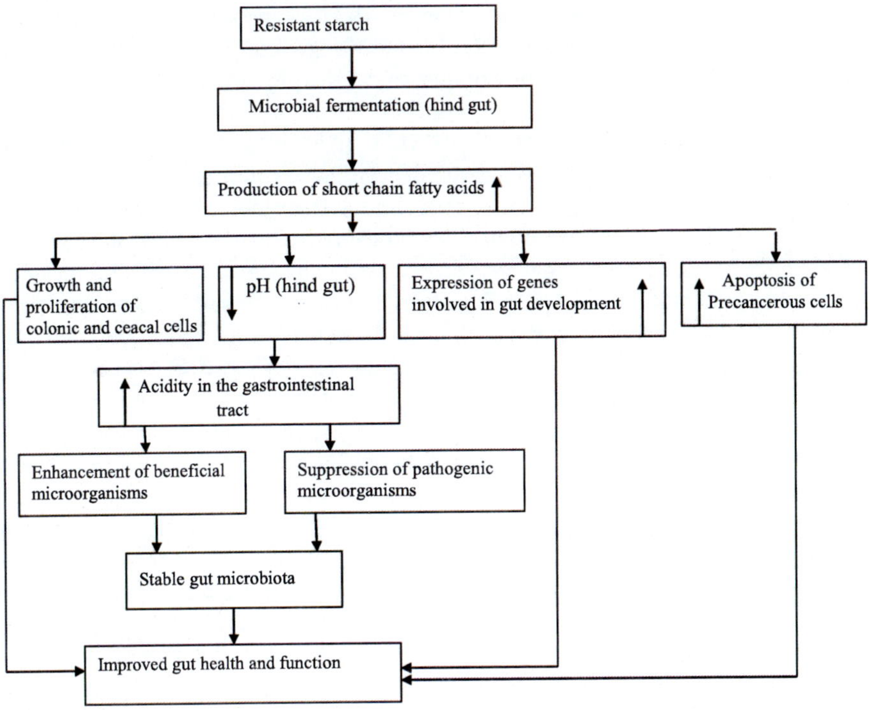

Fig. 1 Diagrammatic illustration of how resistant starch improves gut health (source, Regassa et al. 2018)

studies concluded that all the categories of RS may show close association with various bacterial species including *lactobacillus* and *Bifidobacteria* providing adequate defense from different food processing treatments and storage conditions making RS a "culture protagonist" (Conway 2001).

Anti-Cancerous Properties

Many epidemiological studies have associated the consumption of diets rich in fiber with the protection against colorectal cancer (Zhang and Hamaker 2010). It has been reported that the populations with higher dietary fibre intake could reduce the risk of colorectal cancer by upto 40% (Bingham et al. 2003). Recently, resistant starch has received much attention as a dietary fiber and is thought to be protective against colorectal cancer (Higgins and Brown 2013). Aberrant crypt foci (ACF) are precursor lesions of colorectal cancer (Nugent 2005). It has also been reported that the RS dosage suppressed the formation of colonic aberrant crypt foci(ACF) during the promotion phase to agenotoxic carcinogen in the middle and distal colon, indicating that proper utilization of RS may retard growth of neoplastic lesions in the colon. The other possible mode of action by RS consumption involves modifying the expression path of oncogenes to abscond the growth phase of the cell cycle thereby allowing for differentiation of tumour cells to produce normal matured cells (Young et al. 1995). Adults with pre neoplastic lesions in their colon may therefore benefit from dietary RS (Liu and Xu 2008). A study conducted by Bingham et al. (2003) invariably showed the reductive effect of dietary fibre on the prevalence of colorectal cancer on populations where in the average dose of dietary fibre was doubled. Various studies examining the protective effect of resistant starch on colonic cancer have been conducted on both animals generally focusing on mice and rats as well as human subjects. In animal models tumours induced by mutagens like dimethyl hydrazine or azoxy methane or by using genetic models of intestinal tumours, a diet containing approximately 5.6% high amylose RS3 has been shown to be effective in controlling the cluster formation of abnormal mass of glands in the internal surface of colon and rectum of the animals (Perrin et al. 2001). Biopsy studies conducted on the subjects showed strong effect of butyrate on the development and proliferation of the abnormal cell mass in the intestinal areas possibly because of the inherent ability of resistant starch to lessen the bile acid concentration (Grubben et al. 2001). Various other possible mechanisms involve the acidification of the digestible material limiting the escalation of pathogenic microorganisms as well as restraining the absorption and assimilation of toxic components like ammonia. This valuable effect of SCFA restricts the colonocyte contact with the mutagenic agents (Topping 2007). The effective role of SCFA in modifying various physiological mechanisms including enhanced colonic muscle activity and muscle oxygenation plus the improved colonic mucosal activity reducing the possible risks of cell aberrations and lesions. The effect of RS was shown to be predominant on the sites wherein the maximum fermentation of the starch took place (Young and Le Leu 2004) but the protective effect was found to be ineffective towards the later stages of colonic cancer development. Different studies have shown contradictory results regarding the effect of resistant starch or its adjuncts on the colonic function. In two separate experiments carried by Thorup et al. (1995) as well as Cassand et al. (1997), RS was found to exert a protective effect against the development of aberrant tumours. However the reports published by negated any of these protective effects. Similarly, Sakamoto and co-researchers (1996) found a meagre effect on tumour development after the subjects were fed diets containing 10% RS. However various studies extensively support the repressing activity of butyrate on the abnormal cell proliferation at physiological concentrations and increase in colorectal cancer cells apoptosis of CRC cells (Hague et al. 1993). Some studies have also reported that the rapid tumor growth depends on glucose availability. The consumption of RS has been associated with the reduction of the energy capacity of cells in preneoplastic lesions. As a result, RS enhances resistance to colonization, inhibits bacterial translocation, and finally boosts the chemical and enzymatic functions, involved in the protection of gastrointestinal tract (Roberfroid 2005). The increased motility of the large intestine reduces the time that stool remains in the colon and relieves constipation symptoms. Increased stool transit time and constipation are among the risk factors for colorectal cancer because fecal contact with the colon and rectal walls for prolonged periods of time increases the chances of developing this disease (Roberfroid 2005; Mauro et al. 2013). RS is potential prebiotic agent because it acts as a substrate for growth and development of colonic microbiota (Haralampu 2000). It undergoes fermentation in the large intestine and leads to the production of short-chain fatty acids (SCFA) such as acetate, propionate, and butyrate and also decreases colon pH (Englyst et al. 1992). As butyrate is one of the main energy substrates for large intestinal epithelial cells and inhibits the malignant transformation of such cells in vitro; this makes easily fermentable RS fractions especially interesting in preventing colonic cancer (Asp and Bjorck 1992). As observed in the various studies, the butyrates can have an inhibitory effect on the growth and proliferation of tumor cells in vitro by arresting initiation, progression, and growth of colon tumors (Sharma et al. 2008).

Hypoglycemic Properties

Diabetes mellitus (DM) is now emerging as a global public health disorder and is spreading like an epidemic all over the world. Data has shown the prevalence of diabetes for different age groups from 2.8% in 2000 to double with expected increase up to almost 4.4%, i.e. 366 million by the end of 2030 (Wild et al. 2004). At global level round about 425 million people suffer from diabetes with an incidence of one in every eleventh individual however this figure is very well expected to surpass 642 million mark by the end of 2040. DM as such has engulfed the general populace of whole world but the situation in developing countries like India that contains world's largest diabetic population is totally grim. In due course of time many drugs have been developed to treat and control diabetes including biguanides, sulfonyl ureas and glucosidase inhibitors which ultimately at the end use, result in side effects like liver and kidney dysfunction as well as hypoglycemia (Posuwan et al. 2013). Therefore prevention being better than cure ultimately relies on to devise the approach to focus on how to thwart diabetes on roots rather than to cure. Obesity, one of the most prevalent and chronic metabolic disorder in the world is induced mainly due to energy imbalance. Today at about 60% of the global population with type 2 diabetes are suffering from obesity in south Asian regions. Obesity, considered as the mother of diseases leads a raised fasting glucose level associated with hyperglycemia as well as markedly dilated adipose tissues, insulin resistance, higher amount of insulin in blood circulation (hyperinsulinemia), higher cholesterol in circulating fluids (dyslipedemia) plus hypertension that enhance the incidence of cardiovascular diseases(Cornier et al. 2008). About 80% of the population with DM type 2 also suffer from obesity leading to gain of excessive weight (20-30 %), fatty liver and other related complications that enhance mortality rate to almost three folds (Maggio and Pi-Sunyer 2003). Resistant and slowly digestible starch has shown an inverse relationship with the risk of developing type 2 diabetes (Chandalia et al. 2000). The inherent property of resistant starch to undergo slow and steady digestion implies for its applications in controlled glucose release responses(Sajilata et al. 2006) and thus ultimately by the way of negative feedback triggers the utilization of excessive fat and hence a lower insulin response is expected (Nugent 2005). In order to maintain a normal and regular glucose response a balance is therefore required between consumption of dietary sources containing glucose and end removal of glucose from the system. In such a case resistant starch offers a reliable window to help control blood glucose levels due to a reduced influx of glucose because of slow digestion and reduction in the absorbable glucose in the human gastrointestinal system. A study conducted by Shi et al. (2017) clearly demonstrated the inhibitory effect of resistant starch on hyperglycemic status associated with enhanced levels of gut microflora population which lead to high secretions of short chain fatty acids via resistant starch fermentation secreting propionate and butyrate, which may have an inhibitory effect on obesity induced diabetes. Resistant starch (RS) therefore functions as a new food ingredient that has a low glycemic index. RS itself contains almost zero calories (Alphons 1998) and when used as a low-calorie food additive that in turn can control weight effectively. The ingestion of RS can decrease insulin secretion and control postprandial blood glucose to prevent diabetes (Weickert and Mohlig 2005). The metabolism of RS occurs 5–7 h after consumption, in contrast to normally cooked starch, which is digested almost immediately. Digestion over a 5–7 h period reduces postprandial glycemia and insulinemia and has the potential for increasing the period of satiety (Raben et al. 1994; Reader 1997). As RS has a low glycemic response, adding it as an ingredient to foods will help lower the overall GI value of the food (particularly if it is replacing existing readily absorbed forms of carbohydrate) (Jyoshna and Hymavathi 2017). In a study on humans, it was reported that the consumption of RS3 lowered serum glucose and insulin levels in comparison with other carbohydrates(simple sugars, oligosaccharides, and common starch). The investigation further showed that food rich in RS reduced postprandial blood glucose and might help in providing improved metabolic control in type II diabetes (Reader 1997). Another study showed that RS reduces levels of glucose dependent insulinotropic polypeptide m-RNA along the jejunum and ileum in both normal and type 2 diabetes rats (Shimada et al. 2008). Chemically-modified resistant starches (RS4) have also been reported to produce different glucose responses. In another human intervention study 11 human participants were fed with two test meals containing 1–2% acetylated potato starch and beta cyclodextrin enriched potato starch (2–3%), and it was observed that participants who were fed with diet containing 2–3% beta cyclodextrin enriched potato starch showed lower body glucose levels. This may be due to the more distal absorption of beta cyclodextrin in the intestine or to delayed gastric emptying (Raben et al. 1997). Raben et al. (1994) investigated the effect of RS on postprandial plasma concentration of glucose lipids and harmones and on subjective satiety and palatability ratings. In this study 10 healthy human participants were fed with test meals consisting of 50 g pre gelatinized starch (0% RS) or 50 g raw potato starch (54% RS) together with 500 g artificially sweetened syrup. The postprandial plasma concentrations of glucose, lactate, insulin, gastric inhibitory polypeptide (GIP), glucagon-like peptide-1, and epinephrine were significantly lower in those participants who were fed with meals containing raw potato starch (54% RS). This study suggests that replacement of digestible starch with resistant starch resulted in significant reduction in postprandial glycemia and insulinemia and in the subjective sensation of satiety. In another study, Behall et al. (2006) reported that the reduction in glycemic response was enhanced by combining resistant starch and soluble fiber. Consumption of foods containing moderate amounts of these fibers may improve glucose metabolism in both normal and overweight persons. The results showed that high β-glucan-high resistant starch treatment decreased glucose level by 59% and insulin level by 38% when compared with the low resistant starch treatment. Furthermore, the maize and resistant starch enriched breads have been reported to reduce the postprandial glycemic response in rats (Brites et al. 2011). In this study resistant starch was incorporated in to formulations of breads at 20% of the inclusion rate of wheat and maize flours. 36 rats were divided in to 4 groups and fed either wheat bread, RS wheat bread, maize bread and RS maize bread. The maize bread group presented higher body weight gain and cholesterol level, lower fecal pH, and postprandial blood glucose response than the wheat bread group. The RS-wheat bread group showed significant reductions in feed intake, fecal pH, postprandial blood glucose response, and total cholesterol. The RS-maize group displayed significant reductions of body weight gain, fecal pH, and total cholesterol levels; however, for the glycemic response, only a reduction in fasting level was observed. These results suggest that maize bread has a lower glycemic index than wheat bread, and the magnitude of the effect of resistant starch on glycemic response depends on type of bread.

Hypolipidemic Properties

Resistant starch (RS) is an attractive source of non-caloric food with a more palatable character than insoluble dietary fiber. RS intake seems to improve the blood lipid profile by decreasing serum cholesterol concentrations in normal or hypercholesterolemic conditions in animals (De deckere et al. 1993; Levrat et al. 1996; Park et al. 2004). Therefore incorporation of RS in routine diets could offer an effective strategy for public cardiometabolic health promotion (Nichenametla et al. 2014). The effect of resistant starch on hypolipidemic actions in healthy overweight subjects was examined by Park et al. (2004). Healthy overweight subjects (over 120% of their ideal body weights) were fed with regular meals containing either of 24 g resistant corn starch or regular corn starch for 21 days. The results showed no significant changes in their weights or other physical parameters for the relatively acute period of intakes, however there were significant lowering effects of serum LDL-cholesterol in subjects fed with resistant corn starch supplemented diets in comparison with regular corn starch enriched diets. Various other studies have also concluded the beneficial effect of resistant starch on the metabolism of different blood lipids like low density lipoproteins (LDL), high density lipoproteins (HDL), triglycerides, cholesterol and total lipids (Nugent et al 2005). Studies conducted on rat subjects concluded a heavy reduction in the blood cholesterol levels up to 32%

and up to 42% in triglyceride levels suggesting RS to be more effective in lowering blood lipids than synthetic drugs like cholestyramine (Younes et al. 1995). A study conducted by Cheng and Lai (2000) found a marginal decreasing effect of rice resistant starch on serum cholesterol as well as hepatic cholesterol of animal subjects compared to control receiving corn starch. The researchers ascribed this effect to the increase in propionate levels formed as result of fermentation of rice resistant starch with its major effect in lowering hepatic cholesterol. A study conducted by So et al. (2007) investigating the beneficial effect of RS on lowering of blood lipids subjected mice to dietary interventions wherein one group was fed with diet containing high resistant starch content and vice versa. A reduced fat deposition, serum lipids and plasma leptons were found in the subjects fed with high RS diet. Conclusion were drawn suggesting a possible link between the fermentation of RS leading to SCFA production and protein receptors involved in adipocyte differentiation and proliferation has been sought to be the possible mechanism. Even though RS has been found to be very much beneficial in reducing lipids, the consumption of RS in healthy individuals is yet controversial. Different studies conducted by Stewart et al. (2010) and Heijnen et al. (1996) didn't find any remarkable effect of RS consumption on the reduction of serum cholesterol and adipose fat deposition whereas earlier Behall et al. (1989) found a noteworthy decrease in blood lipids and total fat content on the consumption of RS. Mathe et al. (1993) with their studies on obese rats found RS efficient in decreasing serum cholesterol. Zhou et al. (2015) found a direct influence of RS consumption on the enhanced population of gut micro flora that undoubtly increased the SCFA levels inhibiting the biosynthesis of cholesterol. Similarly, in another human study, resistant starch type 4 (RS4) consumption improved dyslipidemia and body composition that resulted in 7.2% lower mean total cholesterol, 5.5% lower non-HDL, and 12.8% lower HDL cholesterol in individuals with metabolic syndrome (Nichenametla et al. 2014). In a study on animals, the rats were fed with diets supplemented with RS (raw potato starch and high amylose corn starch) to study the effects on ceacal digestion and lipid metabolism in comparison with rats adapted to semi purified diets. The results showed significant ceacal hypertrophy (240% after 7 day of the fiber consumption) and short-chain fatty acids accumulation (especially propionic and butyric acids) in rats fed with RS diets. Apart from this the cholesterol absorption was reduced to 14% of intake in rats fed with resistant starch enriched diets compared with 47% absorption in control rats. RS diets were also effective in lowering plasma cholesterol (−31 and −27%, respectively) and triglycerides (−28 and −22%, respectively) (Lopez et al. 2001). In another study on rats, RS diets led to a marked rise in ceacal size and the ceacal pool of short-chain fatty acids (SCFA), as well as SCFA absorption; cholestyramine did not noticeably affect ceacal fermentation while on the other hand cholestyramine was particularly effective at enhancing bile acid excretion. RS was more effective in lowering plasma cholesterol (−32%) and triglycerides (−29%). In rats fed the RS diet, there was a lower concentration of cholesterol in all lipoprotein fractions, especially the fraction high-density lipoprotein (HDL), while those fed cholestyramine had only a significant reduction of HDL cholesterol. In contrast to cholestyramine, RS also depressed the concentration of triglycerides in the triglyceride-rich lipoprotein fraction. There was no noticeable synergy between the effects of RS and cholestyramine when both were present in the diet (Younes et al. 1995). Furthermore, Rats susceptible to obesity attained lesser body weight with 4, 12 and 16% RS as distinguished with the 0% RS, but in obese-antagonistic animals, significant effect was observed exclusively at 16% RS. Food with 8% RS reported to reduce adiposity irrespective of phenotype (Belobrajdic et al. 2012).

Mineral Absorption

Resistant starch has an inherent property to escape the normal route of digestion and undergoes fermentation in the distle part of the small intestines to yield short chain fatty acids majority of which consists of propionate and butyrate. The organic acids produced via fermentation of RS and related ingredients have shown a considerable effect on absorption of various important mineral components by altering cellular mechanisms by increasing the cell surface area. Various mineral elements show an enhanced solubility in presence of resistant starch that has been related to higher content of amylose in RS. Increased amylose content engages almost all the hydrogen bonds present hence diminishing mineral binding sites on starch molecules thereby increasing their solubility and intestinal absorption (Dobbing et al. 1989). Among all the minerals the SCFA's mainly assist in calcium absorption by modification of the electrolyte balance of the cells through passive diffusion across the cell membrane. Consumption of RS on regular basis has also shown improvement in magnesium assimilation probably by the mechanism of change in ion solubility due to lower pH. Various in vivo studies concluded the positive effect of resistant starch on mineral absorption. Younes et al. found a considerable improvement in dietary calcium levels of rats on feeding them 15–35% RS. Another study conducted reported a much higher assimilation of calcium and magnesium salts on feeding the rats with potato RS and high amylose starch for 1 week. Increase in the consumption frequency of high amylose starch from various have shown a much higher increase in mineral absorption. Another group of researchers conducted a study to compare the intestinal apparent absorption of calcium, phosphorus, iron, and zinc in the presence of either resistant or digestible starch brought out that a meal containing 16.4% RS resulted in a greater apparent absorption of calcium and iron compared with a completely digestible starch. According to Schulz et al. (1993) some RS (RS2 but not RS3) could improve calcium and magnesium absorption by enhancing mineral solubility in the cecum and/or large intestine. Raw potato starch is considered to be RS2. Feeding raw potato starch enhances fermentation in the distal parts of the digestive tract. As a result, the absorption of calcium, magnesium, iron, zinc and copper was increased owing to hypertrophy of the cecal wall and cecal acidification. Furthermore, reported that inulin and resistant starch ingestion led to considerable caecal fermentation in the three experimental groups of rats compared with the control group diet. Moreover, both carbohydrates significantly increased the intestinal absorption and balance of Ca and Mg, without altering the plasma level of these two minerals. Interestingly, the combination of the studied carbohydrates increased significantly the caecal soluble Ca and Mg concentrations, the apparent intestinal absorption and balance of Ca, and non-significantly the plasma Mg level.

Conclusion

Resistant Starch (RS) is one of nature's most interesting functional ingredient. There is a wide variety of starchy food plants in world that are good sources of RS such as cereal grains, root and tuber crops, legume seeds and unripe bananas. Nutritionists and food scientists are paying increasing attention to RS mainly due to fiber fortification and the potential physiological benefits as well as unique technological properties. Resistant starch positively influences the functioning of the digestive tract, microbial flora, the blood cholesterol level, the glycemic index and assists in the control of diabetes. Apart from the potential health benefits of resistant starch, another positive advantage is its lower impact on the sensory properties of food compared with traditional sources of fibre, as whole grains, fruits or bran. There is a lot still to be discovered in terms of the health benefits of RS. The trials are going on in investigating the effect of RS on the various aspects of humans health.

References

Alphons, G. J. V. (1998). Technological aspects of functional food-related carbohydrates.*Trends in Food Science & Technology, 9*: 328–335.

Asp, N. G., & Björck, I. (1992). Resistant starch. *Trends in Food Science & Technology, 3*: 111–114.

Behall, K. M., Scholfield, D. J., Yuhaniak, I. (1989). Diets containing high amylose vs amylopectin starch: effects on metabolic variables in human subjects. *American Journal of Clinical Nutrition. 49*, 337–44.

Belobrajdic, D. P., King, R. A., Christophersen, C. T. & Bird, A. R. (2012). Dietary resistant starch dose-dependently reduces adiposity in obesity-prone and obesity-resistant male rats. *Nutrition & metabolism, 9*(1): 93.

Bingham, S. A., Day, N. E., Luben, R. (2003). Dietary fibre in food and protection against colorectal cancer. European Prospective Investigation into Cancer and Nutrition (EPIC): An observational study. 362 (9388):1000.

Brites, C. M., Trigo, M. J., Carrapiço, B., Alviña, M. & Bessa, R. J. (2011). Maize and resistant starch enriched breads reduce postprandial glycemic responses in rats. *Nutrition Research, 31*(4): 302–308.

Cassand, P., Maziere, S., Champ, M. (1997). Effects of resistant starch and vitamin A supplemented diets on the promotion of precursor lesions of colon cancer in rats. *Nutrition and Cancer, 27*, 53– 9.

Charrier, J. A., Martin, R. J., McCutcheon, K. L., Raggio, A. M., Goldsmith, F., Goita, M. F., Senevirathne, R. N., Brown, I. L., Pelkman, C., Zhou, J. & Finley, J. (2013). High fat diet partially attenuates fermentation responses in rats fed resistant starch from high-amylose maize. *Obesity, 21*(11): 2350–2355.

Champp, M., Langkilde, A. M., Brouns, F., Kettlitz, B. & Le Bail-Collet, Y. (2003). Advances in dietary fibrecharacterisation. 2. Consumption, chemistry, physiology and measurement of resistant starch; implications for health and food labelling. *Nutrition Research Reviews, 16*(2): 143–161.

Charalampopoulos, D., Wang, R., Pandiella, S. S. & Webb, C. (2002). Application of cereals and cereal components in functional foods: A review. *International Journal of Food Microbiology, 79*: 131–141.

Cheng, H. H., & Lai, M. H. (2000). Fermentation of resistant rice starch produces propionate reducing serum & hepatic cholesterol in rats. *Journal of Nutrition, 130*(8), 1991–1995.

Colonna. P., Leloup. V.M., Buleon. A. (1992). Eur. ./. ('lirr. C'Ulr... 46, 17–32.

Conway, P. L. (2001). Prebiotics and human health: The state of the art and future perspectives. *Scandinavian Journal of Nutrition. 45*, 13–21.

Cornier, M. A., Dabelea, D., Hernandez, T. L., Lindstrom, R. C., Steig, A. J., Stob, N. R., … Eckel, R. H. (2008). The metabolic syndrome. *Endocrine Reviews, 29*, 777–822.

Coşkun, T. (2006). Pro-, pre-vesinbiyotikler. *çocuksağlığıvehastalıklarıdergisi, 49*: 128–148.

Cummings, J. H., Beatty, E. R., Kingman, S. M., Bingham, S. A., Englyst H. N. (1996). Digestion and physiological properties of resistant starch in the human large bowel. Br J Nutr. 75(5), 733–47.

De Deckere, E. A., Kloots, W. J. & Van Amelsvoort, J. M. (1993). Resistant starch decreases serum total cholesterol and triacylglycerol concentrations in rats. *The Journal of nutrition, 123*(12): 2142–2151.

Dobbing, J. (1989). Dietary Starches and Sugars in Man: A Comparison. Springer-Verlag, New York, pp 3–7.

Eerlingen, R. C., Crombez, M., Delcour, J. A. (1993). Enzyme-resistant starch. I. Quantitative and qualitative influence of incubation time and temperature of autoclaved starch on resistant starch formation. Cereal Chemistry, 70 (3), 339–344.

Eerlingen, R. C., Crombez, M., Delcour, J. A. (1993). Enzyme-resistant starch. I. Quantitative and qualitative influence of incubation time and temperature of autoclaved starch on resistant starch formation. *Cereal Chemistry, 70*(3), 339–344.

Englyst, H. N. & Cummings J. H. (1987). Digestion of polysaccharides of potato in the small intestine of man. Am. J. Clin. Nutr, 45: 423–431.

Englyst, H. N., Kingman, S. M. & Cummings, J. H. (1992). Classification and measurement of nutritionally important starch fractions. Eur. J. Clin.Nutr. 46 (Suppl. 2): S33–S50.

FAO-WHO (Food and Agricultural organization- World Health Organization). (2002). *Diet, nutrition and prevention of chronic diseases*, vol. 916. Report of a joint FAO-WHO Expert Consultation. WHO technical report series.

Fausto, F. D., Kacchi, A. I, & Mehta, D. (1997). Starch products in confectionery. *Bev Food World, 24*(4): 4–16.

Ferguson, L. R., Tasman Jones, C., Englyst, H., & Harris, P. J. (2007). Comparative effect of three resistant starch preparations on transit time and short chain fatty acid production in rats. *Nutrition and Cancer, 36*, 230–237.

Fuentes-Zaragoza, E., Riquelme-Navarrete, M. J., Sánchez-Zapata, E. & PérezÁlvarez, J. A. (2010). Resistant starch as functional ingredient: A review. *Food Research International, 43*: 931–942.

Grubben, M., Van den Braak, C. C. M., Essenberg, M., Olthof, M., Tangerman, A., Katan, M. B., Nagengast, F. M. (2001). Effect of resistant starch on potential biomarkers for colonic cancer risk in patients with colonic adenomas, a controlled trial. *Digestive Diseases and Sciences, 46* (4), 750–756.

Hague, A., Manning, A. M., Hanlon, K. A., Huschtscha, L. I., Hart, D., & Paraskeva, C. (1993). Sodium butryrate induces apoptosis in human pathways: Implications for the possible rate of dietry fibre in the prevention of large-bowel cancer. *International Journal of Cancer., 55*, 498–505.

Haralampu S. G. (2000). Resistant starch, A review of the physical properties and biological impact of RS3. Carboh. Polymers. 41, 285–292.

Heijinen, M. L., Van Amelsvoort, J. M., Deurenberg, P., & Beynen, A. C. (1996). Neither raw nor retrograded resistant starch lowers fasting serum cholesterol concentration in healthy normo-lipededemic subjects. *American Journal of Clinical Nutrition, 64*(3), 312–318.

Higgins, J. A. & Brown, I. L. (2013). Resistant starch: a promising dietary agent for the prevention/treatment of inflammatory bowel disease and bowel cancer. *Current opinion in gastroenterology, 29*(2): 190–194.

Jenkins, D. J., Vuksan, V., Kendall, C. W., Wursch, P., Jeffcoat, R., Waring, S., Mehling, C. C., Vidgen, E., Augustin, L. S. & Wong, E. (1998). Physiological effects of resistant starches on fecal bulk, short chain fatty acids, blood lipids and glycemic index. *Journal of the American College of Nutrition, 17*(6): 609–616.

Jyoshna, E. & Hymavathi, T. V. (2017). Review of studies on effect of resistant starch supplementation on glucose and insulin. *Journal of Pharmacognosy and Phytochemistry, 6*(3), 55–58.

Kalmokoff, M., Zwicker, B., O'hara, M., Matias, F., Green, J., Shastri, P., Green-Johnson, J. & Brooks, S. P. J. (2013). Temporal change in the gut community of rats fed high amylose cornstarch is driven by endogenous urea rather than strictly on carbohydrate availability. *Journal of applied microbiology, 114*(5): 1516–1528.

Kovatcheva-Datchary, P., Egert, M., Maathuis, A., Rajilić-Stojanović, M., De Graaf, A.A., Smidt, H., De Vos, W. M. & Venema, K. (2009). Linking phylogenetic identities of bacteria to starch fermentation in an in vitro model of the large intestine by RNA-based stable isotope probing. *Environmental microbiology, 11*(4): 914–926.

Leeman, A. M., Karlsson, M. E., Eliasson, A. C. & Björck, I.M. (2006). Resistant starch formation in temperature treated potato starches varying in amylose/amylopectin ratio. *Carbohydrate Polymers, 65*(3), 306–313.

Leitch, E. C. M., Walker, A. W., Duncan, S. H., Holtrop, G. & Flint, H. J. (2007). Selective colonization of insoluble substrates by human faecal bacteria. *Environmental microbiology, 9*(3): 667–679.

Levrat, M. A., Moundras, C., Younes, H., Morand, C., Demigné, C. & Rémésy, C. (1996). Effectiveness of resistant starch, compared to guar gum, in depressing plasma cholesterol and enhancing fecal steroid excretion. *Lipids, 31*(10): 1069–1075.

Liu, R. & Xu, G. (2008). Effects of resistant starch on colonic preneoplastic aberrant crypt foci in rats. *Food and chemical toxicology, 46*(8): 2672–2679.

Lopez, H. W., Levrat-Verny, M. A., Coudray, C., Besson, C., Krespine, V., Messager, A., Demigné, C. & Rémésy, C. (2001). Class 2 resistant starches lower plasma and liver lipids and improve mineral retention in rats. *The Journal of nutrition, 131*(4): 1283–1289.

Luckett, C.R., Wang, Y. J. (2012). Application of enzyme-treated corn starches in breakfast cereal coating. J Food Sci. 77(8), 1750–3841.

Lunn, J., Buttriss J. L. (2007). Carbohydrates and dietary fibre. Journal compilation. *British Nutrition Foundation Nutrition Bulletin, 32*, 21–64.

Maggio, C. A., & Pi-Sunyer, F. X. (2003). Obesity and type 2 diabetes. *Endocrinology and Metabolism Clinics of North America, 32*(4), 805–822, viii.

M. Chandalia, A. Garg, D. Lutjohann, N. (2000) Engl. J. Med. 48, 342, 1392–1398.

Martinez, I., Kim, J., Duffy, P. R., Schlegel, V. L., & Walter, J. (2010). Resistant starch type 2 and 4 have different effects on the composition of the focal microbiota in human subject. *PLoS One, 5*(11), e15046.

Marteau, P. R., de Vrese M., Cellier C. J., Schrezenmeir J. (2001). Protection from gastrointestinal diseases with the use of Probiotics. Am. J. Clin. Nutr. (Suppl.) 73,430–436.

Martínez, I., Kim, J., Duffy, P. R., Schlegel, V. L. & Walter, J. (2010). Resistant starches types 2 and 4 have differential effects on the composition of the fecal microbiota in human subjects. *PloS one, 5*(11), e15046.

Mathe, D., Riottot, M., Rostaqui, N. (1993). Effect of amylo-maize starch on plasma lipoproteins of lean and obese zucker rats. *Journal of Clinical Biochemistry and Nutrition. 14*, 17–21.

Mauro, M. O., Monreal, M. T. F. D., Silva, M., Pesarini, J. R., Mantovani, M. S., Ribeiro, L. R., Dichi, J. B., Carreira, C. M. & Oliveira, R. J. (2013). Evaluation of the antimutagenic and anticarcinogenic effects of inulin in vivo. *Genetics and Molecular Research, 12*(3): 2281–2293.

Mikulikowa, D., Masar, S., & Kraic, J. (2008). Biodiversity of legume health promoting starch. *Starch-Stärke, 60*, 426–432.

Mudgil, D., Barak, S., (2013). Composition, properties and health benefits of indigestible carbohydrate polymers as dietary fiber, A review. International Journal of Biological Macromolecules. 61, 1–6.

Nichenametla, S. N., Weidauer, L. A., Wey, H. E., Beare, T. M., Specker, B. L. & Dey, M. (2014). Resistant starch type 4-enriched diet lowered blood cholesterols and improved body composition in a double blind controlled cross-over intervention. *Molecular nutrition & food research, 58*(6): 1365–1369.

Nugent, A. P. (2005). Health properties of resistant starch. *British Nutrition Foundation, Nutrition Bulletin, 30*: 27–54.

Nordgaard, I., Mortensen, P. B.(1995). Digestive processes in the human colon. *Nutrition. 11*(1), 37–45.

Park, O. J., Ekang, N., Chang, M. J. & Kim, W. K. (2004). Resistant starch supplementation influences blood lipid concentrations and glucose control in overweight subjects. *Journal of nutritional science and vitaminology, 50*(2): 93–99.

Perrin, P., Pierre, F., Patry, Y., Champ, M., Berreur, M., Pradal, G., Bornet, F., Meflah, K., Menanteau, J. (2001). Only fibres promoting a stable butyrate producing colonic ecosystem decrease the rate of aberrant crypt foci in rats. *Gut, 48*(1), 53–61.

Perez-Alvarez, J. A. (2008). Losalimentos Del Dienestar: A Limenpacion Delsiglo. XXI. *Aalmentos, 3*, 54.

Posuwan, J., Prangthip, P., Leardkamolkarn, V., Yamborisut, U., Surasiang, R., Charoensiri, R. (2013). Long-term supplementation of high pigmented rice bran oil (Oryza sativa L.) on amelioration of oxidative stress and histological changes in streptozotocin induced diabetic rats fed a high fat diet, Riceberry bran oil. Food Chemistry, 138, 501–508.

Raben, A., Andersen, K., Karberg, M. A., Holst, J. J. & Astrup, A. (1997). Acetylation of or betacyclodextrin addition to potato beneficial effect on glucose metabolism and appetite sensations. *The American journal of clinical nutrition, 66*(2): 304–314.

Raben, A., Tagliabue, A., Christensen, N. J., Madsen, J., Holst, J. J. & Astrup, A. (1994). Resistant starch: The effect on postprandial glycemia, hormonal response, and satiety. *The American journal of clinical nutrition, 60*(4): 544–551.

Reader, A. (1997). Response of resistant starch in a food bar vs two commercially available bars in persons with type II diabetes mellitus. *Diabetes, 46*: 254.

Regassa, A., Charles, M. (2018). Application of resistant starch in swine and poultry diets with particular reference to gut health and function. *Animal Nutrition, 4*, 305–310.

Roberfroid, M. B. (2005). Introducing inulin-type fructans. *British Journal of Nutrition, 93*(S1): S13–S25.

Rochfort, S., & Panozzo, J. (2007). Phytochemicals for health, the role of pulses. *Journal of Agricultural and Food Chemistry, 55*: 7981–7994.

Rodríguez-Cabezas, M. E., Camuesco, D., Arribas, B., Garrido-Mesa, N., Comalada, M., Bailón, E., Cueto-Sola, M., Utrilla, P., Guerra-Hernández, E., Pérez-Roca, C. and Gálvez, J. (2010). The combination of fructooligosaccharides and resistant starch shows prebiotic additive effects in rats. *Clinical Nutrition, 29*(6): 832–839.

So, P. W., & Yu, W. S. (2007). Impact of resistant starch on body fat patterning and central appetite regulation. *PLoS One, 2*(12), e1309.

Stewart, M. L., Nikhanj, S. D., Timm, D. A., Thomas, W., & Slavin, J. L. (2010). Evaluation of the effect of four fibers on laxation, gastrointestinal tolerance and serum markers in healthy humans. *Annals of Nutrition and Metabolism, 56*, 91–98.

Sajilata, M. G., Singhal, R. S. & Kulkarni, P. R. (2006). Resistant starch – A review. *Comprehensive Reviews in Food Science and Food Safety, 5*: 1–17.

Sakamoto, J., Nakaji, S., Sugawara, K. (1996). Comparison of resistant starch with cellulose diet on 1,2-dimethylydrazine-induced colonic carcinogenesis in rats. Gastroenterology. 110 (1), 116–20.

Scourboutakos, M. (2010). 1 + 1 = 3… Synbiotics: Combining the power of pre-and probiotics. *Journal of Food Science Education, 9*(1): 36–37.

Sharma, A., Yadav, B. S. & Ritika. (2008). Resistant starch: physiological roles and food applications. *Food Reviews International, 24*(2): 193–234.

Shimada, M., Mochizuki, K. & Goda, T. (2008). Dietary resistant starch reduces levels of glucose-dependent insulinotropic polypeptide mRNA along the jejunum-ileum in both normal and type 2 diabetic rats. *Bioscience, biotechnology, and biochemistry, 72*(8): 2206–2209.

Shin, S. I., Choi, H. J., Chung, K. M., Hamaker, B. R., Park, K. H., & Moon, T. W. (2003). Slowly digestible starch from debranched waxy sorghum starch: Preparation and properties. *Cereal Chemistry, 81*, 404–408.

Tharanathan, R. N. (2002). Food-derived carbohydrates: Structural complexity andfunctional diversity. *Critical Reviews in Biotechnology, 22*(1), 65–84.

Thorup, I., Meyer, O., Kristiansen. E. (1995). Effect of potato starch, cornstarch and sucrose on aberrant crypt foci in rats exposed to azoxymethane. *Anticancer Research 15*, 2101–5.

Topping, D. L., Segal, I., Regina, A., Conlon, M. A., Bajka, B. H., Toden, S., … Bird, A. R. (2007). Resistant starch and human health. In *Dietray fibre: New frontiers for food and health.* Wageningen: Wageningen Academic Publishers.

Wang, X., Conway, P. L., Brown, I. L. (1999). In vitro utilization of amylopectin and high-amylose maize (Amylomaize) starch granules by human colonic bacteria. *Journal of Applied Microbiology. 87*, 631–9.

Weickert, M. O. & Mohlig, M. (2005). Impact of cereal fibre on glucose regulating factors. *Diabetologia, 48*: 2343–2353.

Wepner, B., Berghofer, E., Miesenberger, E. & Tiefenbacher, K. (1999). Citrate starch: Application as resistant starch in different food systems. *Starch, 51*(10): 354–361.

Wild, S., Roglic, G., Green, A., Sicree, R. and King, H. (2004). Global prevalence of diabetes: Estimates for the year 2000 and projections for 2030. Diabetes Care. 27, 1047–1053.

Yao, N., Paez, A. V. & White, P. J. (2009). Structure and function of starch and resistant starch from corn with different doses of mutant amylose-extender and floury-1 alleles. J. Agric. Food Chem. 57: 2040–2048.

Younes, H., Levrat, M. A., Demigne, C., & Remesy, C. (1995). Resistant starch is more effective than cholestyramine as a lipid-lowering agent in the rat. *Lipids, 30*(9), 847–853.

Young, G. P., Le Leu, R. K. (2004). Resistant starch and colorectal neoplasia. *Journal of the Association of Official Analytical Chemists International. 87*(3),775–86.

Young, G. P., & Gibson, P. R. (1995). Butyrate and the human cancer cell. In J. H. Cummings, J. L. Rombeau, & T. Sakata (Eds.), *Physiological and clinical aspects of short chain fatty acids* (p. 319). Cambridge, UK: Cambridge University Press.

Younes, H., Levrat, M. A., Demigne, C., & Remesy, C. (1995). Resistant starch is more effective than cholestyramine as a lipid-lowering agent in the rat. *Lipids*, 30(9), 847–853.

Zhang, G. & Hamaker, B. R. (2010). Cereal carbohydrates and colon health. *Cereal chemistry, 87*(4): 331–341.

Zhou, Y., Meng, S., Chen, D., Zhu, X., & Yuan, H. (2015). Structure characterisation and hypoglicemic effect of dual modified resistant starch from indica rice starch. *Carbohydrate Polymers, 103*, 81–86.

Zhu, C. L. & Zhao, X. H. (2013). In vitro fermentation of a retrograded maize starch by healthy adult fecal extract and impacts of exogenous microorganisms on three acids production. *Starch-Stärke, 65*(3–4): 330–337.

Recent Advances in the Application of Starch and Resistant Starch and Slowly Digestible Starch

Mudasir Ahmad, Sayeed Rukhsaar, Adil Gani, Bilal Ahmad Ashwar, Touseef Ahmed Wani, Umar Shah, and Faiza Jhan

Starch Edible Films and Coatings

Introduction

Any type of material used for enrobing (i.e., coating or wrapping) various food to extend shelf life of the product that may be eaten together with food with or without is considered an edible film or coating. Edible films provide replacement and/or fortification of natural layers to prevent moisture losses, while selectively allowing for controlled exchange of important gases, such as oxygen, carbon dioxide, and ethylene, which are involved in respiration processes. A film or coating can also provide surface sterility and prevent loss of other important components.

Edible films and coating are currently gaining large interests among consumers and Industrialists, because of the health and environmental issues raised over the use of plastics throughout the whole world due to lack of recycling facilities or infrastructure, non-recyclability, non-renewability, non-biodegradability or incorporation of toxic additives. This leads to the increased awareness among consumers for the use of non-plastics and current trends indicate steady growth will occur in

M. Ahmad · A. Gani (✉) · B. A. Ashwar · T. A. Wani
Department of Food Science and Technology, University of Kashmir, Srinagar, Jammu and Kashmir, India

S. Rukhsaar
Amity Institute of Food Technology, Amity University, Noida, Uttar Pradesh, India

U. Shah
School of Molecular and Life Science, Faculty of Science and Engineering, Curtin University, Bentley, WA, Australia

F. Jhan
Department of Food Science and Technology, University of Kashmir, Srinagar, Jammu and Kashmir, India

A. Gani, B. A. Ashwar (eds.), *Food biopolymers: Structural, functional and nutraceutical properties*, https://doi.org/10.1007/978-3-030-27061-2_4

the use of biodegradable plastics with increasing availability of suitable materials and because of societal and legislative pressure (Rasato 2009). The considerable improvement of alternative protection methods, for example the use of biopolymers which we can eat, from renewable sources or industrial by-products has been achieved (Campos et al. 2011). The use of starch for edible films and coatings is a subject of great interest due to their high biodegradability and characteristics to increase the shelf life of foods (Baldwin et al. 1995). This section will cover film forming ability of starch, physicochemical properties of these films and coating and their applications in food industry. Other aspects related to the starch-based films or packages, such as the effect of the addition of other components lipid, hydrocolloids, fillers, or active compounds will also be discussed.

Film Forming Ability of Starch and Manufacturing

Starch, is one of the most abundant natural polysaccharides having wide-ranging functionality, available at relatively low cost and great ability to form transparent, tasteless and odorless films, with very good oxygen barrier properties, which is very useful for food preservation purposes. Nevertheless, structure of starch presents some challenges to its potential applications in the packaging industry. Firstly, the starch has high degree of hydrophilicity due to the presence of three hydroxyl groups per D-glycosylic unit. Therefore, water is an effective plasticizer for this polysaccharide and any gain or loss of moisture may result in significant changes in physical and mechanical properties of starch-based materials. Secondly, native starch is not thermoplastic, because it undergoes pyrolysis before the melting point of crystalline regions in starch is reached. Hence, starch cannot be melt-processed via conventional plastic equipment without plasticizers addition (Mohammadi et al. 2013). In order to overcome these limitation starch has been either blended with other materials or chemically or physically modified to achieve balanced characteristics desired for film making properties (Shrestha et al. 2014). Consequently, starch compatibility with other biodegradable polymers, such polycaprolactone or polylactic acid, cellulose, among others, as well as composites processing with micro- and nano-fillers have been broadly studied (Andersons et al. 2016). Starch films may be obtained by two technological processes either by extrusion or casting methods. Casting typically involves the gelatinization of an aqueous starch suspension and its subsequent dehydration under controlled conditions. It has been studied extensively and mostly a related to coating on different surfaces like perishable fruits and vegetables, meat sausage or on seeds and other food stuffs. This method has several limitations like: restricted size and amount of the produced films, long production times and large volumes of solvent disposal used in the process.

The starch material films are brittle with no flexibility and extensibility. To overcome this problem plasticizers are generally added to film formulations to aid processing operations and to modify final product properties. Plasticizers are low molecular mass organic compounds, they reduce intermolecular forces and increase polymer chains mobility and must be compatible with the film-forming polymers; (Fan et al. 2018), oligosaccharides, polyols and lipids are chemically compatible (Lopez et al. 2008; Sothornvit et al. 2005). Excessive addition of plasticizing agents will result in rubbery, sticky films. Thus, it is necessary to optimize the amount of plasticizing agent used in a formulation to attain desired film properties. Starch in presence of a plasticizer (water, glycerol, sorbitol, etc.), high temperature and shearing melts and fluidizes that enables its use in injection, extrusion and blowing equipment, such as those for synthetic plastics. Thermoplastic starch can be prepared by altering the moisture/plasticizer content, amylose/amylopectin ratio of raw material and the temperature and pressure in the extruder, such starch can have different viscosity, mechanical strength, water solubility and water absorption (Mohanty et al. 2000). Large amount of researches have been performed on the plasticization of thermoplastic starch using glycerol (Forssell et al. 1997), sorbitol (Gaudin et al. 1999), urea, formamide (Ma et al. 2004), dimethyl sulfoxide (Nakamura and Tobolsky 1967) and low molecular weight sugars (Kalichevsky et al. 1993). Thermoplastic starch cannot be used in many applications due to its inferior mechanical and water barrier properties, its generally suggested to blend starch with other biodegradable polymers compatible with starch such as Poly lactic acid (PLA), polyvinyl alcohol (PVOH) and others. The starch-co-polymers films have shown better mechanical and water barrier properties. They find use mainly in soluble compostable foams, expanded trays, shape molded parts, and expanded layers as a replacement of polystyrene.

Starch as an Active Packaging Films and Coatings

The development of coatings based on starch has brought a significant increase in their applications and in the amount of products that can be treated, extending the shelf life of fruits and vegetables due to the selective permeability of these polymers to O_2 and CO_2 or development of films loaded with antioxidants and antimicrobial agents. Edible coatings create a passive modified atmosphere, which can influence various changes in fresh and minimally processed foodstuff in some areas such as, color, firmness, sensory quality, microbial growth inhibition, ethylene production and volatile compounds as a result of anaerobic processes (Oms-Oliu et al. 2008). Starch films incorporated with tea polyphenol (TP) were developed to produce active food packaging. TP addition led to an important improvement in antioxidant capability, as well as inhibition efficiency against the microorganisms of *S. aureus* and *E. coli*. However, a decrease in mechanical properties of films was observed (Fenga et al. 2018). The biodegradable films with antioxidant properties were developed, using cassava starch combined with biomass or biomass extract of microalgae *Heterochlorella luteoviridis* and *Dunaliella tertiolecta*. The addition of biomass increases film elongation, rupture and reduced the tensile strength and Young's modulus. The film containing 2.0% of *H. luteoviridis* extract presented the lowest water vapor permeability and good mechanical characteristics (Carissimi et al.

2018). Similarly, biodegradable active films using glycerol plasticized oxidized corn starch with and without ethyl lauroyl arginate (LAE) imparted antimicrobial properties to the films and greatly reduced the total viable counts in salmon (Moreno et al. 2017).

Extracts have also been added in biodegradable films to be used for active, non-edible packaging. The antioxidant activity of the film is related to its ability to release the entrapped active polyphenols. Jimenez et al. (2013) observed that starch-sodium caseinate films incorporated with oleic acid that lost part of their antioxidant capacity, while films incorporated with α- tocopherol showed an increased antioxidant activity. The effectiveness of an edible coating to protect fruits and vegetables depends on the film properties and functionality of some compounds (plasticizers, antimicrobials, antioxidants) within the matrix (Cerqueira et al. 2009). The loss of these molecules affects the thickness of the film (Park 1999) and the solubility of film in water as it is necessary to avoid the dissolution of the coating (Ozdemir and Floros 2008).

Starch Nanocomposites Films

Apart from the lipids or antimicrobial components, addition to biopolymer based films for improvement of mechanical, chemical or functional properties, a novel technique based on the use of very small particles are gaining remarkable popularity in food developments recently. Nowadays nanotechnology is applied with great results in many research areas. One of these fields of application is nanocomposite packaging films containing nanoparticles. A nanoparticle is an ultrafine particle in the nanometer size (Hosokawa et al. 2008), which is able to form nano biocomposite.

Nanocomposites" mean polymeric composite materials that are filled with nano sized rigid particles (Gopalan Nair and Dufresne 2003). These nanocomposite materials have superior mechanical, barrier, and thermal properties when compared with conventional composites, they have also advantage of good recyclability, transparency and low weight (Oksman et al. 2006; Sorrentino et al. 2007). Nanocomposite films containing AgNPs concentrations greater than 71.5 ppm inhibited the growth of *E. coli* ATCC and *Salmonella* spp., which are responsible for most foodborne disease and also caused a slight increase in film thickness and opacity, while decrease in water vapor permeability with increasing AgNPs concentration was observed by Ortega et al. (2017). Since starch is a biodegradable and nontoxic polymer so it has been widely used in non-food applications including papers, textiles, plastics, cosmetics and pharmaceuticals. However, some researchers suggested starch nano-particles as reinforcement filler in the polymeric composites. A wide range of polymeric matrixes including both natural and synthetic polymers have been suggested for the composites with SNPs (Gopalan Nair and Dufresne 2003; Anglès and Dufresne 2000). The incorporation of starch nano-particles into the synthetic polymer matrix, has not only improved the physical properties but the biodegradability of the composite also enhances (Kristo and Biliaderis 2007). However there are safety issues related to nanomaterials as their size may allow them to penetrate into cells, which can be toxic to human body but there is limited scientific data about their eventual toxicological effects.

Encapsulation of Bioactive Compounds

Encapsulation is defined as a technology of packaging solids, liquids, or gaseous materials in sealed capsules/shells (shown in Fig. 1) that can release their contents at controlled rates under specific conditions (Desai and Park 2005). This technology has wide applications both in food and pharmaceutical industry for coating several ingredients such as flavors, antioxidants, colors, acidulants, probiotics, polyunsaturated fatty acids (PUFAs), enzymes, vitamins, drugs, etc. (Can-Karaca et al. 2015; Pegg and Shahidi 2007). Encapsulation has several roles to play like: (1) protection of the core material from degradation by reducing its reactivity to its outside environment such as light, moisture, and oxygen (2) reduction of the evaporation or transfer rate of the core material to the outside environment; (3) To achieve the slow or controlled release in gastrointestinal conditions (5) to mask an unwanted flavor or taste of the core material (4) To increase the solubility and bioavailability of active ingredients by forming nanocapsules (5) to help separate the components of the mixture that would otherwise react with one another (6) protection of heat sensitive bioactive ingredients during thermal processing of foods (7) Development of functional foods by fortification of micro or nano capsules containing active ingredients. Large number of wall materials have been used so far for the encapsulation which includes proteins, carbohydrates, lipids, and so on (Wani et al. 2016). The encapsulating material must be foodgrade, biodegradable, and stable in food systems during processing, storage, and consumption. The selection of the wall

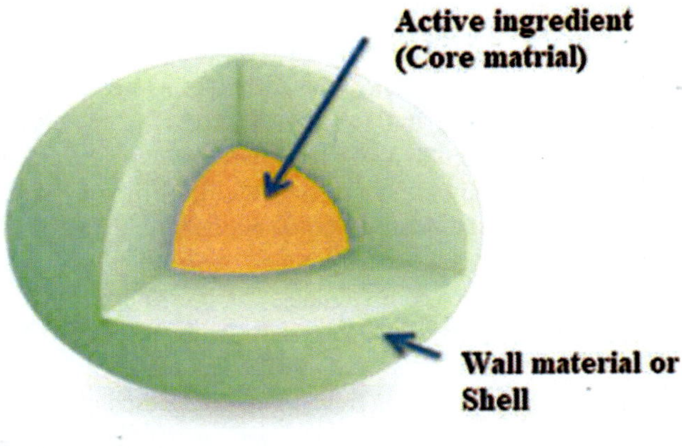

Fig. 1 Representation of encapsulation technology

material depends upon the type of the food ingredients and the encapsulation methods, and can be used either individually or in combination. Various encapsulation methods include electrospinning, gelation, extrusion, co-precipitation, coacervation, spray drying, freeze drying, layer-by-layer deposition, emulsion system, and others (Wani et al. 2016). Different encapsulation methods result in different nano/micro structures, which affects encapsulation efficiency and release properties of core ingredients. The below sections will describe the applications of native, modified and resistant starch for encapsulation of various bioactive ingredients including polyphenols, plant extracts, probiotics, vitamins and etc.

Native and Modified Starch as Encapsulating Agent

Starch is a renewable biopolymer and widely distributed in nature with a huge diversity of industrial applications and among them its use as wall material for encapsulation of bioactive compounds is of great interest. This technique involves the packaging of sensitive functional ingredients within the matrix of wall materials and ensures controlled release under specific conditions. It is widely used to safeguard the bioactive compounds from harsh environmental and gastrointestinal conditions. Functional ingredients can be encapsulated by various wall materials like starch, dextrins, β-glucan, gum arabic, carboxymethyl cellulose, alginate, chitosan, xanthan, pectin, zein and whey protein isolates; (Wani et al. 2016; Shah et al. 2016; Gani et al. 2018; Ahmad et al. 2018), but starch is the predominant one. There has been increasing interest of using starch in native and modified forms to encapsulate food ingredients such as flavours, lipids, polyphenols, carotenoids, vitamins, enzymes, and probiotics because starch is generally recommended as safe, available at cost, non-allergenic and bland in taste. Some studies have observed higher encapsulation efficiency of starch and provided better protection of food ingredients (e.g., flaxseed oil and flavours) against unfavourable conditions than protein and gum arabic based systems (Charve and Reineccius 2009; Tonon et al. 2012).

Some native starch granules have surface pores as well as channels in the granules (Huber and BeMiller 2000; Han et al. 2015) to encapsulate a range of bioactive components including catechin, gallic acid, caffeine, ibuprofen (a water insoluble drug), and amino acids. Other studies on encapsulation using starch include the encapsulation of soy-based ferulic acid derivatives that have antioxidant properties (Compton et al. 2007). Thymol and nisin encapsulated into a starch-based O/W emulsion and found to have enhanced antibacterial efficacy in cantaloupe juice (Sarkar et al. 2017). β carotene was encapsulated in Starch hydrogels and studied in a simulated gastrointestinal tract including oral, gastric, and intestinal phases. Compared with the oil-in-water emulsion systems with WPI (14%), the filled starch hydrogels showed higher bio- accessibility of β-carotene. This was attributed to the prevention of the fat droplets from aggregation by the starch hydrogels during oral and gastric digestion. Catechin and folic acid were encapsulated in starch and studied for release behavior under simulated gastrointestinal conditions, the results reported higher release in intestinal section and protection from adverse gastric conditions (Ahmad et al. 2019a; Ahmad et al. 2018). Starch (mainly its linear glucose homo-polymer fraction named amylose) complexes can retain guest molecules within or in between hollow helices in a form termed V-amylose of nanometric dimensions (Lalush et al. 2005; Lesmes et al. 2008). There has been increasing interest of using starch based delivery systems to encapsulate various food ingredients as reflected by the increasing number of publications during the last few years (Hasanvand et al. 2015; Nielsen et al. 2016; Ahmad et al. 2017, 2019a).

Native starch is commonly modified to have the desired properties for different applications including encapsulation and controlled release by physical, chemical and enzymatic methods (Fathi et al. 2014; Wurzburg 1986). The effect of the partial or total substitution of gum arabic (GA) with native (NA), octenyl succinic anhydride (OSA) and succinylated (SUC) sorghum starches on microencapsulation of nutmeg oleoresin was studied using spray drying technique. Results obtained suggested that both NA and OSA starches could be considered as a good alternative to gum arabic for microencapsulation. Some studies have reported higher encapsulation in starch derivatives like the values of encapsulation for anthocyanins were reported higher in starch derivatives with respect to hydrolyzed native normal and waxy maize starches. Another study showed oxidized tapioca starch has potency to substitute for alginate as encapsulation material (Palupi and Praptiningsih 2016).

Natural starch is predominantly hydrophilic, which limits its application for encapsulating hydrophobic food bioactives. Hydrophobic starch derivatives like Octenyl succinylated starch (OS-starch) have therefore been developed to overcome this drawback. The amphiphilic character of OS-starch makes it an interesting wall material for encapsulating hydrophobic bioactive ingredients. Octenyl succinylated starch (OS-starch) finds wide application in the food industry mainly as emulsifier, encapsulating agent and fat replacer. OS-starch has been used for encapsulation of coenzyme Q10, propolis and conjugated linoleic acid (CLA) (Cheuk et al. 2015; Da-Silva et al. 2013; He et al 2016). The studies have suggested OS-starch as an efficient encapsulating agent (Cheuk et al. 2015; Da-Silva et al. 2013; He et al. 2016).

Starch-based ingredients are also considered as excellent materials for flavor encapsulation over the years. The four most commonly used starch-based ingredients in flavour encapsulation are porous starches, cyclodextrins (CDs), starches and (Octenyl succinic anhydride) OSA starches. Porous starch has been reported to have potential to encapsulate flavors by using a simple plating procedure (Zhao et al. 1996). Numerous pores on the surface and interior exhibit intensive and strong adsorption ability to trap flavor substances tightly inside the starch molecules and can release those adsorbed substances completely. These functions allow porous starch to adsorb volatiles for protection and slow release. CD's have been used for encapsulation of lemon oil, nutraceutical monoterpenes, caraway fruit oil, garlic oil and pine for protection (Bhandari et al. 1998, 1999; Lee et al. 2005). The CD's have unique structures with hydrophobic cavity and hydrophilic surfaces and form inclusion complexes with materials of low water solubility to encapsulate core ingredients entirely or partly in the internal cavities. Hydrolyzed starches like maltodextrin have been used as wall materials for encapsulations of turmeric oleoresin, cumin oleoresin, rose oil, etc. (Wang et al. 2014). The native and hydrolyzed starches have poor emulsifying properties that poses some limitations for its use in encapsulation process. However the chemical modification namely esterification reaction can introduce the functional hydrophobic and hydrophilic groups onto starch involving in effective emulsification properties (Henry and Alistair 1995). There are several possible mechanisms for OSA starches to stabilize emulsion systems like charged OSA starches can stabilize the emulsion by electrostatic stabilization. Several reports on OSA starch production via esterification in different sources of starches from lima bean starch, waxy maize, potato starch and amaranth starches have been used for encapsulation of volatile flavors, such as orange oil, meat flavor and black pepper oleoresin (Wang et al. 2014). The major problem associated with the use of OSA starch in encapsulation is the poor protection of core ingredients against oxidation. To overcome this, OSA starch is frequently used in combination with hydrolyzed starches or other carriers to obtain optimal emulsification and oxidation resistant properties (Reineccius 2004).

Resistant Starch as Encapsulating Agent

Resistant starch (RS) is the fraction of starch that escapes digestion due to its resistance to digestive enzymes in gastrointestinal conditions and does not release monosaccharide within the small intestine. However, it is fermented to a large extent by microbiota in the colon, resulting in the production of short-chain fatty acids (Bird et al. 2007). The starch may be physically inaccessible, retrograded, or chemically modified to resist the digestive enzymes RS occurs naturally in some foods or can be prepared as result of several processing conditions like hydrothermal treatments, annealing, partial gelatinization and ecrystallization, autoclaving, pullulanase debranching, temperature-cycled retrogradation, phosphorylation, hydroxypropylation, acetylation, oxidation, and citric acid modification (Ashwar et al. 2018; Sang et al. 2010). Generally, RS is subdivided into five categories like (1) RS1, which is physically entrapped starch found in cereals within the whole or partially milled seeds due to lack of cell-wall-degrading enzymes in gastrointestinal tract (Raigond et al. 2015). (2) RS2 is non gelatinized native starch granule found in raw potato, green banana, and native high-amylose maize starch; It is protected from digestion by their compact conformation or structure comprising of amylopectin crystals of uncooked native starch granules and is usually in B- or C-type crystalline polymorph (Englyst and Cummings 1987a, b). (3) RS3 is retrograded starch can be produced by gelatinization (i.e., a process of disruption of the granule structure by heating with an excess of water) and retrogradation (i.e., a process of slow recrystallization of starch molecules upon cooling or dehydration) during food processing (Englyst et al. 1992); (4) RS4 is chemically modified starch and prepared by etherification, esterification, and cross-linking, (Lunn and Buttriss 2007); (5) RS is an amylose-lipid complexed starch which is formed from high AM starches that require higher temperatures for gelatinization and are more susceptible to retrograde

(Cummings and Stephen 2007). Apart from positive effects on human health to control obesity, cardiovascular disease, diabetes, and colon cancer (Situ et al.), it is also a good source of encapsulating materials due to its resistance to enzymatic digestion by pancreatic enzymes in mono gastric animals like human. Resistant starch has been used to encapsulate bioactive compounds that are unstable to light, heat, oxygen and upper intestinal physiological conditions.

Resistant starch (RS) has some specific advantages over native or modified starch for its use as encapsulating material, such as less solubility, higher crystallinity and stability in high processing temperatures. RS in combination with milk protein has been used for encapsulation of fish oil and showed improved oxidative stability (Chung et al. 2010). Enhanced oxidative stability of fish oil was reported by encapsulation in culled banana resistant starch-soy protein isolate based microcapsules in functional bakery products (Ayesha, Nasrin and Kumar 2015). The addition of resistant starch-pectin-iron and pectin-iron microparticles to infant powdered milk models showed the increased iron absorption in rats with no significant effects on calcium absorption and sensory evaluation of milk powder. The research suggests the application of the resistant starch as well as pectin for possible fortification and improved delivery of Iron through various food systems (Moslemi et al. 2018). Besides this, the microcapsules of resistant starch containing ascorbic was utilized for development of films and resulted in decreased water vapor permeability (WVP) values with increasing concentrations of microcapsules in the films (Martínez-Ortiza et al. 2017). The resistant starch (RS) prepared by heating starch suspensions or heating followed by microfluidization and their functionality as co-encapsulants in sodium caseinate based fish oil microcapsules were examined and better oxidative stability of fish was achieved (Chung et al. 2010). In our previous study on encapsulation of caffeine in resistant starch, β-glucan and β-cyclodextrin other polysaccharides β-glucan showed maximum decline in the release of caffeine followed by resistant starch and β-cyclodextrin under mimicked stomach conditions whereas RS provided more slow release in intestinal conditions (Noor et al. 2018). Apart from encapsulation of bioactive compounds, essential oils and minerals, the encapsulation of probiotics have also been reported well. The three probiotic strains namely *Lactobacillus casei*, *Lactobacillus brevis* and *Lactobacillus plantarum* were microencapsulated with resistant starch and showed better survival ability than that of free cells in simulated gastrointestinal conditions (Ashwar et al. 2018). Ying et al. (2013) incorporated *L. rhamnosus* in physically modified resistant starch (RS) formulations, whey protein-based systems, and a combined system made of whey protein and physically modified resistant starch. The protective efficiency of single and combined systems was compared in citrate buffer and commercial apple juice (pH 3.5). During 5 weeks of storage at 4 °C and 25 °C under acidic conditions, the whey protein-based formulations provided significantly higher protection of bacteria than the starch formulations. The resistant starch despite some advantages over native starch may some limitations in carrying capacity of core ingredients and its use as encapsulation material has not been studied much.

Table 1 Functional application of starch

Functions	Foods
Adhesion	Battered and breaded foods
Binding	Formed meat, snack seasoning
Clouding	Beverages
Crisping	Fried and baked food, snacks
Dusting	Chewing gum, bakery products
Emulsion stabilisation	Beverages, creamers
Encapsulation	Flavours, beverage clouds
Expansion	Snacks, cereals
Fat replacement	Ice cream, salad dressing, spreads
Foam stabilisation	Marshmallows
Gelling	Gum drops, gelly gum centers
Glazing	Bakery, snacks
Moisture retention	Cakes, meats
Thickening	Gravies, pie fillings, soups

Functional Ingredient in Food Systems

Besides meeting nutritional requirement for humans, starch is widely used as thickening and gelling agents. They are used as flow agents and dusting powders, in dry mixes (puddings and gravies), and as starch blends in salad dressings. Native potato and waxy maize starches are used to provide desired product viscosity during filling of cans prior to retorting. Unmodified tapioca and waxy maize starches are used in packed foods like salad dressings and buttermilk due to improved stability. Native high-amylose starches are used in batters and jelly gum candies. Generally, native or unmodified form, starches have narrow applications in food industry as they produce undesirable pastes and gels. Foods made from such starches have inadequate shelf stability and poor process tolerance during manufacturing. Prolonged holding, followed by repeated freezing and thawing leads to deterioration of products made from native starch. However, these limitations have been overcome to some extent by introducing several modifications to starch. Due to such reasons, the applications of starch as an functional ingredient has been widely in various food formulations for gelling, thickening, increasing stability and replacing costly ingredients. Food applications of starch are discussed below and also given in Table 1.

Food Applications of Starch

Starch as a Thickener

The usage of starch as a thickening agent makes it the most important functional ingredient of the food industry. The reasoning involves the nonspecific linkage of conformational disordered strands. The change in the viscosity of starch solutions is due to the physical entanglement of the chains. In diluted solutions the movement of molecules is free but with the increasing concentration, a change in viscosity is noticed. In concentrated solutions, the movement of molecules gets restricted, thus thickening the solution (Sworn 2003). Starch being the abundant and cheap hydrocolloid is the commonly used thickener. The added advantage of starch is that it does not impart any taste of its own if used at a concentration of 2–5%. Starch is used to thicken many foods such as gravies, soups, puddings, sauces and pie fillings. Modified waxy maize, tapioca and potato starches are preferred for this purpose because of their enhanced stability to textural changes during marketing and distribution.

Starch as a Gelling Agent

A gel like structure is formed when specific association takes place at junctions in conformational ordered chains. The process of gelation involves the formation of three dimensional network that gets formed by the association of dispersed polymer chains. The three dimensional structure contains the solvent in the interstices. There is a formation of junction zones which form the regions of association. These junction zones may be formed by two or three polymer chains (Oakenfull and Glicksman 1987). The formation of gel involves the association of randomly dispersed polymer segments in dispersion in such a way so as to form a three-dimensional network that contains solvent in the interstices. The associated regions known as 'junction zones' may be formed by two or more polymer chains. The gelation process is essentially the formation of these junction zones (Oakenfull and Glicksman 1987). A variety of starches are used for gelling. Dent corn starch is used as a gelling agent in pies, puddings, and cheese products. It provides shorter texture and retains shape of the finished product. Blends of gelling and non-gelling starches can be used to provide intermediate properties, such as slight cut ability in products like salad dressings.

Starch as a Fat Mimetic

Nowadays consumers are becoming more inclined towards eating healthy food. People are well aware about the adverse health effects associated with overconsumption of fats and oils. To overcome this problem, food manufacturers have developed products with reduced or no fat. Fat as a food component enhances texture, flavour and keeping quality of food products. Therefore its replacement may affect the quality of the final product. To deal with such problems, a combination of non-fat ingredients with different functional properties can be used to replace the quality attributes when fat is removed. One such ingredient is starch, which needs to be modified by hydrolysis to be used as a fat replacer. Maltodextrin is a well-recognized and commercially available fat replacer which is used in low- fat mayonnaise, low- fat butter spreads, margarines, low-fat milk products like ice-creams and cheese (Sajilata and Singhal 2005).

Starch as Texture Improver

Modified starches can be used as texture improvers in food industries as it provides desired chewiness to foods, crispiness to biscuits and crackers, quality enhancement of extruded products and resistance to viscosity breakdown in canned foods. It can affect mouth feel. Starch hydrolysates are used for improving mouth feel where cohesiveness or pastiness is to be avoided. High amylose cornstarch is used in extruded snacks to obtain evenly browned crisp product.

Resistant Starch as a Functional Ingredient

Dietary carbohydrates can be categorized into "available" and "resistant" based on their chemically distinct structures and varied metabolic and gastrointestinal properties: (1) 'available or digestible carbohydrates' are digested by the enzymes present in the digestive tract and absorbed in the small intestine to provide energy and carbohydrates for metabolism, for example, sucrose, lactose, HMO, and starchy vegetables, (2) 'resistant or indigestible carbohydrates' escapes digestion in the small intestine and are poorly absorbed and metabolized, for example, dietary fibers present in cereals, fruits, vegetables, and FOS such as inulin (Englyst et al. 2007). Dietary fiber is the most prominent and universally agreed- upon resistant carbohydrate. Besides dietary fiber, there are other sources of naturally occurring and artificially developed resistant carbohydrates such as gums, polydextrose, fructans, resistant maltodextrins and resistant starch.

Resistant starch (RS) is that portion of starch which resists digestion or is incompletely digested by the enzyme pancreatic amylase and in the intact form or as a product of partial hydrolysis reaches the colon for further anaerobic microbial fermentation. Therefore, RS may be defined as the sum of starch and the product of starch digestion that resists digestion in the small intestine of a normal human being. Naturally occurring RS is present in cereals, starchy foods, and heated starch (Charalampopoulos et al. 2002). RS has been classified into four forms (Englyst and Cummings 1987a, b; Eerlingen and Delcour 1995; Brown 1996; Haralampu 2000) as already mentioned in the chapter.

Sources of Resistant Starch

Resistant starch (RS) is found basically in all starchy foods but its quantity may vary depending upon the biological origin of the plant, structure of starch granules, and changes proceeding processing operations. The RS content in foods may vary to about 20% of the product weight or higher (Brighenti et al. 1998). The RS content of various foods has been provided in Table 2.

Food Applications of RS

RS is considered to be safe for human consumption. It is a valuable food ingredient and has enjoyed considerable interest particularly from researchers and food developers. This is due to two reasons, the first being its potential physiological effects (increased laxation, lowering cholesterol levels, reduced risk of cancers, lowering postprandial glucose response, prevention of gall stone formation and others) and the second being its unique functional properties. It can be rightly referred to as a functional fiber as it permits high nutritional claim and desirable physiochemical properties such as water-binding capacity, swelling, gel- formation, viscosity enhancement. As the consumption of processed food is increasing at an alarming rate, with fewer vegetables and fruits, the proportion of RS in food has gone down. Therefore, incorporating RS in food products has become increasingly important. RS containing foods exhibit better sensory properties compared with traditional fiber- containing foods (bran and gums) which are denser, coarser and less palatable than the former.

The low water- holding capacity of RS is responsible improving the texture of the finished product (Yue and Waring 1998). It does not impart its own flavor to the product and provides improved expansion, crispiness, better mouth feel, and color to the final product (Sajilata et al. 2006). They may be used to increase the fiber content of low carbohydrate foods consumed by people following strict diet regimes, providing all the essential fiber without increasing the caloric value of food (Nugent 2005). RS has a lower calorific value (8 kJ/g) and can therefore be substitutive of digestible carbohydrates, lowering the energy value of the final product. RS also finds use in fermented foods such as sausages. The addition of RS in pastas, beverages and baked products enhances health benefits and textural properties of the products.

Resistant Starch as Prebiotic

The recent upsurge in urbanization and industrialization has caused a drastic and dramatic change in the lifestyle and eating habits of people globally. This has led to many lifestyle- related diseases like cancer, diabetes, cardiovascular diseases, immunodeficiency, obesity, etc. Therefore, to tackle these conditions, people are now becoming conscious and are switching to much healthier natural products that can improve their metabolism and make them healthier and fitter. Lately, prebiotics have received much attention from the researchers as well as from the prebiotic ingredient manufacturers because of the advantageous effects conferred by them on human health. As the demand for healthier and wholesome food is increasing, the food and beverage manufacturers are focusing on research and development activities to boost the nutritional and functional profile of foods by incorporating prebiotic rich ingredients into their products such as health drinks, dairy products,

Table 2 Approximate total dietary fibre, starch and resistant starch of some food sources (g/100 g as eaten)

	Source	Total starch	Total dietary fiber	Resistance starch
Legumes	Red kidney beans	42.6	36.8	24.6
	Lentils	53.3	33.1	25.4
	Black eyed peas	53.9	32.8	17.7
Cereal grains	Barley	55.2	17.0	18.2
	Corn	77.9	19.6	25.2
	White Rice	95.1	1.5	14.1
	Wheat	50.8	17.0	13.6
	Oats	43.4	37.7	7.2
Flours	Corn	84.3	2.8	11.0
	Wheat	68.8	12.1	1.7
	Rice	86.9	5.2	1.6
	Potato	81.0	2.1	1.7
Grain-based food products	Spaghetti	73.0	5.6	3.3
	Rolled oats	56.0	10.0	8.5
Cereal products	Crisp bread	67.4	n/a	1.4
	White bread	46.7	n/a	1.9
	Granary bread	44.1	n/a	6.0
	Extruded oat cereal	57.2	n/a	0.2
	Puffed wheat cereal	67.0	n/a	1.2
	Oat porridge	9.0	n/a	0.3
	Cooked spaghetti	n/a	n/a	2.9
	Cooked Rice	n/a	n/a	3.7
Potato products	Boiled potatoes	n/a	n/a	2.0
	Chips	29.5	n/a	4.8
	Mashed potatoes	n/a	n/a	2.4

Adapted from Lunn and Buttriss 2007; *n/a* not available

breakfast cereals, nutrition bars, beverages, bakery products, mineral supplements, weight loss products, infant foods and others.

Prebiotics

The word prebiotic is derived from the Greek phrase "pre- bios" meaning "before life". The earliest definition of prebiotic that had a semantic analogy with the term probiotic, was put forward in 1995 as, "non-digestible food ingredients that beneficially affect the host by selectively stimulating the growth and/or activity of one or a limited number of bacteria in the colon, thus improving host health" (Gibson and

Roberfroid 1995). Ever since this definition has continued to evolve in light of research suggesting that prebiotics extent their impact on the microbiota that resides outside the colon. The definition of prebiotics has been updated by the International Scientific Association for Probiotics and Prebiotics (ISAPP) as, "a substrate that is selectively used by a host microorganism to produce a health benefit" (Gibson et al. 2017).

The prebiotics limit the growth of harmful bacteria and promote the growth and survival of health-promoting advantageous microbiota (probiotics) such as *Lactobacilli, Bifidobacteria, Straptococcus* including yeast such as *Saccharomyces boulardii.* The consumption of prebiotics improves gut health by stimulating the growth of favorable microflora, and promotes overall health of human beings by developing and strengthening the immune system, improving digestion of food and enhancing the absorption and bioavailability of nutrients, lowering the risk of diseases like cancer, cardiovascular diseases, obesity, diabetes etc., alleviating symptoms of Irritable Bowel Syndrome (IBS), lactose intolerance, and allergies (e.g., atopic dermatitis).

Prebiotic rich foods such as cereals, fruits, vegetables and other edible plants have been a significant part of the human diet since time immemorial. The most potential natural sources of prebiotics are the carbohydrate rich foods such as bananas, berries, asparagus, chicory, leeks, onions, garlic, wheat, oats, barley, linseed, artichokes, soya beans, green vegetables and legumes. Prebiotics occur in various forms such as lactose, mannan-oligosaccharides (MOS), chicory fructans, fructo-oligosaccharide (FOS), oligofructose, inulin, galacto-oligosaccharide (GOS), xylo-oligosaccharide (XOS), human milk oligosaccharides (HMO), and others. Oligosaccharides have been identified as the main source of prebiotics (Slavin 2013; Gibson et al. 2017; Rajendran et al. 2017). Resistant starch, pectin and beta-glucans also fit in the definition of prebiotics (Arena et al. 2014; Gómez et al. 2014; Zaman and Sarbini 2016). The recommended daily intake of a prebiotic is 5–8 g, while exceeding 20 g/day may cause intestinal discomfort due to distension. This section deals with the potential of resistant starch as a functional pre-biotic.

Significance and Application of RS as a Prebiotic

Nowadays, the viewpoints on nutrition have changed and people are switching from carbohydrate-rich foods to meals with low caloric value. An increased intake of foods rich in dietary fiber has also seen an upsurge in the recent years as it is considered to be indispensable for proper functioning of the organism. For this reason, resistant starch is presently attracting widespread interest leading to an increased demand for development of economical and robust RS for industrial and research applications. RS, a natural food component, is believed to be neutral to the organism (Annison and Topping 1994). Unlike other dietary fiber components, it's neither a fixed entity nor is it intrinsically indigestible. It increases the food weight without increasing the caloric value (Ranhotra et al. 1996). Consumption of food products containing a part of non-digestible starch decreases levels of glucose and insulin in blood, compared to foods containing same levels of completely digesting starch (Bornet et al. 1989). RS is utilized and acted upon by the microflora of the large intestine, which causes it to ferment and results in the production of hydrogen, methane, and SCFA (mainly propionic, acetic and butyric acid) thereby lowering lumen pH. The resulting pH helps in selective screening of intestinal microflora, eliminating pathogenic microorganisms and stimulating growth of beneficial bacteria. The acids formed in the large intestines demonstrate health-promoting activity. The amount of butyric acid formed upon fermentation of RS is higher than that produced as a result of non-starch polysaccharide fermentation and plays an important role in prevention of anal and colorectal cancers.

In order to be identified as prebiotic RS has to fulfill the following criteria:

I. Resistance to gastric acidity, hydrolysis by mammalian enzymes and absorption in the upper gastrointestinal tract;
II. Fermentable by probiotic microbiota, especially lactobacilli and bifidobacteria,
III. Selectively stimulate activity and/or growth of gut bacteria that contribute to host health and wellbeing.

Criteria I

Structural characteristics are responsible for the resistance of RS. The ratio of amylose to amylopectin being one of the main characteristics that contributes to the resistance of starch. The entangled arrangement between these molecules (amylase and amylopectin) holds the integrity of starch granules. It has been demonstrated that enzymatic digestibility of starch decreases as amylose to amylopectin ratio increases (Jane 2006). The amylase-amylopectin interaction makes the starch inaccessible to hydrolytic enzymes by preventing the starch from swelling. Another starch characteristic that could influence its digestibility is the lipid content. The amylase-lipid complexes exhibit resistance to hydrolysis by amylase enzyme. Formation of such complexes restricts swelling of starch and retard the enzymatic hydrolysis of starch granules (Jane and Robyt 1984; Morrison 2000). In addition, the presence of lipid on the surface of starch granules can contribute to their resistance (Morrison 1981, 1995).

Criteria II

For a RS to be regarded as a prebiotic, it must be used as a substrate for fermentation by gut microflora and confer health benefits to the host. Bacterial fermentation that occurs in the human gut are of two types viz., saccharolytic and proteolytic. As the names suggest, the saccharolytic bacteria break down the carbohydrates and are dominant in proximal colon, whereas, proteolytic bacteria are responsible for protein breakdown and mostly occur in distal part of the colon. Both types of fermentation produce different end products which can further be analyzed to determine the efficacy of RS as prebiotic. The two types of fermentation produce different by products which can be tested to determine the RS efficacy as prebiotic candidates.

Criteria III

Sometimes, the RS might not fulfill Criteria II of being fermentable by beneficial probiotic bacteria. The fermentation, in some cases, might not be specific towards the beneficial bacteria. The RS may also be fermented by a wide array of intestinal microbiota including the pathogens. This causes complications when investigating a putative prebiotic carbohydrate, making Criteria III most difficult to fulfill (Gibson et al. 2004). Such investigations require extensive quantitative microbiological analysis of colonic bacteria such as bifidobacteria, lactobacilli, enterobacteria, clostridia, eubacteria and the total aerobes/anaerobes for which molecular-based microbiological methodologies have been developed and accepted as a reliable tool (Sarbini et al. 2013). Also, the production of gas, enzymes and organic acids has been used as indicators to monitor stimulation of bacterial activity (Sarbini and Rastall 2011). Figure 2 demonstrates the process of RS to be qualified as a prebiotic. Table 3 enlists the methodologies developed to test the potential of RS as a prebiotic.

Starch as Nano Delivery System for Bioactive Compounds

Nanotechnology has emerged as one of the most innovative technologies in the last two decades and focuses on the characterization, fabrication and manipulation of matter structure at dimensions of roughly 1–100 nm (Gallocchio et al. 2015). The fabrication of foods at the nano scale, for example liposomes, micelles, nanoparticles, nanoemulsions, nanofibers, nanotubes and nanocomposites have been exploited for increasing shelf life, improving flavor profiles, enhancement of nutritional value of bioactive components and improving properties of packaging materials (Dufresne 2014; Gallocchio et al. 2015). Recently it is reported that nano-structured particles were used to develop more effective delivery vehicles to target sites for increasing efficiency, bioavailability and absorption of nutraceuticals (Ahmad et al 2019a, 2019b; 2019c; Ahmad et al 2018; Ahmad et al 2017).

The nanoparticles can be produced from proteins (such as, gelatin and milk proteins), polysaccharides (such as chitosan, sodium alginate and starch), and synthetic polymers. Among them starch being the second most abundant biodegradable material can be commercially more important for the food industry (Le-Corre et al. 2010). The starch structure has been under research for years, but because of its complexity, the universally accepted model is still lacking (Buléon et al. 1998). However, the predominant model for starch is a concentric semicrystalline multiscale structure that allows the production of new nanoelements: (1) starch

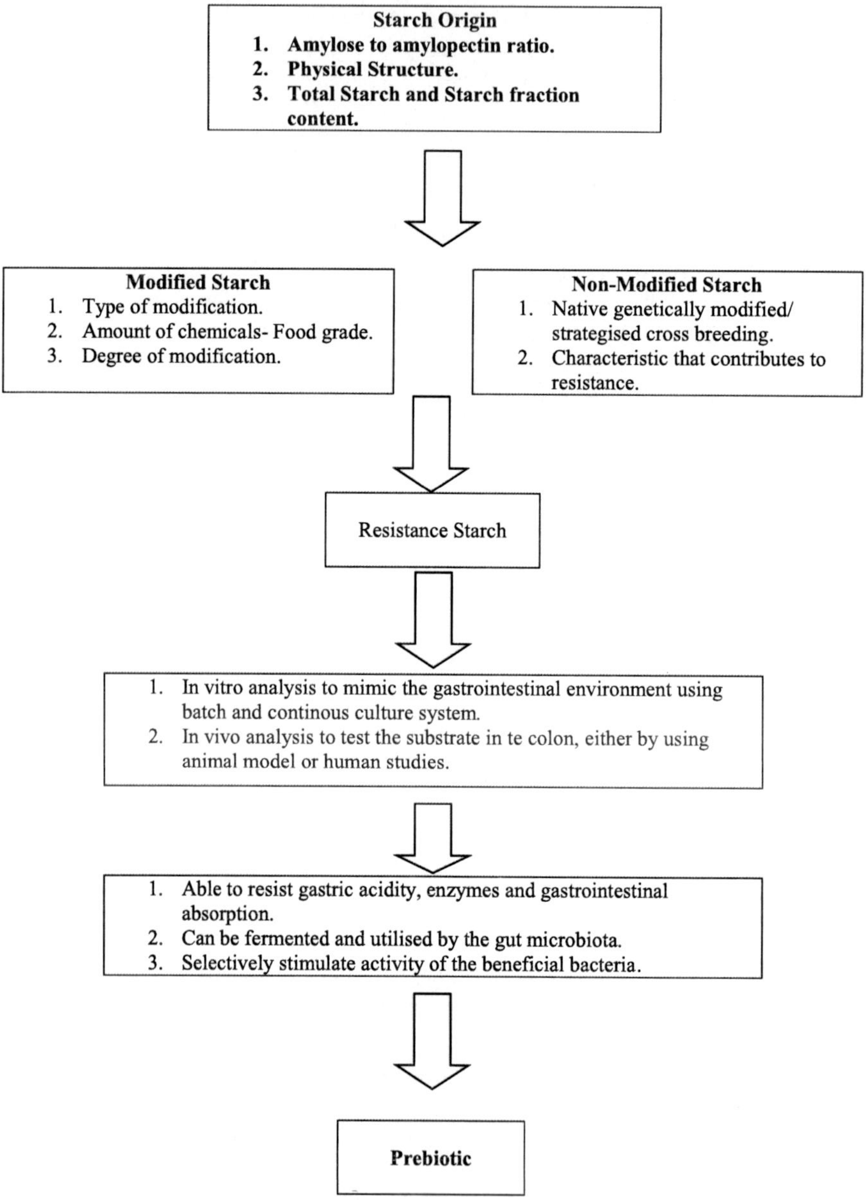

Fig. 2 Process of RS to be qualified as a prebiotic

Table 3 Methodologies adopted to test the potential of RS as prebiotic

Criteria	In vitro test	In vivo test
Resistance towards gastric acidity. Hydrolysis by mammalian enzymes and gastrointestinal absorption	Pre-treatment of the starch substrate with various acids and enzymes that mimic the gastrointestinal condition	Measuring substrate recovery in feces of rat
		Intubation into the gastrointestinal system of anesthetised rats
		Direct recovery of undigested molecules in the distal ileum following oral administration of substrate
Able to be fermented and utilised by gut microbiota	Fermentation of carbohydrates by batch or continuous model with fecal bacteria	Animal model—Rats are fed with food or drink fortified with prebiotic for a duration of time. The animal will then be anesthetised and killed to remove the colon for further analysis
		Indirect method—Collection of breath air at regular time intervals to measure the concentration of gases, i.e. Hydrogen
		Direct method—Collection of feces to measure the recovery of test substrate
Selectively stimulate activity and/or growth of one or limited number of gut microbiota	Same as above, with further analysis includes the enumeration of bacteria, e.g. fluorescent in situ hybridisation and culture dependant methodology	Same as above, with further analysis includes the enumeration of bacteria, e.g. fluorescent in situ hybridisation and culture dependant methodology

nanocrystals resulting from the disruption of amorphous domains from semicrystalline granules by acid hydrolysis and (2) starch nanoparticles produced from gelatinized starch. The starch nanoparticles with particle sizes in the range of 10–1000 nm have been extensively studied as controlled release nanocarriers. For instance, propyl starch nanoparticles loaded with different types of drug (flufenamic acid, testosterone, and caffeine) showed enhanced effectiveness upon permeation studies on human skin (Santander-Ortega et al. 2010). Dialdehyde starch nanoparticles (DASNPs) conjugated with 5-fluorouracil (5-Fu) were found to have enhanced breast cancer cell (MF-7) inhibition in vitro as compared to free 5-Fu (Xiao et al. 2012). The nanoparticles encapsulated lipophilic bioactive compounds can provide physiological benefits in combating diet-related non-communicable diseases, including heart diseases, obesity, hypertension, cancer, and bone health (Gallocchio et al. 2015). Targeted release and protection of bioactive compounds is possible by incorporating them into nanoparticles which can improve their bioavailability. Previous studies have reported that nanoparticles, as delivery systems for polyphenols, can hinder oxidation/degradation, thus improving the stability and bioaccessibility of the phenolic phytochemicals and controlling the release rate of the active ingredient (Onoue et al. 2011). Recently, the starch nano particles have been prepared from maize starch, cassava starch, wheat starch and others (Kim et al. 2017). Starch polymeric nanoparticles using nanoprecipitation method was prepared by Raveendran et al. (2016) in the size range between 250 and 300 nm with smooth surface having spherical cavity inside which contains the anticancer drug (Doxorubicin). The cassava starch nanoparticles were produced by subjecting cassava starch to sulphuric acid hydrolysis and used as a carrier system for curcumin delivery (Rajeswari et al. 2011). Starch nanoparticles from cassava starch, were then characterized by using transmission electron microscopy (TEM), atomic force microscopy (AFM) and Fourier transform spectroscopy (FTIR). The incorporation of ZnO in the genipin crosslinked aminated starch coated iron oxide nanoparticles was used in controlling the release of curcumin from the system (Saikia et al. 2017). Starch Nanoparticles were also prepared by emulsification diffusion method for the delivery of gemcitabine hydrochlorid for the treatment of pancreatic cancer by Khaira, Sharma and Saini (2014). Novel starch-based nanoparticles from three sources: horse chestnut (HSC), water chestnut (WSC) and lotus stem (LSC) were prepared for nano-encapsulation of catechin with the particle size 322.7, 559.2 and 615.6 nm respectively. The results showed the bioactive properties were retained at higher level in encapsulated catechin compared to free catechin upon in-vitro digestion (Ahmad et al. 2019a). The use of starch nanoparticles currently receives much attention because of the abundant availability of starch, its low cost, renewability, biocompatibility, biodegradability, and nontoxicity (Rajeswari et al. 2011). The latter properties make these nanoparticles excellent candidates for implant materials and bioactive compound deliveries. From the literature surveyed, it was concluded that not much has been reported on the production of starch nanoparticles and its utilization as carrier of bioactive compounds for the targeted delivery at their absorption sites.

The effects of amylopectin/amylose ratio and non-solvent type on starch nanoparticle formation including the average particle size, polydispersity index, size distribution, and nanoparticle morphology. The study showed that high amylose starch including Hylon V and Hylon VII gave uniform, spherical and small nanoparticles in the size range of 20e60 and 15e50 nm, respectively. Hylon VII and Hylon V formed more spherical nanoparticles with methanol and ethanol as the non-solvents, and the particle size was less than 50 nm which is very useful for many applications such as delivery of bioactives. Stability of curcumin in the presence of 1 mg/ml native starch nanoparticles was much higher (83.7 ± 3.1%) than curcumin in phosphate buffered saline at pH 7.0 (5.5 ± 1.5%) over 10 days at ambient temperature (Sadeghi et al. 2017).

Natural starch is predominantly hydrophilic, which limits its application for encapsulating hydrophobic food bioactives. Hydrophobic starch derivatives have therefore been developed to overcome this drawback. Dialdehyde starch (Yu et al. 2007) and propyl starch (Jain et al. 2011) nanoparticles have been fabricated for encapsulation and delivery of lipophilic pharmaceutical agents. Octenyl Succinic Anhydride (OSA) modified starches have also been used to encapsulate food and flavour ingredients (Qi and Xu 1999). OSA-starches are surface active molecules that absorb to oilewater interfaces and facilitate the formation of small oil droplets during homogenization, and increase their long term stability during storage. Acetylation of starch has been shown to increase its hydrophobicity, reduce swelling ability, and enhance resistance to enzymatic hydrolysis, which are useful attributes for development of delivery systems for pharmaceutical agents (Xu et al. 2009) and food ingredients (Pu et al. 2011).

Other Applications of Starch

Starches from different sources exhibit uniqueness in functional and physiochemical characteristics. These attributes, in association with the non-toxic, biocompatible and biodegradable nature of starch, makes it a versatile ingredient that can be used in food and biomedical applications. Properties of starch such as gelatinization, swelling index, mechanical behavior, enzymatic digestibility, and solubility have been greatly exploited to tailor its use in a particular biomedical or pharmaceutical application. Native starches, their characteristics, and interactions with other ingredients during product development have been studied and reported from time to time. Due to an increased demand of value-added products, researchers have shown keen interest in isolation and characterization of starches from novel sources. However to be applicable industrially, starch in its native form has certain limitations. These include high tendency towards retrogradation and low thermal resistance, shear stress resistance, thermal decomposition. Starches when modified physically and chemically exhibit improved stability, clarity and resistance to retrogradation and can be used to tailor specific applications. Different treatments have been adopted for production of modified starches commercially, including grafting, self-association, crosslinking, substitution and others (Shefer et al. 1992; Van Soest and Vliegenthart 1997; Malafaya et al. 2006). The mode of treatment for starch modification varies with the application of starch, whether for food or pharmaceutical use. The applications of starch as a matrix for controlled drug release, as a hydrogel for wound dressing and in bone fixation are presently widely recognized and different types of modified starches like hydroxyethyl starch (HES), cadexomer iodine (CI), and oxidized starch (OS) are being used for biomedical purposes.

Modified Starches Used in Biomedical Applications

(a) Hydroxyethyl starch (HES): hydroxyethyl starch (HES) or hetastarch is a naturally derived nonionic polysaccharide obtained by hydrolysis of amylopectin following stabilization by hydroxyl ethylation. The hydroxyethyl is substituted at carbon 2, 3 and 6 in the glucose ring. The degree and type of hydroxylation determines the pharmacokinetics of HES. It is usually produced by the reaction of starch with ethylene oxide at temperatures below 50 °C in aqueous slurries in the presence of a swelling inhibiting salt. Hetastarch contains molecules of variable molecular weights and therefore can be divided into three classes by their average molecular weights: low molecular weight (70–130 kD), medium molecular weight (about 200 kD) and high molecular weight (450–480 kD). Examples of commercially available starches are Pentaspan, Hespan, EloHAES, Hextend, HAES-steril, and Voluven. An intravenous solution of hydroxyethyl starch is used to prevent shock and is used as plasma expander following hypovolemia caused by traumatic injuries, surgery, or other problem like malperfusion, as well as for acute normovolaemic hemodilution and preoperative autologous blood donation (Gallandat Huet et al. 2000).

(b) Cadexomer iodine (CI): It is a hydrophillic modified starch polymer produced by the crosslinking of dextrin with epichlorohydrin. It contains 0.9% iodine arranged in the form of small beads. The iodine is immobilized in the microsphere's helical matrix. Cadexomer iodine has a high absorptive capacity and about 7 g of fluid can be absorbed by 1 g of CI (Torres et al. 2013). It is widely used as a biomaterial for dressing wounds. It acts as an effective debriding agent and antiseptic for exudative wound (Moberg et al. 1983; Laudanska and Gustavson 1988; Holloway et al. 1989). When formulated as a topical wound dressing, the iodosorb aborbs the exudate and particulate matter from the wound surface. As the dressing becomes moist, the size of micropores is increased and iodine is slowly released into the wound bed.

(c) Oxidized starch (OS): OS production involves the introduction of carboxyl and carbonyl groups by means of subsequent starch depolymerization. It is achieved by reacting starch with an oxidizing agent under controlled temperature and pH of 2–7. This results in the hydroxyl groups (primarily at C-2, C-3 and C-6 positions) being transformed to carbonyl and/or carboxyl groups (Wang and Wang 2003). Bromine, chromic acid, hydrogen peroxide, hypochlorite, nitrogen dioxide, oxygen, ozone, and permanganate have been used as oxidants to prepare OS. The desired characteristics of OS include high stability, film forming ability, binding properties and low viscosity (Lawal et al. 2005; Sangseethong et al. 2009). Oxidized starch is used to produce nontoxic, soluble and biodegradable PVA-OS fibers that may be used as a potential biomaterial in wound dressing, tissue engineering and drug delivery (Wang et al. 2011).

Compatibility of Starch and Starch Based Products for Biomedical Applications

The potential of starch and starched based products as medical polymer materials has been widely recognized during recent years due to some advantages biomaterials. These include biodegradability, nontoxicity, low prices, acceptable mechanical properties and most importantly, biocompatibility. Biocompatibility may be defined as the ability of a material to perform with an appropriate host response in a specific situation (Donaruma 1988). For a material to be biocompatible, it has to fulfill the following criteria:

1. Biocompatibility depends not only on the materials but also on conditions in which the material is used.
2. Many applications require that the material reacts with the tissues instead of being ignored by them.
3. Some applications require that the material must degrade over time.

Starch and starch based products owe their biocompatibility to the presence of structural components that are biocompatible. These include products obtained from partial hydrolysis and starch polymer molecules. The biocompatible nature of starch-based materials makes them an important ingredient for biomedical applications.

Biomedical and Pharmaceutical Uses of Starch

Modified Starches in Drug Delivery

Native starches undergo extensive modification (esterification, crosslinking, etherification, grafting etc.) in order to be used as carriers for the controlled release of drugs for better patient compliances. The process of modification makes the native starch a potential biomaterial for developing drug delivery systems by offering more reactive sites for carrying bioactive compounds and can be easily metabolized in the human body. Starch based hydrogels or microspheres are suitable for drug delivery (Reis et al. 2008; Rosado Balmayor et al. 2008). Some starches have a good potential as drug release retardants. These include short chained amylose (amylodextrin), pregelatinized starch, substituted amylase, and cross linked amylase. Starch-urea-borate has been used as an efficient rate controlling matrix for controlled release of glyclazide, diclofenac and carbamazepine. Calcium starch (a modified starch prepared by cross linking the alkali gelatinized starch with calcium chloride), has been reported to be used for controlled release of glyclazide, diltiazem hydrochloride and diclofenac.

Modified Starch as Excipient in Pharmaceutical Formulations

Excipients, often referred to as "diluents", "fillers" or "bulking agents", are additives used to process therapeutically active ingredients into final pharmaceutical dosages suitable for administration to patients (Wade and Weller 1994). It is formulated alongside the active ingredient of medication for the purpose of providing better stabilization, bulking up drug formulations containing otherwise, very small percentage of active ingredients, or to confer therapeutic enhancement on the active ingredient such as enhancing solubility, facilitating drug absorption and lowering viscosity. Starches, due to their poor flowability, cannot be utilized as excipients in their native form. Modified starches as excipients are the non-active ingredients that are vital for the manufacturing of solid drugs such as powders and tablets as they facilitate powder flowability, exhibit non-stick properties, prevent aggregation or denaturation of drug over the period of expected shelf life. In recent times, the most preferred excipient in pharmaceutical industry is pregelatinized starch, marketed as Starch 1500. Modified rice starch, starch acetate and acid hydrolyzed dioscorea starch were established as multifunctional excipient in the pharmaceutical industry recently.

Starch as Plasma Volume Expander

Plasma volume expanders are agents that boost the volume of plasma by increasing the osmotic pressure. They are used for the treatment of patients suffering from hemorrhage, shock and cancer. Acetylated and hydroxyethyl starch (HES) are mainly used as plasma volume expanders.

Starch as Tablet Disintegrant

Modified starches are most commonly employed disintegrating agents in use today. They are used in immediate release tablet formulations where the potent drug should be available to the absorptive area instantaneously. Starches exhibit a great affinity for water resulting in instant expansion and disintegration of tablets. Sodium carboxymethyl starch (marketed as sodium starch glycolate) is commonly used as a disintegrant. Starches from various novel sources have been modified for the same and have been made commercially available. These include low substitute CMS like explotab, primogel and deprogel. These modified starches swell to the extent of 200–300% in water, in comparison to natural starches that increase in volume be only 10–25%.

Starch in Bone Tissue Engineering

Several starch based biodegradable polymers have been used for bone tissue engineering (BTE) and in the development of BTE scaffolds (Gomes et al. 2003). Scaffold, which is an important component in the engineering of hard tissues, permits the flow of culture media while providing cells with all the nutrients and simultaneously removing metabolites resulting from cell activities. Several methodologies have been developed for manufacturing BTE scaffolds including compression molding, injection molding, solvent casting and extrusion using blowing agents, particle leaching, and rapid prototyping. Starch–polycaprolactone composites (SPCL) that possess adequate porosity and required mechanical properties to support cell adhesion and proliferation, have been widely used as scaffolds. Starch based bone cements provide structural support and degrade from the site of application and can combine with bioactive components which induce bone growth in cement-bone interface (Boesel et al. 2004).

Miscellaneous Uses in Pharmaceutical Industry

Starch acts as an excellent absorbent and is used as a dusting powder. For this purpose it's used either alone or mixed with zinc oxide or other similar substances. Starch mucilage may be employed as an emollient, suspending and emulsifying agent. Starch powder is used on surgical gloves. Ointments containing starches may be used as protective application in skin diseases.

Conclusion

Starch has been widely studied throughout the world and as a result it has gained various applications in both food and pharmaceutical industries. Besides meeting the basic nutrition, it is being used for edible coatings, films, nanocomposites, carrier for drug and micronutrient delivery. Native starch has however some technological limitations to be used for encapsulation or film making purpose. Therefore, it is suggested to blend starch with other biodegradable polymers compatible with starch for achieving desirable film characteristics and OS-starch is suggested as an efficient encapsulating agent due to its amphillic nature. Compared to other modifications, the starch at nano-level may show some interesting properties that can be useful in biomedical or drug delivery systems.

Acknowledgements Mr. Mudasir Ahmad is highly thankful to Indian Council of Medical Research (ICMR), Government of India for providing Senior Research Fellowship [3/1/2/81/2018-(Nutri)].

References

Ahmad, M., Ashraf, B., Gani, A., & Gani, A. (2018). Microencapsulation of saffron anthocyanins using β glucan and β cyclodextrin: Microcapsule characterization, release behaviour & antioxidant potential during in-vitro digestion. *International Journal of Biological Macromolecules, 109*, 435–442.

Ahmad, M., Mudghil, P., Gani, A., hameed, F., masoodi, F. A., & Maqsood, S. (2019a). Nano-encapsulation of catechin in starch nanoparticles: Characterization, release behavior and bioactivity retention during simulated in-vitro digestion. *Food Chemistry, 270*(1), 95–104.

Ahmad, M., Gani, A., Hamed, F., Maqsood, S. (2019b). Comparative study on utilization of micro and nano sized starch particles for encapsulation of camel milk derived probiotics (Pediococcus acidolactici), LWT. (2019b). https://doi.org/10.1016/j.lwt.2019.04.078.

Ahmad, M., Mudgil, P., Maqsood, S. (2019c). Camel whey protein microparticles for safe and efficient delivery of novel camel milk derived probiotics, LWT. https://doi.org/10.1016/j.lwt.2019.03.008.

Ahmad, M., Qureshi, S., Maqsood, S., Gani, A., & Masoodi, F. A. (2017). Micro-encapsulation of folic acid using horse chestnut starch and β-cyclodextrin: Microcapsule characterization, release behavior & antioxidant potential during GI tract transition. *Food Hydrocolloids, 66*, 154–160.

Andersons, J., Modniks, J., Joffe, R., Madsen, B., & N€attinen, K. (2016). Apparent interfacial shear strength of short-flax-fiber/starch acetate composites. *International Journal of Adhesion and Adhesives, 64*, 78–85.

Anglès, M. N., & Dufresne, A. (2000). Plasticized starch/tunicin whiskers nanocomposites. 1. Structural analysis. *Macromolecules, 33*, 8344–8353.

Annison, G., & Topping, D. L. (1994). Nutritional role of resistant starch: Chemical structure vs physiological function. *Annual Review of Nutrition, 14*, 297–320.

Arena, P. M., Caggianiello, G., Fiocco, D., Russo, P., Torelli, M., Spano, G., & Capozzi, V. (2014). Barley β-glucans-containing food enhances probiotic performances of beneficial bacteria. *International Journal of Molecular Sciences, 15*(2), 3025–3039.

Ashwar, B. A., Gani, A., Gani, A., Shah, A., & Masoodi, F. A. (2018). Production of RS4 from rice starch and its utilization as an encapsulating agent for targeted delivery of probiotics. *Food Chemistry, 239*, 287–294.

Baldwin, E. A., Nisperos-Carriedo, M. O., & Baker, R. A. (1995). *Horticultural Science, 30*, 35–38.

Bhandari, B. R., D'Arc, B. R., & Padukka, I. (1999). Encapsulation of lemon oil by paste method using β-cyclodextrin: Encapsulation efficiency and profile of oil volatiles. *Journal of Agricultural and Food Chemistry, 47*, 5194–5197.

Bhandari, B. R., D'Arc, B. R., & Thi Bich, L. L. (1998). Lemon oil to β-cyclodextrin ratio effect on the inclusion efficiency of β-cyclodextrin and the retention of oil volatiles in the complex. *Journal of Agricultural and Food Chemistry, 46*, 1494–1499.

Bird, A. R., Vuaran, M., Brown, I., & Topping, D. L. (2007). Two high-amylose maize starches with different amounts of resistant starch vary in their effects on fermentation, tissue and digesta mass accretion, and bacterial populations in the large bowel of pigs. *The British Journal of Nutrition, 97*, 134–144.

Boesel, L. F., Mano, J. F., & Reis, R. L. (2004). Optimization of the formulation and mechanical properties of starch based partially degradable bone cements. *Journal of Materials Science: Materials in Medicine, 15*, 73–83.

Bornet, F. R., Fontvieille, A. M., Rizkalla, S., Colonna, P., Blayo, A., Mercier, C., & Slama, G. (1989). Insulin and glycemic responses in healthy humans to native starches processed in different ways: Correlation with in vitro alpha-amylase hydrolysis. *The American Journal of Clinical Nutrition, 50*, 315–323.

Brighenti, F., Cristina, M., & Baggio, C. (1998). Resistant starch in Italian diet. *British Journal of Nutrition, 80*, 333–341.

Brown, I. (1996). Complex carbohydrates and resistant starch. *Nutrition Reviews, 54*, S115–S119.

Buléon, A., Colonna, P., Planchot, V., & Ball, S. (1998). Starch granules: Structure and biosynthesis. *International Journal of Biological Macromolecules, 23*(2), 85–112.

Campos, C. A., Gerschenson, L. N., & Flores, S. K. (2011). Development of edible films and coatings with antimicrobial activity. *Food and Bioprocess Technology, 4*(6), 849–875.

Can-Karaca, A., Low, N. H., & Nickerson, M. T. (2015). Potential use of plant proteins in the microencapsulation of lipophilic materials in foods. *Trends in Food Science and Technology, 42*(1), 5–12.

Carissimi, M., Flôres, S. H., & Rech, R. (2018). Effect of microalgae addition on active biodegradable starch film. *Algal Research, 32*, 201–209.
Cerqueira, M. A., Lima, Á. M., Souza, B. W. S., Teixeira, J. A., Moreira, R. A., Vicente, A. A. (2009). Functional polysaccharides as edible coatings for cheese, J. Agric. Food Chem. https://doi.org/10.1021/jf802726d.
Charalampopoulos, D., Wang, R., S Pandiella, S., & Webb, C. (2002). Application of cereals and cereal components in functional foods: A review. *International Journal of Food Microbiology, 79*, 131–141.
Charve, J., & Reineccius, G. A. (2009). Encapsulation performance of proteins and traditional materials for spray dried flavors. *Journal of Agricultural and Food Chemistry, 57*, 2486–2492.
Chung, C., Sanguansri, L., & Augustin, M. A. (2010). Resistant starch modification: Effects on starch properties and functionality as co-encapsulant in sodium caseinate-based fish oil microcapsules. *Journal of Food Science, 75*(9), E636–E642.
Compton, D. L., Kenar, J. A., Laszlo, J. A., & Felker, F. C. (2007). Starch-encapsulated, soy-based, ultraviolet-absorbing composites with feruloylated monoacyl-and diacylglycerol lipids. *Industrial Crops and Products, 25*, 17–23.
Cummings, J. H., & Stephen, A. M. (2007). Carbohydrate terminology and classification. *European Journal of Clinical Nutrition, 61*, 5–18.
Da-Silva, F. C., Da-Fonseca, C. R., De-Alencar, S. M., Thomazini, M., Balieiro, J. C. D. C., Paola Pittia, Carmen Sílvia Favaro-Trindade, (2013). Assessment of production efficiency, physicochemical properties and storage stability of spray-dried propolis, a natural food additive, using gum Arabic and OSA starch-based carrier systems. *Food and Bioproducts Processing, 91*(1), 28–36.
Desai, K. G. H., & Park, H. J. (2005). Recent developments in microencapsulation of food ingredients. *Drying Technology, 23*, 1361–1394.
Donaruma, L. G. (1988). Definitions in biomaterials, D. F. Williams, Ed., Elsevier, Amsterdam, 1987, 72 pp. *Journal of Polymer Science Part C: Polymer Letters, 26*, 414–414.
Dufresne, A. (2014). Crystalline starch based nanoparticles. *Current Opinion in Colloid & Interface Science, 19*, 397–408.
Eerlingen, R., & Delcour, J. (1995). Formation, analysis, structure and properties of type III enzyme resistant starch. *Journal of Cereal Science, 22*, 129–138.
Englyst, H. N., & Cummings, J. H. (1987a). Digestion of polysaccharides of potato in the small intestine of man. *The American Journal of Clinical Nutrition, 45*, 423–431.
Englyst, H. N., & Cummings, J. H. (1987b). Resistant starch, a new food component: A classification of starch for nutritional purposes. In *Cereals in a European context. First European conference on food science and technology* (pp. 221–233). Chichester: Ellis Horwood.
Englyst, H. N., Kingman, S., & Cummings, J. (1992). Classification and measurement of nutritionally important starch fractions. *European Journal of Clinical Nutrition, 46*, S33–S50.
Englyst, K. N., Liu, S., & Englyst, H. N. (2007). Nutritional characterization and measurement of dietary carbohydrates. *European Journal of Clinical Nutrition, 61*, 19–39.
Fan, J., Li, J., Huang, Z., Wen, P. H., Bailey, C. G. (2018). Grain size effects on indentation-induced plastic deformation and amorphization process of polycrystalline silicon, Comput. Mater. Sci. https://doi.org/10.1016/j.commatsci.2017
Fathi, M., Martín, Á., & McClements, D. J. (2014). Nanoencapsulation of food ingredients using carbohydrate based delivery systems. *Trends in Food Science & Technology, 39*, 18–39.
Fenga, M., Yua, L., Zhua, P., Zhoua, X., Liua, H., Yanga, Y., … Chen, P. (2018). Development and preparation of active starch films carrying tea polyphenol. *Carbohydrate Polymers, 196*(15), 162–167.
Forssell, P. M., Mikkia, J. M., Moates, G. K., & Parker, R. (1997). Phase and glass transition behavior of concentrated barley starch–glycerol–water mixtures, a model for thermoplastic starch. *Carbohydrate Polymers, 3*, 275–282.
Gallandat Huet, R. C. G., Siemons, A. W., Baus, D., van Rooyen-Butijn, W. T., Haagenaars, J. A., van Oeveren, W., & Bepperling, F. (2000). A novel hydroxyethyl starch (Voluven®) for effective perioperative plasma volume substitution in cardiac surgery. *Canadian Journal of Anesthesia, 47*, 1207–1215.

Gallocchio, F., Belluco, S., & Ricci, A. (2015). Nanotechnology and food: Brief overview of the current scenario. *Procedia Food Science, 5*, 85–88.
Gani, A., Shah, A., Ahmad, M., Ashwar, B. A., & Masoodi, F. A. (2018). β-d-glucan as an enteric delivery vehicle for probiotics. *International Journal of Biological Macromolecules, 106*, 864–869.
Gaudin, S., Lourdin, D., Le Botlan, D., Ilari, J. L., & Colonna, P. (1999). Plasticization and mobility in starch-sorbitol film. *Journal of Cereal Science, 29*(3), 273–284.
Gibson, G. R., Probert, H. M., Loo, J. V., Rastall, R. A., Roberfroid, M. B. (2004). Dietary modulation of the human colonic microbiota: Updating the concept of prebiotics. *Nutrition Research Reviews, 17*, 259–275.
Gibson, G. R., Hutkins, R., Sanders, M. E., Prescott, S. L., Reimer, R. A., Salminen, S. J., … Reid, G. (2017). Expert consensus document: The International Scientific Association for Probiotics and Prebiotics (ISAPP) consensus statement on the definition and scope of prebiotics. *Nature Reviews Gastroenterology & Hepatology, 14*(8), 491.
Gibson, G. R., & Roberfroid, M. B. (1995). Dietary modulation of the human colonic microbiota: Introducing the concept of prebiotics. *The Journal of Nutrition, 125*, 1401–1412.
Gomes, M. E., Sikavitsas, V. I., Behravesh, E., Reis, R. L., & Mikos, A. G. (2003). Effect of flow perfusion on the osteogenic differentiation of bone marrow stromal cells cultured on starch-based three-dimensional scaffolds. *Journal of Biomedical Materials Research Part A, 67A*, 87–95.
G6mez, B., Gull6n, B., Remoroza, C., Schols, H. A., Parajó, J. C., Alonso, J. L. (2014). Purification, characterization, and prebiotic properties of pectic oligosaccharides from orange peel wastes. *Journal of Agricultural and Food Chemistry, 62*, 9769–9782.
Gopalan Nair, G., & Dufresne, A. (2003). Crab shell chitin whisker reinforced natural rubber nanocomposites. 2. Mechanical behavior. *Biomacromolecules, 4*, 666.
Han, S., Choi, S. H., Kim, W., Kim, B. Y., & Baik, M. Y. (2015). Infusion of catechin into native corn starch granules for drug and nutrient delivery systems. *Food Science and Biotechnology, 24*, 2035–2040.
Haralampu, S. G. (2000). Resistant starch - a review of the physical properties and biological impact of RS3. *Carbohydrate Polymers, 41*, 285–292.
Hasanvand, E., Fathi, M., Bassiri, A., Javanmard, M., & Abbaszadeh, R. (2015). Novel starch based nanocarrier for vitamin D fortification of milk: Production and characterization. *Food and Bioproducts Processing, 96*, 264–277.
He, H., Hong, Y., Gu, Z., Liu, G., Cheng, L., & Li, Z. (2016). Improved stability and controlled release of CLA with spray-dried microcapsules of OSA-modified starch and xanthan gum. *Carbohydrate Polymers, 147*, 243–250.
Henry, F. Z., & Alistair, M. S. (1995). *Food polysaccharides and their applications*. New York, NY: Marcel Dekker.
Holloway, G. A., Johansen, K. H., Barnes, R. W., & Pierce, G. E. (1989). Multicenter trial of cadexomer iodine to treat venous stasis ulcer. *Western Journal of Medicine, 151*, 35–38.
Hosokawa, M., Nogi, K., Makio, N., & Yokoyama, T. (2008). *Nanoparticle technology handbook*. Amsterdam: Elsevier.
Huber, K. C., & BeMiller, J. N. (2000). Channels of maize and sorghum starch granules. *Carbohydrate Polymers, 41*, 269–276.
Jain, R., Dandekar, P., Loretz, B., Melero, A., Stauner, T., Wenz, G., Koch, M., Lehr, C., (2011). Enhanced cellular delivery of idarubicin by surface modification of propyl starch nanoparticles employing pteroic acid conjugated polyvinyl alcohol. *International Journal of Pharmaceutics, 420*, 147–155.
Jane, J. (2006). Current understanding on starch granule structures. *Journal of Applied Glycoscience, 53*, 205–213.
Jane, J., & Robyt, J. (1984). Structure studies of amylose-V complexes and retro-graded amylose by action of alpha amylases, and a new method for preparing amylodextrins. *Carbohydrate Research, 132*, 105–118.

Jimenez, A., Fabra, M. J., Talens, P., & Chiralt, A. (2013). Physical properties and antioxidant capacity of starch-sodium caseinate films containing lipids. *Journal of Food Engineering, 116*, 695–702.

Kalichevsky, M. T., Jaroszkiewicz, E. M., & Blanshard, J. M. V. (1993). A study of the glass transition of amylopectin-sugar mixtures. *Polymer, 34*(2), 346–358.

Khaira, R., Sharma, J., & Saini, V. (2014). Development and characterization of nanoparticles for the delivery of gemcitabine hydrochloride. *The Scientific World Journal, 2014*, 1. https://doi.org/10.1155/2014/560962

Kim, J. H., Park, D. H., & Kim, Y. (2017). Effect of heat-moisture treatment under mildly acidic condition on fragmentation of waxy maize starch granules into nanoparticles. *Food Hydrocolloids, 63*, 59–66.

Kristo, E., & Biliaderis, C. G. (2007). Physical properties of starch nanocrystal-reinforced pullulan films. *Carbohydrate Polymers, 68*, 146–158.

Lalush, I., Bar, H., Zakaria, I., Eichler, S., & Shimoni, E. (2005). Utilization of amylase-lipid complexes as molecular nanocapsules for conjugated linoleic acid. *Biomacromolecules, 6*(1), 121–130.

Laudanska, H., & Gustavson, B. (1988). In-patient treatment of chronic varicose venous ulcers. A randomized trial of cadexomer iodine versus standard dressings. *The Journal of International Medical Research, 16*, 428–435.

Lawal, O. S., Adebowale, K. O., Ogunsanwo, B. M., Barba, L. L., & Ilo, N. S. (2005). Oxidized and acid thinned starch derivatives of hybrid maize: Functional characteristics, wide-angle X-ray diffractometry and thermal properties. *International Journal of Biological Macromolecules, 35*, 71–79.

Le-Corre, D., Bras, J., & Dufresne, A. (2010). Starch nanoparticles: A review. *Biomacromolecules, 11*, 1139–1153.

Lesmes, U., Barchechath, J., & Shimoni, E. (2008). Continuous dual feed homogenization for the production of starch inclusion complexes for controlled release of nutrients. *Innovative Food Science and Emerging Technologies, 9*, 507–515.

Lopez, O. V., García, M. A., & Zaritzky, N. E. (2008). Film forming capacity of chemically modified corn starches. *Carbohydrate Polymers, 73*, 573–581.

Lunn, J., & Buttriss, J. L. (2007). Carbohydrates and dietary fibre. *Nutrition Bulletin, 32*, 21–64.

Ma, X., Yu, J., & Jin, F. (2004). Urea and formamide as a mixed plasticizer for thermoplastic starch. *Polymer International, 53*, 1780–1785.

Malafaya, P. B., Stappers, F., & Reis, R. L. (2006). Starch-based microspheres produced by emulsion crosslinking with a potential media dependent responsive behavior to be used as drug delivery carriers. *Journal of Materials Science: Materials in Medicine, 17*, 371–377.

Martínez-Ortiza, M. A., Vargas-Torresa, A., Román-Gutiérrezb, A. D., Chavarría-Hernándeza, N., Zamudio-Floresc, P. B., Meza-Nietoa, M., & Palma-Rodrígueza, H. M. (2017). Partial characterization of chayotextle starch-based films added with ascorbic acid encapsulated in resistant starch. *International Journal of Biological Macromolecules, 98*, 341–347.

Moberg, S., Hoffman, L., Grennert, M. L., & Holst, A. (1983). A randomized trial of cadexomer iodine in decubitus ulcers. *Journal of the American Geriatrics Society, 31*, 462–465.

Mohammadi, N. A., Moradpour, M., Saeidi, M., & Alias, A. K. (2013). Thermoplastic starches: Properties, challenges, and prospects. *Starch - Stärke, 65*, 61–72.

Mohanty, A. K., Misra, M., & Hinrichsen, G. (2000). Biofibres, biodegradable polymers and biocomposites: An overview. *Macromolecular Materials and Engineering, 276-277*, 1–24.

Moreno, O., Gil, A., Atares, L., & Chiralt, A. (2017). Active starch-gelatin films for shelf-life extension of marinated salmon. *LWT - Food Science and Technology, 84*, 189–195.

Morrison, W. (1995). Starch lipids and how they relate to starch granule structure and functionality. *Cereal Foods World, 40*, 437–38, 440–1,443–6.

Morrison, W. (2000). Starch lipids, starch granule structure and properties. *Wheat Structure*, 261–270. https://doi.org/10.1533/9781845698478.6.261

Morrison, W. R. (1981). Starch lipids: A reappraisal. *Starch - Stärke, 33*, 408–410.

Moslemi, M., Hosseini, H., RNeyestan, T., Akramzadeh, N., & Fard, R. M. N. (2018). Effects of non-digestive polymers used in iron encapsulation on calcium and iron apparent absorption in rats fed by infant formula. *Journal of Trace Elements in Medicine and Biology, 50*, 393–398.

Nakamura, S., & Tobolsky, A. V. (1967). Viscoelastic properties of plasticized amylose films. *Journal of Applied Polymer Science, 11*, 1371–1386.

Nasrin, T. A. A., Anal, A. K. (2015). Enhanced oxidative stability of fish oil by encapsulating in culled banana resistant starch-soy protein isolate based microcapsules in functional bakery products, J. Food Sci. Technol. https://doi.org/10.1007/s13197-014-1606-1.

Nielsen, C. K., Kjems, J., Mygind, T., Snabe, T., Schwarz, K., Serfert, Y., & Meyer, R. L. (2016). Enhancing the antibacterial efficacy of isoeugenol by emulsion encapsulation. *International Journal of Food Microbiology, 229*, 7–14.

Noor, N., Shah, A., Gani, A., Gani, A., & Masoodi, F. A. (2018). Microencapsulation of caffeine loaded in polysaccharide based delivery systems. *Food Hydrocolloids, 82*, 312–321.

Nugent, A. P. (2005). Health properties of resistant starch. *Nutrition Bulletin, 30*, 27–54.

Oakenfull, D., & Glicksman, M. (1987). Gelling agents. *C R C Critical Reviews in Food Science and Nutrition, 26*, 1–25.

Oksman, K., Mathew, A. P., Bondeson, D., & Kvien, I. (2006). Manufacturing process of cellulose whiskers/polylactic acid nanocomposites. *Composites Science and Technology, 66*, 2776–2784.

Oms-Oliu, G., Soliva-Fortuny, R., & Martin-Belloso, O. (2008). Edible coatings with antibrowning agents to maintain sensory quality and antioxidant properties of fresh-cut pears. *Postharvest Biology and Technology, 50*, 87–94.

Onoue, S., Ochi, M., & Yamada, S. (2011). Development of (−)-epigallocatechin-3-gallate (EGCG)-loaded enteric microparticles with intestinal mucoadhesive property. *International Journal of Pharmaceutics, 410*, 111–113.

Ortega, F., Giannuzzi, L., Arce, V. B., & García, M. A. (2017). Active composite starch films containing green synthetized silver nanoparticles. *Food Hydrocolloids, 70*, 152–162.

Ozdemir, M., Floros, J. D. (2008). Optimization of edible whey protein films containing preservatives for water vapor permeability, water solubility and sensory characteristics, J. Food Eng. https://doi.org/10.1016/j.jfoodeng.2007.09.028.

Palupi, N. W., & Praptiningsih, Y. (2016). Oxidized tapioca starch as an alginate substitute for encapsulation of antioxidant from coffee residue. *Agriculture and Agricultural Science Procedia, 9*, 304–308.

Park, H. J. (1999). Development of advanced edible coatings for fruits, Trends Food Sci. Technol. (1999). https://doi.org/10.1016/S0924-2244(00)00003-0.

Pegg, R. B., & Shahidi, F. (2007). Encapsulation, stabilization, and controlled release of food ingredients and bioactives. In M. S. Rahman (Ed.), *Handbook of food preservation* (pp. 509–568). Boca Raton, FL: CRC Press.

Pu, H., Chen, L., Li, X., Xie, F., Yu, L., & Li, L. (2011). An oral colon targeting controlled release system based on resistant starch acetate: Synthetization, characterization, and preparation of filmcoating pellets. *Journal of Agricultural and Food Chemistry, 59*, 5738–5745.

Qi Z. H., Xu A. (1999). Starch based ingredients for flavor encapsulation. Cereal Foods World 44:460–465.

Raigond, P., Ezekiel, R., & Raigond, B. (2015). Resistant starch in food: A review. *Journal of the Science of Food and Agriculture, 95*(10), 1968–1978.

Rajendran, C. K., Subin, R., Okolie, C. L., Udenigwe, C. C., & Mason, B. (2017). Structural features underlying prebiotic activity of conventional and potential prebiotic oligosaccharides in food and health. *Journal of Food Biochemistry, 41*(5), e12389.

Rajeswari, L. S., Moorthy, S. N., & Rajasekhran, K. N. (2011). Preparation of cassava starch nanoparticles and their application as a carrier system for curcumin delivery. *International Journal of Nanotechnology and Applications, 5*(3), 193–201.

Ranhotra, G. S., Gelroth, J. A., & Glaser, B. K. (1996). Energy value of resistant starch. *Journal of Food Science, 61*, 453–455.

Rasato, D. (2009). *The emerging world of bioplastics: An industry 'father' looks forward.* Retrieved from http://www.omnexus.com/resources/print.aspx?id=22050

Raveendran, S. S., Subramani, N., Rahman, S. A., & Palsamy, S. (2016). Formulation and characterization of starch nanoparticles for controlled release of doxorubicin. *Nano Science and Nano Technology: An Indian Journal, 10*(1), 30–35.

Reineccius, G. A. (2004). The spray drying of food flavors. *Drying Technology, 22*, 1289–1324.

Reis, A. V., Guilherme, M. R., Moia, T. A., Mattoso, L. H. C., Muniz, E. C., & Tambourgi, E. B. (2008). Synthesis and characterization of a starch-modified hydrogel as potential carrier for drug delivery system. *Journal of Polymer Science Part A: Polymer Chemistry, 46*, 2567–2574.

Rosado Balmayor, E., Tuzlakoğlu, K., Marques, A., Azevedo, H., & Reis, R. (2008). A novel enzymatically-mediated drug delivery carrier for bone tissue engineering applications: Combining biodegradable starch-based microparticles and differentiation agents. *Journal of Materials Science: Materials in Medicine, 19*(4), 1617–1623.

Sherwin Y. Cheuk, Frederick F. Shih, Elaine T. Champagne, Kim W. Daigle, James A. Patindol, Christopher P. Mattison, Stephen M. Bou. (2015). Nano-encapsulation of coenzyme Q10 using octenyl succinic anhydride-modified starch. *Food Chemistry, 174*, 585–590. https://doi.org/10.1016/j.foodchem.2014.11.031

Sadeghi, R., Daniella, Z., Uzun, S., & Kokini, J. (2017). Effects of starch composition and type of non-solvent on the formation of starch nanoparticles and improvement of curcumin stability in aqueous media. *Journal of Cereal Science, 76*, 122–130.

Saikia, C., Das, M. K., Ramteke, A., & Maji, T. K. (2017). Evaluation of folic acid tagged aminated starch/ZnO coated iron oxide nanoparticles as targeted curcumin delivery system. *Carbohydrate Polymers, 157*, 391–399.

Sajilata, M. G., & Singhal, R. S. (2005). Specialty starches for snack foods. *Carbohydrate Polymers, 59*, 131–151.

Sajilata, M. G., Singhal, R. S., & Kulkarni, P. R. (2006). Resistant starch—A review. *Comprehensive Reviews in Food Science and Food Safety, 5*, 1–17.

Sang, Y., Seib, P. A., Herrera, A. I., Prakash, O., & Shi, Y. (2010). Effects of alkaline treatment on the structure of phosphorylated wheat starch and its digestibility. *Food Chemistry, 118*, 323–327.

Sangseethong, K., Lertphanich, S., & Sriroth, K. (2009). Physicochemical properties of oxidized cassava starch prepared under various alkalinity levels. *Starch - Stärke, 61*, 92–100.

Santander-Ortega, M. J., Stauner, T., Loretz, B., Ortega-Vinuesa, J. L., Bastos-González, D., Wenz, G., … Lehr, C. M. (2010). Nanoparticles made from novel starch derivatives for transdermal drug delivery. *Journal of Controlled Release, 141*(1), 85–92.

Sarbini, S. R., Kolida, S., Gibson, G. R., & Rastall, R. A. (2013). In vitro fermentation of commercial α-gluco-oligosaccharide by faecal microbiota from lean and obese human subjects. *British Journal of Nutrition, 109*, 1980–1989.

Sarbini, S. R., & Rastall, R. A. (2011). Prebiotics: Metabolism, structure and function. *Functional Food Reviews, 3*, 93–106.

Sarkar, P., Bhunia, A. K., & Yao, Y. (2017). Impact of starch-based emulsions on the antibacterial efficacies of nisin and thymol in cantaloupe juice. *Food Chemistry, 217*, 155–162.

Shah, A., Gani, A., Ahmad, M., Ashwar, B. A., & Masoodi, F. A. (2016). β-Glucan as an encapsulating agent: Effect on probiotic survival in simulated gastrointestinal tract. *International Journal of Biological Macromolecules, 82*, 217–222.

Shefer, A., Shefer, S., Kost, J., & Langer, R. (1992). Structural characterization of starch networks in the solid state by cross-polarization magic-angle-spinning carbon-13 NMR spectroscopy and wide angle x-ray diffraction. *Macromolecules, 25*, 6756–6760.

Shrestha, A. K., Halley, P. J., Halley, P. J., & Averous, L. (Eds.). (2014). *starch polymers: From genetic engineering to green applications* (pp. 105–143). San Diego, CA: Elsevier.

Slavin, J. (2013). Fiber and prebiotics: Mechanisms and health benefits. *Nutrients, 5*, 1417–1435.

Sorrentino, A., Gorrasi, G., & Vittoria, V. (2007). Potential perspectives of bio-nanocomposites for food packaging applications. *Trends in Food Science and Technology, 18*, 84.

Sothornvit, R., Krochta, J. M., & Han, J. H. (Eds.). (2005). *Innovations in food packaging* (pp. 403–433). London: Academic Press.

Sworn, G. (2003). Hydrocolloid thickeners and their applications. In G. O. Philips & P. A. Williams (Eds.), *Gums and stabilizers for the food industry* (Vol. 12, pp. 13–22). Oxford: Royal Society of Chemistry.

Tonon, R. V., Pedro, R. B., Grosso, C. R. F., & Hubinger, M. D. (2012). Microencapsulation of flaxseed oil by spray drying: Effect of oil load and type of wall material. *Drying Technology, 30*, 1491–1501.

Torres, F. G., Commeaux, S., & Troncoso, O. P. (2013). Starch-based biomaterials for wound-dressing applications. *Starch - Stärke, 65*, 543–551.

Van Soest, J., & Vliegenthart, J. (1997). Crystallinity in starch plastics: Consequences for material properties. *Trends in Biotechnology, 15*, 208–213.

Wade, A., & Weller, P. (1994). *Handbook of pharmaceutical excipients*. London, UK: American Pharmaceutical Association and the Pharmaceutical Press.

Wang, H., Wang, W., Jiang, S., Jiang, S., Zhai, L., & Jiang, Q. (2011). Poly (vinyl alcohol)/oxidized starch fibres via electrospinning technique: Fabrication and characterization. *Iranian Polymer Journal, 20*, 551–558.

Wang, X., Yuan, Y., & Yue, T. (2014). The application of starch-based ingredients in flavor encapsulation December. *Starch-Starke, 67*, 3–4. https://doi.org/10.1002/star.201400163

Wang, Y. J., & Wang, L. (2003). Physicochemical properties of common and waxy corn starches oxidized by different levels of sodium hypochlorite. *Carbohydrate Polymers, 52*, 207–217.

Wani, T. A., Gani, A., Wani, S. M., Wani, I. A., Masoodi, F. A., Nissar, N., & Shagoo, M. A. (2016). Suitability of different food grade materials for the encapsulation of some functional foods well reported for their advantages and susceptibility. *Critical Reviews in Food Science and Nutrition, 56*(15), 2431–2454.

Wurzburg, O. B. (1986). *Modified starches: Properties and uses*. Boca Raton, FL: CRC Press.

Xiao, S. Y., Liu, X. M., & Tong, C. Y. (2012). Dialdehyde starch nanoparticles as antitumor drug delivery system: An in vitro, in vivo, and immunohistological evaluation. *Chinese Science Bulletin, 57*(24), 3226–3232.

Xu, W., Yang, W., & Yang, Y. (2009). Electrospun starch acetate nanofibers: Development, properties, and potential application in drug delivery. *Biotechnology Progress, 25*, 1788–1795.

Ying, D. Y., Schwander, S., Weerakkody Sanguansri, L., Gantenbein-Demarchi, C., & Augustin, M. A. (2013). Microencapsulated Lactobacillus rhamnosus GG in whey protein and resistant starch matrices: Probiotic survival in fruit juice. *Journal of Functional Foods, 5*, 98–105.

Yu, D., Xiao, S., Tong, C., Chen, L., & Liu, X. (2007). Dialdehyde starch nanoparticles: Preparation and application in drug carrier. *Chinese Science Bulletin, 52*, 2913–2918.

Yue, P., & Waring, S. (1998). Resistant starch in food applications. *Cereal Foods World, 43*, 690–695.

Zaman, S. A., & Sarbini, S. R. (2016). The potential of resistant starch as a prebiotic. *Critical Reviews in Biotechnology, 36*(3), 1–7.

Zhao, J., Madson, M. A., & Whistler, R. L. (1996). Cavities in porous corn starch provide a large storage space. *Cereal Chemistry, 73*, 379–380.

Part II
Non-starch Polysaccharides: Structure, Functions, Bioactivity and Applications

Beta-Glucans

Nusrat Jan, Touseef Ahmed Wani, F. A. Masoodi, Adil Gani, and H. R. Naik

Introduction

β-Glucans belongs to a group of non-starch polysaccharides that are unbranched homopolymers of D-Glcp, joined by β (1→3) and β (1→4) linkages ([4)-β-D-Glcp-(1→4)-β-D-Glcp-(1→3)-β-D-Glcp-(1→]n). β-glucans are primarily isolated from cell walls of fungi like mushrooms and yeast, cereals like oats and barley and numerous bacteria (Novak and Vetvicka 2009). β-glucans from mushrooms consists of glucopyranose molecules coupled by (1→3) β-linkages and small number of branches bound by (1→6) β-linkages, whereas glucopyranose molecules in cereal β-glucans are joined by (1→3) β- and (1→4) β-linkages (Laroche and Michaud 2007). Depending on the source of extraction, β-glucans strongly differ in biological and physiochemical properties (Auinger et al. 2013). Further, degrees of purification, as well as the extraction method, also influence the physiological activity of β-glucans (Vetvicka and Vetvickova 2007).

β-glucan, being as dietary fibre, is related to a variety of health benefits for the consumer and has been approved by both the U.S. Food and Drug Administration (FDA) (2008) and the scientific panel of the EU (2010). Due to the fascinating biological properties of β-glucans, they are used for various medical and pharmaceutical applications like lowering blood total cholesterol and blood lipid profiles, moreover as in maintaining body weight (Behall et al. 2004; Liatis et al. 2009). β-glucans have been reported to have synergistic effects such as antioxidant, antigenotoxic, and antimutagenic activities (Kogan et al. 2008). Immune-stimulating activity of oat β-glucans and mushroom β-glucans has additionally been reviewed (Daou and Zhang 2012; Ooi and Liu 2000). Lentinan, schizophyllan, and krestin

N. Jan (✉) · H. R. Naik
SKUAST-K, Srinagar, Jammu and Kashmir, India

T. A. Wani · F. A. Masoodi · A. Gani
University of Kashmir, Srinagar, Jammu and Kashmir, India

A. Gani, B. A. Ashwar (eds.), *Food biopolymers: Structural, functional and nutraceutical properties*, https://doi.org/10.1007/978-3-030-27061-2_5

polysaccharides from mushrooms demonstrated antitumor activities and have massive markets in East Asian countries, including Japan, and Korea (Ooi and Liu 2000). β-glucans prevent diabetes mellitus by controlling blood glucose levels and high blood pressure (Chen and Raymond 2008). In addition, β-glucans promote wound healing and alleviate ischaemic heart injury.

Apart from aforementioned health benefits, β-glucan may even be a valuable ingredient in food applications. It is perceived as a non-caloric thickener for various foods because of its viscosity (Limberger-Bayer et al. 2014). Moreover, it is utilized in foams and emulsions as a stabilizing agent (Lazaridou and Biliaderis 2007), a texturizing agent (Kodama et al. 2015), a fat replacer (Pintado et al. 2016) and as a wall material for controlled release and targetted delivery of nutraceutical compounds (Wani et al. 2016).

The acceptance of β-D-glucan as functional, bioactive ingredients has increased its popularity over the last two decades (Lazaridou and Biliaderis 2007). However, the extraction and purification of β-D-glucan involves a complex process, thus it needs a special attention to take advantage of its recovery and functional properties. In this context, this chapter outlines classification, isolation, purification, modifications, neutraceutical properties and recent advances in the application of beta-glucan.

Classification, Isolation and Purification

Classification

The solubility of β-glucans is related to the degree of polymerization (DP). When DP > 100, β-Glucans are completely insoluble in water. Solubility will increase as DP decreases. Thus, on the basis of their solubility, β-Glucans can be classified as; (a) alkali insoluble, acetic acid insoluble (1→3)-β-glucan; (b) alkali-soluble (1→3)-β-glucan; and (c) highly branched (1→6)-β-glucan (Zekovic et al. 2005). These properties impart a characteristic of insolubility for most β-glucans, limiting application and extrapolation of in vitro experimental data in human beings.

Isolation

β-D-glucan, along with non-starch polysaccharides, starch, protein complex and lipids, resides in the endosperm cell walls. This creates difficulty in extraction of β-D-glucan (Brennan and Cleary 2005). To evaluate the physicochemical properties of β-D-glucan, it needs an optimal extraction procedure to acquire the greatest level of yield and purity and least change in molecular structure of β-D-glucan chains, such as, alteration of extraction condition leads to change in β-D-glucan molecular weight (Wang et al. 2003). To extract β-D-glucan, many methods have been suggested such as hot water extraction (Benito-Román et al. 2011), acidic extraction (Ahmad et al. 2010), alkaline extraction (Ahmad et al. 2010; Din et al. 2009), enzymatic extraction (Ahmad et al. 2010; Li et al. 2006), ultrasound-assisted extraction (UAE) (Benito-Román et al. 2013), and microwave extraction (Ookushi et al. 2006). Among these methods, UAE represents larger potency as this technique is a cost-effective, quick, easy and efficient method that lessens the extraction time compared to conventional methods (Maran and Priya 2014).

Hot-Water Extraction

Water extraction is the most generally used traditional method for extracting polysaccharide because of its potency in getting polysaccharides and its low cost. Extraction time, extraction temperature, and also the ratio of water volume to raw material weight play a key role in the extraction process of polysaccharide. Usually, the lipid component and some other low-molecular weight substances in the raw material are removed firstly by using 75–95% (V/V) ethanol through a reflux divider (Li et al. 2013a, b). Then the residue is subsequently dried and extracted with boiling water many times. The extracting liquids are separated and collected by centrifugation. Various scientists adopted this method for the extraction of β-glucan. In a study, Morgan and Ofman (1998) attempted the hot water extraction process with the incorporation of a modification of freeze–thaw cycles. Using this technique, greater recovery of β-glucan along with elevated level of purity can be obtained.

However, polysaccharide extraction from fungus remains a serious issue, as a result of they are heterogeneous with less solubility in water, and also the yield of water extraction is often quite low. One of the most promising approaches is pressurized water extraction (PWE) because it presents necessary significances over traditional hot-water extractions. The solvents are at high temperature and high pressure throughout the PWE method. There is larger solubility and larger diffusion rate at high temperature, while the solvents remain below its boiling point at high pressure (Villares et al. 2012). For example, Palanisamy et al. (2014) reported that an optimal extraction condition for cultivated mushrooms (*Agaricus bisporus*, *Lentinus edodes* and *Pleurotus ostreatus*) is 200 °C, and 5 cycles of 5 min each at 10.3 MPa. Similarly, there is maximum recovery for extraction of α-(1,4)-glucans, and β-(1,6)-glucans at a pressure of 10.1 MPa for 70 min at 28 °C (Lo et al. 2007).

Alkaline Extraction

Non-cereal sources may vary slightly in β-glucan extraction because the compositions of these sources are variable in nature. This condition is found in case of yeast, molds, and mushrooms from which β-glucan can be extracted. Alkaline treatment favors this kind of extraction as this kind of β-glucan has different chemical composition. To extract β-glucan, the cell mass is treated with a high concentration of alkali (1 M KOH) at low temperature for a long period of time with continuous stirring. Then the supernatant and residue is separated by an exhaustive centrifugation process. The residue material still contains β-glucan, thus to completely separate β-glucan, it is resuspended twice in the alkaline media under the same conditions of temperature. All of the alkali extracts is also precipitated using ammonium sulfate by keeping the material overnight at a temperature of 4 °C or below followed by use of centrifugation at high speed under low temperature to recover β-glucan. This technique is highly effective to recover higher amounts of β-glucan from the mushroom cell wall along with high levels of purity can be achieved (Nguyen et al. 1998).

Using alkali as a solvent, β- glucan extraction from mycelium of *P. chrysogenum* has also been reported; the mycelium was first freeze-dried and then ground into powder form. Then the material was treated with alkaline solution for the extraction of β-glucan. Maximum recovery can be obtained at a concentration of 1 M NaOH at 40 °C as NaOH is a more effective alkali in this form. β-glucan can be extracted by the treatment of freeze-dried powder of mycelia with the said concentration of alkali. β-glucan contains protein impurities, thus it must be separated by the sevage method that works at low pH. After protein has been excluded, the pH of the media is again adjusted toward neutral for separating β-glucan. Further purification is also obtained through dialysis against distilled water (Wang et al. 2007).

Acidic Extraction

Several reports are available on various extraction methods for β-glucan however; extraction under acidic conditions has been reported only by a few workers. Ahluwalia and Ellis (1984) have focussed the determination of β-glucan and starch in barley and its malt. They adopted a simple and quick method for the extraction of β-glucan and starch by using perchloric acid (50 mM). Then the extract was directly used without the drying process to determine the β-glucan and starch contents within the extract solution. For β-glucan determination, the extract was hydrolyzed to glucose by incubating with sodium acetate buffer and "treated cellulose." Glucose released was measured enzymatically using glucose-6-phosphate dehydrogenase and results were expressed as β-glucan (%) on dry weight basis. Moreover, Babu (2015) investigated that with defatted oat flour, a lower yield of β-glucan was achieved under acidic extraction compared to alkaline and enzymatic extractions.

Enzymatic Extraction

To improve purity and extraction within the extraction media, many enzymes are used to remove impurities which then separated in a series of steps by using centrifugation process. For that purpose a large number of enzymes like endo-β-(1→3)(1→4) glucanase, feruloyl esterase, endo-xylanases, xyloacetylesterase, and arabinofuranosidase could also be used. Some indigenous enzymes need to be deactivated as they will decrease the quantity and quality of extracted β-glucan during the extraction process. Kanauchi and Bamforth (2001) employed a method of β-glucan extraction which is based on the β-glucan solubilization from cell wall by using enzymes from outer sources. Lichenase, xylanase, and esterases are some other enzymes which will improve recovery of β-glucan. To increase the solubilization of β-glucan, these enzymes are sometimes used as the only treatment, however in other cases these could also be used along with acid, alkali, or heat treatment. Skendi et al. (2003) used different varieties of oats for β-glucan extraction by using heat stable enzyme, termamyl, along with hot water treatment and found greater recovery of β-glucan by this technique. Additionally, the researchers also used iso-propanol and petroleum ether as defatting media that resulted in higher yield. Soluble and insoluble substances can be separated from each other using high-speed centrifuges during this extraction method. Most of the protein impurities can become insoluble at their isoelectric point or by use of protein-degrading enzymes and are excluded during the centrifugation process. After removal of impurities, solubilized β-glucan was precipitated by either ethanol or calcium sulfate.

Another enzymatic extraction method for β-glucan extraction included boiling and refluxing of grounded oat powder with ethanol. After defatting, all of the material was suspended in water and heat treated to further deactivate the indigenous enzymes. α-amylase (termamyl) preparation that is stable at high temperature was incorporated in this medium to degrade starch impurities. Enzyme pancreatin was used for removal of protein impurities. Finally, β-glucan was precipitated by slow addition of ethanol until the concentration reached 50% in the extraction media. High-speed centrifugation at low temperature was carried out to recover β-glucan (Dongowski et al. 2005). In a similar study, Papageorgiou et al. (2005) recovered β-glucan from barley and oats by employing an enzymatic process.

Ultrasound-Assisted Extraction

Ultrasonic-assisted extraction (UAE) is a quick and efficient technique that was adopted for polysaccharide extraction. The flow diagram for the extraction of β-glucan from barley by UAE is given in Fig. 1. As is well-known, the application of UAE may provide several benefits including the reduction of solvents, temperature, and extraction time (Rahmanian et al. 2015). It has been determined that the extraction rate of UAE is correlated to ultrasound power per unit volume of liquid (Cheung and Wu 2013), and the kinetics of UAE strongly depend on the fungal materials, particularly the microstructure and morphology of the fungal cell and aggregation or dispersion of solid particles in the extracting liquid (Cheung et al. 2013). For example, an optimum UAE extraction condition of polysaccharide from *Ganoderma lucidum* (Fr.) Karst was reported to be an ultrasonic frequency of 8 kHz, extraction temperature 95 °C, extraction time 3 h, and the ratio of water to raw material was 12 (Chen et al. 2010). Similarly, an optimal extraction condition has been investigated for *A. bisporus* polysaccharide under the conditions of ultrasonic power 230 W, extraction temperature 70 °C, extraction time 62 min, and

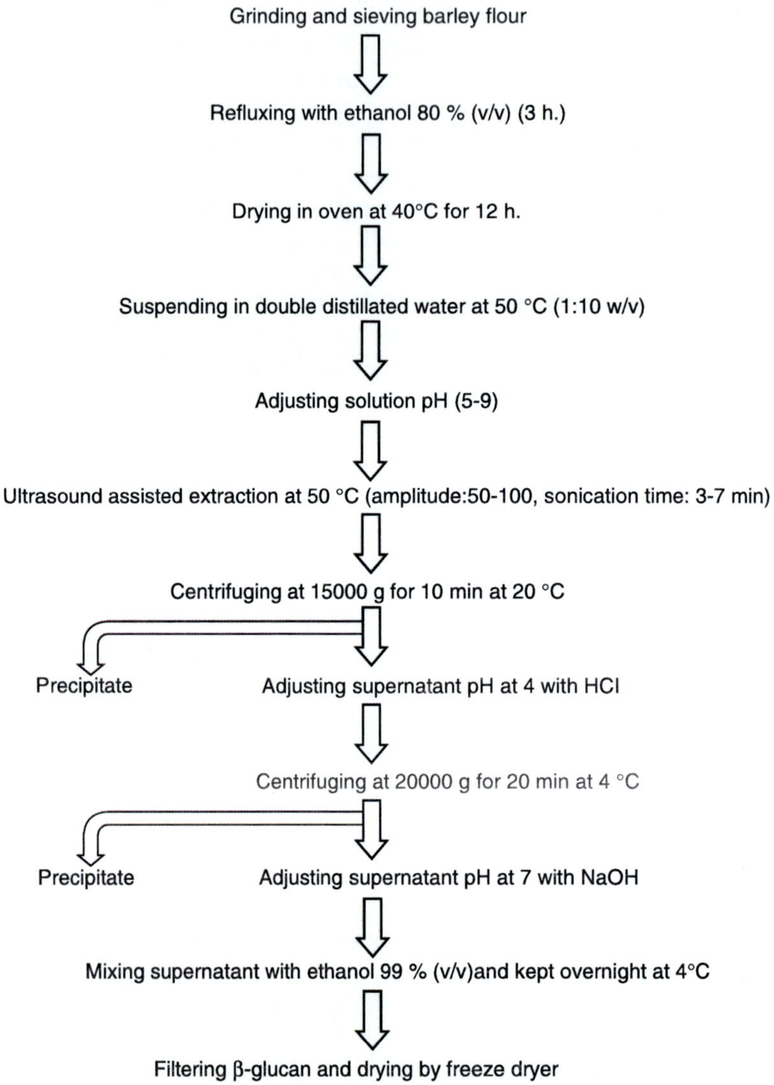

Fig. 1 Schematic outline for UAE method and purification procedures of barley β-D-glucan

water-material ratio 30 (mL:g), and the greater recovery of *A. bisporus* polysaccharide was obtained at 6.02% (Tian et al. 2012). Moreover, the yield of polysaccharide from *L. edodes* has been determined under the conditions of ultrasound power 340 W, water-material ratio 30:1 (mL:g), and ultrasound time 14 min was 14.39%, which is significantly higher than water heating extraction (Ke 2014).

However, the utilization of ultrasounds for β-glucan extraction is very limited in the literature. Izydorczyk et al. (2000) have employed sonication for physical disruption before the extraction (pre-treatment) showing that this pretreatment will improve solubilisation of β-glucan. Ultrasound assisted extraction have also been reported to be an efficient process to extract high molecular weight β-glucans from barley. The maximum extraction yield (66.1%) was reported to be achieved when delivering the maximum amount of energy (962.5 kJ/L), leading at the same time to the lowest MW (269 kDa). A reduction in the intensity of the treatment (energy output of 170 kJ/L) decreases the extraction yield up to 44.3% but increases the β-glucan MW to 461 kDa (Benito-Román et al. 2013). Recently, Sourki et al. (2017) reported that with increasing sonication time up to 5.5 min, β-d-glucan flow behaviour index (n) and its extraction yield significantly increased.

Microwave Extraction

Microwave energy is a non-ionizing type of electromagnetic radiation with wavelengths typically in the range of 1 mm to 1 m, corresponding to 0.3–300 GHz. MAE was employed for the polysaccharide extraction from the fruiting body of *Hericium erinaceum* mushroom. Comparison of the results obtained by MAE with those obtained by conventional hot-water extraction revealed that the extractability of the former for 5 min at 140 °C was almost equivalent to that of the latter extraction for 6 h at 100 °C (Ookushi et al. 2006). Besides, microwave treatments have also been found to increase the solubilization of polysaccharide. It is reported that microwave heating in a high-pressure vessel improves the dispersion of β-glucan solution in water without polymer degradation (Wang et al. 2002). A water insoluble hyperbranched β-glucan isolated from sclerotia of *Pleurotus tuberregium* was treated by microwave heating to produce an aqueous solution, and the soluble polysaccharide existed in a sphere-like conformation in 0.02% aqueous sodium azide solution (Tao and Xu 2008).

Purification of Beta-Glucan

During β-glucan extraction, water or other solvents are used. However, impurities (protein-based or non-starch polysaccharides) are present in solutions during these treatments, thus, polysaccharide enriched extracts obtained from the extraction procedure are further purified. Ion-exchange chromatography and gel filtration chromatography are the most common and convenient methods for purifying polysaccharide. In general, the impurities are removed from the polysaccharide through ethanol precipitation, neutral polysaccharides are separated from acidic ones using ion-exchange chromatography through DEAE-cellulose columns, and the neutral polysaccharides are then separated into α-glucan and β-glucan using gel filtration and affinity chromatography (Wasser 2002). Using cetyl trimethyl ammonium bromide or cetyl pyridinum chloride, acidic and neutral polysaccharide might even be separated which then form a precipitated complex with the acidic polysaccharide (Yang and Zhang 2009). Polysaccharide with a broad polydispersity can be fractionated by stepwise precipitation or preparative gel permeation chromatography, through which polysaccharides with different molecular weights and low polydispersity can be achieved (Zhang et al. 2007).

Structural and Solution Behavior

Structural Features

Polysaccharides derived from different sources are usually composed of various monosaccharides, and many of them have hyperbranched structures. To determine the structure of glucans from different resources, a series of analytical methods have been developed. The chemical structure was analyzed by Fourier transform infrared spectroscopy (FITR), nuclear magnetic resonance spectroscopy (NMR), gas chromatography (GC), gas chromatography-mass spectrometer (GC-MS), and high-performance liquid chromatography (HPLC).

The primary structure of cereal β-glucans consists of a linear chain of glucopyranosyl monomers joined by a mixture of single β-(1→3) linkages and consecutive β-(1→4) linkages (Fig. 2) (Skendi et al. 2003; Tosh et al. 2004). (1→3), (1→4)-β-D-glucan-4-glucanohydrolase, also known as lichenase, acts on (1→3), (1→4)-β-D-glucan from cereal grains produces oligosaccharides which can be characterised using methylation analysis. Using this analysis for oat β-glucans, composition was predominantly β-(1→3)-linked cellotriosyl (3-O-β-cellobiosyl-D-glucose, DP3) and cellotetraosyl (3-O-cellotriosyl-D-glucose, DP4) units with a small number of regions containing 4-8 consecutive (1→4)-linked units (Wood et al. 1991a, b). Endo-1→4-β-glucanases and one non-cellulolytic β-glucanase isolated from *Trichoderma reesei* were found to specifically hydrolyse cellotetraosyl units and higher homologues, allowing for isolating cellotriosyl units which will be determined by lichenase hydrolysis (Ajithkumar et al. 2006). Water-soluble and -insoluble β-glucans from oat and barley were studied by digesting with lichenase and analysing oligosaccharides using HPAEC-PAD analysis which revealed that in oat β-glucans, the trisaccharide and tetrasaccharide cellulosic units accounted for 95% of molecule (Johansson et al. 2000). In one study, cellulosic oligomers released by lichenase showed cellotriosyl and cellotetraosyl units accounted for 91-92% for barley and 92-94% for oats (Papageorgiou et al. 2005) and in another study barley β-glucans cleaved by lichenase had 93% of residues as tri- or tetrasaccharide cellulosic units (Cyran et al. 2002). Furthermore, in cereal β-glucans, HPLC can be

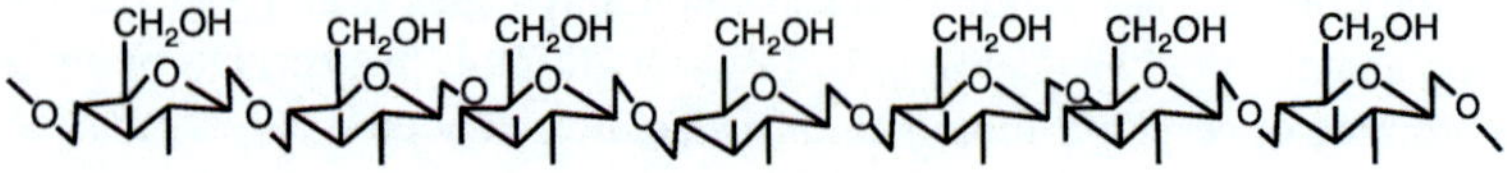

Fig. 2 General structure of cereal β-glucans

used to determine the molar ratio of tri- to tetrasaccharide (DP3:DP4). This molar ratio varies, but is typically about 2.1 for oats, 3.2 for barley and 3.5 for wheat (Wood et al. 1991a, b), and the ratio always is higher for barley than oats (Cui et al. 2000; Tosh et al. 2004). HPAEC-assisted lichenase treatment can be used to determine barley β-glucan structure and molar ratio of 1→4 to 1→3 linkages for barley varieties ranged from 2.26 to 2.35 (Yoo et al. 2007). High-performance size-exclusion chromatography (HPSEC) can be used to determine the molar mass of β-glucans. Molecular weight is important in influencing β-glucan solubility as 500,000 g/mol molecular weight has been demonstrated for soluble β-glucans of both oat and barley and <200,000 g/mol for insoluble β-glucans (Johansson et al. 2004). Molecular weight of β-glucans varies greatly with oats (0.065–3 × 106) typically higher than barley (0.15–2.5 × 106) or wheat (0.25–0.7 × 106) (Andersson et al. 2004; Beer et al. 1997; Lazaridou et al. 2003).

Likewise, the structure of mushroom β-glucans consists of linear β-(1→3)-linked backbones with β-(1→6)-linked side chains of varying length and distribution (Fig. 3), and can form complex tertiary structures stabilized by interchain hydrogen bonds (Brown and Gordon 2005). Both schizophyllan from mushroom *Scizophyllum commune* and scleroglucan from mushroom *Sclerotium glucanicum* have a β-(1→3)-linked backbone, with on average one β-(1→6)-glucose substitution every three backbone residues. Epiglucan from mushroom *Epicoccum nigrum* also has a backbone of β-(1→3)-linked glucose residues, but now with two β-(1→6) substitutions on average every three residues, while the β-(1→3)-linked backbone of lentinan from mushroom *L. edodes* has two β-(1→6) side chains every five residues (Schmid et al. 2001). Using infrared spectroscopy, nuclear magnetic resonance (NMR), methylation and monosaccharide composition analysis, the structural characterization of β-glucan from mushroom *G. frondosa* was determined to be a β-D-(1-3)-linked glucan backbone with a single β-D-(1-6)-linked glucopyranosyl residue branched at C-6 on every two residues (Fang et al. 2012). Similarly, Dong et al. (2012) determined the chemical structure of β-glucan (GLSA50-1B) from mushroom *G. lucidum* using sugar compositional analysis, methylation analysis, partial

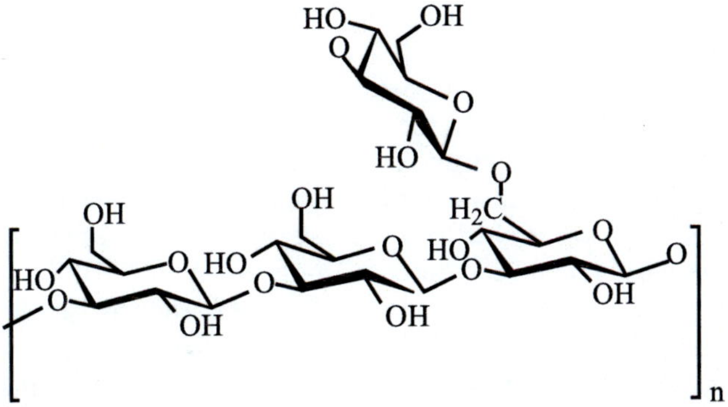

Fig. 3 General structure of mushroom (1→3) β-glucans with ramifications β-(1→6)

acid hydrolysis, acetolysis, and NMR and ESI-MS spectroscopy. The results indicated that GLSA50-1B was elucidated to be a completely unique β-D-glucan featured by a 1,6-linked β-D-glucopyranosyl backbone with different length of branches consisting of terminal and 1,4-linked glucopyranosyl residues, connected to O-4 of alternative glucose residues in the backbone.

Solution Behaviour

Solubility in water is one of the most important characteristic of β-glucan. In water, solubility of β-glucans is dependent, above all, on their structure, and this is associated with their origin as well (Rop et al. 2009). It is well documented that the solubility of β-glucan increases with temperature. In general, water solubility chain conformation, and addition of suitable ionic groups with appropriate degree of substitution can change the bioactivities of polysaccharides (Tao et al. 2006). Soluble β-glucans are considered to be stronger immunostimulators than insoluble β-glucans. Zhang et al. (2001) have reported that water-insoluble polysaccharides exhibit very little bioactivity, whereas their soluble derivatives show high antitumor and/or antiviral activities. Similarly, Xiao et al. (2004) reported that soluble high molecular weight yeast β-glucan may improve IFNγ-producing cells in a dose-dependent manner, whereas the insoluble β-glucan had no impact.

The viscosity is another important characteristic feature of β-glucan. The viscosity of β-glucan is dependent on the molecular weight, molecular structure, and food matrix (Kerckhoffs et al. 2003; Bae et al. 2009). The range of relative MW of β-glucans is quite broad and fluctuates (depending on origin) from tens to thousands of kilodaltons (kDa) (Rop et al. 2009). High molecular weight β-glucans are most effective as they are able to form viscous solutions. These are mainly found in native forms of oat or barley (≥1000 kDa) (Wilson et al. 2004), while food processing or extraction results in reducing β-glucan molecular weight (Rieder et al. 2015; Tosh et al. 2010).

Oat and mushroom β-glucans within the diet have been reported to reduce human plasma lipid due to its capability to extend food viscosity (Bae et al. 2009; Andersson et al. 2010). Studies reported that β-glucans from each oat and mushroom shiitake exhibited sheer thinning behaviour where the viscosity decreased with higher speed. They additionally determined that the mushroom diet was better than the oat diet as a part of a high fat diet as a result of it reduced body weight gain, total fat mass, plasma triacylglycerol and increased fat faecal excretion (Handayani et al. 2012). Moreover, viscosity is controlled by β-glucan concentration in solution and β-glucan MW, and thus the glycaemic response was considerably related to values of concentration MW (Wood et al. 2000). In a study, the relationship of viscosity and MW of cereal β-glucan in postproandial blood glucose and insulin response and serum cholesterol has been reported (Wood 2007). Furthermore, it has been reported that endogenous enzymes are to blame for the changes in viscosity. There are a minimum of two endogenous glucan hydrolase enzymes that cause degradation of β-glucans that ultimately modify the viscosity property of β-glucan. Generally these enzymes initiate depolymerization reactions, resulting in low viscosities (Ahmad and Anjum 2010; Burkus and Temelli 2000).

Moreover, final viscosity of the β-glucan can be affected by the extraction conditions. At very high pH, extraction of oat β-glucan can cause changes in β-glucan which will lower its viscosity property. The interaction between high pH and medium temperature is extremely unfavourable for the viscous property of β-glucan. This is due to the sensitive nature of 1→3 linkages at higher pH levels or medium temperatures.

Modifications

β-Glucan is taken into account as a very important bioactive compound for human health, but its low solubility has led to the development of chemical modification technologies to improve bioavailability. To modify β-glucan, various methods including physical and chemical crosslinking reactions are laid out to increase their functional and technological properties (Ahmad et al. 2015). β-glucans in this respect can be chemically modified to obtain various derivatives with potential industrial or medicinal applications (Synytsya and Novak 2013). Chemical modifications including sulfonylation, carboxymethylation, phosphorylation and acetylation are regarded as important tools to improve the properties, bioactivities and applications in the structure of polysaccharides, (Tranquilan-Aranilla et al. 2012; Ye et al. 2012; Jindal et al. 2013). Chemical derivatization of β-D-glucans improved antioxidant, antitumoral, anticoagulating and immunomodulating activities. Additionally, chemical characteristics can be changed by modifying β-D-glucans which sufficiently enhance bioactivity, thereby widening their range of applications (Magnani et al. 2009; Tranquilan-Aranilla et al. 2012; Jindal et al. 2013). Thus, chemical modification can be a useful tool in developing β-D-glucans as therapeutic agents (Sarangi et al. 2006; Zong et al. 2012; Zhang et al. 2013).

Chemical Modification by Sulfonylation

Sulfonated polysaccharides are complex molecules that exert excellent physicochemical properties and bioactivity. The sulfonate group is attached to the hydroxyl group on the monosaccharide unit, thus, improves the biological activities compared to the parent non-sulfonated polysaccharides (Raveendran et al. 2013). Sulfonylation is obtained through the reaction between pyridine and chlorosulfonic acid or pyridine and sulfur trioxide, indimethyl sulfoxide (Carvalho et al. 2013). Sulfonylation methods employing pyridine-chlorosulfonic acid producing greater yields of sulfonated polysaccharides with a large degree of substitution, and more interestingly, are simply executed (Lu et al. 2012). The chemical modification of polysaccharides through sulfonylation has been broadly studied due to the properties obtained of the modified products. Sulfonylation protocols have been developed for various polysaccharides including curdlan, dextran, and botryosphaeran (Jindal et al. 2013; Brandi et al. 2011), mannogalactan (Gracher et al. 2010), carboxymethyl cellulose (Shao et al. 2014).

Various biomacromolecules have been investigated in an attempt to obtain antithrombotic agents that may function as alternatives to heparin. In this respect, sulfonated polysaccharides are therefore of great interest (Gracher et al. 2010). Brandi et al. (2011) sulfonated and resulfonated botryosphaeran (1→3; 1→6)-β-D-glucan) and demonstrated that the polysaccharide have a potential of thrombin inhibition similar to that of heparin. Accordingly, the sulfonated botryosphaeran showed anticoagulant activity in vitro. Lin et al. (2004) investigated a sulfonated (1→3)-β-D-glucan in vitro and in vivo and exhibited potentiated antitumor activity on Sarcoma 180 cells derived by *Poria cocos* and their results verified antiproliferative activity significantly higher than the unmodified glucan. Cui et al. (2008) showed high antioxidant potential of a sulfonated (1→6)-β-D-glucan obtained by *Pueraria lobata* (Willd.) Ohwi. Additionally, sulfonated polysaccharides have been found for improving their solubility in water. They have been used as texturizing agents in foods, as food supplements, and in the development of new drugs (Jindal et al. 2013).

Chemical Modification by Carboxymethylation

In recent years, carboxymethylation of several natural polymers including cellulose, starch, dextran, chitin and chitosan have been investigated. Carboxymethylation takes place by substituting hydroxyl groups on the polysaccharide with carboxyl groups. Derivatives obtained by carboxymethylation exhibit properties with excellent applications in the chemical, pharmaceutical, food and cosmetic industries (Ye et al. 2012). Carboxymethylation is typically performed under alkaline conditions through suspension of the biopolymer in chloroacetic acid (Carvalho et al. 2013). Carboxymethylated β-glucan from corn kernels obtained a molecule with improved solubility, decreased viscosity, and enhanced antioxidant activity (Chen et al. 2013a, b). Moreover, similar to sulfonated polysaccharides, carboxymethylation will trigger enhancements in the physicochemical and biological properties of these polysaccharides.

Derivatization through carboxymethylation on a (1→3)-β-D-glucan from *Poria cocos* has been demonstrated and it has been verified that it significantly increased solubility and a dose-dependent elimination of free radicals (Wang et al. 2009). Carboxymethylation of (1→3)-β-D-glucan from baker's *Saccharomyces cerevisiae* yeast has been observed and administered with cyclophosphamide (an alkylating antineoplastic agent), showed increased inhibition of Lewis carcinoma cells, and inhibited pulmonary metastases, thereby improving the efficacy of cyclophosphamide action (Kogan et al. 2002). Similarly, carboxymethylated (1→3)-β-D-glucan from *Ganoderma lucidum* has been evaluated and verified decreasing cell metabolism in human cervical carcinoma cells which was primarily based on activation of the immune system with consequent action on the morphology, metabolism and viability of the tumor cells (Wiater et al. 2012). Besides, Krizkova et al. (2002) also observed that caboxymethylation of yeast β-glucan exhibits very high antioxidative activity, as well as expressive antimutagenic effects.

Chemical Modification by Phosphorylation

Phosphorylation is achieved by replacing a hydroxyl group on the monosaccharide residues with a phosphate group. Compared to the unmodified molecule, the phosphorylated glucan has improved anti-inflammatory, anti-viral, and anti-proliferative biological activities (Ye et al. 2013; Chen et al. 2009; Chen et al. 2014a, b; Huang and Zhang 2011).

Phosphorylation of (1→3)-β-D-glucan from *Lachnum sp*. has been investigated and was found significantly more efficient in inhibiting proliferation of murine sarcoma S-180 cells compared to parent unmodified-polysaccharide (Ye et al. 2013). Similarly, phosphorylated (1→3)-β-D-glucan from *Poria cocos* exposed to S-180 sarcoma cells has been described and it has been observed that the derivatized polysaccharide with high molecular weights showed strong antitumor activity at low concentrations compared to the parent (1→3)-β-D-glucan (Chen et al. 2009). In vivo experiments showed that derivatization through phosphorylation of a (1→4)-β-D-glucan from *Rhizoma panacis japonici* found that the glucan greatly inhibited the proliferation of H-22 tumor cells (Chen et al. 2014a, b)

Similarly, phosphorylation also improved the solubility of a (1→3)-β-D-glucan (Chen et al. 2009). According to the authors, the introduction of phosphate groups improved the solubility of the molecule without changing the glucan structure and conferred protection in a model of periodontitis induced by *Escherichia coli* in rats. In vitro and in vivo experiments demonstrated that (1→3)-β-D-glucan produced by *Poria cocos* following phosphorylation improved the solubility of the molecule and exhibited high degree of inhibition of Sarcoma 180 (S-180) tumor cells in a dose-dependent manner (Huang and Zhang 2011).

Chemical Modification by Acetylation

Acetylation is a method of substituting hydroxyl groups with O-acetyl groups (Popescu et al. 2012). This process has triggered much attention in different commercial sectors, however, there are few studies reporting these types of modifications in association with enhanced biological activity.

As there are few studies on acetylation of glucans, derivatization by acetylation on other polysaccharides are recognized with respect to their biological activities.

In one study, acetylation modified the oligosaccharide, chitotriose, isolated from chitosan observed antioxidant activity in a dose-dependent manner related to the degree of substitution in the molecule (Li et al. 2013a, b). Similarly, porphyran (an algal *Porphyra haitanensis*) polysaccharide made up of disaccharide units consisting of 3-β-D-galactosyl residues alternating with 4-linked 3,6-anhydro-α-L-galactose units also possesses antioxidant activity (Zhang et al. 2009). Acetylation of a heteropolysaccharide from the fruiting body of *Ganoderma atrum* has been reported and exhibited immune-modulating effects and high antioxidant activity than the parent un-modified polysaccharide (Chen et al. 2014a, b). Acetylation of a polysaccharide (repeating disaccharide units are [β-D-GlcpA-(1→4)-α-L-Rhap3s] and [α-L-IdopA-(1→4)-α-L-Rhap3s] from the alga *Ulva pertusa* has been reported to control hyperlipidemia and it was revealed that the strongest anti-hyperlipidemic activity, reducing total and LDL cholesterol, was obtained at a dose of 125 mg/kg animal body weight (Qi et al. 2012). Therefore, antioxidant and immunomodulating activities represent the main beneficial effects in tissue culture from acetylation of polysaccharides. However, there are scarce reports available for the testing of glucans in vivo and proposing mechanisms to explain these effects.

Nutraceutical Properties

There are innumerable therapeutic effects of β-glucans, varying from general health to specific therapeutic benefits. Numerous reports from in vitro and in vivo experiments have shown some biological properties of β-glucans including anti-oxidant and immunomodulatory (Maity et al. 2014; Nandi et al. 2014), anti-hypercholesterolemic and hypoglyceamic (Pomeroy et al. 2001; Charles 2005) and anticancerous properties (Hong and Jung 2014). Table 1 illustrates the sources and bioactivities of some β-glucans. Some of the important health promoting properties of β-glucans is discussed as under:

Anti-Oxidant Property

Increasing evidence reports that oxidative stress may cause cell injury which might trigger both the physiological process of ageing (Harman 2006) and many pathological progressions that eventually results in serious health issues (Halliwell 2003). There is a great deal of analysis showing that β-glucan from yeast and mushrooms are potential antioxidant agents.

The antioxidant properties of β–glucan isolated from edible mushrooms such as *Agaricus bisporus*, *Pleurotus ostreatus* and *Coprinus attrimentarius* were demonstrated using different assays like DPPH (2,2-diphenyl-1-picrylhydrazyl), reducing

Table 1 Sources and bioactivity of some β-glucans

Source	Type	Bioactivity	References
Agaricus bisporus	β-glucan	Antioxidant, immune-modulating effect	Wang et al. (2013) Kozarski et al. (2011)
Scizophyllum commune	Schizophyllan	Antioxidant, anticancer activities	Du and Xu (2014) Wasser (2002)
Poria cocos	β-glucan	Antitumor activities	Chen et al. (2009)
Lentinus edodes	Mannoglucan, polysaccharide-protein complex, glucan, lentinan	Immunomodulating, anticancer and antitumor activities	Maity et al. (2014)
Ganoderma lucidum	β-d-glucan	Immunostimulatory, antioxidant activities	Wang et al. (2014) Ma et al. (2013)
Pleurotus tuberregium	β-glucan	Hyperglycemia and oxidative stress; hepatoprotective, anti-breast cancer	Huang et al. (2012) Zhang et al. (2006) Zhang et al. (2001)
Hordeum vulgare L.	Barley β-glucan	Immunestimulatory, hypocholesterolemic	Chan et al. (2007) Choi et al. (2010)
Avena sativa L.	Oat β-glucan	Immunestimulatory, hypocholesterolemic, antitumor	Udayangani et al. (2017) Theuwissen and Mensink (2007) Choromanska et al. (2015)

power, metal chelating ability and ABTS (2,2-Azino-bis, 3-ethylbenzothiazoline-6-sulfonic acid) and all these activities varied significantly ($p \leq 0.05$) among all the β-glucan, however the β-glucan from *Coprinus attrimentarius* exhibited the highest values for all antioxidant activities (Khan et al. 2017). Maity et al. (2014) evaluated the antioxidant properties of a soluble β-glucan derived from the fruiting bodies of mushroom *Entoloma lividoalbum* by alkaline extraction. The β-glucan was found to possess hydroxyl and superoxide radical-scavenging activities with effective concentration of 480 and 150 μg/mL and the reducing power was found to be 0.5 at 480 μg/mL. The antioxidant properties of hot-water polysaccharide extracts of fruiting bodies of *A. bisporus, Agaricus brasiliensis, Phellinus linteus* and *G. lucidum* containing β-glucan were compared, and the order was found to be *G. lucidum, P. linteus, A. brasiliensis, A. bisporus* (Kozarski et al. 2011).

Sener et al. (2005) reported that yeast β-glucan treatment replenished the glutathione level in tissue and suppressed myeloperoxidase activity, indicating a protective effect against this oxidative injury in vivo. Additionally, Sener et al. (2007) evaluated that the yeast β-glucan treatment ameliorated the nicotine-induced oxidative damage in the kidney. It not only increased glutathione levels, decreased both malondialdehyde level and myeloperoxidase activity, but also restored the morphological changes in the damaged tissues, and improved the impairments in renal function.

Immuno-Modulatory Property

β-glucans are known to have various immunomodulatory properties. In vitro and in vivo reports determined that the immune-stimulating property of β-glucan depends on structure, molecular weight and number of branches (Miura et al. 2003; Brown and Gordon 2003). Bohn and Miller (1995) reported that the immune-stimulating effect of β-glucan is probably related to the activation of cytotoxic macrophages and T-helper and natural killer (NK) cells and with the promotion of T lymphocyte differentiation and activation, for the alternative complement pathway. β-glucans have also been observed as modulators of both humoral and cellular immunity (Falch et al. 2000; Tokunaka et al. 2000; Kubala et al. 2003). β-(1→3)-D-glucans from fungi were reported to be capable of having beneficial effects in pre-inflammatory responses, suggesting that β-glucan can be a modulator of the anti-inflammatory response as interleukin mediators (Luhm et al. 2006). β-1,3-glucan from *Saccharomyces cerevisiae* showed in vitro enhanced Th2 immune response and inhibited Th1 by stimulating the release of anti-inflammatory cytokines such as IL-10 and TGF-b (Chen et al. 2013a, b), whereas mushroom β-1,3-glucan showed in vivo decreased Th2 immune response induced by OVA and improved Th1 type immune response (Wu et al. 2014).

Nowadays, β-glucans are broadly studied in aquaculture and food industry partly because of their immunostimulatory activities (Meena et al. 2013; Ringo and Song 2016). The immune function of β-glucans has been evaluated by administered zebrafish larvae with oat derived β-glucan, and then challenged the fish with pathogenic bacteria *Edwardsiella tarda*. The authors concluded that the β-glucan-treated group had significantly higher survival rates than the control group (Udayangani et al. 2017). In vitro experiment studied the immunostimulant activity of β-glucan isolated from the fruiting bodies of *Russula albonigra* mushroom showed macrophage activation by nitric oxide (NO) production as well as spleenocytes and thymocytes proliferation (Nandi et al. 2014). Similarly, Sun et al. (2013) extracted β-glucan from the fruiting bodies of *Trametes robiniophila* exhibited immunomodulatory activity.

Anti-Hypercholesterolemic Property

Now-a-days, heart disease is the major cause of human death around the world. In addition to pharmacotherapy, reducing serum cholesterol levels through making dietary changes including lowering total fat intake and dietary cholesterol is one way to decrease the risk of developing the disease. Interest in dietary fiber has increased due to its beneficial effects on blood cholesterol levels. Introducing β-glucan rich fiber in the daily diet is now regarded as an effective measure to decrease serum lipid levels.

Among the cereal grains, oats and barley have been reported to be the most effective in reducing serum total cholesterol and low density lipoprotein (LDL)-cholesterol in humans and animals (Pomeroy et al. 2001). Oats has been reported to reduce both serum TC and LDL-C compared with the control (Theuwissen and Mensink 2007) and barley has also been observed to have a similar effect (Behall et al. 2004). A study has conducted that barley β-glucan significantly decreased liver weight, but not adipose tissue weight (Choi et al. 2010). Another study has revealed that in mice fed a high-fat-diet, barley significantly lowered adipose tissue (Yokoyama and Shao 2006). However, oat and barley β-glucans do not increase triacylglycerol (TAG) (Colhoun et al. 2004).

Mushroom β-glucans are also considered as an important component for reducing hyperlipidaemia (Chen and Seviour 2007). Several researches investigated that feeding of *Pleurotus* mushroom exhibited a hypocholesterolemic effect (Bobek et al. 1996a, b; Gunde-Cimerman and Plemenitas 2001). Consumption of *P. ostreatus* mushroom soup significantly decreased triacylglycerol concentration and oxidized LDL levels (Schneider et al. 2011), and hypercholesterolemic mice fed with 5% fruiting bodies of *P. ostreatus* mushroom lowered plasma total cholesterol by 28%, LDL-cholesterol by 55%, triglyceride by 34%, nonesterified fatty acid by 30%, and total live cholesterol levels by >34% (Hossain et al. 2003).

Hypoglyceamic Property

Diabetes is a group of metabolic disorders which includes alterations in carbohydrate, fat, and protein metabolisms associated with absolute or relative deficiencies in insulin secretion and insulin action. To treat diabetes mellitus, different hypoglycemic drugs have been manufactured but they have various side effects (May et al. 2002). Therefore, it is very important to have a hypoglycemic drug with minimal side effects in clinical experience. Demand for natural products like β-glucan with anti-diabetic activity has been increased.

Oat β-glucan intake has been reported to be beneficial for healthy subjects and patients with type-2 diabetes (Charles 2005). Two oat products namely flaked oats (muesli) and boiled oat flakes (oat porridge) has been investigated for the postprandial effect in healthy subjects. The results revealed that both products had a similar

GI effect as white bread, while intake of boiled oat kernels tested at the same time gave lower glucose and insulin responses (Granfeldt et al. 1995). Tapola et al. (2005) studied volunteers with type-2 diabetes fed on oat bran flour, oat bran crisp, and a glucose load providing 12.5 g glycemic carbohydrate (series 1) and 25 g glucose load alone, and 25 g glucose load with 30 g oat bran flour (series 2). In both series oat bran products rapidly lowered postprandial glucose concentrations than after the 12.5 g or 25 g glucose load during the 1st h, but the glucose concentration was greater at 120 min after the oat bran products ingestion than after the glucose load. This decrease of glucose absorption will decrease insulin release and thereby attenuate pancreatic insulin response.

Mushrooms are also exemplary anti-diabetic sources of natural medicines. Mushroom β-glucans lowers blood glucose concentrations by delaying stomach emptying so that dietary glucose is absorbed gradually (Lo et al. 2006) hence decrease appetite and reduce food intake. Orally ingested fruiting bodies and the acidic polysaccharide of both *Tremella mesenterica* and *Tramella aurantia* reduced the blood glucose concentrations in induced diabetic rats (Kiho et al. 1995). Hot water extracted β-glucans from *Agaricus blazei* was evaluated and showed anti-hyperglycemic, anti-hypertriglyceridemic activity indicating overall anti-diabetic activity in diabetic rats (Kim et al. 2005). β-glucan from *G. lucidum* is reported to increase body weight and serum insulin levels, but decrease fasting blood glucose in streptozotocin induced diabetic mice (Li et al. 2011).

Anticancerous Property

Cancer as broad groups of diseases involving deregulated cell growth is diverse, complex, and only partially understood, which makes it difficult to cure (Stosic-Grujicic et al. 2011). Many anti-cancer therapies are available like chemotherapy and anticancer drugs which are known to be cytotoxic to cancer cells, but at the same time, they are toxic to normal cells and harmful to the immune system (Yang et al. 2007). Therefore, there is a need for new and natural anticancer compounds of low or without toxicity to normal cells. β-glucan obtained from cereals, fungi, seaweeds, yeasts and bacteria is considered as one of the active compounds responsible for the immune effects (Chan et al. 2009; Hong and Jung 2014; Jung et al. 2008; Cheung et al. 2002).

Due to the properties of mushroom β-glucans as immune stimulants, they have been widely employed in tumor therapies. The mushroom β-glucans that have shown significant anticancer property against several human cancers in clinical trials includes lentinan from *Lentinus edodes*, D-fraction from *Grifola frondosa*, schizophyllan from *Schizophyllum commune*, polysaccharide-K (PSK) from *Trametes versicolor*, and polysaccharide-peptide (PSP), also from *Trametes versicolor* (Wasser 2002). It has been demonstrated that a polysaccharide-peptide complex (PSP) isolated from *Trametes versicolor* significantly reduced proliferation of MAD-MB-231 breast cancer cells as compared with the control (Chow et al. 2003).

It was also reported that a protein bound polysaccharide (PBP) extracted from *Phellinus linteus* had an anti-proliferative effect for SW-480 human colon cancer cells (Li et al. 2004a, b).

Antitumor activities of low molecular weight β-glucan obtained from oats were also investigated in cancer cells: Me45, A431 and normal HaCaT and murine macrophages P388/D1. The low molecular weight β-glucan from oat significantly deceased cancer cells viability, while for the normal cells it was non-toxic. Immunocytochemical ABC analysis showed that β-glucan induced strong expression of caspase-12 in both cancer cell lines, while in HaCaT cells ABC reaction was significantly lower and in P388/D1 cell line ABC reaction was negative (Choromanska et al. 2015). Similarly, Murphy et al. (2004) reported that oat β-glucan administration decreased lung tumor foci, increased macrophage cytotoxicity, and decreased the metastatic spread of injected melanoma cells.

Recent Advances in Application of Beta-Glucan

β-glucans have great potential in a wide variety of fields due to their conformation arrangement, gel-forming capacity and potent pharmacological properties.

Applications in Foods

β-Glucans have traditionally been used in foods as ingredients to technically improve their physical properties (rheological, texture), but can also improve the sensorial characteristics and provide important health benefits due to their biological functions (Kagimura et al. 2015; Laroche and Michaud 2007; Thammakiti et al. 2004).

Physiochemical properties of β-glucans show great tendency to use this valuable ingredient for preparation of various foods such as, ice creams, sauces, and salad dressings (Wood 1986). Due to stabilizing and hydrocolloid nature of β-glucan which makes it an alternative to guar gum, xanthan gum, locust bean gum, and gum arabic, and thus its major use in food could be for frozen ready-to-eat foods (Lyly et al. 2004). β-glucan has vital properties as a thickening agent for beverages, sauces, ice creams, and salad dressings (Temelli et al. 2004; Wood 1986). There are reports that β-glucan sourced from barley and mushrooms has great tendency to enrich snack products that have exerted better physical and textural characteristics when compared with a control recipe. Such β-glucan-containing products may increase the possibility of controlling glycemic responses to an appropriate condition (Brennan et al. 2013; Tosh 2013). Using a response surface methodology technique, effects of resistant starch, β-glucan, and simple starch on sausages has been studied and every treatment influenced most of the physical and sensory parameters. β-glucan in combination with resistant starch produced a softer texture in sausages that is needed for great consumer acceptance. Incorporation of these dietary fiber sources additionally improved the color and other sensory parameters (Amini Sarteshnizi et al. 2015).

β-glucan extracted from spent brewer's yeast may be used to develop a low-fat mayonnaise product. In this case, replacement of fat is dependent on the texture-modifying properties of β-glucan. Several levels of β-glucan have a potential to exhibit acceptable quality parameters when incorporated to a mayonnaise product. However, selective organoleptic properties related to color and appearance may have undesirable effects on the quality of mayonnaise (da Silva Araújo et al. 2014). Apart from spent yeast β-glucan, cereal β-glucan also imparts similar fat replacer properties. Cereal β-glucan was explored for the development of a low-fat cheese product resulted in improved texture in the final product (Lazaridou et al. 2014). In meat products, the effectiveness of β-glucan as a fat replacer was also reported and observed the applicability of β-glucan in color, emulsion stability, microstructure, and other textural characteristics in cooked meat batters (Álvarez and Barbut 2013).

Curdlan gum, a bacterial-insoluble linear (1→3)-β-D-glucan, is widely used in various types of foods including jellies, edible fiber, and noodles. Due to its moisture retention capacity and lack of taste, color, and odor, curdlan can be effectively used in baked, meat, and dairy products, among others (Divyasri et al. 2014; Shih et al. 2009).

Applications in Medicine

Food applications are not the only area in which β-glucan from different sources can be used. There are some instances where β-glucan was used for pharmaceutical products, showing its effectiveness as biological material. This is based on its natural applications against anti-inflammatory and antitumor activities (Ishibashi et al. 2004; Li et al. 2004a, b).

β-glucans have been used in various clinical trials to test their general effects on health and clarify the mechanisms responsible (Tungland and Meyer 2002). Venkatachalam et al. (2013) reported that cyclic glucans have larger inner cavity diameter, therefore, can be used as wound dressing material. Moreover, Kofuji et al. (2010) recognized a transparent wound dressing sheet by forming a complex between β-glucan and chitosan on evaluating in wounds created on the dorsal surfaces of mice. The β-glucan-chitosan complex sheets exhibited pharmacological properties comparable or superior to a commercial wound dressing product, thus can be used as promising new wound dressing product. It is found that β-glucan collagen matrix (BGC) had the ability to effectively treat partial-thickness burns in children. BGC also had the ability to significantly decrease post injury pain with affective acceleration of wound healing (Delatte et al. 2001). β-glucans are also recognized as therapeutic agents in the treatment of burn injuries (Toklu et al. 2006) and wound healing (Berdal et al. 2007). In one study, protective effect of β-glucan treatment on burn-induced remote organ injury was investigated by administration of β-glucan in the rats and findings were found effective against burn-induced oxidative tissue damage (Toklu et al. 2006). In another study, wound-healing effects of the poly(lactic-co-glycolic acid) membranes containing β-glucan has been investigated, the results revealed that the membranes accelerated wound healing by improving the interaction, proliferation of cells, and angiogenesis. Therefore, these membranes can be useful as a skin substitute for improving wound healing (Kim et al. 2012). Furthermore, Sakurai et al. (2001) reported that curdlan sulfates with DS from 1.7 to 8.7 mol% could form complexes with polycytidylic acid and the complex polynucleotide chain showed a significant resistance against enzymatic hydrolysis. It has been demonstrated that these water-soluble sulfated curdlan-nucleotide complexes could find application in gene technology in medicine.

Applications in Cosmetics

β-glucans have increasingly being used as essential ingredient in cosmetics and drugs produced by industries for the manufacturing of products such as facial creams and antiaging lotions, gels, makeup, and hair care products, in addition to those applied for the treatment of acne, cellulitis, dermatitis, and minor burns.

β-glucans usage as an active ingredient has been expanded in the manufacturing of these and other cosmetic products and is associated with numerous advantages that these biopolymers manifest on the skin. The β-glucans have a high ability to regenerate cells that produce collagen and immune cells, have high antioxidant potential, high capacity for hydration and skin penetration, and healing power of wounds and burns, besides increased collagen synthesis that confers protection against ultraviolet radiation (Kagimura et al. 2015; Zhu et al. 2016; Du et al. 2014). Some β-glucans have also been demonstrated to possess antinociceptive (pain reduction) activity (Moreno et al. 2016), and this property can also be exploited in cosmetic creams and products applied to inflamed and healing wounds, bruises, insect bites, and hematomas, among others. Additionally, healing effect has also been observed in glucan-based ointments or lotions for the treatment of eczema, psoriasis and other dermatitis ailments owing to their film forming capacity over the injured area, which improves hydration, wound healing ability, and protection of the site against external agents (Zhu et al. 2016).

Pillai et al. (2005) clinically evaluated efficacy of oat β-glucan in human skin models for reducing fine-lines and wrinkles and observed that β-glucan can deeply penetrated the skin into the epidermis and dermis. The research supports the use of β-glucan in the care and maintenance of healthy skin and the cosmetic treatment of the signs of aging. It has been shown that scleroglucan polysaccharide at a low concentration is useful in the field of skin and dermatological applications, thus, the glucan composition revealed beneficial effects including moisturization of skin or mucosa, and has an anti-aging and revitalizing effect on the skin (Baschong et al. 2009). Mushroom glucan has found greater moisture retention and has xerophthalmia relieving property (Chen 2014). Schizophyllan, a β-glucan reduces inflammation, irritation and other damage due to UV and toxic environment exposure of the skin (Kumari et al. 2008). The skin health promotion effects of natural β-glucan obtained from cereals and microorganisms has also discussed (Du et al. 2014).

β-1, 6-branched-β-1, 3-glucan obtained from liquid mycelia culture of *S. commune* have shown that it can defer skin aging, impart skin whitening effect and effectively cure skin damage (Park et al. 2001). Ketkeaw et al. (2012) found that a β-glucan from Hevea latex improved production of fibroblast collagen as compared to control. The authors concluded that Hevea β-glucan has possible application for treatment of collagen loss in aging skin.

Conclusion

β-glucans are a group of dietary fibers or polysaccharides derived from natural sources with proven pharmacological applications. The bioavailability of β-glucan is very low as it has low solubility in aqueous solutions. In order to improve its solubility, several derivation procedures including sulfonylation, carboxymethylation, phosphorylation and acetylation have been applied that in turn can alter their biological activities such as antioxidation and anticoagulation. Owing to its antioxidant, anti-hypercholesterolemic, immuno-modulatory, hypoglyceamic and anti-cancerous properties, β-glucan could offer tremendous applications in food, medical and cosmetic sectors.

References

Ahluwalia, B., & Ellis, E. E. (1984). A rapid and simple method of determination of starch and β-glucan in barley and malt. *Journal of the Institute of Brewing, 90*, 254–259.

Ahmad, A., & Anjum, F. M. (2010). *Perspective of β-Glucan: Extraction and Utilization*. Saarbrücken, Germany: VDM Publishers.

Ahmad, A., Anjum, F. M., Zahoor, T., Nawaz, H., & Ahmed, Z. (2010). Extraction and characterization of β-d-glucan from oat for industrial utilization. *International Journal of Biological Macromolecules, 46*(3), 304–309.

Ahmad, N. H., Mustafa, S., & Che Man, Y. B. (2015). Microbial polysaccharides and their modification approaches: A review. *International Journal of Food Properties, 18*, 332–347.

Ajithkumar, A., Andersson, R., Siika-Aho, M., Tenkanen, M., & Aman, P. (2006). Isolation of cellotriosyl blocks from barley β-glucan with endo-1,4-β-glucanase from *Trichoderma reesei*. *Carbohydrate Polymers, 64*(2), 233–238.

Álvarez, D., & Barbut, S. (2013). Effect of inulin, β-glucan and their mixtures on emulsion stability, color and textural parameters of cooked meat batters. *Meat Science, 94*(3), 320–327.

Amini Sarteshnizi, R., Hosseini, H., Bondarianzadeh, D., Colmenero, F. J., & Khaksar, R. (2015). Optimization of prebiotic sausage formulation: Effect of using β-glucan and resistant starch by D-optimal mixture design approach. *LWT - Food Science and Technology, 62*(1, Part 2), 704–710.

Andersson, A. A. M., Armo, E., Grangeon, E., Fredriksson, H., Andersson, R., & Aman, P. (2004). Molecular weight and structure units of (1→3, 1→4)-β-glucans in dough and bread made from hull-less barley milling fractions. *Journal of Cereal Science, 40*, 195–204.

Andersson, K., Svedberg, K., Lindholmb, M., Ostec, R., & Hellstranda, P. (2010). Oats (*Avena sativa*) reduce atherogenesis in LDL-receptor-deficient mice. *Atherosclerosis, 212*, 93–99.

Auinger, A., Riede, L., Bothe, G., Busch, R., & Gruenwald, J. (2013). Yeast (1,3)-(1,6)-β-glucan helps to maintain the body's defence against pathogens: A double-blind, randomized, placebo-controlled, multicentric study in healthy subjects. *European Journal of Nutrition, 52*, 1913–1918.

Babu, L. R. (2015). Green extraction techniques, structural analysis and antioxidant activites of β-glucan present in oats. *International Journal of Latest Trends in Engineering and Technology, 5*(4), 125–135.

Bae, I., Lee, S., Kim, S., & Lee, H. (2009). Effect of partially hydrolyzed oat β-glucan on the weight gain and lipid profile of mice. *Food Hydrocolloids, 23*, 2016–2021.

Baschong, W., Monglat, S., & Ochs, D. (2009). *Glucan compositions*. US Patent 0156563 A1.

Beer, M. U., Wood, P. J., & Weisz, J. (1997). Molecular weight distribution and (1→3) (1→4)-β-D-glucan content of consecutive extracts of various oats and barley cultivars. *Cereal Chemistry, 74*, 476–480.

Behall, K. M., Scholfield, D. J., & Hallfrisch, J. (2004). Diets containing barley significantly reduce lipids in mildly hypercholesterolemic men and women. *The American Journal of Clinical Nutrition, 80*, 1185–1193.

Benito-Román, O., Alonso, E., & Cocero, M. J. (2013). Ultrasound-assisted extraction of β-glucans from barley. *LWT - Food Science and Technology, 50*(1), 57–63.

Benito-Román, O., Alonso, E., & Lucas, S. (2011). Optimization of the β-glucan extraction conditions from different waxy barley cultivars. *Journal of Cereal Science, 53*(3), 271–276.

Berdal, M., Appelbom, H. I., Eikrem, J. H., Lund, A., Zykova, S., & Busund, L. T. (2007). Aminated β-1,3-D-glucan improves wound healing in diabetic db/db mice. *Wound Repair and Regeneration, 15*, 825–832.

Bobek, P., Ozdin, L., & Kajaba, I. (1996a). Dose-dependent hypocholesterolaemic effect of oyster mushroom (*Pleurotus ostreatus*) in rats. *Physiological Research, 46*, 327–329.

Bobek, P., Ozdın, L., & Kuniak, L. (1996b). Effect of oyster mushroom (*Pleurotus ostreatus*) and its ethanolic extract in diet on absorption and turnover of cholesterol in hypercholesterolemic rat. *Food/Nahrung, 40*, 222–224.

Bohn, J. A., & Miller, J. N. (1995). (1→3)-β-D-Glucans as biological response modifiers: A review of structure–functional activity relationships. *Carbohydrate Polymers, 28*, 3–14.

Brandi, J., Oliveira, E. C., Monteiro, N. K., Vasconcelos, A. F. D., Dekker, R. F. H., Barbosa, A. M., … Silva, J. M. L. C. (2011). Chemical modification of botryosphaeran: Structural characterization and anticoagulant activity of a water-soluble sulfonated (1→ 3)(1→ 6)-β-D-glucan. *Journal of Microbiology and Biotechnology, 21*, 1036–1042.

Brennan, C. S., & Cleary, L. J. (2005). The potential use of cereal (1→3, 1→4)-β-d-glucansas functional food ingredients. *Journal of Cereal Science, 42*(1), 1–13.

Brennan, M. A., Derbyshire, E., Tiwari, B. K., & Brennan, C. S. (2013). Integration of β-glucan fibre rich fractions from barley and mushrooms to form healthy extruded snacks. *Plant Foods for Human Nutrition, 68*(1), 78–82.

Brown, G. D., & Gordon, S. (2003). Fungal β-glucans and mammalian immunity. *Immunity, 19*, 311–315.

Brown, G. D., & Gordon, S. (2005). Immune recognition of fungal β-glucans. *Cellular Microbiology, 7*, 471–479.

Burkus, Z., & Temelli, F. (2000). Stabilization of emulsions and foams using barley β-glucan. *Food Research International, 33*, 27–33.

Carvalho, J. C. T., Perazzo, F. F., Machado, L., & Bereau, D. (2013). Biologic activity and biotechnological development of natural products. *BioMed Research International, 2013*, 1–4.

Chan, G., Chan, W., & Sze, D. (2009). The effects of β-glucan on human immune and cancer cells. *Journal of Hematology & Oncology, 2*(1), 25.
Chan, W. K., Law, H. K., Lin, Z. B., Lau, Y. L., & Chan, G. C. (2007). Response of human dendritic cells to different immunomodulatory polysaccharides derived from mushroom and barley. *International Immunology, 19*, 891–899.
Charles, S. B. (2005). Dietary fiber, glycemic response, and diabetes. *Molecular Nutrition & Food Research, 49*, 560–570.
Chen, C., Wu, W., Xu, X., Zhang, L., Liu, Y., & Wang, K. (2014a). Chain conformation and anti-tumor activity of derivatives of polysaccharide from *Rhizoma Panacis Japonici*. *Carbohydrate Polymers, 105*, 308–316.
Chen, J., & Raymond, K. (2008). β-glucans in the treatment of diabetes and associated cardiovascular risks. *Vascular Health and Risk Management, 4*, 126–1272.
Chen, J., & Seviour, R. (2007). Medicinal importance of fungal β-(1/3), (1/6)-glucans. *Mycological Research, 111*, 635–652.
Chen, S., Chen, H., Tian, J., Wang, Y., Xing, L., & Wang, J. (2013a). Chemical modification, antioxidant and α-amylase inhibitory activities of corn silk polysaccharides. *Carbohydrate Polymers, 98*, 428–437.
Chen, S. N. (2014). *Composite glucan and method for preparing the same*. US Patent 20140031542 A1.
Chen, X., Wang, W., Li, S., Xue, J., Fan, L., & Sheng, Z. (2010). Optimization of ultrasound-assisted extraction of Lingzhi polysaccharides using response surface methodology and its inhibitory effect on cervical cancer cells. *Carbohydrate Polymers, 80*, 944–948.
Chen, X., Xu, X., Zhang, L., & Zeng, F. (2009). Chain conformation and anti-tumor activities of phosphorylated (1→3)-β-D-glucan from *Poria cocos*. *Carbohydrate Polymers, 78*(3), 581–587.
Chen, Y., Dong, L., Weng, D., Liu, F., Song, L., & Li, C. (2013b). 1,3-β-glucan affects the balance of Th1/Th2 cytokines by promoting secretion of anti-inflammatory cytokines in vitro. *Molecular Medicine Reports, 8*, 708–712.
Chen, Y., Zhang, H., Wang, Y., Nie, S., Li, C., & Xie, M. (2014b). Acetylation and carboxymethylation of the polysaccharide from *Ganoderma atrum* and their antioxidant and immunomodulating activities. *Food Chemistry, 156*, 279–288.
Cheung, N. K., Modak, S., Vickers, A., & Knuckles, B. (2002). Orally administered beta-glucans enhance anti-tumor effects of monoclonal antibodies. *Cancer Immunology, Immunotherapy, 51*(10), 557–564.
Cheung, Y., Siu, K., & Wu, J. (2013). Kinetic models for ultrasound-assisted extraction of watersoluble components and polysaccharides from medicinal fungi. *Food and Bioprocess Technology, 6*, 2659–2665.
Cheung, Y., & Wu, J. (2013). Kinetic models and process parameters for ultrasound-assisted extraction of water-soluble components and polysaccharides from a medicinal fungus. *Biochemical Engineering Journal, 79*, 214–220.
Choi, J. S., Kim, H., & Jung, M. H. (2010). Consumption of barley beta-glucan ameliorates fatty liver and insulin resistance in mice fed a high-fat diet. *Molecular Nutrition & Food Research, 54*, 1004–1013.
Choromanska, A., Kulbacka, J., Rembialkowska, N., Pilat, J., Oledzki, R., Harasym, J., & Saczko, J. (2015). Anticancer properties of low molecular weight oat beta-glucan – An in vitro study. *International Journal of Biological Macromolecules, 80*, 23–28.
Chow, L., Lo, W., Loo, C. S., Hu, W., & Sham, X. J. (2003). Polysaccharide peptide mediates apoptosis by up-regulating p21 gene and down-regulating cyclin D1 gene. *The American Journal of Chinese Medicine, 31*, 1–9.
Colhoun, H. M., Betteridge, D. J., & Durrington, P. N. (2004). Primary prevention of cardiovascular disease with atorvastatin in type 2 diabetes in the Collaborative Atorvastatin Diabetes Study (CARDS): Multicentre randomised placebocontrolled trial. *Lancet, 364*, 685–696.
Cui, H., Liu, Q., Tao, Y., Zhang, H., Zhang, L., & Ding, K. (2008). Structure and chain conformation of a (1→6)-α-D-glucan from the root of *Pueraria lobata* (Willd.) Ohwi and the antioxidant activity of its sulfated derivative. *Carbohydrate Polymers, 74*(4), 771–778.

Cui, W., Wood, P. J., Blackwell, B., & Nikiforuk, J. (2000). Physicochemical properties and structural characterization by two dimensional NMR spectroscopy of wheat β-D-glucan-comparison with other β-D-glucans. *Carbohydrate Polymers, 41*, 249–258.

Cyran, M., Izydorczyk, M. S., & Macgregor, A. W. (2002). Structural characteristics of water-extractable nonstarch polysaccharides from barley malt. *Cereal Chemistry, 79*, 359–366.

da Silva Araújo, V. B., de Melo, A. N. F., Costa, A. G., Castro-Gomez, R. H., Madruga, M. S., de Souza, E. L., & Magnani, M. (2014). Followed extraction of β-glucan and mannoprotein from spent brewer's yeast (*Saccharomyces uvarum*) and application of the obtained mannoprotein as a stabilizer in mayonnaise. *Innovative Food Science and Emerging Technologies, 23*, 164–170.

Daou, C., & Zhang, H. (2012). Oat β-Glucan: Its role in health promotion and prevention of diseases. *Comprehensive Reviews in Food Science and Food Safety, 11*, 355–365.

Delatte, S. J., Evans, J., Hebra, A., Adamson, W., Othersen, H. B., & Tagge, E. P. (2001). Effectiveness of β-glucan collagen for treatment of partial thickness burns in children. *Journal of Pediatric Surgery, 36*, 113–118.

Din, A., Anjum, F. M., Zahoor, T., & Nawaz, H. (2009). Extraction and utilization of barley β-glucan for the preparation of functional beverage. *International Journal of Agriculture and Biology, 11*(6), 737–740.

Divyasri, D., Gunasekar, V., Benny, I. S., & Ponnusami, V. (2014). A review on industrial applications of curdlan. *International Journal of ChemTech Research, 6*, 3000–3003.

Dong, Q., Wang, Y., Shi, L., Yao, J., Li, J., & Ma, F. (2012). A novel water-soluble β-D-glucan isolated from the spores of *Ganoderma lucidum*. *Carbohydrate Research, 353*, 100–105.

Dongowski, G., Drzikova, B., Senge, B., Blochwitz, R., Gebhardt, E., & Habel, A. (2005). Rheological behaviour of β-glucan preparations from oat products. *Food Chemistry, 93*, 279–291.

Du, B., Bian, Z. X., & Xu, B. J. (2014). Skin health promotion effects of natural beta-glucan derived from cereals and microorganisms: A review. *Phytotherapy Research, 28*, 159–166.

Du, B., & Xu, B. (2014). Oxygen radical absorbance capacity (ORAC) and ferric reducing antioxidant power (FRAP) of β-glucans from different sources with various molecular weight. *Bioactive Carbohydrates and Dietary Fibre, 3*, 11–16.

European Food Safety Authority. (2010). Scientific Opinion on the substantiation of a health claim related to oat beta-glucan and lowering blood cholesterol and reduced risk of (coronary) heart disease pursuant to Article 14 of Regulation (EC) No 1924/2006. *EFSA Journal, 8*, 1885.

Falch, B. H., Espevik, T., Ryan, L., & Stokke, B. T. (2000). The cytokine stimulating activity of (1→3)-β-D-glucans is dependent on the triple helix conformation. *Carbohydrate Research, 329*, 587–596.

Fang, J., Wang, Y., Lv, X., Shen, X., Ni, X., & Ding, K. (2012). Structure of a β-glucan from *Grifola frondosa* and its antitumor effect by activating Dectin-1/Syk/NF-kB signaling. *Glycoconjugate Journal, 29*, 365–377.

Gracher, A. H. P., Cipriani, T. R., Carbonero, E. R., Gorin, P. A. J., & Iacomini, M. (2010). *Thrombosis Research, 126*, 180–187.

Granfeldt, Y., Hagander, B., & Bjoerck, I. (1995). Metabolic responses to starch in oat and wheat products. On the importance of food structure, incomplete gelatinisation, or presence of viscous dietary fiber. *European Journal of Clinical Nutrition, 49*, 189–199.

Gunde-Cimerman, N., & Plemenitas, A. (2001). Hypocholesterolemic activity of the genus pleurotus (Jacq.: Fr.) P. Kumm. (*Agaricales S. I., Basidiomycetes*). *International Journal of Medicinal Mushrooms, 3*, 395–397.

Halliwell, B. (2003). Free radicals and other reactive species in disease. *eLS*. https://doi.org/10.1038/npg.els.0002269

Handayani, D., Meyer, B. J., Chen, J. Z., Tang, P., Kwok, P. C. L., & Chan, H. K. (2012). The comparison of the effect of oat and shiitake mushroom powder to prevent body weight gain in rats fed high fat diet. *Food and Nutrition Sciences, 3*, 1009–1019.

Harman, D. (2006). Free radical theory of aging: An update. *Annals of the New York Academy of Sciences, 1067*, 10–21.

Hong, J. H., & Jung, H. K. (2014). Antioxidant and antitumor activities of beta-glucan-rich exopolysaccharides with different molecular weight from *Paenibacillus polymyxa* JB115. *Journal of Korean Society for Applied Biological Chemistry, 57*(1), 105–112.

Hossain, S., Hashimoto, M., Choudhury, E. K., Alam, N., Hussain, S., & Hasan, M. (2003). Dietary mushroom (*Pleurotus ostreatus*) ameliorates atherogenic lipid in hypercholesterolaemic rats. *Clinical and Experimental Pharmacology and Physiology, 30*, 470–475.

Huang, H. Y., Korivi, M., Chaing, Y. Y., Chien, T. Y., & Tsai, Y. C. (2012). *Pleurotus tuberregium* polysaccharides attenuate hyperglycemia and oxidativestress in experimental diabetic rats. *Evidence-based Complementary and Alternative Medicine, 2012*, 856381.

Huang, Q., & Zhang, L. (2011). Preparation, chain conformation and anti-tumor activities of water-soluble phosphated (1→3)-alpha-D-glucan from *Poria cocos* mycelia. *Carbohydrate Polymers, 83*, 1363–1369.

Ishibashi, K. I., Miura, N. N., Adachi, Y., Tamura, H., Tanaka, S., & Ohno, N. (2004). The solubilization and biological activities of *Aspergillus* β-(1→3)-d-glucan. *FEMS Immunology and Medical Microbiology, 42*, 155–166.

Izydorczyk, M. S., Storsley, J., Labossiere, D., MacGregor, A. W., & Rossnagel, B. G. (2000). Variation in total and soluble β-glucan content in hulless barley: Effects of thermal, physical, and enzymic treatments. *Journal of Agricultural and Food Chemistry, 48*, 982–989.

Jindal, M., Rana, V., Kumar, V., Singh, R. S., Kennedy, J. F., & Tiwary, A. K. (2013). Sulfation of *Aegle marmelos* gum: Synthesis, physico-chemical and functional characterization. *Carbohydrate Polymers, 92*, 1660–1668.

Johansson, L., Tuomainen, P., Ylinen, M., Ekholm, P., & Virkki, L. (2004). Structural analysis of water-soluble and -insoluble β-glucans of whole-grain oats and barley. *Carbohydrate Polymers, 58*, 267–274.

Johansson, L., Virkki, L., Maunu, S., Lehto, M., Ekholm, P., & Varo, P. (2000). Structural characterization of water soluble β-glucan of oat bran. *Carbohydrate Polymers, 42*, 143–148.

Jung, H. K., Park, S. C., Park, B. K., & Hong, J. H. (2008). Physiological activities of a β-glucan produced by *Panebacillus polymyxa*. *Biotechnology Letters, 30*, 1545–1551.

Kagimura, F. Y., da Cunha, M. A. A., Barbosa, A. M., Dekker, R. F. H., & Malfatti, C. R. M. (2015). Biological activities of derivatized D-glucans: A review. *International Journal of Biological Macromolecules, 72*, 588–598.

Kanauchi, M., & Bamforth, C. W. (2001). Release of β-glucan from cell walls of starchy endosperm of barley. *Cereal Chemistry, 78*(2), 121–124.

Ke, L. Q. (2014). Optimization of ultrasonic extraction of polysaccharides from *Lentinus edodes* based on enzymatic treatment. *Journal of Food Processing and Preservation, 39*, 254–259.

Kerckhoffs, D., Hornstra, G., & Mensink, R. (2003). Cholesterol-lowering effect of beta-glucan from oat bran in mildly hypercholesterolemic subjects may decrease when beta-glucan is incorporated into bread and cookies. *American Journal of Clinical Nutrition, 78*, 221–227.

Ketkeaw, R., Oungbho, K., & Wititsuwannakul, R. (2012). The β-glucan from *Hevea brasiliensis* latex and its possible application in anti-aging cosmeceuticals. In *38th congress on science and technology of Thailand*.

Khan, A. A., Gani, A., Masoodi, F. A., Mushtaq, U., & Naik, A. S. (2017). Structural, rheological, antioxidant, and functional properties of β–glucan extracted from edible mushrooms *Agaricus bisporus, Pleurotus ostreatus* and *Coprinus attrimentarius*. *Bioactive Carbohydrates and Dietary Fibre, 11*, 67–74.

Kiho, T., Morimoto, H., Sakushima, M., Usui, S., & Ukai, S. (1995). Polysaccharides in fungi. XXXV. Anti diabetic activity of an acidic polysaccharide from the fruiting bodies of *Tremella aurantia*. *Biological and Pharmaceutical Bulletin, 18*(12), 1627–1629.

Kim, H. L., Lee, J. H., Lee, M. H., Kwon, B. J., & Park, J. C. (2012). Evaluation of electrospun (1,3)-(1,6)-β-D-glucans/biodegradable polymer as artificial skin for full thickness wound healing. *Tissue Engineering Part A, 18*, 2315–2322.

Kim, Y. W., Kim, K. H., Choi, H. J., & Lee, D. S. (2005). Antidiabetic activity of β-glucans and their enzymatically hydrolyzed oligosaccharides from *Agaricus blazei*. *Biotechnological Letters, 27*, 483–487.

Kodama, S., Shoda, T., Machmudah, S., Wahyudiono, W., Kanda, H., & Goto, M. (2015). Enhancing pressurized water extraction of β-glucan from barley grain by adding CO_2 under hydrothermal conditions. *Chemical Engineering and Processing: Process Intensification, 97*, 45–54.

Kofuji, K., Huang, Y., Tsubaki, K., Kokido, F., Nishikawa, K., & Isobe, T. (2010). Preparation and evaluation of a novel wound dressing sheet comprised of β-glucan-chitosan complex. *Reactive and Functional Polymers, 70*, 784–789.

Kogan, G., Pajtinka, M., Babincova, M., Miadokova, E., Rauko, P., Slamenova, D., & Korolenko, T. A. (2008). Yeast cell wall polysaccharides as antioxidants and antimutagens: Can they fight cancer? *Neoplasma, 55*, 387–393.

Kogan, G., Xandula, J., Korolenko, T. A., Falameeva, O. V., Poteryaeva, O. N., Zhanaeva, S. Y., … Kaledin, V. I. (2002). *International Immunopharmacology, 2*, 775–781.

Kozarski, M., Klaus, A., Niksic, M., Jakovljevic, D., Helsper, J. P. F. G., & Van Griensven, L. J. L. D. (2011). Antioxidative and immunomodulating activities of polysaccharide extracts of the medicinal mushrooms *Agaricus bisporus, Agaricus brasiliensis, Ganoderma lucidum* and *Phellinus linteus*. *Food Chemistry, 129*, 1667–1675.

Krizkova, L., Durackova, Z., Sandula, J., Slamenova, D., Sasinkova, V., & Sivonova, M. (2002). Fungal beta-(1-3)-D-glucan derivatives exhibit high antioxidative and antimutagenic activity in vitro. *Anticancer Research, 23*, 2751–2756.

Kubala, L., Ruzickova, J., Nickova, K., Sandula, J., Ciz, M., & Lojek, A. (2003). The effect of (1→3)-β-D-glucans, carboxymethylglucan and schizophyllan on human leukocytes in vitro. *Carbohydrate Research, 338*, 2835–2840.

Kumari, M., Survase, S. A., & Singhal, R. S. (2008). Production of schizophyllan using Schizophyllum commune NRCM. *Bioresource Technology, 99*, 1036–1043.

Laroche, C., & Michaud, P. (2007). New developments and prospective for β-(1-3)-glucans. *Recent Patents on Biotechnology, 1*, 59–73.

Lazaridou, A., & Biliaderis, C. G. (2007). Molecular aspects of cereal β-glucan functionality: Physical properties, technological applications and physiological effects. *Journal of Cereal Science, 46*(2), 101–118.

Lazaridou, A., Biliaderis, C. G., & Izydorczyk, M. S. (2003). Molecular size effects on rheological properties of oat beta-glucans in solution and gels. *Food Hydrocolloids, 17*, 693–712.

Lazaridou, A., Serafeimidou, A., Biliaderis, C. G., Moschakis, T., & Tzanetakis, N. (2014). Structure development and acidification kinetics in fermented milk containing oat β-glucan, a yogurt culture and a probiotic strain. *Food Hydrocolloids, 39*, 204–214.

Li, B., Dobruchowska, J. M., Gerwig, G. J., Dijkhuizen, L., & Kamerling, J. P. (2013a). Structural investigation of water-soluble polysaccharides extracted from the fruit bodies of *Coprinus comatus*. *Carbohydrate Polymers, 91*, 314–321.

Li, C., Ha, T., Kelley, J., Gao, X., Qiu, Y., & Kao, R. L. (2004a). Modulating Toll-like receptor mediated signaling by (1→3)-β-d-glucan rapidly induces cardioprotection. *Cardiovascular Research, 61*, 538–547.

Li, F., Zhang, Y., & Zhong, Z. (2011). Antihyperglycemic effect *of Ganoderma lucidum* polysaccharides on streptozotocin-induced diabetic mice. *International Journal of Molecular Sciences, 12*, 6135–6145.

Li, G., Kim, D., Kim, T., & Park, B. (2004b). Protein-bound polysaccharide from *Phellinus linteus* induces G2/M phase arrest and apoptosis in SW480 human colon cancer cells. *Cancer Letters, 216*, 175–181.

Li, K., Liu, S., Xing, R., Qin, Y., & Li, P. (2013b). Preparation, characterization and antioxidant activity of two partially N-acetylated chitotrioses. *Carbohydrate Polymers, 92*(2), 1730–1736.

Li, W., Cu, S. W., & Kakuda, Y. (2006). Extraction, fractionation, structural and physical characterization of wheat β-d-glucans. *Carbohydrate Polymers, 63*(3), 408–416.

Liatis, S., Tsapogas, P., Chala, E., Dimosthenopoulos, C., Kyriakopoulos, K., Kapantais, E., & Katsilambros, N. (2009). The consumption of bread enriched with β-glucan reduces LDL-cholesterol and improves insulin resistance in patients with type 2 diabetes. *Diabetes & Metabolism, 35*, 115–120.

Limberger-Bayer, V. M., de Francisco, A., Chan, A., Oro, T., Ogliari, P. J., & Barreto, P. L. M. (2014). Barley β-glucans extraction and partial characterization. *Food Chemistry, 154*, 84–89.

Lin, Y., Zhang, L., Chen, L., Jin, Y., Zeng, F., Jin, J., … Cheung, P. C. K. (2004). Molecular mass and antitumor activities of sulphated derivatives of α-glucan from *Porio cocos* mycelia. *International Journal of Biological Macromolecules, 34*, 231–236.

Lo, H. C., Tsai, F. A., Wasser, S. P., Yang, J. G., & Huang, B. M. (2006). Effects of ingested fruiting bodies, submerged culture biomass, and acidic polysaccharide glucuronoxylomannan of *Tremella mesenterica* Retz.: Fr. on glycemic responses in normal and diabetic rats. *Life Sciences, 78*(17), 1957–1966.

Lo, T. C. T., Tsao, H. H., Wang, A. Y., & Chang, C. A. (2007). Pressurized water extraction of polysaccharides as secondary metabolites from *Lentinula edodes*. *Journal of Agricultural and Food Chemistry, 55*, 4196–4201.

Lu, X., Mo, X., Guo, H., & Zhang, Y. (2012). Sulfation modification and anticoagulant activity of the polysaccharides obtained from persimmon (*Diospyros kaki* L.) fruits. *International Journal of Biological Macromolecules, 51*(5), 1189–1195.

Luhm, J., Langenkamp, U., Hensel, J., Frohn, C., Brand, J. M., Hennig, H., … Mueller, A. (2006). β-(1→3)-D-Glucan modulates DNA binding of nuclear factors kB, AT and IL-6 leading to an anti-inflammatory shift of the IL-Ib/IL-1 receptor antagonist ratio. *BMC Immunology, 7*, 1–15.

Lyly, M., Salmenkallio-Marttila, T., Suortti, K., Autio, K., Poutanen, K., & Lahteenmaki, L. (2004). The sensory characteristics and rheological properties of soups containing oat and barley β-glucan before and after freezing. *Lebensmittel-Wissenschaft und Technologie, 37*, 749–761.

Ma, C. W., Feng, M., Zhai, X., Hu, M., You, L., Luo, W., & Zhao, M. (2013). Optimization for the extraction of polysaccharides from *Ganoderma lucidum* and their antioxidant and antiproliferative activities. *Journal of the Taiwan Institute of Chemical Engineers, 44*, 886–894.

Magnani, M., Calliari, C. M., Macedo, F. C., Jr., Mori, M. P., Cólus, I. M. S., & Castro-Gomez, R. J. H. (2009). Optimized methodology for extraction of (1→3)(1→6)-β-D-glucan from *Saccharomyces cerevisiae* and in vitro evaluation of the cytotoxicity and genotoxicity of the corresponding carboxymethyl derivative. *Carbohydrate Polymers, 78*, 658–665.

Maity, P., Samanta, S., Nandi, A. K., Sen, I. K., Paloi, S., Acharya, K., & Islam, S. S. (2014). Structure elucidation and antioxidant properties of a soluble β-d-glucan from mushroom *Entoloma lividoalbum*. *International Journal of Biological Macromolecules, 63*, 140–149.

Maran, J. P., & Priya, B. (2014). Ultrasound-assisted extraction of polysaccharide from *Nephelium lappaceum* L. fruit peel. *International Journal of Biological Macromolecules, 70*, 530–536.

May, L. D., Lefkowitch, J. H., Kram, M. T., & Rubin, D. E. (2002). Mixed hepatocellular cholestatic liver injury after pioglitazone therapy. *Annals of Internal Medicine, 136*, 449–452.

Meena, D. K., Das, P., Kumar, S., Mandal, S. C., Prusty, A. K., & Singh, S. K. (2013). Beta-glucan: An ideal immunostimulant in aquaculture (a review). *Fish Physiology and Biochemistry, 39*, 431–457.

Miura, N. N., Adachi, Y., Yadomae, T., Tamura, H., Tanaka, S., & Ohno, N. (2003). Structure and biological activities of β-glucans from yeast and mycelial forms of *Candida albicans*. *Microbiology and Immunology, 47*, 173–182.

Moreno, R. B., Ruthes, A. C., Baggio, C. H., Vilaplana, F., Komura, D. L., & Iacomini, M. (2016). Structure and antinociceptive effects of β-D-Glucans from *Cookeina tricholoma*. *Carbohydrate Polymers, 141*, 220–228.

Morgan, K. R., & Ofman, D. J. (1998). Glucagel, a gelling β-glucan from barley. *Cereal Chemistry, 75*, 879–881.

Murphy, E. A., Davis, J. M., Brown, A. S., Carmichael, M. D., Mayer, E. P., & Ghaffar, A. (2004). Effects of moderate exercise and oat β-glucan on lung tumor metastases and macrophage antitumor cytotoxicity. *Journal of Applied Physiology, 97*, 955–959.

Nandi, F. A. K., Samanta, S., Maity, S., Sen, I. K., Khatua, S., Devi, K. S. P., & Islam, S. S. (2014). Antioxidant and immunostimulant β-glucan from edible mushroom *Russula albonigra* (Krombh.). *Carbohydrate Polymers, 99*, 774–782.

Nguyen, T. H., Fleet, G. H., & Rogers, P. L. (1998). Composition of the cell walls of several yeast species. *Applied Microbiology and Biotechnology, 50*, 206–212.

Novak, M., & Vetvicka, V. (2009). Glucans as biological response modifiers. *Endocrine, Metabolic & Immune Disorders - Drug Targets, 9*, 67–75.

Ooi, V. E. C., & Liu, F. (2000). Immunomodulation and anti-cancer activity of polysaccharide-protein complexes. *Current Medicinal Chemistry, 7*, 715–729.

Ookushi, Y., Sakamoto, M., & Azuma, J. (2006). Optimization of microwave-assisted extraction of polysaccharides from the fruiting body of mushrooms. *Journal of Applied Glycoscience, 53*, 267–272.

Palanisamy, M., Aldars Garcıa, L., Gil Ramırez, A., Ruiz Rodrıguez, A., Marın, F. R., & Reglero, G. (2014). Pressurized water extraction of β-glucan enriched fractions with bile acids binding capacities obtained from edible mushrooms. *Biotechnology Progress, 30*, 391–400.

Papageorgiou, M., Lakhdara, N., Lazaridou, A., Biliaderisd, C. G., & Izydorczyk, M. S. (2005). Water extractable (1→3) (1→4)-β-d-glucans from barley and oats: An intervarietal study on their structural features and rheological behaviour. *Journal of Cereal Science, 42*, 213–224.

Park, K. M., Park, B. H., So, S., Kim, M. S., Kim, J. S., & Kim, Y. T. (2001). *Composition for external application containing a beta-1,6-branched-beta-1,3-glucan*. US Patent 0029253 A1.

Pillai, R., Redmond, M., & Roding, J. (2005). Anti-wrinkle therapy: Significant new findings in the non-invasive cosmetic treatment of skin wrinkles with beta-glucan. *International Journal of Cosmetic Science, 27*(5), 292.

Pintado, T., Herrero, A. M., Jiménez-Colmenero, F., & Ruiz-Capillas, C. (2016). Emulsion gels as potential fat replacers delivering β-glucan and healthy lipid content for food applications. *Journal of Food Science and Technology, 53*(12), 4336–4347.

Pomeroy, S., Tupper, R., Cehun-Aders, M., & Nestel, P. (2001). Oat β-glucan lowers total and LDL-cholesterol. *Australian Journal of Nutrition and Dietetics, 58*, 51–55.

Popescu, C., Larsson, P. T., Olaru, N., & Vasile, C. (2012). Spectroscopic study of acetylated kraft pulp fibers. *Carbohydrate Polymers, 88*, 530–536.

Qi, H., Liu, X., Zhang, J., Duan, Y., Wang, X., & Zhang, Q. (2012). Synthesis and antihyperlipidemic activity of acetylated derivative of ulvan from *Ulva pertusa*. *International Journal of Biological Macromolecules, 50*(1), 270–272.

Rahmanian, N., Jafari, S. M., Wani, T. A. (2015). Bioactive profile, dehydration, extraction and application of the bioactive components of olive leaves. *Trends in Food Science & Technology, 42*, 150–172.

Raveendran, S., Yoshida, Y., Maekawa, T., & Kumar, D. S. (2013). Pharmaceutically versatile sulfated polysaccharide based bionano platforms. *Nanomedicine: Nanotechnology, Biology, and Medicine, 9*(5), 605–626.

Rieder, A., Ballance, S., Lovaas, A., & Knutsen, S. H. (2015). Minimizing molecular weight reduction of β-glucan during barley bread making. *LWT - Food Science and Technology, 64*(2), 767–774.

Ringo, E., & Song, S. K. (2016). Application of dietary supplements (synbiotics and probiotics in combination with plant products and β-glucans) in aquaculture. *Aquaculture Nutrition, 22*, 4–24.

Rop, O., Mlcek, J., & Jurikova, T. (2009). Beta-glucans in higher fungi and their health effects. *Nutrition Reviews, 67*, 624–631.

Sakurai, K., Mizu, M., & Shinkai, S. (2001). Polysaccharide-polynucleotide complexes. 2. Complementary polynucleotide mimic behavior of the natural polysaccharide schizophyllan in the macromolecular complex with single-stranded RNA and DNA. *Biomacromolecules, 2*, 641–650.

Sarangi, I., Ghosh, D., Bhutia, S. K., Mallick, S. K., & Maiti, T. K. (2006). Anti-tumor and immunomodulating effects of *Pleurotus ostreatus* mycelia-derived proteoglycans. *International Immunopharmacology, 6*(8), 1287–1297.

Schmid, F., Stone, B. A., McDougall, B. M., Bacic, A., Martin, K. L., & Brownlee, R. T. (2001). Structure of epiglucan a highly side-chain/branched (1→3; 1→6)-β-glucan from the micro fungus *Epicoccum nigrum* Ehrenb. ex Schlecht. *Carbohydrate Research, 331*, 163–171.

Schneider, I., Kressel, G., Meyer, A., Krings, U., Berger, R. G., & Hahn, A. (2011). Lipid lowering effects of oyster mushroom (*Pleurotus ostreatus*) in humans. *Journal of Functional Foods, 3*, 17–24.

Sener, G., Toklu, H., Ercan, F., & Erkanlı, G. (2005). Protective effect of β-glucan against oxidative organ injury in a rat model of sepsis. *International Immunopharmacology, 5*, 1387–1396.

Sener, G., Toklu, H. Z., & Cetinel, S. (2007). β-Glucan protects against chronic nicotine-induced oxidative damage in rat kidney and bladder. *Envrionmental Toxicology and Pharmacology, 23*, 25–32.

Shao, L. L., An, Q. F., Ji, Y. L., Zhao, Q., Wang, X. S., Zhu, B. K., & Gao, C. J. (2014). Preparation and characterization of sulfated carboxymethyl cellulose nanofiltration membranes with improved water permeability. *Desalination, 338*, 74–83.

Shih, I. L., Yu, J. Y., Hsieh, C., & Wu, J. Y. (2009). Production and characterization of curdlan by *Agrobacterium sp. Biochemical Engineering Journal, 43*, 33–40.

Skendi, A., Biliaderis, C. G., Lazaridou, A., & Izydorczyk, M. S. (2003). Structure and rheological properties of water soluble β-glucans from oat cultivars of *Avena sativa* and *Avena bysantina*. *Journal of Cereal Science, 38*, 15–31.

Sourki, A. H., Koocheki, A., & Elahi, M. (2017). Ultrasound-assisted extraction of β-d-glucan from hull-less barley: Assessment of physicochemical and functional properties. *International Journal of Biological Macromolecules, 95*, 462–475.

Stosic-Grujicic, S., Mijatovic, S., & Maksimovic-Ivanic, D. (2011). An evidence-based perspective of *Coriolus Versicolor* (multicolored polypore mushroom) for cancer patients. In C. S. William (Ed.), *Evidence-based anticancer materia medica* (pp. 265–302). Dordrecht, The Netherlands: Springer.

Sun, Y., Sun, T., Wang, F., Zhang, J., Li, C., Chen, X., & Sun, S. (2013). A polysaccharide from the fungi of Huaier exhibits anti-tumor potential and immunomodulatory effects. *Carbohydrate Polymers, 92*, 577–582.

Synytsya, A., & Novak, M. (2013). Structural diversity of fungal glucans. *Carbohydrate Polymers, 92*, 792–809.

Tao, Y., & Xu, W. (2008). Microwave-assisted solubilization and solution properties of hyperbranched polysaccharide. *Carbohydrate Research, 343*, 3071–3078.

Tao, Y. Z., Zhang, L. N., & Cheung, P. C. K. (2006). Physicochemical properties and antitumor activities of water-soluble native and sulfated hyperbranched mushroom polysaccharides. *Carbohydrate Research, 341*, 2261–2269.

Tapola, N., Harvonen, H., Niskamen, L., Mikola, M., & Sarkkinen, E. (2005). Glycemic responses of oat bran products in type 2-diabetic patients. *Nutrition, Metabolism and Cardiovascular Diseases, 15*, 255–261.

Temelli, F., Bansema, C. B., & Stobbe, K. S. (2004). Development of an orange-flavored barley β-glucan beverage with added whey protein isolate. *Journal of Food Science, 69*(7), 237–242.

Thammakiti, S., Suphantharika, M., Phaesuwan, T., & Verduyn, C. (2004). Preparation of spent brewer's yeast β-glucans for potential applications in the food industry. *International Journal of Food Science and Technology, 39*, 21–29.

Theuwissen, E., & Mensink, R. P. (2007). Simultaneous intake of beta-glucan and plant stanol esters affects lipid metabolism in slightly hypercholesterolemic subjects. *The Journal of Nutrition, 137*, 583–588.

Tian, Y., Zeng, H., Xu, Z., Zheng, B., Lin, Y., & Gan, C. (2012). Ultrasonic-assisted extraction and antioxidant activity of polysaccharides recovered from white button mushroom (*Agaricus bisporus*). *Carbohydrate Polymers, 88*, 522–529.

Toklu, H. Z., Sener, G., Jahovic, N., Uslu, B., Arbak, S., & Yegen, B. C. (2006). Beta-glucan protects against burn-induced oxidative organ damage in rats. *International Immunopharmacology, 6*, 156–169.

Tokunaka, K., Ohno, N., Adachi, Y., Tanaka, S., Tamura, H., & Yadomae, T. (2000). Immunopharmacological and immunotoxicological activities of a water soluble (1→3)-β-glucan, CSBG from *Candida albicans*. *International Journal of Immunopharmacology, 22*, 383–394.

Tosh, S. M. (2013). Review of human studies investigating the post-prandial blood-glucose lowering ability of oat and barley food products. *European Journal of Clinical Nutrition, 67*(4), 310–317.

Tosh, S. M., Brummer, Y., Miller, S. S., Regand, A., Defelice, C., Duss, R., & Wood, P. J. (2010). Processing affects the physicochemical properties of β-glucan in oat bran cereal. *Journal of Agricultural and Food Chemistry, 58*(13), 7723–7730.

Tosh, S. M., Brummer, Y., Wood, P. J., Wang, Q., & Weisz, J. (2004). Evaluation of structure in the formation of gels by structurally diverse (1→3)(1→4)-beta-D-glucans from four cereal and one lichen species. *Carbohydrate Polymers, 57*, 249–259.

Tranquilan-Aranilla, C., Nagasawa, N., Bayquen, A., & Dela Rosa, A. (2012). Synthesis and characterization of carboxymethyl derivatives of kappa-carrageenan. *Carbohydrate Polymers, 87*(2), 1810–1816.

Tungland, B. C., & Meyer, D. (2002). Non-digestible oligo- and polysaccharides (dietary fiber): Their physiology and role in human health and food. *Comprehensive Reviews in Food Science and Food Safety, 1*, 90–109.

Udayangani, R. M. C., Dananjaya, S. H. S., Fronte, B., Kim, C. H., Lee, J., & De Zoysa, M. (2017). Feeding of nano scale oats beta-glucan enhances the host resistance against Edwardsiella tarda and protective immune modulation in zebrafish larvae. *Fish and Shellfish Immunology, 60*, 72–77.

US Food and Drug Administration. (2008). Food labeling: Health claims; soluble fiber from certain foods and risk of coronary heart disease. *Federal Register, 73*, 47828–47829.

Venkatachalam, G., Narayanan, S., & Doble, M. (2013). Applications of cyclic β-glucans. In *Cyclic β-glucans from microorganisms: Production, properties and applications*. Berlin. Germany: Springer.

Vetvicka, V., & Vetvickova, J. (2007). Physiological effects of different types of β-glucan. *Biomedical Papers of the Medical Faculty of the University Palacky, Olomouc, Czech Republic, 151*, 225–231.

Villares, A., Mateo-Vivaracho, L., & Guillamon, E. (2012). Structural features and healthy properties of polysaccharides occurring in mushrooms. *Agriculture, 2*, 452–471.

Wang, J., Yuan, Y., & Yue, T. (2014). Immunostimulatory activities of β-d-glucan from *Ganoderma lucidum*. *Carbohydrate Polymers, 102*, 47–54.

Wang, J. T., Wang, Q., & Han, J. R. (2013). Yield, polysaccharides content and antioxidant properties of the mushroom *Agaricus subrufescens* produced on different substrates based on selected agricultural wastes. *Scientia Horticulturae, 157*, 84–89.

Wang, Q., Wood, P., Huang, X., & Cui, W. (2003). Preparation and characterization of molecular weight standards of low polydispersity from oat and barley(1→3)(1→4)-β-d-glucan. *Food Hydrocolloids, 17*(6), 845–853.

Wang, Q., Wood, P. J., & Cui, W. (2002). Microwave assisted dissolution of β-glucan in water—Implications for the characterisation of this polymer. *Carbohydrate Polymers, 47*, 35–38.

Wang, T., Deng, L., Li, S., & Tan, T. (2007). Structural characterization of a water-insoluble (1→3)-α-d-glucan isolated from the *Penicillium chrysogenum*. *Carbohydrate Polymers, 67*, 133–137.

Wang, Y., Yu, Y., & Mao, J. (2009). Carboxymethylated β-glucan derived from *Porio cocos* with biological activities. *Journal of Agricultural and Food Chemistry, 57*, 10913–10915.

Wani, T. A., Gani, A., Wani, S. M., Masoodi, F. A., Wani, I. A., Baba, N., & Shagoo, M. A. (2016). Suitability of Different Food Grade Materials for the Encapsulation of Some Functional Foods Well Reported for Their Advantages and Susceptibility. *Critical Reviews in Food Science and Nutrition, 56*, 2431–2454.

Wasser, S. (2002). Medicinal mushrooms as a source of antitumor and immunomodulating polysaccharides. *Applied Microbiology and Biotechnology, 60*, 258–274.

Wiater, A., Paduch, R., Choma, A., Pleszczynska, M., Siwulski, M., Dominik, J., … Szczodrak, J. (2012). Biological study on carboxymethylated (1→3)-α-D-glucans from fruiting bodies of *Ganoderma lucidum. International Journal of Biological Macromolecules, 51*, 1014–1023.

Wilson, T. A., Nicolosi, R. J., Delaney, B., Chadwell, K., Moolchandani, V., Kotyla, T., & Ostergren, T. (2004). Reduced and high molecular weight barley β-glucans decrease plasma total and non-HDL-cholesterol in hypercholesterolemic Syrian golden hamsters. *The Journal of Nutrition, 134*(10), 2617–2622.

Wood, P. J. (1986). Oat β-glucan: Structure, location and properties. In F. H. Webster (Ed.), *Oats: Chemistry and technology* (pp. 121–152). St. Paul, MN: American Association of Cereal Chemists.

Wood, P. J. (2007). Cereal β-glucans in diet and health. *Journal of Cereal Science, 46*, 230–238.

Wood, P. J., Beer, M. U., & Butler, G. (2000). Evaluation of the role of concentration and molecular weight of oat b-glucan in determining effect of viscosity on plasma glucose and insulin following an oral glucose load. *British Journal of Nutrition, 84*, 19–23.

Wood, P. J., Weisz, J., & Blackwell, B. A. (1991a). Molecular characterization of cereal β-D-glucans. Structural analysis of oat β-D-glucan and rapid structural evaluation of β-D-glucans from different sources by high-performance liquid chromatography of oligosaccharides released by lichenase. *Cereal Chemistry, 68*, 31–39.

Wood, P. J., Weisz, J., & Mahn, W. (1991b). Molecular characterisation of cereal β-glucans. II. Size-exclusion chromatography for comparison of molecular weight. *Cereal Chemistry, 68*, 530–536.

Wu, Y. S., Chen, S., Wang, W., Lu, C. L., Liu, C. F., & Chen, S. N. (2014). Oral administration of MBG to modulate immune responses and suppress OVA-sensitized allergy in a murine model. *Evidence-Based Complementary and Alternative Medicine, 2014*, 567427.

Xiao, Z., Trincado, C. A., & Murtaugh, M. P. (2004). β-Glucan enhancement of T cell IFNγ response in swine. *Veterinary Immunology and Immunopathology, 102*, 315–320.

Yang, L., & Zhang, L. (2009). Chemical structural and chain conformational characterization of some bioactive polysaccharides isolated from natural sources. *Carbohydrate Polymers, 76*, 349–361.

Yang, X., Guo, D., Zhang, J., & Wu, M. (2007). Characterization and anti-tumor activity of pollen polysaccharide. *International Immunopharmacology, 7*, 427–434.

Ye, L., Xu, L., & Li, J. (2012). Preparation and anticoagulant activity of a fucosylated polysaccharide sulfate from a sea cucumber *Acaudina molpadioidea. Carbohydrate Polymers, 87*, 2052–2057.

Ye, M., Yuan, R., He, Y., Du, Z., & Ma, X. (2013). Phosphorylation and anti-tumor activity of exopolysaccharide from *Lachnum* YM120. *Carbohydrate Polymers, 97*(2), 690–694.

Yokoyama, W. H., & Shao, Q. (2006). Soluble fibers prevent insulin resistance in hamsters fed high saturated fat diets. *Cereal Foods World, 51*, 16–18.

Yoo, D. H., Lee, B. H., Chang, P. S., Lee, H. G., & Yoo, S. H. (2007). `Improved quantitative analysis of oligosaccharides from lichenase-hydrolyzed water-soluble barley β-glucans by high-performance anion-exchange chromatography. *Journal of Agricultural and Food Chemistry, 55*, 1656–1662.

Zekovic, D. B., Kwiatkowski, S., Vrvic, M. M., Jakovljevic, D., & Moran, C. A. (2005). Natural modified (1→3)-β-glucans in health promotion and disease alleviation. *Critical Reviews in Biotechnology, 25*, 205–230.

Zhang, M., Cheung, P. C. K., Chiu, L. C. M., Wong, E. Y. L., & Ooi, V. E. C. (2006). Cell cycle arrest and apoptosis induction in human breast carcinoma MCF-7 cells by carboxymethylated β-glucan from the mushroom sclerotia of *Pleurotus tuberregium*. *Carbohydrate Polymers, 66*, 455–462.

Zhang, M., Cheung, P. C. K., & Zhang, L. (2001). Evaluation of mushroom dietary fiber (non-starch polysaccharides) from sclerotia of *Pleurotus tuberregium* (Fries) Singer as a potential antitumor agent. *Journal of Agricultural and Food Chemistry, 49*, 5059–5062.

Zhang, M., Cui, S. W., Cheung, P. C. K., & Wang, Q. (2007). Antitumor polysaccharides from mushrooms: A review on their isolation process, structural characteristics and antitumor activity. *Trends in Food Science and Technology, 18*, 4–19.

Zhang, Z., Wang, X., Mo, X., & Qi, H. (2013). Degradation and the antioxidant activity of polysaccharide from *Enteromorpha linza*. *Carbohydrate Polymers, 92*, 2084–2087.

Zhang, Z., Zhang, Q., Wang, J., Shi, X., Song, H., & Zhang, J. (2009). In vitro antioxidant activities of acetylated, phosphorylated and benzoylated derivatives of porphyran extracted from *Porphyra haitanensis*. *Carbohydrate Polymers, 78*, 449–453.

Zhu, F., Du, B., & Xu, B. (2016). A critical review on production and industrial applications of beta-glucans. *Food Hydrocolloids, 52*, 275–288.

Zong, A., Cao, H., & Wang, F. (2012). Anticancer polysaccharides from natural resources: A review of recent research. *Carbohydrate Polymers, 90*(4), 1395–1410.

Pectin

Nairah Noor, Asima Shah, Asir Gani, Adil Gani, Faiza Jhan, Zanoor ul Ashraf, Bilal Ahmad Ashwar, and Tariq Ahmad Ganaie

Introduction

Pectin was discovered in apple juice by Vauque in 1790. The name has been derived from the Greek word *pektikos* meaning to congeal or solidify and has been coined by (Braconnot 1825a, b). However the production of pectin originated in the 1900s in Germany where an apple juice processor tried to cook dried apple pomace, the by-product from apple juice processing. Since then, pectin has been studied and used extensively by scientists from different areas of science. In plant tissues, it is linked to the other cell wall constituents including celluloses or hemicelluloses and plays an important role in the plant growth and development. It forms an important component in the young tissues of fruits and vegetables. It is mostly abundant in the middle lamella layers between adjoining plant cells. Next to this, the primary cell wall also exhibits a high amount of pectin. From the secondary cell wall of plants towards the plasma, the concentration of the pectin goes on decreasing or is even completely absent. Chemically, pectin is an anionic polysaccharide made up of covalently linked α-1, 4-d-galacturonic acid (also known as galactosyluronic acid), galacturonic acid constituting the main monomeric unit of the complex pectin molecule. The sugar acid, D-galacturonic acid (GalA) is an oxidized form of D-galactose, and is present mostly as homogalacturonan (HG) and rhamnogalacturonan I (RG-I) structural elements. These units though linked by a-1/4 galacturonosyl linkage, are often interrupted by L-rhamnose units bearing side-chains which result in discontinuity in the (GalA) polymer chain (Fig. 1).

As far as the sources of pectin are concerned, pectin is available in large number of plant species but its commercial sources are few. The major source of pectin has been apple pomace but nowadays use of citrus peel (Brejnholt 2009) is also

N. Noor · A. Shah · A. Gani (✉) · F. Jhan · Z. ul Ashraf · B. A. Ashwar · T. A. Ganaie
Department of Food Science and Technology, University of Kashmir,
Srinagar, Jammu and Kashmir, India

A. Gani, B. A. Ashwar (eds.), *Food biopolymers: Structural, functional and nutraceutical properties*, https://doi.org/10.1007/978-3-030-27061-2_6

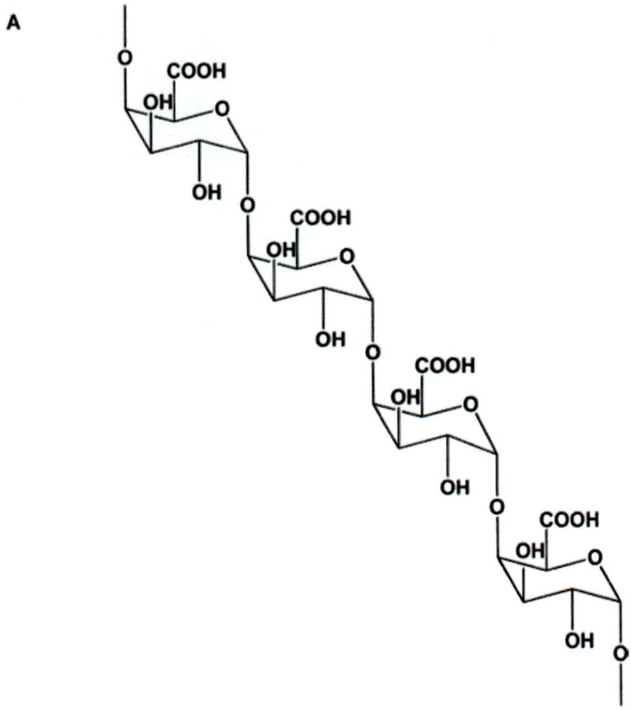

Fig. 1 Pectin structure depicting covalently -1,4-linked d-galacturonic acid (GalA) units. (Source: Chan et al. 2017)

increasing. Commercial pectin is available mostly from citrus peel (85.5%), followed by apple pomace (14.0%) and a little amount is also obtained from sugar beet pulp (0.5%) (Ciriminna et al. 2015). The main source of citrus peel is in turn obtained from juices and essential oil production, while sugar beet pulp is obtained from the sugar industry (Yapo et al. 2007). For proper utilization of pectin, the production process can be divided into three stages: pre-treatment stage, an extraction operation, and a post-extraction stage (Gentilini et al. 2014: Lopes da Silva and Rao 2006; May 1990).

The physicochemical properties of pectin vary from source to source. As pectin is a high molecular weight polysaccharide, its use is very common in the food industry and has mostly been exploited for its physicochemical properties like, gelling, viscosity, emulsification and thickening. The interaction of pectin with water involves the inclusion of water in the capillaries, network voids and junction zones formed by gelling. The most striking feature of the pectin is its gel forming ability. When the pectin molecule is partially dehydrated in presence of acid and sugar it forms gel that is an intermediate between solution and precipitation (May 1990). The properties of pectin which are responsible for its gel strength include molecular weight and degree of esterification. This correlation can be explained that high methoxyl pectins form quick gel solutions gel whereas low methoxyl pectin gels by the so-called egg box structure. In the latter, there is an ionic interaction between calcium ions and oxygen group of the adjoining chains, which results in the cross linking of the chains. There is a positive correlation between molecular weight and gel strength (Whistler and Daniel 1993). This gelling property is employed in jams and jellies while as the other properties allow pectin to be used as a stabilizer in fruit juices and acidified milk drinks, protein rich fruit drinks, and fortified anti-oxidant foods and as a fat replacer in emulsified meat products. Apart from these uses, it has been seen as a target encapsulating agent for the drug delivery, wound healing ability and tissue engineering. Further its use in healthcare, pharmaceuticals and cosmetics is also growing owing to its, non toxicity, biocompatibility and biodegradability. It has been reported to decrease blood fat, inhibit the lipase activity, and reduce the occurrence of heart diseases and growth and proliferation of cancer cells. However the growing demand in the market sometimes requires alterations or modifications in the pectin macromolecule in order to tailor the needs. Accordingly, the backbone of the pectin constituting the hydroxyl and the carboxyl groups and the side chains constituting neutral sugars can be altered to prepare the broad range of derivatives. As in case of acidic or alkaline demethoxylation or amidation, where the high methoxylated pectins can undergo structural modifications in the galacturonic acid by changing ratio of the hydrophilic and hydrophobic groups yielding low-methoxylated (LMP) or amidated (AP) pectins. These modifications can also affect the pectin-water interactions.

Classification, Isolation and Purification

Classification

The pectins employed for industrial use are recognised on the basis of methyl–esterification. The classification of pectin is done on the basis of methyl-esterification of GalA units (also known as degree of methylation) or DM (Deuel 1943). DM can be defined as the percentage of the molar ratio of methyl-esters to that of GalA units (includes both free GalA and substituted GalA). This is the most crucial parameter that affects various physicochemical properties of pectins like gelling, surface tension and emulsion formation (Müller-Maatsch et al. 2014). So accordingly, pectins can be classified as high methoxyl (HM) and low methoxyl (LM) pectins depending upon the degree of methyl esterification. Methyl groups can esterify the carboxyl groups on the continuous poly-(GalA) chain of pectin in varying proportions resulting in certain degree of methyl esterification (DE or DM). If DE is greater than 50%, it is known as high methoxyl (HM) pectin and if DE is lesser than 50%, it is known as low methoxyl (LM) pectins. High methoxyl (HM) forms gel in high co-solute concentration of 55–75% and acidic (pH 2.50–3.50) systems (Oakenfull and Scott 1984) whereas low methoxyl (LM) pectins can form gel in presence of cations over a broad pH range (Hoefler 1991). Apart from this conventional classification,

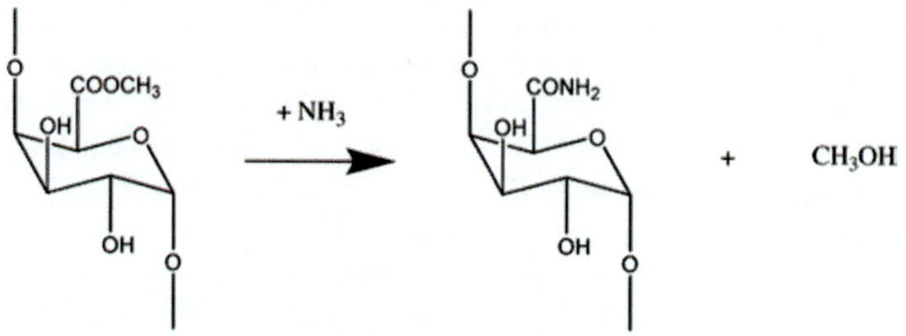

Fig. 2 Ammonia de-esterification of pectin, yielding amidated low methoxypectin (ALMP); Figure adapted from Chan et al. (2017)

commercial pectin has another variant called amidated low methoxy pectin (ALMP). The ammonia de-esterification of HMP results in the formation of ALMP (Fig. 2).

One of the major differences between LMP and ALMP is the presence of primary amide groups which replaces the methyl–ester groups on carboxyl groups in HMP. Here we can also explain the term DAm or Degree of amidation which is expressed as the percentage of the molar ratio of primary amide group present to GalA units (including both free GalA and substituted GalA). ALMP pectin having a DAm value of 15–25 is most commonly used. It was in 1943, that Deuel put forth the idea of calculating the degree of acetyl-esterification (DA). DA can be defined as the percentage which expresses the molar ratio of acetyl-esters present to GalA units (includes both free GalA and substituted GalA). It can be considered as an important indicator of gelling capacity of pectin but is less significant as compared to DM. The total amount of non-esterified mono-, di- and tri- GalA released by the action of endopolygalacturonase on the pectin to the total amount of non-esterified GalA already present in the pectin is expressed as DB (Daas et al. 1999). This allows us to predict the pattern of methyl-esterification. The gel forming ability and the other properties of the gels are largely affected by the pattern of methyl-esterification. Similarly the amount of the non-esterified mono-, di- and tri- GalA released by the action of endopolygalacturonase on pectin to the total amount of GalA present in the pectin is expressed by the term "absolute degree of blockiness" (DBabs), which takes the original total GalA units into consideration (Guillotin et al. 2005).

Isolation

The complexity of the pectic polysaccharides with the other constituents of the cell matrix makes it difficult to extract these in pure form. So the methods employed should be sufficient enough to obtain these high molecular weight compounds in isolated forms. Further the methods employed must have the advantages of not only protecting the integrity of the molecular structures but must be user friendly as well. There are various methods of extraction that have been reported in literature but the most conventional technique employed uses direct boiling and microwave heating

(Fishman et al. 2000). But these methods are associated with the drawbacks like the direct boiling takes about 2 h to extract pectin which can degrade the molecular structure of pectins. Pectin can also be extracted by using different solvents like water, buffers, chelating agents, acids, and/or bases in a sequence mentioned. This not only isolates pectin selectively but also the extracted pectin may have different properties like solubility, gel forming ability and strength etc. It must be noted that the sequence in which the solvents are used affects the purity and the solubility of the pectin. For instance if the solvents used follow the order from water to bases as mentioned above, the extraction of pectin will be selective and the solubility will also be different from the source material. However when the acid is used first, the selective extraction of pectin is not so effective. Similarly if we use the other solvents individually, it not only affects the yield but the properties of pectin as well even if the starting material remains same. The stepwise industrial extraction of pectin is given in (Fig. 3).

Broadly speaking, conventional pectin extraction involves the following steps:

1. Pre-treatment stage.
2. Extraction operation.
3. Post-extraction stage.

Conventional Acid Extraction

Acids are generally cheaper and give the highest yield of pectin among the commercially mineral acids are employed to extract pectin and these include hydrochloric acid, nitric acid and sulphuric acid (Iglesias and Lozano 2004; Mesbahi et al. 2005).The extraction method involving acids yield pectin that is rich in GalA units. The process involves pre-treatments such as washing of the parent material to remove off the dirt and ensure stability during transportation, storage and blanching or drying to inactivate enzymes such as pectin methyl-esterase (PME) and bacteria which can be detrimental in degrading pectin. Following the pretreatments, the peels can be dried to a moisture content of 10–12% to prevent fermentation. The most commonly used sources of pectin include citrus peels, apple pomace, and sugar beet pulp. Currently food and agricultural wastes from industries are also being harnessed for obtaining pectin. This enables us to combat the waste management problems and also widen the availability of the cheap materials for commercial use. Fruit wastes such as banana peels (Emaga et al. 2008), kiwifruit pomace (Yuliarti et al. 2015) papaya peels (Koubala et al. 2014) grapefruit peel (Xu et al. 2014) sugar beet (Maxwell et al. 2016) and many other sources have been exploited as sources of pectin. The pretreated or fresh parent material is then subjected to extraction in demineralized water whose acidity is maintained to pH 1.5–3.0 at 50–100 °C for approximately 3 h (Nussinovitch and Hirashima 2013). The next step involves separation of extracted pectin from the raw materials by filtration or centrifugation. Pectin can also be precipitated by using polyvalent cations such as aluminum or copper salts. Here in this case the cations can be removed by first washing

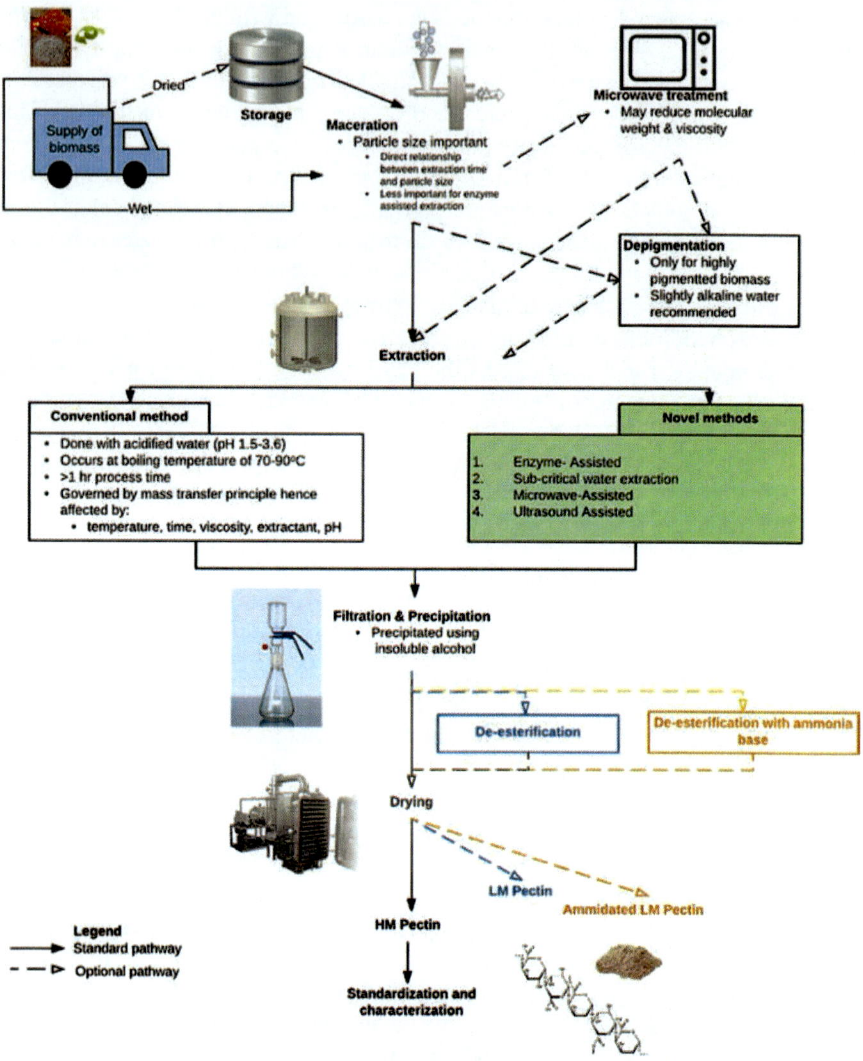

Fig. 3 Schematic representation of Pectin production. Figure adapted from Adetunji et al. (2017)

the product with acidified alcohol followed by an alkaline wash to neutralize the product. This is again followed by alcohol wash and finally the product is press dried. For the modification or de-esterification of pectin, the final step may involve treatment of pectin with alcohol suspension and then drying and milling the pectin to yield powdered pectin. The process is completed by blending the product with different batches in order to standardize structure and functionality. Various factors like temperature, time, viscosity, extractant and pH influences the process as it is governed by mass transfer. This section summarizes some of the many non-conventional sources and technologies employed for the production of pectin.

Enzymatic Methods

Enzymes can be used to catalyse the release of the pectin and this extraction has been found to be comparable to that of acidic extraction. The physicochemical properties of pectin like, viscosity, yield and gel strength are speculated to be similar to conventional extraction. Also as the environmental concerns are growing, large volumes of acidic affluent may further aggravate this. Further the long chain pectins may break down as a result of harsh treatment of acids which may cause loss of functional properties. So enzymatic extraction can offer the advantages of specificity, less energy consumption and affluent management. Many enzymes have been used for this purpose, like cellulases, hemi cellulases and fungal or plant-derived pectin methyl esterases in a study carried out by Dominiak et al. (2014) six commercial cellulases were used for obtaining the pectin from lime peel and it was reported that when lime peel was treated with enzyme preparation Laminex C2K derived from *Penicillium funiculosum* during 4 h treatment (pH 3.5, 50 °C), it released pectin with molecular weight, functional properties like gelling, viscosity similar to that obtained from conventional acid extraction during 8 h treatment (at 70 °C, pH < 2). It was further reported that pectin extraction carried out using enzymes had higher amounts of methylated pectin whereas pectins that were extracted using acids, had lower amounts pf methylated pectins.

Non-Conventional Methods

The success of the pectin extraction process is dependent upon the two factors (1) The rate at which the hydrolysis of the protopectin (i.e. pectin bound to the other polysaccharides in the cell matrix) takes place and (2) How fast the resultant pectin gets solubilised i.e. the diffusion of dissolved pectin into the extracting solvent (Cho and Hwang 2000; Minkov et al. 1996). Various mathematical models have been developed that take into account the above mentioned factors i.e. hydrolysis and solubilisation/degradation and have been found to follow a first-order chemical reaction. The equation of hydrolysis for a spherical pectin-bearing particle is given by:

dCdt¼kCð1 εÞ (1)

where, C is the concentration of protopectin, k is the rate constant of hydrolysis and ε is the porosity of the particle. There are certain factors that are crucial to the successful extraction of pectin but temperature and solvent pH are the governing ones. It has been reported that temperature has an influence on the diffusivity, D, and is accordingly related by stokes-Einstein equation:

$$D \infty [T/\eta]$$

where T, is the absolute temperature and h, is the dynamic viscosity coefficient (Loncin and Merson 1979). Overall the factors that influence both the quality and the quantity of the extracted pectin can be categorised as intrinsic, extrinsic and techno functional ones. Intrinsic factors are related to matrix of polysaccharide like particle size and moisture content. Extrinsic factors are based on the nature of solvent like its molecular weight, polarity, solid-liquid ratio, toxicity and volatility. The techno functional factors like temperature, pressure, and optimal operating conditions are equally important. High temperatures can result in excessive depolymerisation and de-esterification of the protopectin, which can result in pectin extract having inferior functional properties although yield per unit may be high at elevated temperatures. Another important parameter that needs to be taken into consideration is the pH of the extraction solvent while extracting pectin. Generally pH range between 1 and 3 has been reported by most of the workers (Jensen et al. 2013). In order to adjust the pH, small amount of base such as NaOH can be used although time-temp combinations are also taken into account. Low acidic pH favours the hydrolysis of the protopectin whereas high pH conditions can cause pectin degradation owing to β-eliminations (Kravtchenko et al. 1992a, b). So what we need is to strike a balance between having a superior quality but low yield at high pH and having an inferior quality extract but with higher yield at a lower pH. Talking about particle size, El-Nawawi and Shehata (1987) reported that for smaller particle size higher yields were obtained and the reason was that for increased surface area, the area of contact between the solid and solvent increased and distance for diffusion for the solute decreased. But for the particles which are too fine, problems may arise as the fine particle size may jam within the cervices of adjacent coarser ones, which may again pose a problem to the flow of the solvent. It was further reported by El-Nawawi and Shehata (1987) that agitation of the solvent had no significant effect on the yield of pectin. The workers explained this owing to the outward diffusion of the solute within the solid matrix and the diffusion of solute from matrix surface to the solvent. If it would have been vice-versa then agitation may have played a role in increasing the extraction yield following the eddy diffusion induced by agitation. Another significant factor attributing to the rate of mass transfer is time as with time the concentration gradient decreases the rate of mass transfer decreases. As the duration of the extraction increases, more and more pectin starts accumulating in the solution. This makes the solution more viscous thereby declining the rate (Richardson et al. 2002). Further the optimal duration for the process is important as an extended period of extraction may cause the thermal degradation of the polysaccharide but also may stop the pectin dissolution. Thus control of the above mentioned parameters is very important to maintain the superior quality of the extracted pectin. The role of the above mentioned factors have been studied in detail by El-Nawawi and Shehata (1987).With respect to increasing both the quality and the quantity of the extracted pectin, a number a non-conventional techniques have been sorted to overcome the shortcomings of the conventional extraction techniques. These aim to increase the yield in an environment friendly and energy conserving way. A few of them have been described below:

Microwave-Assisted Extraction (MAE)

Microwaves are the electromagnetic waves in the range between 300 MHz and 300 GHz on the electromagnetic spectrum. These consist of two oscillating perpendicular fields: electric and magnetic fields. The frequency bands at 0.915 and 2.450 GHz have been mostly utilized at homes in ovens and operate at 2.450 GHz (wavelength, 12.2 cm; energy 0.94 J/mol). At such frequencies, rotation of the polar molecules of a microwave-irradiated matrix and/or solvent takes place which in turn creates heat. Here the transfer of energy can be explained on the basis of dipole rotation involving the reversal of dipoles in polar molecules; and ionic conduction, involving displacement of charged ions present in the solvent (Routray and Orsat 2012). But the generation of heat can take place only if the material absorbs some energy i.e. it possesses dielectric losses. The positive side of using microwave assisted extraction is that as compared to the conventional heating, there is no temperature gradient. With respect to the pectin modifying effects of microwave assisted extraction, various works have been carried as in case of pectin extracted from grapefruit by (Bagherian et al. 2011). The workers reported that using microwave power (0.45, 0.63, and 0.9 kW) not only increased yield of pectin but also the values for DE increased generally with increasing microwave power and time (70.96–80.89%). However the technology bears the cost constraint and cannot be employed on commercial scale. A lot of investigation needs to be carried to bring it to use.

Enzyme-Assisted Extraction (EAE)

There is a growing pressure on food and pharmaceutical industries to replace extraction methods based on solvents with some novel methods in order to prevent the presence of trace chemical solvents in products from solvent-based extraction processes. For this purpose, the use of enzymes is finding its way to extract and modify various polysaccharides owing to their selective reactivity and ability to catalyse reactions. Enzymes have a widespread use in food processing sector such as clarification of juices. Similarly the use of enzymes for extraction of seaweed hydrocolloids, carrageenan, alginate and agar modified the physicochemical properties of the marine polysaccharide (Rhein-Knudsen et al. 2015). The network of the polysaccharides in the plant cell is comprised of cellulose, hemicellulose (like xyloglucan), and pectin and protein. With respect to pectin, the cellulose/xyloglucan network is implanted in its matrix along with a protein network. So the cell wall along with the various cell components can be degraded using enzymes such as cellulases, hemicellulases and proteases with minimal pectinolytic activity. So we can have two approaches as regards to enzyme assisted extraction i.e.: (1) Degrading pectin using enzymes followed by isolation of pectin fragments e.g. galacturonic acid, and (2) using enzymes capable of deconstructing plant cell wall and isolating pectin (Panouille et al. 2006). However the method requires a good deal of knowledge both of the catalytic action of the selected enzymes and the optimum conditions for their use in EAE (Ptichkina et al. 2008; Puri et al. 2012). The former method has been used to study de-esterification of commercial HM pectin into LM using a synergistic application of high hydrostatic pressure and pectin methyl esterase (PME) enzyme by Zhao et al. (2015). Similarly, Zykwinska et al. (2008) also reported that LM pectin can be obtained from HM pectin source by incorporating PME alongside cellulases and proteases into the Enzyme assisted extraction process.

Electro-Magnetic Field Assisted Extraction (EME)

This technology has been employed to extract pectin from citrange by Zouambia et al. (2014). In their work, extraction was carried out in magnetizable and enameled containers that were placed on an induction plate. The apparatus contained 9 levels where each level provided for a known temperature and power. The time was varied for 10, 20, 30, 40, 50, 60 and 90 min. The yield of pectin was found to be remarkably high and it was only 2.09 min that gave half the amount of pectin extracted by conventional heating for 90 min. So a considerable reduction in extraction time is an added benefit. Hence, this technology can be projected as a novel and promising method for the extraction of pectin.

Subcritical Water Extraction (SWE)

Subcrtical water extraction involves the use of liquid water at an elevated pressure which without a change in phase can reach to the temperatures higher than its normal boiling point. It is also known as pressurized hot water extraction (PHWE) and superheated water extraction (SHWE) (Zakaria and Kamal 2016) and the water may reach to a temperature of 160 °C which enhances the extraction rate. The high temperature of water can offer a wide range of advantages including low viscosity, higher rate of mass transfer, low surface tension and high diffusion. It even alters the physicochemical properties of the subcritical water like dielectric constant and solubility (Azmir et al. 2013). SWE results in reduction of dielectric constant of water from 79 at 25 °C to 43 at 160 °C and to 33 at temperatures around 200 °C which enables it to extract both ionic and non-ionic compounds. The solubility of pectin extracted by solvent having low dielectric constant is enhanced and this in turn increases the yield (Brunner 2009; Chen et al. 2015). There are certain reports of low yields at high temperatures, but in case of SWE the yields at 120 °C are high which highlight the significant increase in mass transfer and prevention of hydrolysis encountered upon conventional extraction. An increase in the yield was observed by Liu et al. (2016) in case of soy hulls by synergistic application of phosphate-assisted subcritical water and ultrasonic treatment. Further at times some other solvent other than water is used and the process is known as accelerated solvent extraction (ASE). Again the most important property i.e. solubility of water is also enhanced at subcritical conditions. Further studies in this regard are still needed. Owing to the higher quality of pectin extracts, short duration of process and acid free solvent, SWE is becoming the most sorted technique for extracting pectin (Ueno et al. 2008).Also the GRAS status of solvent being employed, the technique finds more usage in food and agro based industries. However the process needs to be standardised and kept in control owing to the susceptibility of the pectin chains to degradation (Khajavi et al. 2005). Also the cost considerations prove to be a hindrance for using this technique on commercial scale. Various factors play a key role in governing the extraction yield in case of SWE and these include: temperature, time, pressure, particle size etc. For temperatures as high as 170 °C, investigations have been done to study the effect of such high temperatures on extracting pectin. Another important factor that must be taken into consideration during optimization of the process is time, as the long duration process may result in the degradation of the pectin chains. Pressure not only keeps the water away from changing its phase but also affects the yield of pectin. This factor has been studied by Chen et al. (2015) and others who investigated SWE at pressures up to 10 MPa. The yield of pectin can also be influenced by particle size of the parent material.

Ultra Sound Assisted Extraction (UAE)

Ultra sound waves are the sound waves having frequencies from 20 kHz to1–16 kHz. A major difference between ultra sound and electromagnetic waves is that ultrasound waves require a medium (solid, liquid or gas) for their propagation. This involve cycles of expansion and compression and while passing through the liquid medium, these lead to the formation and collapse of bubbles, that ultimately result in the formation of microjets. This phenomenon through which these bubbles grow and collapse is known as cavitation. The process of cavitation can last for 400 ms and within this, high temperatures and pressures about 5000 °C and 1000 atm can be attained (Luque-Garcıa and Luque de Castro 2003). These high-speed jets and waves are then used to swell the materials because of the intermixing of solvents and pore enlargement. This also increases the molecular mobility and subsequently increases the yield. It has been reported that an increase beyond the 65 W can increase the bubble formation which can decrease the ultrasound energy transmitted into the system. As the ultrasound can break down the polymers, so it can have implications on physicochemical properties of pectin side chains of the pectin molecules are more prone to the degradation by ultrasound and is witnessed by the fact that there is reduction in the (Gal + Ara/Rha) ratio which corresponding to the neutral side chains. Again the increased intensity and time duration decreased the ratio. Zhang et al. (2013) reported negative effects on the gelling properties of the sonicated pectin. Pectin gel in such a case was reported to have weak molecular structure but were more transparent. Also the rate of gelation decreased with the increasing intensity and time of ultrasonication. Of course, UAE has a number of advantages such as it reduces the extraction time, lowers the usage of solvent and also saves the energy. Many workers have also reported this technique in being most effective for mixing and micromixing, for having higher mass and energy transfer rates most notably reducing the concentration gradients and extraction temperature.

Besides it also increases production, reduces equipment size and has very few process steps. Apart from these, it also improves the bioactive profile of polysaccharides, by modifying their molecular properties. As we know that unmodified pectin is not easily absorbed by the body, so modification of pectin using UAE can break these molecules into smaller fragments. Shear forces are responsible for decreasing the molecular weight of pectin and this force is generated as a result of relative motion between solvent and polymer chains, as the bubbles collapse. The effect is more pronounced in long chains as compared to chains below a certain critical size. Various researchers have reported the synergistic application of the ultrasound and the other novel techniques for obtaining better yield of pectin e.g. the workers in china reported that an ultrasound-/microwave- assisted acid method' (abridged UMAAM) not only helped to achieve higher yields of pectin but also the pectin obtained was having much higher MW (636.20 vs. 295.95 kDa) in comparison to pectin obtained by traditional means. It has been reported that grapefruit pectin obtained from UAHE had lower molecular weight, viscosity and degree of esterification as compared to the conventional heating extraction (CHE).On the contrary it had higher degree of branching and purity and yield comparable to CHE. Two modes of the UAE system exists in two forms: namely bath and probe units. In the former, there are limitations of reproducibility and lack of uniformity in distribution of ultrasound energy. On the other hand, ultrasonic probes are much efficient as the ultrasound energy can be focussed on specific sample zones which can result in better cavitation. Probes are better suited to laboratory works.

Purification of Pectin

The industrial production of pectins mostly consists of HMP. Although LMP occurs naturally in plants but can also be obtained from HMP by its de-esterification using acid, alkaline and/or enzymes. Short extraction time at a temperature close to boiling yields rapid-setting HMP as the high temperature- short extraction time combinations reduce de-esterification. This is reverse in case of long extraction time with low temperatures which favour de-esterification to yield slow-setting HMP or even LMP (Nussinovitch and Hirashima 2013). Therefore, by selecting the suitable time-temperature combinations, the pectin with desired properties can be obtained (Chan and Choo 2013). In case of acid de-esterification, inorganic acids such as hydrochloric acid is used to bring out the de-esterification. The method has a major disadvantage that it can lead to the pectin degradation via hydrolysis and β-elimination. In case of hydrolysis, which is a pH-dependent reaction, and occurs mainly in acidic environment (Krall and McFeeters 1998; Smidsrød et al. 1966), neutral polysaccharides (ballast) are removed. A lesser portion of proteins and phenolic compounds are also removed (Constenla and Lozano 2003; Kravtchenko et al. 1992a, b). The pH of the plant cell wall lies in the range of 4–6 and as the pH decreases, the rate of the reaction of acid hydrolysis increases (Brett and Waldron 1990). But for the pH range of 4–6, acid hydrolysis was found negligible by Fraeye et al. (2007). Again, lower the DM, lower will be the amount of methyl-esterfied target groups and faster will be the rate of the pectin hydrolysis. Owing to this, a number of the properties are influenced including gel strength resulting in the formation of weaker gel. For obtaining LMP without any main-chain damage, a controlled or reduced temperature is sufficient. And in case of β-elimination, a double bond is formed between C4 and C5 in the GalA units as a result of the cleavage of glycosidic bonds (BeMiller 1986). In case of β-elimination, which is temperature-dependent, unlike of acid hydrolysis which is pH dependent, the elimination occurs only in presence of a methyl group at C6 (Keijbets and Pilnik 1974). Again the rate of β -elimination decreases as the DM decrease. A kinetic study was carried out by Kravtchenko et al. (1992a, b) and the workers proposed that as the temperature increased, the rate of β-elimination reaction also increased. They reported similar values of activation energy for β-elimination at pH values ranging from 4.5 to 11.0. Moreover, the β-elimination reaction takes place under alkaline conditions, or even at near neutral pH (Lopes da Silva & Rao 2006; Taylor 1982). The reaction rate is very low at weak acidic pH.

Another most common industrial methods to de-esterify pectin involves the use of ammonia in alcohol at relatively low temperatures. This technique is called as alkaline de-esterification (ALMP). Besides ammonia, hyrroxylamine and other ammonia derivatives can also be used to carry out the aminolysis (Racape et al. 1989). The benefits of using ALMP include the production of relatively high pectin molecular weights. In this process, the saponification of the methyl-esterified GalA residues takes place at a rate which is slower than with sodium hydroxide in water at the same pH. Therefore, with the increase in time of reaction, DAm increases, DM decreases and the GalA content remains constant (Renard and Thibault 1996). It may also be noted that there is no major change in the molecular weight pectin derived from this de-esterification which signifies that the reaction does not lead to drastic depolymerisation (Anger and Dongowski 1988). Also the composition of the neutral sugar and the number of side chains remains constant (Reitsma et al. 1986). When enzyme de-esterification is followed, a block distribution of unesterified GalA units is obtained (Denès et al. 2000; Savary et al. 2003). In such case, the de-esterification of the pectin is carried out using pectinmethyl-esterase (PME), the most commonly used enzyme. The de-esterification of pectin by PME can be explained via two mechanism: (1) PME follows a single chain-reaction by acting on a single chain of pectin to produce block pattern of carboxylates and (2) It produces shorter blocks of de-esterified pectins by having multiple attachments. Overall enzyme de-esterification is an efficient and environmentally safe method for obtaining LMP.

Structure and Solution Behaviour

The chemical and structural properties of pectins are still a subject of study since its discovery owing to the homogeneity of this polymer family. With respect to its basic structure, this heteropolysaccharide is made up of atleast 17 different

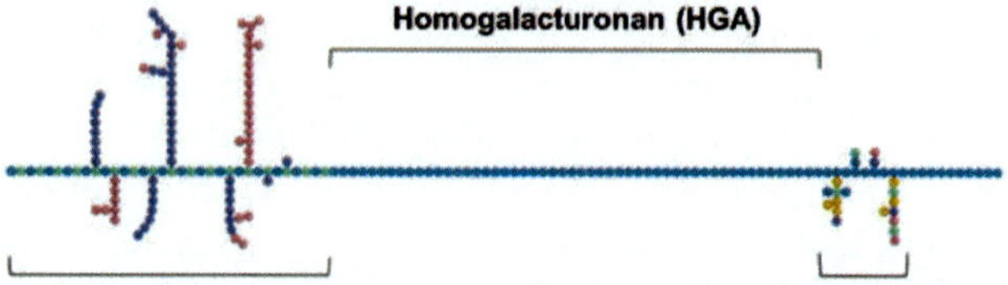

Fig. 4 Pectin chain comprising of covalently linked (i) homogalacturonan (HGA), (ii) rhamnogalacturonan I (RG-I) and (iii) rhamnogalacturonan II (RG-II). The diagram shown here is intended only to illustrate some of the major domains found in most pectins rather than indicate definitive structures

monosaccharides, of which GalA is in abundance. This is followed by L-arabinose, D-galactose, L-rhamnose and others (Kaya et al. 2014). The different monosaccharides are connected through 20 different linkages. In case of GalA monomers, methyl esterification or O-acetyl-esterification of the carboxylic groups and hydroxyls groups may take place (Fig. 4).

In the latter, the O-acetyl-esterification occurs mainly at the O-3 position and occasionally at the O-2 position. The composition of the native pectins includes polysaccharide domains: homogalacturonan (HGA), rhamnogalacturonan I (RG-I), rhamnogalacturonan II (RG-II), xylogalacturonan, and apiogalac-turonan. The latter three have been classified as substituted HGA (Caffall and Mohnen 2009; Harholt et al. 2010). Very less is known about their linkages and remains a matter of debate. However various studies over the time have revealed the covalent linkage of these polysaccharides. Besides these, the ionic cross linking takes place between pectin strands to form pectic networks branching throughout the primary cell walls (Willats et al. 2003). Structure of the pectin has been studied under three models as proposed: (a) Alternating HGA and RGI segments on the pectin backbone (Schols and Voragen 1996) (b) the pectin backbone comprising RGI segment with HG being linked as its side chain (Vincken et al. 2003) and (c) Alternating and perpendicularly linked-HGA strands and a RGI domain on the pectin backbone (Yapo 2011).

Pectin Polysaccharide Structure

Homogalacturonan (HGA)

HGA is an abundant and most significant polysaccharide consisting of a linear homopolymer of α-1,4-linked GalA (Fig. 4i). It is estimated that the number of GalA units in HGA are around 200 units and is known as the smooth region. It forms about 60–65% of total pectin and is synthesized in the Golgi apparatus and from there it is transferred to the middle lamella and then to primary cell walls. Finally it is deposited as 70–80% of GalA units methyl-esterified and could be

O-acetyl-esterified at O-2 or O-3. Sometimes one or two α-1, 2-linked l-rhamnopyranose units can join the smooth region and can be seen in most of the pectin structures. Additionally at C-2 or C-3 positions, GalA units may be substituted with xylose or apiose residues, forming domains known as xylogalacturonan or apiogalacturo-nan, respectively. As a result of these modifications, the functional properties of HGA are altered.

Rhamnogalacturonan I (RG-I)

In case of RG-I, 100 repeating units of the dis-accharide α-1,2-linked-l-rhamnose -α-1,4-d-GalA are linked. In other words, every twenty-fifth GalA of the main chain is substituted with rhamnose and the backbone comprising of the repeating units of α-1,2-linked-l-rhamnose -α-1,4-d-GalA (Fig. 4ii). More that 30% of the pectin is comprised of the rhamnose residues (Renard et al. 1995). It has been estimated that 50–78% of the rhamnose is RGI. But the GalA residues of RG-I can be O-acetyl-esterified and not methyl-esterfied (Caffall and Mohnen 2009). It has been reported that 20–80% of rhamnose residues have neutral side chains substituted at the C-4 position. There can even be the attachments of side chains of neutral sugars at the C-2, C-3 and/or C-4 positions of rhamnose residues. The major part of these side chains include arabinose, forming galactan, arabinogalactans, ara-binan and galactose. The covalent linkage of the other sugars like glucose, mannose, fucose, xylose, and glucuronic acid can also be found as side chains. The neutral side chains can be made up of a single glycosyl residue up to 50 or more, which can result in polysaccharide having high branching and complexity (Voragen et al. 2009).The massive branching characteristics of RG-I is the reason that it is called as "hairy region".

Rhamnogalacturonan II (RG-II)

One of the most striking features of the RG-II is that it is the only domain of pectin that does not have structural diversification. It consists of around nine α-1, 4-linked GalA units having four different polymeric side chains. Some of the side chains may be methyl- esterfied and these contain eleven rare sugars that include 2-O-methyl-d-xylose, 3-deoxy-d-manno-octulosonic acid, 3-deoxy-d-lyxoheptulosaric acid, 2-O-methyl-l-fucose, 3-C-carboxy-5-deoxy-l-xylose (aceric acid) and apiose. Its cleavage by endopolygalactouronase signifies its covalent bonding to HGA and it also corsslinks the two pectin molecules (apiosyl residues) within the cell wall by borate ester links. RG-II has whole togetherly different structure than RG-I athough their names resemble in sequence and it is also included in the hairy region. The ends of RG-II are believed to be glycosidically linked to the HGA domains (Fig. 4iii).

Solution Behaviour

When pectin is homegenoulsy dispersed in water, the molecules of pectin which are far apart from each other do not interact with each other, and the effect of this dispersion is seen only as a slight increase in viscosity. This increase of viscosity may be attributed to the increased proportion of hydrated molecules which causes the distortion in the velocity pattern of the liquid (Guimarães et al. 2009). As we increase the concentration of pectin, the viscosity of the solution increases and this pattern has been observed for pectin from citrus peel (Sousa et al. 2015), tamarillo fruit pulp (do Nascimento et al. 2016), dragon fruit peel (Muhammad et al. 2014), mango pulp (Iagher et al. 2002), apple pomace and cacao pod husks (Vriesmann and Petkowicz 2013). An increase in the concentration results in the decrease of the intermolecular distances between pectin molecules which provides a platform for intermolecular interactions such as hydrogen bonding. When a graph is plotted between shear stress and shear rate of the pectin solutions, it is observed that pectin solutions demonstrate a Newtonian behaviour (Fig. 5), but only up to a certain concentration. At concentrations upto 3%, these dispersions exhibit characteristics of Newtonian liquids and when the concentration of pectin is increased, these Newtonian liquids may display shear thinning behaviour i.e. with the increase in shear rate the viscosity of the solution decreases.

Here the most important factor is the molecular weight of pectin which determines the concentration of pectin at which the transformation from Newtonian to shear thinning takes place. However many studies have reported an increase in pseudoplasticity with an increase in pectin concentration (Muhammad et al. 2014; do Nascimento et al. 2016; Sato et al. 2008; Sousa et al. 2015).

Gel Forming Ability of Pectin

Pectin gel consists of a three-dimensional (3D) network of polymer chains with solvent and solutes trapped within. The formation of the gel is dependent upon the molecular structure, the behaviour of junction zones which form the site of polymer crosslinking, and the intermolecular forces which are responsible for holding the network together (Axelos and Thibault 1991a). In case of high molecular weight polysaccharides like pectins, complexity arises at the junction zones. As such a number of individually weak interactions like hydrogen bonds and electrostatic interactions between molecular structures come into play. The solution behaviour of the pectin polysaccharides like HMP and LMP is different although the macromolecular properties like, composition, size and conformation have the same effect on the characteristics of the gel.

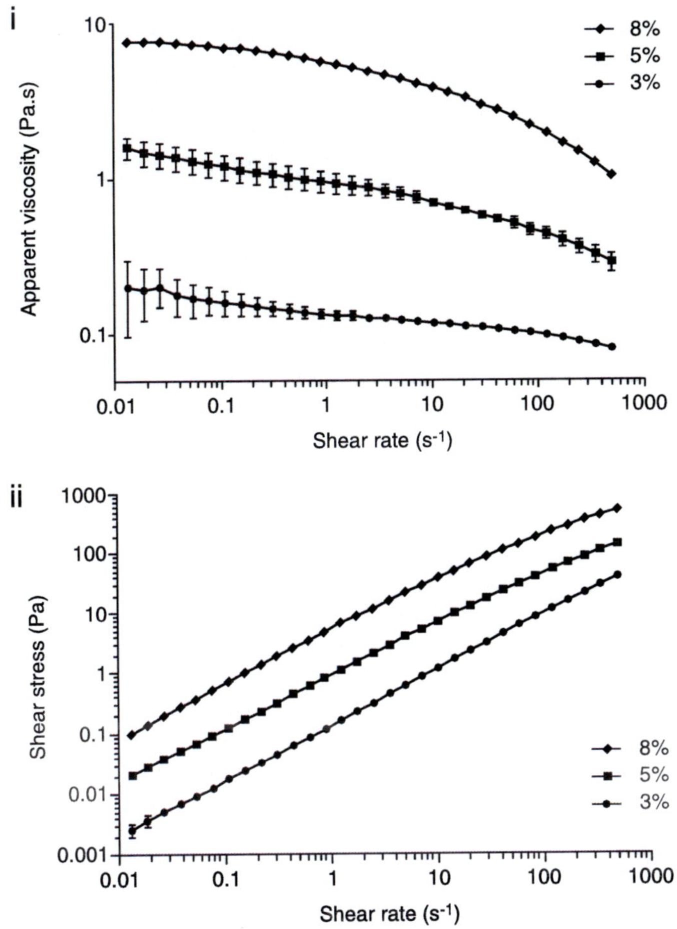

Fig. 5 The impact of shear rate on the (i) apparent viscosity and (ii) shear stress of 3%, 5% and 8% of high methyl-esterified pectin from tamarillo pulp forming the aqueous dispersion in water. Figure adapted from do Nascimento et al. (2016)

High Methoxy Pectin (HMP)

High methoxy gel pectins are formed at low pH in the presence of a high sucrose concentration or similar co-solute. The conditions of low pH results in protonation of carboxylate which reduces the electrostatic repulsion. Further the high concentration of sucrose reduces the water activity that triggers the interchain reactions as compared to chain-solvent interaction (Thibault and Ralet 2003). X-ray diffraction (XRD) studies have revealed that intermolecular hydrogen bonds and hydrophobic bonding between methyl esters are the main forces behind the stabilization of the HMP gels (Walkinshaw and Arnott 1981). The conformation of the adjacent GalA

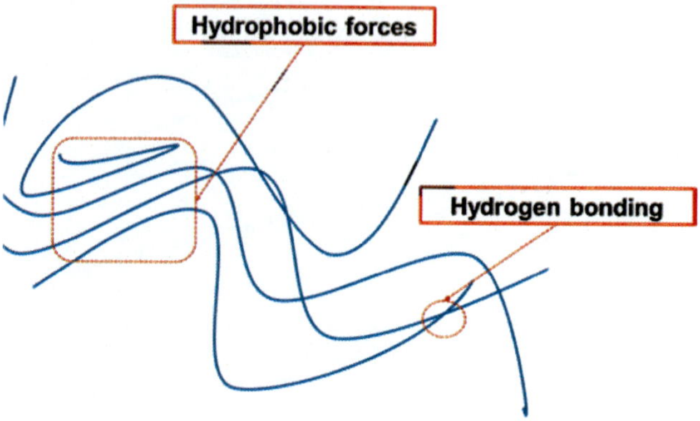

Fig. 6 The intermolecular hydrogen bonds and hydrophobic interactions between methyl esters governing the HMP gelation

units promotes the hydrogen bonding between the pectin molecules Although the hydrogen bonds are weak, still these offer a good thermodynamic stability to the gel and is the only force that is maintains the gel structure. This interaction is however incapable to overcome the entropic barrier to gelation. Another factor i.e. hydrophobic interaction is as a result of the interaction between the water molecules and the non-polar methyl esters of pectin (Thakur et al. 1997) (Fig. 6), the free energy contribution of the junction zones by the hydrophobic interaction is half to that contributed by hydrogen bonding.

The surface area that is exposed to water gets reduced as a result of coalescence of methyl esters, which as a matter of fact reduces the overall entropy. It has been observed that for HMP the hydrophobic interaction result in increased viscosity of the gels. The effect of methylation on the gel properties of HMP have been studied recently and (Löfgren et al. 2005) demonstrated that pectins having a block distribution of methyl esters have faster rate of gel formation as compared to the pectins having random methyl ester distribution and same DM. In preparation of HMP gels a number of additives are used such as sucrose, glucose, fructose and many polyols e.g. glycerol (Evageliou et al. 2000). The sugar molecules play an important role of stabilizing the junction zones by triggering the hydrophobic interactions between methyl-ester groups. it can be explained on the basis of "Preferential hydration" theory that states that macromolecules are surrounded by the heterogenous distribution of the water These solutes have a wider influence on the hydrophobic interactions as they undergo interactions with the water molecules (Matubayasi 1994). For instance sucrose can increase the hydrophobic interaction between two methyl esters by 67%. pH is another factor that influences the gel formation of HMP. When the pH is lowered, protonation of carboxylic groups present in the anionic pectin takes place which reduces the electrostatic repulsions occurring between pectin chains (Thakur et al. 1997). Not only this, inter and intramolecular hydrogen bonds along with secondary alcohol groups are also formed at low pH. In a study carried out by Seshadri et al. (2003), it was shown that HMP gels are time-dependent. The study involved subjecting the solutions of native pectin mixture (1.15% apple pectin, 41.40% sucrose, pH 1.50) having high viscosity at time zero to stress at different gelation times. It was observed that with the increase in gelation time, the magnitude of the stress response to the oscillatory sweeps increased display in the elastic nature of the solution. Also the storage modulus (G'), having dependency on frequency became less evident specifying the change from viscous to rubbery nature. This behaviour can be explained on the basis of network formation. In the beginning the pectin behaves as single and dispersed molecule. During shearing, owing to the high mobility of the pectin dispersion, there is quick relaxation of pectin molecules. This keeps the building up of excessive stress at low oscillation frequencies. With the passage of time, as result of hydrogen bonding and hydrophobic interactions association of pectin molecules begins. This forms the 3Dnetwork which blocks the molecules to interact with the superimposed flow. As a matter of fact there is a shift from viscous region to rubber plateau.

Da Silva et al. (1995) reported that gelation of HMP is highly temperature dependent. At 5 °C gelation of HMP was performed and it resulted in a weak gel due to slow gelation. The reason for this being the lack of hydrophobic interaction at low temperature. At temperatures up to 30 °C Pseudo-equilibrium G increased and then decreased. It was also observed that at an intermediate temperature range, hydrogen bonds and hydrophobic interactions can result in formation of the network having higher elasticity. On heating the structure of HMP-sugar gels, they tend to be irreversible. When the gel was heated and cooled an unusual 'thermal annealing' behaviour was found (Evageliou et al. (2000). This showed its dependency on pH. In presence of 65 wt% sucrose and at pH 4.7, a HMP (of DM = 70%, 0.5%) gel was fully reversible with no detectable thermal hysteresis on heating. Upon heating, there was slight reduction in G' and G'' (pH 3.0 and 3.5) Despite having pH difference, stability remained virtually constant at 30 °C.

Low Methoxy Pectin (LMP)

The stabilization of LMP gels is brought by ionic cross-linkages via calcium bridges between two carboxylates from two different chains. There is seldom any hydrogen bonding or any hydrophobic interaction. The structure hence formed is given the term "egg-box" model (Gidley et al. 1979; Grant et al. 1973).The egg box model consists of junction zones that are formed by the sequential, side by side units of GalA, in which the ionic and electrostatic ionic bonding of carboxyl groups results in interlocking of GalA monomers in adjacent chains. This forms an egg-box structure and was first explained for alginate. Various workers have studied the egg-box models for alginate and pectin (Braccini and Pérez 2001). The "egg-box" model is best suited to explain the gelation mechanism of alginate but in case of LMP, a model termed as ashifted "egg-box" can be used to understand the gelation (Low et al. 2015). It is termed as "ashifted egg-box" as one of the chains is shifted with respect to the other. The workers stated that there is two-step mechanism involved, the first step comprising of dimerization of two macromolecules and the second step involving the aggregation of these dimmers (Axelos and Thibault 1991b).The unbranched, non esterfied blocks of galacturonan are held together by calcium ions. The important parameters that modify the intra and intermolecular interactions include the distribution of non-methyl-esterified carboxyl groups and the charge density. Following the changes in conformation, the interchain hydrogen bonding takes place by placement of the carboxyl group outward from the polymer backbone (Gilsenan et al. 2000; Mansel et al. 2015). The major forces responsible for the stabilization of the egg-box structure are stabilization of the egg-boxes formed between the two neighbouring chains are electrostatic interactions, hydrogen bonding and to a lesser extent by van der Waals interactions. Overall the stabilization of the egg-boxes is brought by the forces in the order electrostatic interactions ˃hydrogen bonding ˃Van der Waals interactions. The stability of the egg-boxes is dependent upon six consecutive carboxyl groups that is present on the interior of each interacting chain (Powell et al. 1982). The formation of the gel is hindered by the presence of methyl ester groups in the primary backbone. The transformation of LMP-calcium mixtures from a state of liquid to gel with increasing G' takes place as the DM decreases henceforth increasing the gel strength. The probability of the formation of egg box structure is increased as the DM increases which increases the length of the residues of non-mehtylated GalA. The stronger gel is formed with the formation of greater amount of "egg-boxes". Also the impact of calcium ions on the gel strength was studies and it was reported by Dobies et al. (2004) that as the concentration of the calcium ions increased, the value of G' increased which in turn increased the strength of the LMP-calcium gel. This mechanism of action lies in the fact that higher concentration of the calcium ions permits the formation of additional cross-links of non-methylated GalA, which in turn create a more elastic and denser gel network (Basak and Bandyopadhyay 2014). The addition of calcium or iron ions cab be used as an alternative to thickening agents in fluids as these increase the pseudoplasticity. On the contrary addition of the magnesium ions can result in the formation of weak pectin gels as a consequence of decreasing G'. The dispersion of LMP in aqueous calcium solution is carried out at 70 °C. This is followed by cooling the mixture slowly. At an elevated temperature, hydrophobic interactions favour the egg-box structure via calcium bridging. Upon cooling there is an increased interdimer associations that support the gel formation. The gel formation is further encouraged by the electrostatic interactions of the single dissociated carboxyl groups of pectin chains with calcium. Also just like in HMP gelation, addition of sugar increases the inter-chain reactions in LMP gels. But with the increase in the concentration of sugar, hydrogen bonding may be promoted as a result of dehydrating effect of sugars which will in turn decrease the cross-linking by calcium ions. This will overall have an adverse effect on LMP gels (Fu and Rao 2001).

Amidated Low Methoxy Pectin (ALMP)

Löfgren et al. (2006) reported that ALMP S generally have less sensitivity towards calcium whereas (Capel et al. 2006) reported that the process of amidation may slight influence the sensitivity towards calcium. Also the gels formed by ALMP and LMP at a pH of 3 and 7 appear to be similar in kinetic behaviour and gel microstructure. ALMP gels are formed as a result of hydrogen bonding between amide groups at low pH. These gels are thermally reversible and sensitive. With or without 30% sucrose, these gels show a rapid formation when G' > G". When heated, they can be solidified again whereas in case of pectin gel, the liquid state remains unaltered (Chen et al. 2014a, b).

Factors Affecting the Gelation Process

Overall the process of gelation is affected by both extrinsic and intrinsic factors. Extrinsic factors include concentration of calcium, concentration of pectin, pH, ionic strength and temperature whereas the molecular weight, number and the pattern of free carboxyl groups and the types of pectin form the intrinsic factors (Gigli et al. 2009). The underlying effect of molecular weight is well understood in terms of rigidity of the gel as the lower the molecular weight, less will be the junction zones per molecules and less will be the extent of cross linking and accordingly weaker will be the gel. This parameter is mainly dependent upon the origin of pectin but handling procedures like extraction can also have a significant effect on the molecular weight of the pectin. In a study carried out by Seshadri et al. (2003) the workers reported an increase in the yield of pectin when the extraction was carried out by ultrasound although the rheological properties of pectin were inferior. Further, by increasing the intensity and time of sonication, reduction in the gel strength was reported. The underlying mechanism of this result was attributed to chain degradation that was promoted by cavitation that brought the reduction in the molecular weight of pectin. Another important factor, that is the presence of sugar monomers such as rhamnose. This sugar monomer is not compatible with the configuration of the junction zones that are formed by the GalA monomers. As a result it hinders the molecular orientation that is required for the formation of the junction zones (Axelos and Thibault 1991a). As a consequence, the geling properties of the polymer in solution are affected. Thus rhamnose, which is the abundant sugar in rhamnogalactouranan forms the hairy region by dismantling the backbone of galactouranan and this hairy region further limits the inter-chain association. However, these rhamnose interruptions vary within the same plant species or in different pectins from different plant species. This phenomenon of rhamnose displaying poor gelling properties can be observed in sugar beet pectin, which is mostly composed of high rhamnose content. But again due to its emulsifying properties and its ability to form covalently cross-linked hydrogels, it is used in various commercial applications. Apart from sugar beet pectin, other commercial sources like apple pectin gel displays visco-elastic properties. Also citrus pectin gel is more elasto-brittle in nature. Its noteworthy to mention that inter-chain association is limited by the presence of neutral sugars but as these branches are often removed during purification or extraction, so these have less significance in commercial pectins.

Rheology of Mixed Pectin Systems

In order to enhance the textural and rheological properties of the pectin, synergistic approaches may be exploited. These synergistic mixed gels are finding a great use in industry. This involves mixing of HMP, LMP and/or ALMP which can form stronger mixed gel. In such mixed systems, the value of G' is higher as reported by Löfgren and Hermansson (2007) who reported higher G' values for mixed HMP/LMP as compared to HMP gels. They further reported that the presence of calcium and sucrose during the association of LM pectin has a marked effect on the storage modulus of mixed gels. The storage modulus was increased by thirty times. The rheological behaviour of HMP gels having higher concentration of sugar coincides with that of mixed gels. The addition of ALMP to HMP increased the viscosity and reduced the pseudo plasticity (Sato et al. 2008). The food industry employs two major components i.e. polysaccharides and proteins in order to manipulate the structural and textural stability. When the different or even the same components/biopolymers (carbohydrate–carbohydrate or protein-protein) are mixed, it can cause the phase separation. This separation can be divided into two cases: segregative, in which there is an unmixing of two phases and associative, which can form the complexes between the two biopolymers. The latter has been exploited widely in cosmetic, pharmaceutical, biotechnological, medical and food industries. The stability and rheology of the dairy products can be altered by generating he pectin-casein system in which pectin can either prevent the association of casein micelles or can lead to the aggregation (Maroziene and de Kruif 2000). The interaction of the pectin and milk protein can be brought by the calcium bridges between the two as these are generally not considered compatible at neutral pH. Similarly, Sadahira et al. (2016) reported that there was an increase in the foam stability and continuous phase viscosity as a result of the formation of complexes between egg white and pectin. Also when pectin was incorporated in low fat spreads with fish gelatin, there was an increase in the melt ability and consistency of such systems. The more the content of pectin in such spreads, more better was the performance in relation to melt in mouth and instant in-mouth flavour release effects.

Modification of Pectin

Pectin can be modified in order to achieve desired formulations such as gels, 3-D matrices, films and micro-/nano-particles. This can be studied by dividing the whole domain into two categories: (1) Developing new technologies for pectin isolation, purification, and characterization and (2) Modification of pectin macromolecules using physical, chemical and enzymatic methods. In terms of developing new technologies, various methods have been put forward by the workers and these are summarized below:

Developing New Technologies

Pulsed Electric Field Assisted Extraction (or PEF)

PEF has received a good response recently and has been employed in food and agro based industries food industry to inactive enzymes and pathogenic microorganisms in order to improve the quality and shelf life stability of milk, fruit juice, or wine (Jaeger et al. 2009). In a study by Ma and Wang (2013), PEF technology was used to modify the sugar beet pectin. The treatment chamber consisted of two parallel stainless-steel plate electrodes and a tubular Teflon. In order to enhance the electric field, Teflon was placed parallel to the planar electrode. The Pectin solutions of 25 gLA were pumped by peristaltic pumps. The pulse intensities 18, 24 and 30 kV cm were applied respectively. After PEF treatments, the samples were oven-dried and finely ground and analysed for the modifications. It has been reported that physicochemical properties of the chitosan and corn-starch showed variations with regards to electric field intensity, pulse width, and number of pulses (Han et al. 2009). In case of pectin, PEF increased the DE of pectin derivatives significantly with the intensity varying from 18 to 30 $kVcm^{-1}$. It also increased the thermal stability of the modified pectin as compared to the native one.

Enzyme-Assisted Extraction (or EAE)

This method has already been discussed above. The method has been found to be effective in extracting as well as modifying the pectin. With respect to modifications, a study of de-esterification of commercial HM pectin into LM using a synergistic application of high hydrostatic pressure and pectin methyl esterase (PME) enzyme was conducted by Zhao et al. (2015). Similarly, Zykwinska et al. (2008) also reported that LM pectin can be obtained from HM pectin source by incorporating PME alongside cellulases and proteases into the Enzyme assisted extraction process.

Microwave-Assisted Extraction (or MAE)

With respect to the pectin modifying effects of microwave assisted extraction, various works have been carried out as in case of pectin extracted from grapefruit by (Bagherian et al. 2011). The workers reported that using microwave power of 0.45, 0.63, and 0.9 kW not only increased the yield of pectin but also the values for DE increased generally with increasing microwave power and time (70.96–80.89%). Similarly Galacturonic acid unit contents were also found to be higher with increasing power and time, thus showing the modifying effect of this novel technique. Another notable feature of the microwave based extraction is that it takes less time approximately 15 min to obtain satisfactory extraction (Fishman et al. 2000). Also the yield efficiency and the quality of pectin obtained is better. In a study carried out by Fishman et al. 2000, orange albedo were subjected to microwave heating under high pressure for pectin extraction and it was observed that there was an increase in molar mass, size and intrinsic viscosity as compared to conventional methods.

Modification of Pectin Macromolecules

The pectin modification techniques can be broadly classified and studied under the subheadings including:

1. Enzymatic methods.
2. Physical methods such as ultrasonication, high pressure treatment, radiation, photolysis and microfluidization.
3. Chemical methods such as alkylation, amidation, quaternization, thiolation, sulfating and oxidation.

Modification of Pectin with Physical Treatment

In case of physical treatment such as ultra sound, a number of studies reveal the improved properties of pectin as a result of disaggregation of pectin polymers. For instance Muñoz-Almagro et al. (2017) reported that with the increase in time and intensity of ultrasound the depolymerisation of citrus and apple pectin increases although there was a reduction in polydisparity. Also there may no alteration in the primary structure of pectin, but this technique offers the advantages of transforming the pectins and formulating the pectin products with anti-oxidant properties (Ogutu and Mu 2017). Another study on physical modification of pectin, involving the technique dynamic high pressure microfluidization (DHPM) for black-cherry tomato pomace by was carried out by (Zhang et al. 2018). This treatment altered the rheological and physiochemical properties of the test material by increasing the average particle size which resulted in the formation of particles foamed filiform microparticles. The study also reported that there was a decrease in the apparent viscosity which however increased the fluidity and consistency for the solutions. Marquis et al. (2015) reports that this technique can provide for the formulation of monodisperse spherical and non spherical polymeric microparticles apart from providing the versatile geometry. In brief, microparticles of pectin hydrogels which vary in shape, (oblate ellipsoid, torus or mushroom type) and morphology can be generated using the technique of microfluidization.

Modification of Pectin with Enzymatic Treatment

A number of enzymes are utilized for modifying the pectin in a more precise way. These are commercially obtained and purified from sources like Aspergillus niger and include the enzymes like galactanase, arabinase, pectin methyl esterase (PME) and pectin acetyl esterase (PAE). The enzymes can produce highly modified pectins which has tremendous commercial applications. For instance pectins having high DM are frequently modified using PMEs (Owen et al. 2017). It is important to mention that the variation in PME can affect the final properties of pectin and further the degree As in case of processive PMEs, blockier distributions are possible because the enzyme is capable of moving along the chain and hydrolysing the consecutive methyl esters. This results in the presence of consecutive unmethylesterfied residues even at high DM. In case of non-processive PME, the enzyme after binding to HG chain removes only one methyl ester and later dissociates again which results in formation of residues that are randomly unmetylesterfied. The resulting pectins have different patterns of DM with peculiar calcium sensitivities, different degrees of network formation and polyelectrolytic effects. Another enzyme, i.e. PAE also functions in modifying the pectin by deacetyation of pectin. In a study carried out by Oosterveld et al. (2000) sugar beet pectin was incubated with the combination of PAE and pectin esterase in presence of Ca2þ ions releasing 13.8% of the acetyl groups 27.2% of methyl ester groups. One of the disadvantages of this treatment in such case the modified pectin obtained is resistant to the attack of pectin esterase due to the presence of acetyl groups which results in its non homgenity. The rheological and conformational properties of the pectin are dependent upon the degree of methoxylation. As a result, pectins having higher degree of methylesterification and galacturonic acid content can provide a good source of substrate for modification. There are few enzymes which allow for the modification of pectin and also helps in assessing the effects of various constituents in RG-I side chains on the gelling capabilities of pectin. These include endo-a-1,5-arabinanase and endo-b-1,4-galactanase. Their enzymatic treatment has various implications on the gelling temperature and gel strength of the debranched pectin. Following the reduction in neutral sugar concentration, it reduces the gelling temperature and gel strength in debranched pectin. The enzymatically extracted pectin possesses a higher degree of esterification and finds a great commercial application as so-called Ultra-Rapid-Set pectin. The unique property of this gel is that it gels rapidly at higher temperature.

Modification of Pectin with Chemical Treatment

A number of chemical methods are available in literature that has been utilized for the extraction of pectin. On such method which has been used widely includes alkylation. This method is broadly divided into two domains methoxylation and long chain. In the former, minimum carbon chains are introduced while the latter involves the introduction of more carbons to the carboxyl groups of pectin. When the pectin is treated with an alkali, it results in the generation of the demethyesterfied residues which are randomly distributed along the polymer backbones that are methylesterfied. These additions can change the structural, functional and physicochemical properties of the pectic polysaccharides. It has been reported by various workers that alkylation enhances the water solubility, gel strength, antibacterial activity and emulsifying cababilty. The synergistic interaction of some factors such as DM, pH, ionic strength Amidation and demethoxylation influences the overall properties of the modified pectins including water binding, emulsifying and interfacial properties. The reason for this can be attributed to the hydrophilic groups at the galacturonic acid which are increased during the alkaline treatment. Similarly upon acidic modification, there occurs the degradation of pectin which modifies its molecular weight and composition subsequently. Fracasso et al. (2018) reported the modification of citrus peels using alkaline and acidic treatment. The modified pectins displayed good water dissolving properties but poor strength gels, pseudoplastic rheological behaviour with less viscous behaviour and good stability.

Miscellaneous Modifications of Pectin by Blending with Proteins

Blending with Proteins

When pectin is blended with whey proteins, the resulting pectin-protein complex forms a gel which can be employed for the production of gel controlled low-fat mayonnaise (Sun et al. 2018). This treatment employs the interaction between the positively charged proteins and negatively charged proteins polysaccharides. Guo et al. (2018) reported an increase in the emulsifying properties of whey protein after they were treated with sugar beet pectin for modification. Similar results were obtained from citrus pectin (Koch et al. 2018). The mechanism can be understood on the basis that higher degree of glycosylation in the protein-pectin conjugates increases the emulsifying properties. Also it has been observed that the neutral region of pectin is more readily conjugated with whey protein isolate in comparison with the galacturonic region of pectin. The film incorporating the protein-pectin blends is usually prepared in presence of transglutaminase. These films are often employed for fresh-cut fruits and vegetables in order to prevent weight loss and microbial growth.

Modification of Pectin by Graft Copolymerization

The use of synthetic polymers for increasing the functionality of the polysaccharides is attracting great attention and has wide applications in immobilization of enzymes, drug delivery systems and bioseparations. This can be easily achieved by the process of graft copolymerization in which the pectin can be modified by using ceric ammonium nitrate, N,N,N0,N0-tetramethylethylenediamine (TEMED) and microwave-induced graft copolymerization of N,N-diethylacrylamide (Is¸õklan and Tokmak 2018). The formulated graft polymers have low critical solution temperatures of 31 °C, high thermal resistivity, biocompatibility as a biomaterial and are thermos-sensitive. Giacomazza et al. 2014 proposed that the development of a gelling system utilizing high esterified pectin and water-soluble maltose-conjugated chitosan that can gel rapidly as a model for tissue engineering. The hydrophobic modification of pectin involves the reaction of pectin with diacyl chlorides glutaryl and sebacoyl chloride) which yields mono-grafted (isolated chains) and bi-grafted (chemical gels) structures. The length of the carbon chains inserted as a result of acylation have a underlying effect on the yield of mono-grafted and bi-grafted structures (Seslija et al. 2018). The heterogeneity of the modified pectin, depicted by the entanglement of sticky polymer microparticles as compared to unmodified pectin is more and is more liable to heat degradation. There is a reduction in the glass transition temperature with the increase in size of the acyl chains on the polymer backbone. There is an enhancement of the hydrophobicity of the pectin as a result of acylation. The pectins modified with phenols are becoming more popular. The structure of pectin has a good amount of ferulic acid units but its abundance is not so high (Morris et al. 2010). When the enzymatic graft of the ferulic acid oxidation products is carried on, the process usually takes place in aqueous medium via catalyzation using Myceliophthora thermophila laccase as the biocatalyst (Karaki et al. 2017).

The reaction between the pectin and the ferulic acid oxidation products involves the alteration of thermal and morphological properties. The hygroscopic behaviour is less when the water activity is less than 0.5 and more when the water activity is above 0.5. Also the viscosity of the feruloylated pectin is less and exhibits a different calcium-dependent gelation behaviour. The antioxidant properties also display a significant improvement. Covalent bonds are found between the pectin carboxyl groups and the laccase-mediated ferulic acid oxidation products. So modification of pectin using phenols can help to achieve pectins with various properties of interest.

Modification of Pectin by Thiolation

Thiolated pectin from citrus has been obtained with thioglycolic acid and has been reported to be as a potent mucoadhesive delivery system for metformin (Sharma and Ahuja 2011).Similarly the use of thiolated lemon pectin for buccal delivery of drugs has also been reported by (Hauptstein et al. 2014).The modification in such cases takes place by conjugation of cysteine through the formation of amide bonds. In this conjugation, the major fraction of sulphur is present in thiol form where as the remaining takes part in disulfide bonding. The content of thiol group has a marked effect on the properties of thiolated pectin including gelation time, swelling ratio, viscosity, polymer concentration, hardness, elastic moduli and H2O2 concentration. The resulting gels also display microgel particles and pseudoplastic properties.

Sun Drying Modification

Sun drying as a technique for modifying pectin has been analysed by Asgar et al. (2003) in case of Japanese persimmon fruits. The workers reported a 44% increase in the yields, and sugar content of cold-water-soluble pectic polysaccharides (CWPs) during sun drying indicating the extensive solubilisation of pectins as a result of sun drying. The presence of galacturonic acid in the hydrolyzate of CWPs from fresh and sun-dried fruits was confirmed by paper chromatography. Table 1 shows that the yield of CWPs was increased about 44% during the sun-drying process.

Applications of Pectin

Delivery of Probiotics

Pectin finds use in the delivery of probiotics owing to its ability to form three dimensional and water insoluble gels (Lee et al. 2009) and exhibiting resistance to GI conditions (Cabrera et al. 2011). Pectin is less sensitive towards chemical agents and shows higher resistance to gastric conditions, than other polysaccharides like alginate. Voo et al. (2011) revealed that Pectin beads encapsulating poultry probiotic cells exhibited higher mechanical strength than those of alginate. Gebara et al. (2013) reported that pectin encapsulated Lactobacillus acidophilus exhibited lesser reduction (1.51 log cycles) than non-encapsulated cells (3.54 log cycles) when incubated in gastric and intestinal juices. Encapsulation of *Lactobacillus rhamnosus* with pectin also improved its viability in gastric environments at very low pH. Due to highly porous nature and weak binding to cross-linking agents, pectin finds less use in encapsulation (Chan et al. 2011). The workers suggested the use of starch or rice bran as filler in order to counter the porosity in pectin and to increase the protective effect for delivery. Chotiko and Sathivel (2016) used a combination of pectin and rice bran extract for probiotic delivery of Lactobacillus plantarum under gastric conditions that resulted in its increased viability. Increasing the rice bran content resulted in an increase in the sphericity and encapsulation efficiency of the pectin-rice bran capsules. Rice bran increased the viscosity of the pectin solutions, which is necessary for forming spherical beads or capsules. The study further revealed the creation of a mesh-like network in the calcium-pectinate capsules resulting in an increase in the diffusion path length of acid and bile in the gastric conditions, thereby reducing their penetration and thus enhancing cell viability. Encapsulation with sugar beet pectin increased the viability of *Lactobacillus salivarius* NRRL B-30514 under simulated gastrointestinal conditions. Khorasani and Shojaosadati (2016) designed various bionano composite formulations of bacterial nanocellulose (Gorrasi et al. 2012) and pectin for enhancing the survivability of probiotic *Bacillus coagulans* under drying and GI conditions. Out of various formulations, 50% BNC and 50% pectin were found to confer the highest survival rate, being 98.15–99.95% and 90.26–95.18% respectively. The study also revealed that the combination of pectin and BNC was more stable and effective in protecting the probiotic cells rather than pectin alone. The molecular interactions between pectin and other polysaccharides improve the resistance towards acidic and enzymatic hydrolysis. Addition of CMC in the pectin-BNC bio-nanocomposite increased the prebiotic count and provided resistance to drying and GI conditions whereas, bio-nanocomposite of starch and pectin conferred the thermal protection (Khorasani and Shojaosadati 2017a). The problem of pectin dissolution in the GI conditions can be checked out by using a composite of pectin and water insoluble polysaccharides. Cellulose protects collapse of pectin in presence of GI fluids, by restricting and reinforcing the polymeric chains to confined domains which decreases the permeability of GI fluids into the biocomposite. Owing to these properties, pectin-biocomposites could thus find an application in probiotic delivery systems (Khorasani and Shojaosadati 2017b).

As a Source of Prebiotics

Studies have suggested that pectin can act as a prebiotic functional package to deliver probiotics. Pectin and pectin derived products have been evaluated for their role as potential prebiotics. Microorganisms like clostridia and bacteroides have been found to increase using pectins, but the oligosaccharides derived from them antagonized their growth. Pectic-oligosaccharides are obtained by enzymatic hydrolysis (Concha and Zúñiga Hansen 2012), acid hydrolysis of the feedstock (Hu et al. 2009), hydrothermal processing (Gomez et al. 2014) and various other techniques. Pectic oligosaccharides show diverse biological activities such as anti-cancer, anti-ulcer, immunomodulation and also can be used as potential prebiotics (Hotchkiss et al. 2003). Pectins obtained from different fruits presented prebiotic potential that is examined by substrate consumption, distribution of metabolic products, and the effects on bacterial numbers. Gullón et al. (2013) revealed an increase in the metabolic products and bacterial count in the intestine using apple pomace pectic-oligosaccharides. Pectic-oligosaccharides obtained by subjecting apple pomace to dynamic high-pressure microfluidization increased the bifidobacteria and lactobacilli count and also led to an increase in the concentration of acetic, lactic, and propionic acid. Pectic-oligosaccharides obtained from Valencia orange albedo have been found to inhibit the invasion of intestinal cells by *Campylobacter jejuni*.

Pectic-oligosaccharides exhibited an anti-invasive effect to *C. jejuni* using an intestinal model of Caco-2 cells. Bacteria need multiple points of attachment to get effectively adhered to cells but pectic-oligosaccharides affect the efficiency of cell adhesion by saturating the receptor sites in cells and thereby form an interfere which prevents cell invasion (Ganan et al. 2010). Chen et al. (2013) revealed a considerable drop in the count of bacteroides and clostridia with the administration of pectic-oligosaccharides. Hydrolyzates comprising of monosaccharides obtained from the enzymatic treatment of citrus pectin also enhanced the growth and acid tolerance of Lactobacillus acidophilus and Bifidobacterium bifidum (Ho et al. 2017).Pectic-oligosaccharides obtained from lemon peel wastes and sugar beet pulp showed better prebiotic activities (Gomez et al. 2016). Michalak et al. (2012) reported the prebiotic activity of pectic-oligosaccharides obtained from potato. The study revealed a rise in the growth of Bifidobacterium longum and Lactobacillus acidophilus and prevent the growth of Clostridium perfringens. Parkar et al. (2010) examined kiwi fruit pectin obtained by the selective re-solubilisation of fruit fiber using chemical treatments and revealed a significant inhibition in the adhesion of *S. typhimurium* and an enhanced adhesion of *L. rhamnosus*.

As an Emylsifying Agent

Pectins obtained from different sources have been evaluated for emulsifying properties. Pectin obtained from okra (*Abelmosus esculentus L. Moench*) possesses hydrophobicity and surface activity at the o/w interface owing to its higher concentration of protein and acetyl groups that enables its use as an emulsifying agent. Emulsions obtained from okra pectin are good in stability due to formation of a rigid interface by the higher protein content. The ability to form good emulsions in acidic environments enables their use in fruit drinks or acidified milk products (Alba et al. 2013). Sugar beet pectin have been evaluated for its emulsifying behaviour and have been found to possesses excellent emulsifying properties. Many factors are responsible for emulsifying properties of sugar beet pectin such as highly branched polysaccharide structures, protein moiety, and acetyl groups, among all these factors protein is of vital importance (Zhang et al. 2014).Further evaluation on the emulsifying properties of sugar beet pectin revealed that not only protein plays a vital role in emulsification but, ferulic acid-araban/galactan-protein complexes and ferulic acid alone are also important in improving emulsifying capacity, emulsifying stability, and surface activity. Moreover, emulsifying properties of sugar beet pectin have also been correlated with the degree of methylation. Highly methylesterified pectin have been shown to reduce the interfacial tension between oil and water phase, owing to its hydrophobicity (due to presence of COOCH3– groups), thus giving pectin its emulsifying properties (Chen et al. 2016). Food-grade nano emulsions were prepared from high methoxylpectin containing essential oils such as oregano, thyme, and lemongrass. Nano emulsions containing lemongrass presented long term stability and absence of creaming over storage (Rosas et al. 2016). Emulsions containing high methoxyl pectin and resveratrol remained stable over seven days of storage and did not show phase separation (Benjasirimongkol and Sriamornsak 2016).On comparison, sugar beet and citrus pectin were found to lower surface tension effectively than apple pectin, whereas the later revealed less suitability for emulsification (Schmidt et al. 2015). Further studies into pectin emulsions can enable their use as delivery systems in food products.

Uses in Active Packaging and Food Preservation

Conventional plastic packaging materials possesses the problem of disposal and accumulation due to deficient recycling (Sutherland et al. 2010). Food packaging materials account for millions of tons of waste in landfills, making it an environmental issue and affecting all life forms. Researchers have come forward to address this problem and have emerged with some alternatives in the form of edible films and coatings for food packaging. Biopolymers such as polysaccharides, proteins, and lipids find wide use in the formulation of packaging films. Due to the ability to form strong insoluble polymer upon reacting with multivalent metal cations like calcium, pectin has been used in edible packaging (Espitia et al. 2014). Medeiros et al. (2012) prepared a nanomultilayer coating using pectin and chitosan. This nanomulti layer system prevented the deterioration of 'Tommy Atkins' mangoes. The antimicrobial and gas barrier properties contributed by chitosan along with low oxygen permeability of pectin helped in extending the shelf life of mangoes. As such pectin does not possess any antimicrobial property but can act as a carrier molecule for antimicrobials, antioxidants and such type of functional compounds. Apple pectin coating in combination with antimicrobials and antioxidants effectively arrested the microbial growth and browning of fresh-cut persimmon (Sanchis et al. 2016). Pectin, as a carrier of essential oils or their active constituents have been reported to preserve the acceptability and sensory attributes of fresh fruits. Application of pectin carrying citral and eugenol as functional compounds helped in effective storage of fresh raspberry for 7 days (Guerreiro et al. 2015). Pectin bio-nanohybrid film manufactured by incorporation of layered double hydroxide salicylate has been shown to extend the shelf-life of fresh apricots by preventing the mould formation, the main factor responsible for spoilage of uncoated fruit (Gorrasi and Bugatti 2016). However, there are some issues related to the mechanical and barrier properties of pectin films that could be addressed by certain modifications. The issue of oxygen permeability has been addressed by use of functionalized pectin. Functionalized pectin applied on polyethylene films reduced the oxygen permeability through the polymer which implies that the application of pectin derivatives could be useful for active food packaging (Calce et al. 2014). Pectin film incorporated with crystalline nanocellulose has been reported to increase tensile strength and reduce water vapour transmission (Chaichi et al. 2017). Mechanical and barrier properties of pectin films can also be improved by incorporation of layered double hydroxide based inorganic nanohybrid fillers. Pectin films incorporated with layered double hydroxide and antimicrobial molecules improved thermal resistance, mechanical, and barrier properties. Pectin composite films produced by incorporating silver nanoparticles resulted in improved thermal stability, mechanical strength, water vapour, and UV-light barrier properties (Shankar et al. 2016).

Nutraceutical Potential

Metabolism of Cholesterol

Cholesterol synthesis begins with acetyl CoA and acetoacetyl-CoA and is converted into two primary bile acids: cholic acid and chenodeoxycholic acid. Excessive cholesterol in the body is removed by the excretion of bile acids and free cholesterol but bile acids get reabsorbed in the small intestine and are transported to liver. Dietary fibers like pectin, psyllium and β–glucan enhance the excretion of bile acids thus reducing the serum cholesterol levels (Theuwissen and Mensink 2008). Dietary fibers lower the cholesterol levels by following possible mechanisms: (1) hindrance of bile salt reabsorption, (2) reduction in the rate of glucose absorption leading to decreased insulin production. Reduced insulin causes a reduction in HMG-CoA activity, thus reducing cholesterol synthesis, and (3) production of short chain fatty acids by colonic fermentation, hindering cholesterol synthesis in the liver (Gunness and Gidley 2010). Cholesterol lowering ability of pectin is due to its viscosity, molecular weight, degree of methoxylation, acetylation and degree of esterification. Another factor responsible for cholesterol-lowering effect of pectin is that it forms a viscous gel and binds cholesterol and bile acids, that promotes their excretion and reduces reabsorption. Also, absorption of cholesterol is hindered by disruption of micelle formation constituted by bile salts, phospholipids and fatty acids. Molecular weight of pectin contributes towards lowering of cholesterol by inhibiting lipid digestion, through viscosity effect. High molecular weight pectin form complex gels that trap water and lipids inside, thus increasing the viscosity of the gastrointestinal phases. Increase in gel viscosity limits the diffusion and the resulting interaction of lipids and lipases. Further, the entrapment of lipids inside the pectin gels make them inaccesible to be acted upon by lipases. This results in reduced lipolytic activity overall (Espinal-Ruiz et al. 2016).The degree of methoxylation alters lipid digestion through its interactive role in cation binding, thus helping in lowering of cholesterol. Calcium ions play a crucial role in lipid digestion. Calcium ions form insoluble soaps with long chain FFAs, thus making them inaccessible for lipases to act upon. High methoxylated pectin (HMP) and medium methoxylatedpectin (MMP) (Degree of Methoxylation = 52%) efficiently enclose the lipid droplets, thus hinder the adsorption of lipases. On the other hand, low methoxylated pectin (LMP) possesses an open structure due to repulsion with negative charge carrying lipid droplets, thus making lipids easily availability for lipases to act upon. Acetylation of pectins have been reported to decrease the interaction of pectin with bile acids

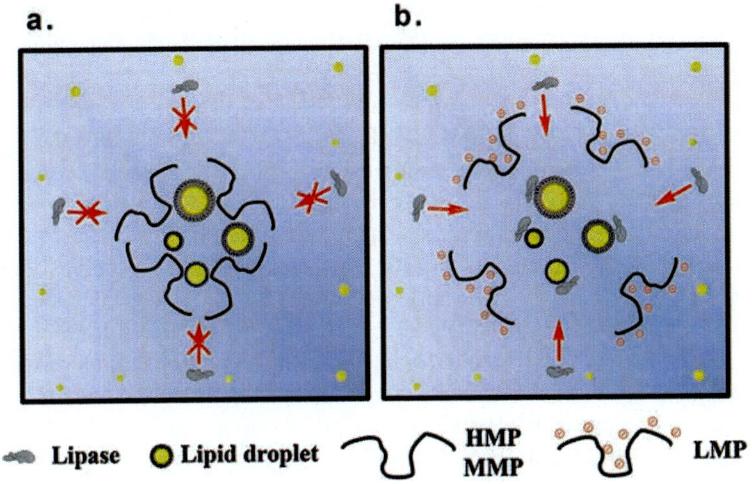

Fig. 7 Displaying the effect of High methoxylated pectin (HMP) and medium methoxylated pectin (MMP) on the inhibition of lipid droplet digestion by pectin. High methoxylated pectin (HMP) and medium methoxylated pectin (MMP) form a closed structure around the lipid droplets (depletion flocculation) which prevents the access of lipase to their surfaces thus preventing the lipid digestion (**a**), whereas low methoxylated pectin (LMP) promotes lipid digestion due to its open structure formed as a result of electrostatic repulsion with negatively charged lipid droplets, allowing the access of lipase to their surfaces (**b**). Adapted from Elsevier 2016

(Dongowski 1995). Pectins with higher degree of esterification i.e. hydrophobic substitutions have been found to lower plasma cholesterol, due to their lesser solubility (Anderson et al. 2009; Gronde et al. 2016). Pectins have been shown to promote droplet flocculation which slows down their digestion than the non-flocculated ones owing to the reduced surface area of lipids for lipases to act upon (Reis et al. 2009) (Fig. 7).

Role in Prevention of Cancer

Researchers have established the role of Pectin and its derivatives in the inhibition of metastasis. Flint et al. (2008) reported that the anti-tumour property of dietary pectin is mainly due to its role in protection of immune system and promotion of probiotic growth. Pectin derived from *Centella asiatica (L.)* Urban, have been reported to modulate the immunological activity of Band T cells owing to its carboxyl and acetyl groups (Wang et al. 2003). Inside the colon, dietary pectin is fermented into short-chain fatty acids that help the gut microflora to grow by lowering pH and by preventing the overgrowth of pathogens and also regulate the apoptotic proteins in the colonic crypts, and enhance growth of colocyte crypts (Louis et al. 2014). The in-vitro and in-vivo studies demonstrated anti-cancerous effects of apple pectin. In vitro studies in 4T1 breast cancer cells revealed that pectic acid (0.1%

concentration) induced apoptosis, inhibited cell growth and attachment, fragmented chromatin and also blocked the sub-G1 phase of the cell cycle. In vivo studies in mice as animal model of breast cancer, showed an inhibition of tumour progression and increased apoptosis (Delphi and Sepehri 2016). Moreover, pectin modified with heat, enzymes, and chemicals possesses better antitumor activity than the native or unmodified pectin. Modified pectins induce programmed cell death by the detachment of cells from its matrix i.e. 'anoikis' (Jackson et al. 2007). Heat modified citrus pectin has been reported to possess apoptotic action over prostate cancer cells (Jackson et al. 2007). Leclere et al. (2015) revealed cell death inducing activity in HepG2 and A549 cells of heat modified citrus pectin. The mechanism involved caspase 3 cleavage which is different from the classical apoptosis. Further research revealed that the cytotoxic activity is related to molecules with molecular weight corresponding to oligogalacturonic acid with lower degree of polymerization. Heat modification of galacturonic acid generates active molecules for instance, 4,5-dihy droxy-2-cyclopenten-1-one (cytotoxic molecule) from polygalacturonc acid through Maillard reactions. This molecule forms covalent adducts on cysteines, leading to protein denaturation and ubiquitination. Protein degradation brought by tubulin adducts results in prevention of mitosis. Ubiquitinylation of proteins inhibits proteosome, causing the activation of caspase 8 and hence apoptosis. Pectin obtained by the enzymatic hydrolysis of Glinus opposite folius showed better immune modulating effects than the native forms (Inngjerdingen et al. 2007). pH modified citrus pectin revealed an inhibitory action on tumor growth, metastases, and angiogenesis (Glinsky and Raz 2009). pH modified pectin increases solubility due to β-elimination causing a significant reduction in the DE and reduces the pectin chain length. Due to acid-hydrolysis and β-elimination certain domains in pectin molecule get enriched that are regarded as its pharmacophores (i.e. structures required for interaction with a biological target to trigger response). These pharmacophores are present in the RG-I domain and are mainly galactans rich in β-galactosides. Gal-3 (Galectin-3), a key target in metastasis recognizes these pharmacophores via a carbohydrate recognition domain. Gal-3 is a lectin that has been reported to show over-expression in metastatic cancers promoting cell migration and survival (Fortuna-Costa et al. 2014). Modified pectins that are rich in RG-I domains present β-galactosides to the carbohydrate recognition domain of Gal-3 efficiently, as they are comprise more of galactosides overxylanan and arabinan residues, which do not resist acid-hydrolysis effectively compared to galactans. When pro-metastatic Gal-3 gets binded, this blocks its interaction to other proteins and peptides, thereby arresting all possible pathways involving cell adhesion, migration and apoptosis. Cheng et al. (2013) reported considerable anti-tumor activity in okra and potato pectins, rich in RG-Idomains. Maxwell et al. (2016) studied apoptosis inducing activity of colon cancer cells in modified sugar beet pectin. Sugar beet pectin was found to be about 7 times rich in RG-I domains compared to citrus pectin. Pectin treated with alkali exhibited an increased ratio of RG-I to HG and also showed an enhancement in cancer cell apoptosis. Further insights into the mechanisms involving pectin could help in the treatment of cancers.

Conclusions

The applicability of pectin in food and agro based industries has increased its market and even the non–food industries are finding its use. The structure and physicochemical properties of pectin have a direct relation with its applications. These properties are largely influenced by the technique of the extraction used. All the techno functional properties like yield efficiency, molecular structure, viscosity, gelling properties have a profound effect of the extraction method. Conventionally pectin extraction process has been well established but with the growing demand, new methods need to be investigated and optimized to keep up with the scenario. Many novel techniques like ultrasound, microwaves, sub critical water and enzyme assisted extractions discussed above have been worked upon and to some extent optimized on lab scale but in an attempt to bring it on a commercial scale, further studies need to be carried upon. Lack of understanding limits the usage of this hydrocolloid so the need of the hour becomes to fully undertake the studies comparable those of Casas-Orozco et al. (2015) and Wiesenborn et al. (1999). Since the structure of pectin is associated with number of applications, which in turn is dependent upon the parent material or source of pectin, the method of extraction and the modification techniques, so it becomes necessary to understand these phenomenon on wider scale. As far as the nutraceutical properties of pectin is concerned, it has been found to act as prebiotic as well as the target delivery vehicle for probiotics. Also as the demand for the minimally processed foods and products is increasing need for developing new packaging materials becomes equally important and in this respect pectin edible films are finding their way into the market. Owing to their low cost, easy availability, biodegradability, pectin films can become a good choice for manufacturers. Also when these edible films are incorporated with some natural bioactive compounds, they were reported with antimicrobial activities against food borne pathogens. Again more studies need to be carried so as to achieve the desired functions. To sum up, pectin is a polysaccharide which could be tailor-made to obtain the novel benefits and bring forward the methods and techniques which could prove exceptional to food and non-food industry.

References

Adetunji, Ridwan, Adekunle, Ademola, Orsat, Valerie, Raghavan, &Vijaya. (2017). Advances in the pectin production process using novel extraction techniques: A review. *Food Hyrdocolloids, 62*, 239–250.

Alba, K., Ritzoulis, C., Georgiadis, N., & Kontogiorgos, V. (2013). Okra extracts as emulsifiers for acidic emulsions. *Food Research International, 54*, 1730–1737.

Anderson, J. W., Baird, P., Davis, R. H., Ferreri, J. S., Knudtson, M., Koraym, A., Waters, V., Christine L Williams, C. L. (2009). Health benefits of dietary fiber. *Nutrition Reviews, 67*, 188–205.

Anger, H., & Dongowski, G. (1988). Amidated pectins—Characterization and enzymatic degradation. *Food Hydrocolloids, 2*(5), 371–379.

Asgar, M. A., Yamauchi, R., & Kato, K. (2003). Modification of pectin in Japanese persimmon fruit during the sun-drying process. *Food Chemistry, 81*, 555–560.

Axelos, M. A. V., & Thibault, J. F. (1991a). The chemistry of low-methoxyl pectin gelation. In R. H. Walter (Ed.), *The chemistry and technology of pectin* (pp. 109–118). San Diego, CA: Academic Press.

Axelos, M. A. V., & Thibault, J. F. (1991b). Influence of the substituents of the carboxyl groups and of the rhamnose content on the solution properties and flexibility of pectins. *International Journal of Biological Macromolecules, 13*(2), 77–82.

Azmir, J., Zaidul, I. S. M., Rahman, M. M., Sharif, K. M., Mohamed, A., Sahena, F., Jahurul, M. H. A., Ghafoor, K., Norulainid, N. A. N., Omar, A. K. M. (2013). Techniques for extraction of bioactive compounds from plant materials: A review. *Journal of Food Engineering, 117*(4), 426–436.

Bagherian, H., Zokaee Ashtiani, F., Fouladitajar, A., & Mohtashamy, M. (2011). Comparisons between conventional, microwave- and ultrasound-assisted methods for extraction of pectin from grapefruit. *Chemical Engineering and Processing: Process Intensification, 50*(11–12), 1237–1243.

Basak, R., & Bandyopadhyay, R. (2014). Formation and rupture of Ca2+ induced pectin biopolymer gels. *Soft Matter, 10*(37), 7225–7233.

BeMiller, J. N. (1986). An introduction to pectins: Structure and properties. In *Chemistry and function of pectins* (pp. 2–12). Washington, DC: American Chemical Society.

Benjasirimongkol, P., & Sriamornsak, P. (2016). Stability study of resveratrol-loaded emulsions using pectin as an emulsifier. *Asian Journal of Pharmaceutical Sciences, 11*, 199–200.

Braccini, I., & Pérez, S. (2001). Molecular basis of Ca2+-induced gelation in alginates and pectins: The Egg-Box model revisited. *Biomacromolecules, 2*(4), 1089–1096.

Braconnot, H. (1825a). Nouvelles observations sur l'acide pectique. *Annales de Chimie et de Physique, 30*, 96–102.

Braconnot, H. (1825b). Recherches sur un nouvel acide universellement répandudans tous les végétaux. *Annales de Chimie et de Physique, 28*(2), 173–178.

Brejnholt, S. M. (2009). Pectin. In *Food stabilisers, thickeners and gelling agents* (pp. 237–265). Oxford, UK: Wiley-Blackwell.

Brett, C., & Waldron, K. (1990). Cell wall structure and the skeletal functions of the wall. In *Physiology and biochemistry of plant cell walls* (pp. 4–57). Dordrecht, The Netherlands: Springer.

Brunner, G. (2009). Near critical and supercritical water. Part I. Hydrolytic and hydrothermal processes. *The Journal of Supercritical Fluids, 47*(3), 373–381.

Cabrera, J. C., Cambier, P., & Cutsem, P. V. (2011). Drug encapsulation in pectin hydrogel beads–A systematic study of stimulated digestion media. *International Journal of Pharmacy and Pharmaceutical Sciences, 3*(5), 292–299.

Caffall, K. H., & Mohnen, D. (2009). The structure, function, and biosynthesis of plant cell wall pectic polysaccharides. *Carbohydrate Research, 344*(14), 1879–1900.

Calce, E., Mignogna, E., Bugatti, V., Galdiero, M., Vittoria, V., & De Luca, S. (2014). Pectin functionalized with natural fatty acids as antimicrobial agent. *International Journal of Biological Macromolecules, 68*, 28–32.

Capel, F., Nicolai, T., Durand, D., Boulenguer, P., & Langendorff, V. (2006). Calcium and acid induced gelation of (amidated) low methoxyl pectin. *Food Hydrocolloids, 20*(6), 901–907.

Casas-Orozco, D., Villa, A. L., Bustamante, F., & Gonzalez, L.-M. (2015). Process development and simulation of pectin extraction from orange peels. *Food and Bioproducts Processing, 96*, 86–98.

Chaichi, M., Hashemi, M., Badii, F., & Mohammadi, A. (2017). Preparation and characterization of a novel bionanocomposite edible film based on pectin and crystalline nanocellulose. *Carbohydrate Polymers, 157*, 167–175.

Chan, E. S., Wong, S. L., Lee, P. P., Lee, J. S., Ti, T. B., Zhang, Z., … Yim, Z. H. (2011). Effects of starch filler on the physical properties of lyophilized calcium-alginate beads and the viability of encapsulated cells. *Carbohydrate Polymers, 83*(1), 225–232.

Chan, S.-Y., & Choo, W.-S. (2013). Effect of extraction conditions on the yield and chemical properties of pectin from cocoa husks. *Food Chemistry, 141*(4), 3752–3758.

Chan, S. Y., Choo, W. S., Young, D. J., & JunLoh, X. (2017). Pectin as a rheology modifier: Origin, structure, commercial production and rheology. *Carbohydrate Polymers, 161*, 118–139.

Chen, H., Fu, X., & Luo, Z. (2016). Effect of molecular structure on emulsifying properties of sugar beet pulp pectin. *Food Hydrocolloids, 54*, 99–106.

Chen, H.-m., Fu, X., & Luo, Z.-g. (2015). Properties and extraction of pectin-enriched materials from sugar beet pulp by ultrasonic-assisted treatment combined with subcritical water. *Food Chemistry, 168*(0), 302–310.

Chen, J., Liang, R. H., Liu, W., Li, T., Liu, C. M., Wu, S. S., W, Z. J. (2013). Pectic oligosaccharides prepared by dynamic high-pressure microfluidization and their in vitro fermentation properties. *Carbohydrate Polymers, 91*, 175–182.

Chen, J., Liu, W., Liu, C.-M., Li, T., Liang, R.-H., & Luo, S.-J. (2014b). Pectin modifications: A review. *Critical Reviews in Food Science and Nutrition, 55*(12), 1684–1698.

Chen, Y., Zhang, J. G., Sun, H. J., & Wei, Z. H. (2014a). Pectin from Abelmoschus esculentus: Optimization of extraction and rheological properties. *International Journal of Biological Macromolecules, 70*, 498.

Cheng, H., Zhang, Z., Leng, J., Liu, D., Hao, M., Gao, X., … Zhou, Y. (2013). The inhibitory effects and mechanisms of rhamnogalacturonan I pectin from potato on HT-29 colon cancer cell proliferation and cell cycle progression. *International Journal of Food Science and Nutrition, 64*(1), 36–43.

Cho, Y.-J., & Hwang, J.-K. (2000). Modeling the yield and intrinsic viscosity of pectin in acidic solubilization of apple pomace. *Journal of Food Engineering, 44*(2), 85–89.

Chotiko, A., & Sathivel, S. (2016). Development of a combined low-methoxyl-pectin and rice-bran-extract delivery system to improve the viability of Lactobacillus plantarum under acid and bile conditions. *LWT – Food Science and Technology, 66*, 420–427.

Ciriminna, R., Chavarría-Hernandez, N., Ines Rodríguez Hernandez, A., & Pagliaro, M. (2015). Pectin: A new perspective from the biorefinery standpoint. *Biofuels, Bioproducts and Biorefining, 9*(4), 368–377.

Concha, J., & Zúñiga Hansen, M. E. (2012). Enzymatic depolymerization of sugar beet pulp: Production and characterization of pectin and pectic-oligosaccharides as a potential source for functional carbohydrates. *Chemical Engineering Journal, 192*, 29–36.

Constenla, D., & Lozano, J. E. (2003). Kinetic model of pectin demethylation. *Latin American Applied Research, 33*, 91–95.

da Silva, J. A. L., Goncalves, M. P., & Rao, M. A. (1995). Kinetics and thermal behaviour of the structure formation process in HMP/sucrose gelation. *International Journal of Biological Macromolecules, 17*(1), 25–32.

Daas, P. J. H., Meyer-Hansen, K., Schols, H. A., De Ruiter, G. A., & Voragen, A. G. J. (1999). Investigation of the non-esterified galacturonic acid distribution inpectin with endopolygalacturonase. *Carbohydrate Research, 318*(1–4), 135–145.

Delphi, L., & Sepehri, H. (2016). Apple pectin: A natural source for cancer suppression in 4T1 breast cancer cells in vitro and express p53 in mouse bearing 4T1 cancer tumors, in vivo. *Biomedicine and Pharmacotherapy, 84*, 637–644.

Denès, J.-M., Baron, A., Renard, C. M. G. C., Péan, C., & Drilleau, J.-F. (2000). Different action patterns for apple pectin methylesterase at pH 7.0 and 4.5. *Carbohydrate Research, 327*(4), 385–393.

Deuel, H. (1943). Pektin als hochmolekularer elektrolyt. *Mitteilungen aus dem Gebiete der Lebensmitteluntersuchung und Hygiene, 34*, 41–51.

do Nascimento, G. E., Simas-Tosin, F. F., Iacomini, M., Gorin, P. A. J., & Cordeiro, L. M. C. (2016). Rheological behavior of high methoxyl pectin from the pulp of tamarillo fruit (Solanum betaceum). *Carbohydrate Polymers, 139*, 125–130.

Dobies, M., Kozak, M., & Jurga, S. (2004). Studies of gelation process investigated by fast field cycling relaxometry and dynamical rheology: The case of aqueous low methoxyl pectin solution. *Solid State Nuclear Magnetic Resonance, 25*(1–3), 188–193.

Dominiak, M., Søndergaard, K. M., Wichmann, J., Vidal-Melgosa, S., Willats, W. G. T., Meyer, A. S., & Mikkelsen, J. D. (2014). Application of enzymes for efficient extraction, modification, and development of functional properties of lime pectin. *Food Hydrocolloids, 40*, 273–282.

Dongowski, G. (1995). Influence of pectin structure on the interaction with bile acids under in vitro conditions. *Zeitschrift für Lebensmittel-Untersuchung und Forschung, 201*, 390–398.

El-Nawawi, S. A., & Shehata, F. R. (1987). Extraction of pectin from Egyptian orange peel. Factors affecting the extraction. *Biological Wastes, 20*(4), 281–290.

Emaga, T. H., Ronkart, S. N., Robert, C., Wathelet, B., & Paquot, M. (2008). Characterisation of pectins extracted from banana peels (Musa AAA) under different conditions using an experimental design. *Food Chemistry, 108*, 463–471.

Espinal-Ruiz, M., Restrepo-Sanchez, L. P., Narvaez-Cuenca, C. E., & McClements, D. J. (2016). Impact of pectin properties on lipid digestion under simulated gastrointestinal conditions: Comparison of citrus and banana passion fruit (Passiflora tripartite var. mollissima) pectins. *Food Hydrocolloids, 52*, 329–342.

Espitia, P. J. P., Du, W., de Jesús Avena-Bustillos, R., Soares, N. D. F. F., & McHugh, T. H. (2014). Edible films from pectin: Physical-mechanical and antimicrobial properties—A review. *Food Hydrocolloids, 35*, 287–296.

Evageliou, V., Richardson, R. K., & Morris, E. R. (2000). Effect of pH, sugar type and thermal annealing on high-methoxy pectin gels. *Carbohydrate Polymers, 42*(3), 245–259.

Fishman, M. L., Chau, H. K., Hoagland, P., & Ayyad, K. (2000). Characterisation of pectin, flash-extracted from orange albedo by microwave heating, under pressure. *Carbohydrate Research, 323*, 126–138.

Flint, H. J., Bayer, E. A., Rincon, M. T., Lamed, R., & White, B. A. (2008). Polysaccharide utilization by gut bacteria: Potential for new insights from genomic analysis. *Nature Reviews Microbiology, 6*(2), 121–131.

Fortuna-Costa, A., Gomes, A. M., Kozlowski, E. O., Stelling, M. P., & Pavao, M. S. (2014). Extracellular galectin-3 in tumor progression and metastasis. *Frontiers in Oncology, 4*, 138. https://doi.org/10.3389/fonc.2014.00138

Fracasso, A. F., Perussello, C. A., Carpiné, D., Petkowicz, C. L. O., & Haminiuk, C. W. I. (2018). Chemical modification of citrus pectin: Structural, physical and rheological implications. *International Journal of Biological Macromolecules, 109*, 784–792.

Fraeye, I., De Roeck, A., Duvetter, T., Verlent, I., Hendrickx, M., & Van Loey, A. (2007). Influence of pectin properties and processing conditions on thermal pectin degradation. *Food Chemistry, 105*(2), 555–563.

Fu, J. T., & Rao, M. A. (2001). Rheology and structure development during gelation of low-methoxyl pectin gels: The effect of sucrose. *Food Hydrocolloids, 15*(1), 93–100.

Ganan, M., Collins, M., Rastall, R., Hotchkiss, A. T., Carrascosa, A. V., & Martinez-Rodriguez, A. J. (2010). *International Journal of Food Microbiology, 137*, 181–185.

Gebara, C., Chaves, K. S., Ribeiro, M. C. E., Souza, F. N., Grosso, C. R. F., & Gigante, M. L. (2013). Viability of Lactobacillus acidophilus La5 in pectin-whey protein microparticles during exposure to simulated gastrointestinal conditions. *Food Research International, 51*(2), 872–878.

Gentilini, R., Bozzini, S., Munarin, F., Petrini, P., Visai, L., & Tanzi, M. C. (2014). Pectins from Aloe Vera: Extraction and production of gels for regenerative medicine. *Journal of Applied Polymer Science, 131*(2). https://doi.org/10.1002/app.39760

Giacomazza, D., Sabatino, M. A., Catena, A., Leone, M., San Biagio, P. L., & Dispenza, C. (2014). Maltose-conjugated chitosans induce macroscopic gelation of pectin solutions at neutral pH. *Carbohydrate Polymers, 114*, 141–148.

Gidley, M. J., Morris, E. R., Murray, E. J., Powell, D. A., & Rees, D. A. (1979). Spectroscopic and stoicheiometric characterisation of the calcium-mediated association of pectate chains in gels and in the solid state. *Journal of the Chemical Society, Chemical Communications*, (22), 990–992. https://doi.org/10.1039/C39790000990

Gigli, J., Garnier, C., & Piazza, L. (2009). Rheological behaviour of low-methoxyl pectin gels over an extended frequency window. *Food Hydrocolloids, 23*(5), 1406–1412.

Gilsenan, P. M., Richardson, R. K., & Morris, E. R. (2000). Thermally reversible acid-induced gelation of low-methoxy pectin. *Carbohydrate Polymers, 41*(4), 339–349.

Glinsky, V. V., & Raz, A. (2009). Modified citrus pectin anti-metastatic properties: One bullet, multiple targets. *Carbohydrate Research, 344*(14), 1788–1791.

Gomez, B., Gullon, B., Remoroza, C., Schols, H. A., Parajo, J. C., & Alonso, J. L. (2014). Purification, characterization, and prebiotic properties of pectic oligosaccharides from orange peel wastes. *Journal of Agricultural and Food Chemistry, 62*(40), 9769–9782.

Gomez, B., Gullon, B., Yanez, R., Schols, H., & Alonso, J. L. (2016). Prebiotic potential of pectins and pectic oligosaccharides derived from lemon peel wastes and sugar beet pulp: A comparative evaluation. *Journal of Functional Foods, 20*, 108–121.

Gorrasi, G., & Bugatti, V. (2016). Edible bio-nano-hybrid coatings for food protection based on pectins and LDH-salicylate: Preparation and analysis of physical properties. *LWT - Food Science and Technology, 69*, 139–145.

Gorrasi, G., Bugatti, V., & Vittoria, V. (2012). Pectins filled with LDH-antimicrobial molecules: Preparation, characterization and physical properties. *Carbohydrate Polymers, 89*, 132–137.

Grant, G. T., Morris, E. R., Rees, D. A., Smith, P. J. C., & Thom, D. (1973). Biological interactions between polysaccharides and divalent cations: The egg-box model. *FEBS Letters, 32*(1), 195–198.

Gronde, T., Hartog, A., van Hees, C., Pellikaan, H., & Pieters, T. (2016). Systematic review of the mechanisms and evidences behind the hypocholesterolaemic effects of HPMC, pectin and chitosan in animal trials. *Food Chemistry, 199*, 746–759.

Guerreiro, A. C., Gago, C. M. L., Faleiro, M. L., Miguel, M. G. C., & Antunes, M. D. C. (2015). Raspberry fresh fruit quality as affected by pectin-and alginate-based edible coatings enriched with essential oils. *Scientia Horticulturae, 194*, 138–146.

Guillotin, S. E., Bakx, E. J., Boulenguer, P., Mazoyer, J., Schols, H. A., & Voragen, A. G. J. (2005). Populations having different GalA blocks characteristics are present in commercial pectins which are chemically similar but have different functionalities. *Carbohydrate Polymers, 60*(3), 391–398.

Guimarães, G. C., Coelho Júnior, M. C., & Garcia Rojas, E. E. (2009). Density and kinematic viscosity of pectin aqueous solution. *Journal of Chemical & Engineering Data, 54*(2), 662–667.

Gullón, B., Gómez, B., Martínez-Sabajanes, M., Yá˜nez, R., Parajó, J. C., & Alonso, J. L. (2013). Pectic oligosaccharides: Manufacture and functional properties. *Trends in Food Science and Technology, 30*, 153–161.

Gunness, P., & Gidley, M. J. (2010). Mechanisms underlying the cholesterol-lowering properties of soluble dietary fibre polysaccharides. *Food & Function, 1*(2), 149–155.

Guo, X., Guo, X., Yu, S., & Kong, F. (2018). Influences of the different chemical components of sugar beet pectin on the emulsifying performance of conjugates formed between sugar beet pectin and whey protein isolate. *Food Hydrocolloids, 82*, 1–10.

Han, Z., Zeng, X. A., Zhang, B. S., & Yu, S. J. (2009). Effects of pulsed electric fields (PEF) treatment on the properties of corn starch. *Journal of Food Engineering, 93*, 318–323.

Harholt, J., Suttangkakul, A., & Vibe Scheller, H. (2010). Biosynthesis of pectin. *Plant Physiology, 153*(2), 384–395.

Hauptstein, S., Hintzen, F., Müller, C., Ohm, M., & Bernkop-schnürch, A. (2014). Development and in vitro evaluation of a buccal drug delivery system based on preactivated thiolated pectin. *Drug Development and Industrial Pharmacy, 40*, 1530–1537.

Ho, Y. Y., Lin, C. M., & Wu, M. C. (2017). Evaluation of the prebiotic effects of citruspectin hydrolysates. *Journal of Food and Drug Analysis, 25*, 550. https://doi.org/10.1016/j.jfda.2016.11.014

Hoefler, A. C. (1991). Other pectin food products. In R. H. Walter & S. Taylor (Eds.), *The chemistry and technology of pectin* (pp. 51–66). San Diego, CA: Academic Press.

Hotchkiss, A. T., Jr., Olano-Martin, E., William, E. G., Gibson, G. R., & Rastall, R. A. (2003). Pectic oligosaccharides as prebiotics. In *Oligosaccharides in food and agriculture* (ACS symposium series) (Vol. 849, pp. 54–62). Washington, DC: American Chemical Society.

Hu, K., Liu, Q., Wang, S., & Ding, K. (2009). New oligosaccharides prepared by acid hydrolysis of the polysaccharides from Nerium indicum Mill and their anti-angiogenesis activities. *Carbohydrate Research, 344*, 198–203.

Iagher, F., Reicher, F., & Ganter, J. L. M. S. (2002). Structural and rheological properties of polysaccharides from mango (Mangifera indica L.) pulp. *International Journal of Biological Macromolecules, 31*(1–3), 9–17.

Iglesias, M. T., & Lozano, J. E. (2004). Extraction and characterization of sunflower pectin. *Journal of Food Engineering, 62*(3), 215–223.

Inngjerdingen, K. T., Patel, T. R., Chen, X., Kenne, L., Allen, S., Morris, G. A., Harding, S. E., Matsumoto, T., Diallo, D., & Yamada, H. (2007). Immunological and structural properties of a pectic polymer from Glinus Oppositifolius. *Glycobiology, 17*(12), 1299–1310.

Is¸õklan, N., & Tokmak, S. (2018). Microwave based synthesis and spectral characterization of thermo-sensitive poly (N,N-diethylacrylamide) grafted pectin copolymer. *International Jouranl of Biological Macromolecules*, 113, 669–680.

Jackson, C. L., Dreaden, T. M., Theobald, L, K., Tran, N. M., Beal, T. L., Eid, M., Gao, M. Y., Shirley, R. B., Stoffel, M. T., Kumar, M. V., & Mohnen, D. (2007). Pectin induces apoptosis in human prostate cancer cells: Correlation of apoptotic function with pectin structure. *Glycobiology, 17*(8), 805–819.

Jaeger, H., Meneses, N., & Knorr, D. (2009). Impact of PEF treatment inhomogeneity such as electric field distribution, flow characteristics and temperature effects on the inactivation of E. coli and milk alkaline phosphatase. *Innovative Food Science & Emerging Technologies, 10*, 470–480.

Jensen, S. V., Sørensen, S. O., & Rolin, C. (2013). US8592575 B2: USPTO.

Karaki, N., Aljawish, A., Muniglia, L., Bouguet-Bonnet, S., Leclerc, S., Paris, C., … Humeau-Virot, C. (2017). Functionalization of pectin with laccase-mediated oxidation products of ferulic acid. *Enzyme and Microbial Technology, 104*, 1–8.

Kaya, M., Sousa, A. G., Crépeau, M.-J., Sørensen, S. O., & Ralet, M.-C. (2014). Characterization of citrus pectin samples extracted under different conditions: Influence of acid type and pH of extraction. *Annals of Botany, 114*(6), 1319–1326.

Keijbets, M. J. H., & Pilnik, W. (1974). Elimination of pectin in the presence of anions and cations. *Carbohydrate Research, 33*(2), 359–362.

Khajavi, S. H., Kimura, Y., Oomori, T., Matsuno, R., & Adachi, S. (2005). Degradation kinetics of monosaccharides in subcritical water. *Journal of Food Engineering, 68*(3), 309–313.

Khorasani, A. C., & Shojaosadati, S. A. (2016). Bacterial nanocellulose-pectin bionanocomposites as prebiotics against drying and gastrointestinal condition. *International Journal of Biological Macromolecules, 83*, 9–18.

Khorasani, A. C., & Shojaosadati, S. A. (2017a). Pectin-non-starch nanofibers biocomposites as novel gastrointestinal-resistant prebiotics. *International Journal of Biological Macromolecules, 94*, 131–144.

Khorasani, A. C., & Shojaosadati, S. A. (2017b). Starch- and carboxymethylcellulose-coated bacterial nanocellulose-pectin bionanocomposite as novel protective prebiotic matrices. *Food Hydrocolloids, 63*, 273–285.

Koch, L., Hummel, L., Schuchmann, H. P., & Emin, M. A. (2018). Improving the emulsifying properties of whey protein isolate-citrus pectin blends by a novel reactive extrusion approach. *Journal of Food Engineering, 223*, 175–188.

Koubala, B. B., Christiaens, S., Kansci, G., Van Loey, A. M., & Hendrickx, M. E. (2014). Isolation and structural characterisation of papaya peel pectin. *Food Research International, 55*, 215–221.

Krall, S. M., & McFeeters, R. F. (1998). Pectin hydrolysis: Effect of temperature, degree of methylation, pH, and calcium on hydrolysis Rates. *Journal of Agricultural and Food Chemistry, 46*(4), 1311–1315.

Kravtchenko, T. P., Arnould, I., Voragen, A. G. J., & Pilnik, W. (1992a). Improvement of the selective depolymerization of pectic substances by chemical b-elimination in aqueous solution. *Carbohydrate Polymers, 19*(4), 237–242.

Kravtchenko, T. P., Berth, G., Voragen, A. G. J., & Pilnik, W. (1992b). Studies on the intermolecular distribution of industrial pectins by means of preparative size exclusion chromatography. *Carbohydrate Polymers, 18*(4), 253–263.

Leclere, L., Fransolet, M., Cote, F., Cambier, P., Arnould, T., Cutsem, P. V., & Michiels, C. (2015). Heat-modified citrus pectin induces apoptosis-like cell death and autophagy in HepG2 and A549 cancer cells. *Public Library of Science, 10*(3), e0115831. https://doi.org/10.1371/journal.pone.0115831

Lee, J. S., Kim, E. J., Chung, D., & Lee, H. G. (2009). Characteristics and antioxidant activity of catechin-loaded calcium pectinate gel beads prepared by internal gelation. *Colloids and Surfaces B: Biointerfaces, 74*(1), 17–22.

Liu, C., Cheng, F. F., Wang, J. M., Wan, Z. L., Sun, Y. E., & Yang, X. Q. (2016). Preparation and characterization of surface-active pectin from soya hulls by phosphate-assisted subcritical water combined with ultrasonic treatment. *International Journal of Food Science and Technology, 51*, 61–68.

Löfgren, C., Guillotin, S., Evenbratt, H., Schols, H., & Hermansson, A.-M. (2005). Effects of calcium, pH, and blockiness on kinetic rheological behavior and microstructure of HM pectin gels. *Biomacromolecules, 6*(2), 646–652.

Löfgren, C., Guillotin, S., & Hermansson, A.-M. (2006). Microstructure and kinetic rheological behavior of amidated and nonamidated LM pectin gels. *Biomacromolecules, 7*(1), 114–121.

Löfgren, C., & Hermansson, A.-M. (2007). Synergistic rheological behaviour of mixed HM/LM pectin gels. *Food Hydrocolloids, 21*(3), 480–486.

Loncin, M., & Merson, R. L. (1979). Equations related to the transfer of mass, heat, and momentum. In M. L. L. Merson (Ed.), *Food engineering. Principles and selected applications* (p. 11e40). New York, NY: Academic Press.

Lopes da Silva, J. A., & Rao, M. A. (2006). Pectins: Structure, functionality, and uses. In A. M. Stephen, G. O. Phillips, & P. A. Williams (Eds.), *Food polysaccharides and their applications* (2nd ed.). Boca Raton, FL: CRC/Taylor & Francis.

Louis, P., Hold, G. L., & Flint, H. J. (2014). The gut microbiota, bacterial metabolites and colorectal cancer. *Nature Reviews Microbiology, 12*(10), 661–672.

Low, Z. W., Chee, P. L., Kai, D., & Loh, X. J. (2015). The role of hydrogen bonding inalginate/poly(acrylamide-co-dimethylacrylamide) and alginate/poly(ethyleneglycol) methyl ether methacrylate-based tough hybrid hydrogels. *RSC Advances, 5*(71), 57678–57685.

Luque-Garcıa, J. L., & Luque de Castro, M. D. (2003). Ultrasound: A powerful tool for leaching. *TrAC Trends in Analytical Chemistry, 22*(1), 41–47.

Ma, S., & Wang, Z. (2013). Pulsed electric field-assisted modification of pectin from sugar beet pulp. *Carbohydrate Polymers, 92*, 1700–1704.

Mansel, B. W., Chu, C.-Y., Leis, A., Hemar, Y., Chen, H.-L., Lundin, L., & Williams, M. A. K. (2015). Zooming in: Structural investigations of rheologically characterized hydrogen-bonded low-methoxyl pectin networks. *Biomacromolecules, 16*(10), 3209–3216.

Maroziene, A., & de Kruif, C. G. (2000). Interaction of pectin and casein micelles. *Food Hydrocolloids, 14*(4), 391–394.

Marquis, M., Davy, J., Cathala, B., Fang, A., & Renard, D. (2015). Microfluidics assisted generation of innovative polysaccharide hydrogel microparticles. *Carbohydrate Polymers, 116*, 189–199.

Matubayasi, N. (1994). Matching-mismatching of water geometry and hydrophobic hydration. *Journal of the American Chemical Society, 116*(4), 1450–1456.

Maxwell, E. G., Colquhoun, I. J., Chau, H. K., Hotchkiss, A. T., Waldron, K. W., Morris, V. J., et al. (2016). Modified sugar beet pectin induces apoptosis of colon cancer cells via an interaction with the neutral sugar side-chains. *Carbohydrate Polymers, 136*, 923–929.

May, C. D. (1990). Industrial pectins: Sources, production and applications. *Carbohydrate Polymers, 12*(1), 79–99.

Medeiros, B. G. D. S., Pinheiro, A. C., Carneiro-da-Cunha, M. C., & Vicente, A. A. (2012). Development and characterization of a nanomultilayer coating of pectin and chitosan—Evaluation of its gas barrier properties and application on 'Tommy Atkins' mangoes. *Journal of Food Engineering, 110*, 457–464.

Mesbahi, G., Jamalian, J., & Farahnaky, A. (2005). A comparative study on functional properties of beet and citrus pectins in food systems. *Food Hydrocolloids, 19*(4), 731–738.

Michalak, M., Thomassen, L. V., Roytio, H., Ouwehand, A. C., Meyer, A. S., & Mikkelsen, J. D. (2012). Expression and characterization of an endo-1,4-β-galactanase from Emericella nidulans and Pichia pastoris for enzymatic design of potentially prebiotic oligosaccharides from potato galactans. *Enzyme and Microbial Technology, 50*(2), 121–129.

Minkov, S., Minchev, A., & Paev, K. (1996). Modelling of the hydrolysis and extraction of apple pectin. *Journal of Food Engineering, 29*(1), 107–113.

Morris, G. A., Ralet, M., Bonnin, E., Thibault, J., & Harding, S. E. (2010). Physical characterisation of the rhamnogalacturonan and homogalacturonan fractions of sugar beet (Beta vulgaris) pectin. *Carbohydrate Polymers, 82*, 1161–1167.

Muhammad, K., Mohd Zahari, N. I., Gannasin, S. P., Mohd Adzahan, N., & Bakar, J. (2014). High methoxyl pectin from dragon fruit (Hylocereus polyrhizus) peel. *Food Hydrocolloids, 42*(Part 2), 289–297.

Müller-Maatsch, J., Caligiani, A., Tedeschi, T., Elst, K., & Sforza, S. (2014). Simple and validated quantitative1H NMR method for the determination of methylation, acetylation, and feruloylation degree of pectin. *Journal of Agricultural and Food Chemistry, 62*(37), 9081–9087.

Muñoz-Almagro, N., Montilla, A., Moreno, F. J., & Villamiel, M. (2017). Modification of citrus and apple pectin by power ultrasound: Effects of acid and enzymatic treatment. *Ultrasonics Sonochemistry, 38*, 807–819.

Nussinovitch, A., & Hirashima, M. (2013). *Cooking innovations: Using hydrocolloids for thickening, gelling, and emulsification*. London, UK: CRC Press.

Oakenfull, D., & Scott, A. (1984). Hydrophobic interaction in the gelation of highmethoxyl pectins. *Journal of Food Science, 49*(4), 1093–1098.

Ogutu, F. O., & Mu, T. H. (2017). Ultrasonic degradation of sweet potato pectin and its antioxidant activity. *Ultrasonics Sonochemistry, 38*, 726–734.

Oosterveld, A., Beldman, G., Searle van Leeuwen, M. J. F., & Voragen, A. G. J. (2000). Effect of enzymatic deacetylation on gelation of sugar beet pectin in the presence of calcium. *Carbohydrate Polymers, 43*, 249–256.

Owen, J. L., Kent, L. M., Ralet, M. C., Cameron, R. G., & Williams, M. A. K. (2017). A tale of two pectins: Diverse fine structures can result from identical processive PME treatments on similar high DM substrates. *Carbohydrate Polymers, 168*, 365–373.

Panouille, M., Thibault, J.-F., & Bonnin, E. (2006). Cellulase and protease preparations can extract pectins from various plant byproducts. *Journal of Agricultural and Food Chemistry, 54*(23), 8926–8935.

Parkar, S. G., Redgate, E. L., Wibisono, R., Luo, X., Koh, E. T. H., & Schroder, R. (2010). Gut health benefits of kiwifruit pectins: Comparison with commercial functional polysaccharides. *Journal of Functional Foods, 2*, 210–218.

Powell, D. A., Morris, E. R., Gidley, M. J., & Rees, D. A. (1982). Conformations and interactions of pectins: II. Influence of residue sequence on chain association in calcium pectate gels. *Journal of Molecular Biology, 155*(4), 517–531.

Ptichkina, N. M., Markina, O. A., & Rumyantseva, G. N. (2008). Pectin extraction from pumpkin with the aid of microbial enzymes. *Food Hydrocolloids, 22*(1), 192–195.

Puri, M., Sharma, D., & Barrow, C. J. (2012). Enzyme-assisted extraction of bioactives from plants. *Trends in Biotechnology, 30*(1), 37–44.

Racape, E., Thibault, J. F., Reitsma, J. C. E., & Pilnik, W. (1989). Properties of amidated pectins. II. Polyelectrolyte behavior and calcium binding of amidated pectins and amidated pectic acids. *Biopolymers, 28*(8), 1435–1448.

Reis, P., Holmberg, K., Watzke, H., Leser, M. E., & Miller, R. (2009). Lipases at interfaces: A review. *Advances in Colloid and Interface Science, 147-148*, 237–250.

Reitsma, J. C. E., Thibault, J. F., & Pilnik, W. (1986). Properties of amidated pectins. I. Preparation and characterization of amidated pectins and amidated pecticacids. *Food Hydrocolloids, 1*(2), 121–127.

Renard, C. M. G. C., Crépeau, M.-J., & Thibault, J.-F. (1995). Structure of the repeating units in the rhamnogalacturonic backbone of apple, beet and citrus pectins. *Carbohydrate Research, 275*(1), 155–165.

Renard, C. M. G. C., & Thibault, J.-F. (1996). Degradation of pectins in alkaline conditions: Kinetics of demethylation. *Carbohydrate Research, 286*(0), 139–150.

Rhein-Knudsen, N., Ale, M. T., & Meyer, A. S. (2015). Seaweed hydrocolloid production: An update on enzyme assisted extraction and modification technologies. *Marine Drugs, 13*(6), 3340–3359.

Richardson, J. F., Harker, J. H., & Backhurst, J. R. (2002). *Coulson and Richardson's chemical engineering volume 2-Particle technology and separation processes* (5th ed.). Oxford, UK: Elsevier.

Rosas, G. M. I., Castro, J. M., Martinez, L. A. O., Trujillo, L. S., & Belloso, O. M. (2016). Long-term stability of food-grade nanoemulsions from high methoxyl pectin containing essential oils. *Food Hydrocolloids, 52*, 438–446.

Routray, W., & Orsat, V. (2012). Microwave-assisted extraction of flavonoids: A review. *Food and Bioprocess Technology, 5*(2), 409–424.

Sadahira, M. S., Rodrigues, M. I., Akhtar, M., Murray, B. S., & Netto, F. M. (2016). Effect of egg white protein-pectin electrostatic interactions in a high sugar content system on foaming and foam rheological properties. *Food Hydrocolloids, 58*, 1–10.

Sanchis, E., Ghidelli, C., Palou, L., Perez-Gago, M. B., Sheth, C., Mateo, M. (2016). Browning inhibition and microbial control in fresh-cut persimmon (Diospyroskaki Thunb. cv. Rojo Brillante) by apple pectin-based edible coatings. *Postharvest Biology and Technology, 112*, 186–193.

Sato, A. K., Oliveira, P., & Cunha, R. (2008). Rheology of mixed pectin solutions. *Food Biophysics, 3*(1), 100–109.

Savary, B. J., Hotchkiss, A. T., Fishman, M. L., Cameron, R. G., & Shatters, R. G. (2003). Development of a Valencia orange pectin methylesterase for generating novel pectin products. In F. Voragen, H. Schols, & R. Visser (Eds.), *Advances in pectin and pectinase research* (pp. 345–361). Dordrecht, The Netherlands: Springer.

Schmidt, U. S., Schmidt, K., Kurz, T., EndreB, H. U., & Schuchmann, H. P. (2015). Pectins of different origin and their performance in forming and stabilizing oil-in-water emulsions. *Food Hydrocolloids, 46*, 59–66.

Schols, H. A., & Voragen, A. G. J. (1996). Complex pectins: Structure elucidation using enzymes. In J. Visser & A. G. J. Voragen (Eds.), *Progress in biotechnology* (pp. 3–19). Amsterdam, The Netherlands: Elsevier.

Seshadri, R., Weiss, J., Hulbert, G. J., & Mount, J. (2003). Ultrasonic processing influences rheological and optical properties of high-methoxyl pectin dispersions. *Food Hydrocolloids, 17*(2), 191–197.

Seslija, S., Spasojevic, P., Panic, V., Dobrzynska-Mizera, M., Immirzi, B., Stevanovic, J., & Popovic, I. (2018). Physico-chemical evaluation of hydrophobically modified pectin derivatives: Step toward application. *International Journal of Biological Macromolecules, 113*, 924–932.

Shankar, S., Tanomrod, N., Rawdkuen, S., & Rhim, J. W. (2016). Preparation of pectin/silver nanoparticles composite films with UV-light barrier and properties. *International Journal of Biological Macromolecules, 92*, 842–849.

Sharma, R., & Ahuja, M. (2011). Thiolated pectin: Synthesis, characterization and evaluation as a mucoadhesive polymer. *Carbohydrate Polymers, 85*, 658–663.

Smidsrød, O., Haug, A., & Larsen, B. (1966). The influence of pH on the rate of hydrolysis of acidic polysaccharides. *Acta Chemica Scandinavica, 20*, 1026–1034.

Sousa, A. G., Nielsen, H. L., Armagan, I., Larsen, J., & Sorensen, S. O. (2015). The impact of rhamnogalacturonan-I side chain monosaccharides on the rheological properties of citrus pectin. *Food Hydrocolloids, 47*, 130–139.

Sun, C., Liu, R., Liang, B., Wu, T., Sui, W., & Zhang, M. (2018). Microparticulated whey protein-pectin complex: A texture-controllable gel for low-fat mayonnaise. *Food Research International, 108*, 151–160.

Sutherland, W. J., Clout, M., Cote, I. M., Daszak, P., Depledge, M. H., Fellman, L., Fleishman, E., Garthwaite, R., Gibbons, D. W., Lurio, J. D., Impey, A. J., Lickorish, F., Lindenmayer, D., Madgwick, J., Margerison, C., Maynard, T., Peck, L. S., Pretty, J., Prior, S., Redford., K. H., Scharlemann, J. P. W., Spalding, M., Watkinson, A. R. (2010). A horizon scan of global conservation issues for 2010. *Trends in Ecology & Evolution, 25*(1), 1–7.

Taylor, A. J. (1982). Intramolecular distribution of carboxyl groups in low methoxylpectins—A review. *Carbohydrate Polymers, 2*(1), 9–17.

Thakur, B. R., Singh, R. K., Handa, A. K., & Rao, M. A. (1997). Chemistry and uses of pectin—A review. *Critical Reviews in Food Science and Nutrition, 37*(1), 47–73.

Theuwissen, E., & Mensink, R. P. (2008). Water-soluble dietary fibers and cardiovascular disease. *Physiology and Behavior, 94*(2), 285–292.

Thibault, J.-F., & Ralet, M.-C. (2003). Physico-chemical properties of pectins in the cell walls and after extraction. In F. Voragen, H. Schols, & R. Visser (Eds.), *Advances in pectin and pectinase research* (pp. 91–105). Dordrecht, The Netherlands: Springer.

Ueno, H., Tanaka, M., Hosino, M., Sasaki, M., & Goto, M. (2008). Extraction of valuable compounds from the flavedo of Citrus junos using subcritical water. *Separation and Purification Technology, 62*(3), 513–516.

Vincken, J.-P., Schols, H. A., Oomen, R. J. F. J., McCann, M. C., Ulvskov, P., Voragen, A. G. J., & Visser, R. G. F. (2003). If homogalacturonan were a side chain of rhamnogalacturonan I. Implications for cell wall architecture. *Plant Physiology, 132*(4), 1781–1789.

Voo, W. P., Ravindra, P., Tey, B. T., & Chan, E. S. (2011). Comparison of alginate and pectin based beads for production of poultry probiotic cells. *Journal of Bioscience and Bioengineering, 111*(3), 294–299. https://doi.org/10.1016/j.jbiosc.2010.11

Voragen, A. J., Coenen, G.-J., Verhoef, R., & Schols, H. (2009). Pectin, a versatile polysaccharide present in plant cell walls. *Structural Chemistry, 20*(2), 263–275.

Vriesmann, L. C., & Petkowicz, C. L. O. (2013). Highly acetylated pectin from cacaopod husks (Theobroma cacao L.) forms gel. *Food Hydrocolloids, 33*(1), 58–65.

Walkinshaw, M. D., & Arnott, S. (1981). Conformations and interactions of pectins: II. Models for junction zones in pectinic acid and calcium pectate gels. *Journal of Molecular Biology, 153*(4), 1075–1085.

Wang, X. S., Dong, Q., Zuo, J. P., & Fang, J. N. (2003). Immunological activities and structure of pectin from Centella asiatica. *Carbohydrate Research, 338*(22), 2393–2402.

Whistler, R., & Daniel, J. (1993). Chapter 3. In O. Fennema (Ed.), *Quı'mica de los alimentos* (p. 42). Zaragoza, Spain: Editorial Acribia.

Wiesenborn, D. P., Wang, J., Chang, K. C., & Schwarz, J. G. (1999). Comparison of continuous and batch processes for pectin extraction from sunflower heads. *Industrial Crops and Products, 9*(3), 171–181.

Xu, Y., Zhang, L., Bailina, Y., Ge, Z., Ding, T., Ye, X., & Liu, D. (2014). Effects of ultrasound and/or heating on the extraction of pectin from grapefruit peel. *Journal of Food Engineering, 126*, 72–81.

Yapo, B. M. (2011). Pectic substances: From simple pectic polysaccharides to complex pectins—A new hypothetical model. *Carbohydrate Polymers, 86*(2), 373–385.

Yapo, B. M., Robert, C., Etienne, I., Wathelet, B., & Paquot, M. (2007). Effect of extraction conditions on the yield, purity and surface properties of sugar beet pulp pectin extracts. *Food Chemistry, 100*(4), 1356–1364.

Yuliarti, O., Goh, K. K. T., Matia-Merino, L., Mawson, J., & Brennan, C. (2015). Extraction and characterisation of pomace pectin from gold kiwifruit (Actinidiachinensis). *Food Chemistry, 187*, 290–296.

Zakaria, S. M., & Kamal, S. M. M. (2016). Subcritical water extraction of bioactive compounds from plants and Algae: Applications in pharmaceutical and food ingredients. *Food Engineering Reviews, 8*, 23–34.

Zhang, L., Shi, Z., Shangguan, W., Yapeng F., Katsuyoshi, N, Phillips, G, O., & Jiang, F. (2014). Emulsification properties of sugar beet pectin after modification with horseradish peroxidase. *Food Hydrocolloids, 43*, 1–7.

Zhang, L., Ye, X., Ding, T., Sun, X., Xu, Y., & Liu, D. (2013). Ultrasound effects on the degradation kinetics, structure, and rheological properties of apple pectin. *Ultrasonics Sonochemistry, 20*(1), 222–231.

Zhang, W., Xie, F., Lan, X., Gong, S., & Wang, Z. (2018). Characteristics of pectin from black cherry tomato waste modified by dynamic high-pressure microfluidization. *Journal of Food Engineering, 216*, 90–97.

Zhao, W., Guo, X., Pang, X., Gao, L., Liao, X., & Wu, J. (2015). Preparation and characterization of low methoxyl pectin by high hydrostatic pressure-assisted enzymatic treatment compared with enzymatic method under atmospheric pressure. *Food Hydrocolloids, 50*, 44–53.

Zouambia, Y., Ettoumi, K. Y., Krea, M., & Moulai-Mostefa, N. (2014). A new approach for pectin extraction: Electromagnetic induction heating. *Arabian Journal of Chemistry, 10*, 480. https://doi.org/10.1016/j.arabjc.2014.11.011

Zykwinska, A., Boiffard, M.-H., Kontkanen, H., Buchert, J., Thibault, J.-F., & Bonnin, E. (2008). Extraction of green labeled pectins and pectic oligosaccharides from plant byproducts. *Journal of Agricultural and Food Chemistry, 56*(19), 8926–8935.

Arabinoxylans

Asima Shah, F. A. Masoodi, Asir Gani, Adil Gani, Zanoor ul Ashraf, Nairah Noor, and Aala Fazli

Occurrence and Structure

Arabinoxylansare non-starch polysaccharides mainly localized in the endosperm cell walls, aleurone layer and the pericarp of cereal grains (Ayala-Soto et al. 2016). They are the pentose sugars made of arabinose and xylose and hence are often referred to as pentosans (Izydorczyk and Bilideris 2000). Arabinoxylans have been found in cereals, like rye, wheat, barley, oats, rice, corn, sorghum, and millets (Izydorczyk and Biliaderis 1995; Roubroeks et al. 2000; Sun et al. 2004). Among cereals rye has been reported as richest sources of arabinoxylans ranging between 8.2 and 10.0% (Nilsson et al. 1997). Sapirstein et al. (2013) reported total pentosan content in four Canadian whole grain wheat cultivars within the range from 5.45 to 6.72%. A positive relationship was found between the amount of arabinoxylan accumulated in wheat and the dry conditions (Coles et al. 1997). The total arabinoxylan content in ten covered and six naked barley cultivars ranged between 4.8 and 9.7% (Oscarsson et al. 1996). Aman (1987) reported an average arabinoxylan content of 6.9% in oats. Arabinoxylans have been reported in other sources such as linseed mucilage, rye grass, pangola grass, psyllum, banana peels and Ispaghula seed husk.

Pentosans from various cereals have similar molecular structures. However, they differ in the degree of polymerization, sequence of glycosidic linkages and their proportions, ratio of arabinose to xylose residues, pattern of substitution of the xylan backbone with arabinose residues and the number of other substituents, such as feruloyl groups or glucuronic acid residues (Izydorczyk and Biliaderis 1995). The pentosan structure mainly consists of β-(1→4)-D-xylopyranose units which forms the backbone to which arabinofuranose units are attached as side branches on C (O)-2 and/or C (O)-3 positions (Carvajal-Millan et al. 2007). The ratio arabinose to xylose

A. Shah · F. A. Masoodi (✉) · A. Gani · Z. ul Ashraf · N. Noor
Department of Food Science and Technology, University of Kashmir, Srinagar, Jammu and Kashmir, India

A. Gani, B. A. Ashwar (eds.), *Food biopolymers: Structural, functional and nutraceutical properties*, https://doi.org/10.1007/978-3-030-27061-2_7

indicates the degree of branching in arabinoxylans. The ratio of Arabinose to xylose may vary from 0.3 to 1.1 depending on the origin of arabinoxylans, being lower in endosperm than in bran (Rattan et al. 1994). A higher ratio has been associated with a lower content of 3-monosubstituted and unsubstituted xylose residues and a higher content of 2-monosubstituted and disubstituted xylose residues (Vinkx et al. 1991; Gruppen et al. 1992). Also, some of the arabinose moieties are ester-linked on C (O)-5 positions to hydroxycinnamic acids such as ferulic acid and p- coumaric acid (Ayala-Soto et al. 2016). In addition to monomeric ferulic acid, dimer and trimer structures of this compound has been reported in arabinoxylans extracted from wheat and corn bran (Lapierre et al. 2001). It has been reported that in water-soluble arabinoxylan, ferulic acid substituents are involved in the oxidative gelation by formation of dehydrodiferulic acid units (Geissmann and Neukom 1973). Rattan et al. (1994) reported that arabinoxylans extracted from several Canadian wheat varieties contained ferulic acid in the range of 0.63–1.37 mg of ferulic acid per gram of purified polymer. Ayala-Soto et al. (2015) reported the ferulic acid content of glucoarabinoxylans from three sorghum varieties in the range of 0.09–0.16 (μg FA/mg). The concentration of ferulic acid and p-coumaric acid of arabinoxylans extracted from different maize fiber sources has been reported in the range of 0.52–1.197, and 0.1–0.22 μg/mg extract, respectively (Ayala-Soto et al. 2014a, b). Dervilly-Pinel et al. (2001) found that the amount of ferulic acid in the wheat, barley, rye and triticale endosperm ranged from 18 to 60 residues per 10,000 xylose units with wheat and barley arabinoxylans containing more ferulic acid than those from rye and triticale. In general, arabinoxylans are heterogeneous molecule consisting of a range of structures with different degrees and substitution patterns. The molecular weight of arabinoxylans extracted from cereals varies with the origin and the method used for their estimation. Fincher and Stone (1986) reported molecular weight of water extractable wheat arabinoxylans in the range of 800,000–5,000,000 Da as estimated by gel filtration chromatography. Gruppen et al. (1992) reported a weight-average molecular weight of 850,000 for alkali extractable wheat arabinoxylans, as determined by laser light scattering analysis. Girhammar and Nair (1992) reported molecular weight of rye arabinoxylans in the range of 519,000–770,000 Da using high performance gel permeation chromatography. The weight average molecular weight of 124,000 Da was obtained for water-extractable arabinoxylan from rye as determined by multiple angle laser light scattering (Nilsson et al. 2000). Girhammar et al. (1986) reported that the molecular weight of water-extractable wheat arabinoxylans range from 65,000 to 66,000 Da as obtained by sedimentation (Fig. 1).

Classification, Isolation and Purification of Pentosans

Arabinoxylan are divided into water-extractable and water-unextractable fractions. The differences in the water extractability of the arabinoxylans is related to the pattern of substitution and the extent of physical entanglement, covalent ester bonding between the –OH group of arabinoxylans and –COOH group of uranic acids and the

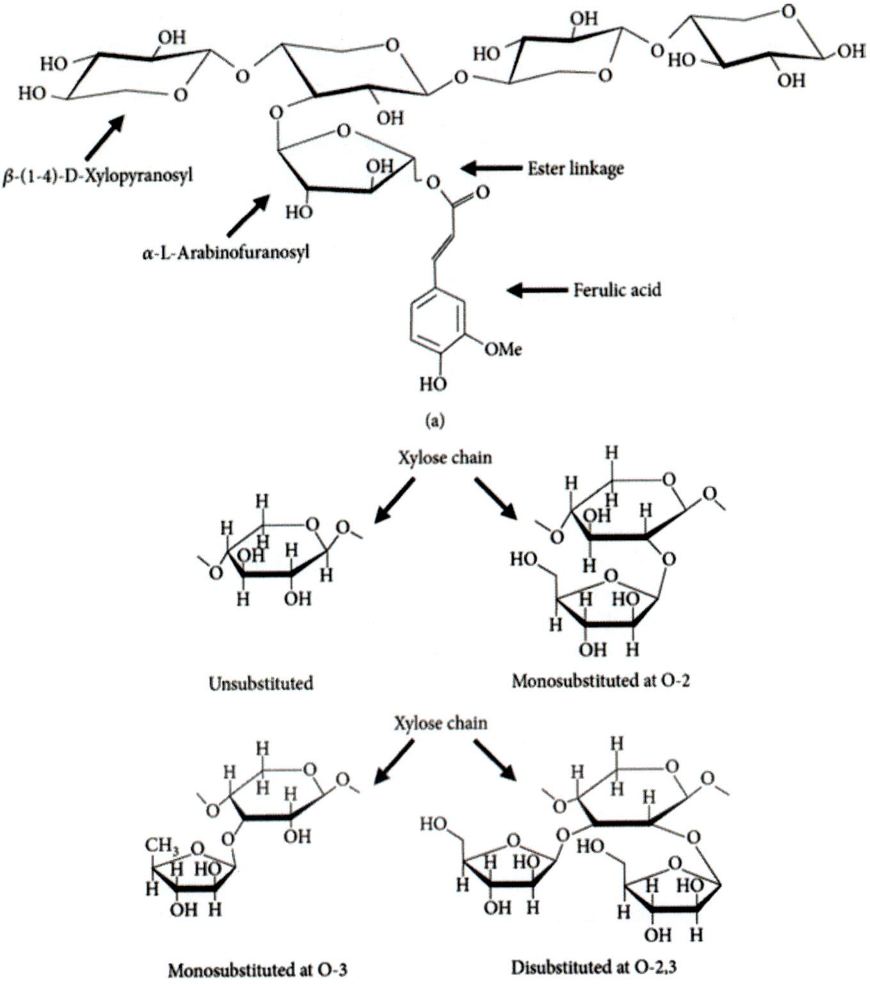

Fig. 1 Chemical structure of arabinoxylan. (Source: Mendez-Encinas et al. 2018)

formation of diferulic acid bridges between adjacent arabinoxylan chains (Geissmann and Neukom 1973; Gruppen et al. 1992). In cereals and millets, water-extractable pentosans mainly consist of arabinose, xylose and glucose in different proportions (Voragen et al. 1987) whereas water-unextractable pentosans consist of arabinose, xylose and glucose with small amounts of other sugars (Shyama and Muralikrishna 2004).

Arabinoxylan chains are linked to each other or other cell wall components through various covalent and non-covalent interactions. As a result of these cross-links, majority of arabinoxylans cannot be extracted with water, and therefore require treatments with alkali solutions to liberate them. Aqueous and alkali extraction are the most common methods used for isolating arabinoxylans from whole grains or specific plant tissues (Gruppen et al. 1992). Fengler and Marquardt (1988) reported use of aqueous extraction for isolating water soluble arabinoxylans from rye flour. Michniewicz et al. (1990) reported centrifugation of flour-water suspensions resulting in a sludge layer which is further purified with hydrolytic enzymes like protease and amylase to remove proteins and starch. Praznik et al. (2009) reported use of hot water extraction in combination with α-amylase for extraction of water soluble pentosans from rye flour. After cellite filtration, the clear filtrate was precipitated in ethanol—acetone solution. Revanappa et al. (2010) reported extraction of pentosans from hemicellulose of wheat varieties by glucoamylase digestion for 4 h. The polysaccharide was then precipitated in ethanol followed by centrifugation and freeze drying. Izydorczyk and Biliaderis (1993) obtained nearly protein free pentosan extracts from wheat flour after adsorption of contaminating proteins in the water extracts on various clays. Chemical, physical and enzymatic methods have been reported to extract arabinoxylans from wheat and rye bran (Figueroa-Espinoza et al. 2004). The extraction of pentosans from rye bran and flour was carried by alkaline method followed by addition of enzymes (α-amylase, amyloglucosidase, and protease). Furthermore, the pentosan was purified by ultra-dia-filtration through a ceramic membrane with a cut-off-level of 8 kDa (Mansberger et al. 2014). Tran and Nordin (1977) reported following methods for extraction of pentosans from wheat flour: (a) Flour extracted with water; (b) Flour extracted with 0.01 M sodium dodecyl sulfate in 0.1 M phosphate buffer pH 7.6; and (c) Flour extracted by stirring in about two volumes of 15% trichloroacetic acid.

Physicochemical Properties of Arabinoxylans

Viscosity

Arabinoxylan exhibit high viscosity in aqueous solution. The factors affecting the viscosity of arabinoxylans includes molecular weight, chain length, degree of substitution and substitution patterns, and ferulic acid cross-links (Izydorczyk and Biliaderis 1995; Dervilly-Pinel et al. 2001). Arabinoxylans with high ferulic acid content, high xylose/arabinose ratio, and low content of doubly substituted xylose on the backbone chain have high intrinsic viscosity (Izydorczyk and Bilideris 2000). The intrinsic viscosity of arabinoxylans was reported in the range of 2.5–3.1 dL/g (Medcalf et al. 1968), 0.8–3.1 dL/g (D'Appolonia and Macarthur 1975), and 2.75–5.48 dL/g (Izydorczyk et al. 1991). Fincher and Stone (1986) reported viscosity of 0.214 dL/g for dextran, and 0.195 dL/g for beet arabinan. Izydorczyk et al. (1991) reported viscosity of guar gum in the range of 2.3–6.75 dL/g and pectin with a low degree of esterification in the range of 3.38 dL/g. Izydorczyk and Biliaderis (1992) reported the intrinsic viscosity to vary between 3.4 and 8.5 dL/g for the different fractions of wheat arabinoxylans obtained by preparative gel permeation chromatography.

Water Holding Capacity

Both water extractable and unextractable pentosans are hydrophilic in nature. It has been reported that 23% of the water in dough is associated with pentosans. Izydorczyk et al. (1991) reported the hydration capacity of cross-linked pentosans and arabinoxylans from wheat varieties in the range of 42.7–59.9 (%) and 75.0–87.2 (%). Ghribi et al. (2015) reported the water holding capacity of 5.14 (g/g) for water-soluble polysaccharide from chickpea flours. Sciarini et al. (2009) reported water retention capacity of 15.20 (g/g) for galactomannans. The water holding capacity of water-soluble polysaccharides from potatoes peels was found to be 4.097 g/g (Jeddou et al. 2016). The water soluble polysaccharides isolated from pistachio and almond displayed a water holding capacity of 1.97 g/g (Sila et al. 2014).

Gelation

Arabinoxylans in the presence of free radical generating agents such as hydrogen peroxide, ammonium persulfate, ferric chloride, and linoleic acid/lipoxygenase are capable of forming three-dimensional gel network. Dimerization of ferulic acid which is linked to the arabinoxylans is considered to be responsible for oxidative gelation. Figueroa-Espinoza and Rouau (1998) reported that ferulic acid sites, phenolic ring, and propenoic chain, can participate in cross-linking reactions. The five main dimers of ferulic acid which have been identified in gelled arabinoxylans are 5-50, 8-50 benzo, 8-O-40, 8-50 and 8-80 forms in which 8-50 and 8-O-40 forms are found to be predominant (Nino-Medina et al. 2009). Izydorczyk et al. (1991) reported an increase in the development of a three-dimensional network when water-soluble pentosans from wheat varieties were treated with H_2O_2. Arabinoxylan gel structures are governed by both covalent and non-covalent linkages. The covalent interactions include di-FA, tri-FA bridges and non-covalent linkages are the weak hydrogen interactions (Vansteenkiste et al. 2004).

Foam Stabilization

Arabinoxylans can stabilize protein foams by (a) increasing the viscosity of interlamellar liquid; (b) reducing the drainage properties of the foam films; (c) intercession of interactions and crosslinks between proteins in the adsorbed layer; and (d) increased surface elasticity (Sarker et al. 1998). Izydorczyk et al. (1990) reported that arabinoxylans stabilize protein films against thermal disruption by increasing the viscosity of the solution and limiting foam formation.

Arabinoxylans and Baking Performance

Due to high water binding capacity and viscosity, arabinoxylans greatly influence the distribution of water and consistency in the dough. Michniewicz et al. (1991) reported that water soluble pentosans increases the water absorption of dough and enhance the loaf volume of bread. Buksa et al. (2010) reported bread with the largest loaf volume was obtained, when the wholemeal rye flour with highest water absorption and soluble pentosans was used. Maeda and Morita (2003) reported an increase in the loaf volume of wheat bread in the range of 3.79–4.38 (cm^3/g) on addition of water soluble pentosans in the range of 0.2–1.0%. However if the concentration of arabinoxylans is greater than optimum then there is increase in the viscosity of dough which in turn decreases the loaf expansion during oven spring (Kulp and Bechtel 1963). The fineness and homogeneity of crumb depends on the foam stability of dough. The arabinoxylans adds to the strength and extensibility of the gluten-starch films which is surrounding the gas bubbles and thereby reducing the rate of carbon dioxide diffusion from the dough (Izydorczyk and Bilideris 2000). Courtin and Delcour (1998) reported that crumbs of bread fortified with arabinoxylans are less firm than those of controls. This positive effect has been related to the increased moisture content of the samples. Water acts as a plasticizer of the gluten-starch composite matrix thereby lowering rigidity of the products. Biliaderis et al. (1995) reported the effects of high and low molecular weight arabinoxylan fractions on bread making quality. The arabinoxylans with higher molecular weight increases the farinograph water absorption to a greater extent than the low molecular weight counterparts. Adding 1–2% pentosans to poor wheat flours markedly improves dough properties (Shelton and D'Appolonia 1985; Rouau 1993). Another important property of arabinoxylans is associated to their role in delaying bread staling. Bread staling is a complex phenomenon involving loss of aroma, deterioration of crust characteristics and increase in crumb firmness (Izydorczyk and Biliaderis 1995; Shelton and D'Appolonia 1985). Biliaderis et al. (1995) reported that after seven day storage period, the arabinoxylans fortified breads exhibited lower crumb firmness than the controls. Kinetic studies indicated that adding pentosans decreased the amount of starch components available for crystallization and thus decreased the bread staling rate. The observed effects were also attributed to a higher moisture content of breads substituted with arabinoxylans and to the plasticizing effects of water (Shelton and D'Appolonia 1985). Also, breads with added arabinoxylans displays increased rates in starch retrogradation mostly because of their high moisture content (Biliaderis et al. 1995). Zeleznak and Hoseney (1986) reported increase in rate of retrogradation as the moisture content increased from 20 to 45% in arabinoxylan containing wheat breads.

Modification of Arabinoxylans

Arabinoxylans have various food applications such as thickener, stabilizer, an emulsifier, film former, and water binder (Ayala-Soto et al. 2016; Saeed et al. 2011). Also, arabinoxylans exhibits their nutritional benefits as soluble and insoluble fiber and the presence of ferulic acid having some antioxidant properties (Katapodis et al. 2003). The degree of branching and spatial arrangement of arabinoxylans recovered by different methods can influence nutraceutical potential (Saulnier et al. 2007). Also, the structural characteristics and molecular weight of arabinoxylans can affect the techno-functional properties of the products (Sun et al. 2011). So in order to acquire the improved functional, technological and nutraceutical properties, various physical, chemical, and enzymatic methods have been used to modify arabinoxylan molecule.

Physical Modification

Physical modification of arabinoxylans includes ultrasonic and microwave assisted modification. Arabinoxylans subjected to ultrasonic power (120–480 W) and treat time (15–45 min) resulted in increased phenolic compounds and antioxidant properties whereas the chemical composition remained unchanged upon the treatment. The molecular weight, main chain composition and monosaccharide constituents of arabinoxylans significantly decreased after ultrasound treatment. Upon ultrasonication arabinoxylans resulted in low viscosity gels (Li et al. 2016). Microwave irradiation is the another method for modification, when applied in the temperature-time range of 160–200 °C for 2–20 min results in production of arabinoxylans with a wide range of molecular weights and some of them being highly feruloylated. These feruloylated arabinoxylans may confer health benefits by preventing oxidation stress and having prebiotic effects (Rose and Inglett 2010).

Chemical Modification

The chemical modifications include carboxymethylation, arabinoxylan Na salt formation, and esterification. Carboxymethylation of arabinoxylan with sodium monochloroacetate resulted in replacement of –OH group with carboxymethyl group. Carboxymethylation of arabinoxylan yielded water soluble derivative with antimicrobial activity. The carboxymethylation of arabinoxylan represents a useful way to get water soluble arabinoxylan derivatives with anionic functions (Erum et al. 2015). Bhatia and Ahuja (2015) reported that carboxymethylation of arabinoxylan from psyllium resulted in increased crystallinity, decreased viscosity and improved thermal stability. In arabinoxylan-Na salt formation, antimicrobial activity was observed to be concentration dependent with partial solubility in water (Erum et al. 2015). In esterification, arabinoxylans are treated with a C2–C4 aliphatic anhydride using methanesulfonic acid as a catalyst or trifluoroacetic anhydride as a promoter. The resulted arabinoxylan esters are amorphous with high molecular weight, soluble in a range of solvents and thermally stable upto 200 °C (Buchanan et al. 2003).

Enzyme Modification

Enzyme hydrolysis is an efficient method for modifying the microstructure of arabinoxylans. Enzymes degrade arabinoxylanin order to achieve better functionality in dough development. Courtin and Delcour (2002) reported use of endoxylanases which are hydrolytic enzyme to improve dough handling loaf volume and oven spring. Yeast fermentation of wheat bran resulted in solubilization of arabinoxylans along with increased levels of level of folates and phenolic. Also, addition of yeast fermented bran into wheat flour resulted in bread with improved volume and crumb softness (Katina et al. 2012). Pitkanen et al. (2011) reported use of two α-L-arabinofuranosidases to modify the arabinose to xylose ratio of arabinoxylans which resulted in a significant effect on the solution properties of arabinoxylans. Pretreatment of arabinoxylans (water-unextratable) from wheat with xylanase, results in oligomers which are better utilized by the gut bacteria, thereby increasing their prebiotic potential (Vardakou et al. 2008).

Nutraceutical Potential of Arabinoxylan

The biological property of arabinoxylans which includes antioxidant, anticancer and prebiotic properties depends on its structural characteristics and content of phenolic acids in particular ferulic acid (Mendez-Encinas et al. 2018). The anticancer activity of arabinoxylans is related to its antioxidant, prebiotic, and gelation properties. It has been reported that highly feruloylated arabinoxylans are less fermented than arabinoxylans without ferulic acid, which has a great advantage of increasing the growth of probiotic bacteria, such as Lactobacillus and Bifidobacterium, which are capable of producing ferulic acid esterases to release ferulic acid from feruloylated arabinoxylans and inhibiting the growth of some non-beneficial bacteria (Snelders et al. 2014; Szwajgier and Jakubczyk 2010). The fermentation of arabinoxylans results in the production of short-chain fatty acids (SCFA) like acetic, propionic, and butyric acids, which has immunomodulatory and anti-inflammatory properties and plays an important role in the maintenance of healthy gut (Bach Knudsen 2015; Riviere et al. 2016). Butyric acid is the main energy source for the epithelial cells of human colon and the proliferation of bacteria producing butyric acid protects the colon from pathogenic microflora besides having immunomodulatory and anti-inflammatory properties (Riviere et al. 2016; Yacoubi et al. 2016). Arabinoxylans, because of these reasons is considered as a prebiotic with excellent health benefits, especially those related with the prevention of colon cancer. Ghoneum (2016) reported that arabinoxylans can stimulate the production of cytokines IL-2 and IL-12 which are the main anticancer cytokines in humans. Arabinoxylans exert antioxidant effects by increasing the activity of endogenous antioxidant enzymes, suppressing lipid peroxidation and protecting against oxidative stress (Noaman et al. 2008). Phenolic acids in wheat bran can prevent the oxidation of fat and protein by the free radical scavenging system and exert anti-inflammatory and anti-cancerous effects (Onipe et al. 2015). Katapodis et al. 2003) reported that a feruloylated oligosaccharide from wheat flour showed a high antioxidant activity as evaluated by DPPH scavenging activity and inhibition of copper-mediated oxidation of human low density lipoprotein. Li et al. (2015) reported that arabinoxylans from wheat endosperm produce nitric oxide (NO) and interleukin 8 (IL-8) from U937 and Caco-2 cell lines reflecting their immunomodulatory potential. Arabinoxylans play vigorous role in the management of diabetes inhuman health. Arabinoxylans lowers blood sugar and can be used as a sweetener, base material for the production of vitamin B_2 (Kaats et al. 2011; Aguedo et al. 2013). Lu et al. (2000) showed that fortification of arabinoxylans in order to increase fiber content in bread, lowered postprandial glucose and insulin responses in humans. Arabinoxylans possess viscosity enhancing properties, which affects the degree of contact of food substrates with the enzymes that digest them in the small intestine. In this way, arabinoxylans delays gastric emptying and reduce small intestine motility, which in turn results in delayed glucose absorption. Arabinoxylans have the ability to reduce the risk of coronary heart disease and paly role in weight management system (Hromadkova et al. 2012).

Recent Advances in Application of Arabinoxylans

Arabinoxylans are receiving increased attention as delivery vehicles due to its three-dimensional porous structure, ability to swell on contact with water and dietary fiber nature (Carvajal-Millan et al. 2007; Hernandez-Espinoza et al. 2012). Berlanga-Reyes et al. (2009) reported use of arabinoxylan gels as a colon-specific protein delivery vehicle. Vansteenkiste et al. (2004) demonstrated that in vitro pepsin proteolysis of BSA (bovine serum albumin) can be prevented by embedding BSA in arabinoxylan gels. Arabinoxylan gels could also be used as matrices for targeted delivery of insulin in the colon (Berlanga-Reyes 2008; Berlanga-Reyes et al. 2009). Iravani et al. (2011) reported encapsulation of caffeine in arabinoxylan hydrogels for its controlled release. A study reported that a probiotic bacteria *Bifidobacterium longum* was entrapped in arabinoxylan gels and their microstructure resembles that of an imperfect honeycomb. This expands the use of arabinoxylans as a wall material for encapsulation of probiotics (Morales-Ortega et al. 2014). Carvajal-Millan et al. (2005)studied the capacity of arabinoxylan gels with different FA contents to load ovalbumin. Gelling capacity of arabinoxylans result to be of great interest for biomedical and pharmaceutical applications

Conclusion

Arabinoxylans are the reservoirs of large amounts of ferulic acid and other phenolic acids which are covalently linked to them. Phenolic acids are involved in defense including protection against fungal pathogen. This chapter describes the functional properties of arabinoxylans and their potential application as antioxidant and anticancer agents. Moreover, its antioxidant activity exerts a protective effect against the free radicals and oxidative stress. This study evaluated the influences of various modifications on the molecular weight of Arabinoxylans. Modification of arabinoxylan resulted in products carboxymethyl arabinoxylan, etherified arabinoxylans and arabinoxylan-Na with altered structural and technofunctional properties. The nutraceutical perspectives were also highlighted as they are helpful in regulating blood cholesterol which thereby protects the body from cardiovascular disorders like atherosclerosis and post-prandial glycemic response reduction. Arabinoxylans are also involved in development of novel colon-targeted delivery systems for the treatment of colon associated diseases, such as inflammatory bowel disease, irritable bowel disease, and particularly colon cancer.

References

Aguedo, M., Vanderghem, C., Goffin, D., Richel, A., & Paquot, M. (2013). Fast and high yield recovery of arabinose from destarched wheat bran. *Industrial Crops and Products, 43*, 318–325.

Alia, E., Sajid, B., & Shazia, S. (2015). Modified and unmodified arabinoxylans from Plantago ovata husk: Novel excipients with antimicrobial potential. *Bangladesh Journal of Pharmacology, 10*, 765–769.

Aman, P., & Graham, H. (1987). Analysis of total and insoluble mixed-linked (1→3), (1→4)-β-D-glucans in barley and oats. Journal of Agricultural and Food Chemistry, 35, 704-709.

Ayala-Soto, F. E., Serna-Saldivar, S. O., Garcia-Lara, S., & Perez-Carrillo, E. (2014a). Hydroxycinnamic acids, sugar composition and antioxidant capacity of arabinoxylans extracted from different maize fiber sources. *Food Hydrocolloids, 35*, 471–475.

Ayala-Soto, F. E., Serna-Saldivar, S. O., Perez-Carrillo, E., & Garcia-Lara, S. (2014b). Relationship between hydroxycinnamic profile with gelation capacity and rheological properties of arabinoxylans extracted from different maize fiber sources. *Food Hydrocolloids, 39*, 280–285.

Ayala-Soto, F. E., Serna-Saldívar, S. O., & Welti-Chanes, J. (2016). Effect of processing time, temperature and alkali concentration on yield extraction, structure and gelling properties n of corn fiber arabinoxylans. *Food Hydrocolloids, 60*, 21–28.

Ayala-Soto, F. E., Serna-Saldivar, S. O., Welti-Chanes, J., & Gutierrez-Uribe, J. A. (2015). Phenolic compounds, antioxidant capacity and gelling properties of glucoarabinoxylans from three types of sorghum brans. *Journal of Cereal Science, 65*, 277–284.

Bach Knudsen, K. E. (2015). Microbial degradation of whole-grain complex carbohydrates and impact on short-chain fatty acids and health. *Advances in Nutrition, 6*, 206–213.

Berlanga-Reyes, C. M. (2008). *Study on the controlled release of insulin from Maize arabinoxylan gels*. Dissertation. Research Center for Food and Development, CIAD, A. C.

Berlanga-Reyes, C. M., Carvajal-Millan, E., Lizardi-Mendoza, J., Márquez-Escalante, J. A., Martínez-López, A. L. (2009). Maize arabinoxylan gels as protein delivery matrices. Molecules, 14, 1475–1482

Bhatia, M., & Ahuja, M. (2015). Psyllium arabinoxylan: Carboxymethylation, characterization and evaluation for nanoparticulate drug delivery. *International Journal of Biological Macromolecules, 72*, 495–501.

Biliaderis, C. G., Izydorczyk, M. S., & Rattan, O. (1995). Effect of arabinoxylans on bread making quality of wheat flours. *Food Chemistry, 5*, 165–172.

Buchanan, C. M., Buchanan, N. L., Debenham, J. S., Gatenholm, P., Jacobsson, M., Sheltona, M. C., … Wood, M. D. (2003). Preparation and characterization of arabinoxylan esters and arabinoxylan ester/cellulose ester polymer blends. *Carbohydrate Polymers, 52*(4), 345–357.

Carvajal-Millan, E., Guilbert, S., Doublier, J. L., Micard, V. (2005). Arabinoxylan/protein gels: Structural, rheological and controlled release properties. Food Hydrocolloids, 20, 53–61.

Carvajal-Millan, E., Rascon-Chu, A., Marquez-Escalante, J., Micard, V., Ponce de Leon, N., & Gardea, A. (2007). Maize extraction gum: Characterization and functional properties. *Carbohydrate Polymers, 69*, 280–285.

Coles, G. D., Hartunian-Sowa, P. D., Jamiesona, P. D., Haya, A. J., Atwel, W. A., Fulcher, R. G. (1997). Environmentally induced variation in starch and non-starch polysaccharide content in wheat. Journal of Cereal Science, 26, 47–54.

Courtin, C. M., & Delcour, J. A. (1998). Physico-chemical and bread-making properties of low molecular weight wheat-derived arabinoxylans. *Journal of Agricultural and Food Chemistry, 46*, 4066–4073.

Courtin, C. M., & Delcour, J. A. (2002). Arabinoxylans and endoxylanases in wheat flour bread-making. *Journal of Cereal Science, 35*, 225–243.

D'Appolonia, B. L., & Macarthur, L. A. (1975). Comparison of starch, pentosans, and sugars of some conventional height and semi dwarf hard red spring wheat flours. *Cereal Chemistry, 52*, 230–239.

Dervilly-Pinel, G., Rimsten, L., Saulnier, L., Andersson, R., & Aman, P. (2001). Water-extractable arabinoxylans from pearled flours of wheat, barley, rye and triticale. Evidence for the presence of ferulic acid dimers and their involvement in gel formation. *Journal of Cereal Science, 34*, 207–214.

Erum, A., Bashir, S., & Saghir, S. (2015). Modified and unmodified arabinoxylans from *Plantago ovata* husk: Novel excipients with antimicrobial potential. *Bangladesh Journal of Pharmacology, 10*, 765–769.

Fengler, A. I., & Marquardt, R. R. (1988). Water-soluble pentosans from Rye: II. Effects on rate of dialysis and on the retention of nutrients by the chick. *Cereal Chemistry, 9*(65), 298–302.

Figueroa-Espinoza, M. C., & Rouau, X. (1998). Effect of cysteinyl caffeic acid, caffeic acid, and l-dopa on the oxidative cross-linking of feruloylated arabinoxylans by a fungal laccase. *Journal of Agricultural and Food Chemistry, 47*, 497–503.

Geissmann, T., & Neukom, H. (1973). A note on ferulic acid as a constituent of the water-insoluble pentosans of wheat flour. *Cereal Chemistry, 50*, 414–416.

Ghoneum, M. (2016). From bench to bedside: The growing use of arabinoxylan rice bran (MGN-3/Biobran) in cancer immunotherapy. Austin. *Immunology, 1*, 1006.

Ghribi, A. M., Sila, A., Gafsi, I. M., Blecker, C., Danthine, S., Attia, H., … Besbes, S. (2015). Structural, functional, and ACE inhibitory properties of water-soluble polysaccharides from chickpea flours. *International Journal of Biological Macromolecules, 75*, 276–282.

Girhammar, U., & Nair, B. M. (1992). Isolation, separation and characterization of water-soluble non starch polysaccharides from wheat and rye. *Food Hydrocolloids, 6*, 285–299.

Girhammar, U., Nakamuram, M., & Nair, B. M. (1986). Water-soluble pentosans from wheat and rye-chemical composition and some physical properties in solution. In G. O. Phillips, D. J. Wedlock, & P. A. Williams (Eds.), *Gums and stabilisers for the food industry 3*. London, UK: Elsevier Applied Science.

Gonzalez-Estrada, R., Calderon-Santoyo, M., & Carvajal-Millan, E. (2015). Covalently cross-linked arabinoxylans films for Debaryomyces hansenii entrapment. *Molecules, 20*, 11373–11386.

Gruppen, H., Hamer, R. J., & Voragen, A. G. J. (1992). Water-unextractable cell wall material from wheat flour. 1. Extraction of polymers with alkali. *Journal of Cereal Science, 16*, 41–51.

Hernandez-Espinoza, A. B., Pinon-Muniz, M. I., Rascon-Chu, A., Santana-Rodriguez, V. M., & Carvajal-Millan, E. (2012). Lycopene/arabinoxylan gels: Rheological and controlled release characteristics. *Molecules, 17*, 2428–2436.

Hromadkova, Z., Kostalova, Z., & Ebringerova, A. (2008). Comparison of conventional and ultrasound-assisted extraction of phenolics-rich heteroxylans from wheat bran. *Ultrasonics Sonochemistry, 15*, 1062–1068.

Hromadkova, Z., Paulsen, B. S., Polovka, M., Kostalova, Z., & Ebringerova, A. (2012). Structural features of two heteroxylan polysaccharide fractions from wheat bran with anti-complementary and antioxidant activities. *Carbohydrate Polymers, 93*(1), 22–30.

Iravani, S., Fitchett, C. S., & Georget, D. M. R. (2011). Physical characterization of arabinoxylan powder and its hydrogel containing a methyl xanthine. *Carbohydrate Polymers, 85*, 201–207.

Izydorczyk, M., Biliaderis, C. G., & Bushuk, W. (1991). Comparison of the structure and composition of water-soluble pentosans from different wheat varieties. *Cereal Chemistry, 68*, 139–144.

Izydorczyk, M. S., & Biliaderis, C. G. (1992). Effect of molecular size on physical properties of wheat arabinoxylan. *Journal of Agricultural and Food Chemistry, 40*, 561–568.

Izydorczyk, M. S., & Biliaderis, C. G. (1993). Structural heterogeneity of wheat endosperm arabinoxylans. *Cereal Chemistry, 70*, 641–646.

Izydorczyk, M. S., & Biliaderis, C. G. (1995). Cereal arabinoxylans: Advances in structure and physicochemical properties. *Carbohydrate Polymers, 23*, 33–48.

Izydorczyk, M. S., Biliaderis, C. G., & Bushuk, W. (1990). Oxidative gelation studies of water-soluble pentosans from wheat. *Journal of Cereal Science, 11*, 153–169.

Izydorczyk, M. S., & Bilideris, C. G. (2000). Structural and functional aspects of cereal arabinoxylans and β-glucans. In G. Doxastakis & V. Kiosseoglou (Eds.), *Novel macromolecules in food systems* (pp. 361–384). Amsterdam, The Netherlands: Elsevier.

Jeddou, K. B., Chaari, F., Maktouf, S., Nouri-Ellouz, O., Helbert, C. B., & Ghorbel, R. E. (2016). Structural, functional, and antioxidant properties of water-soluble polysaccharides from potatoes peels. *Food Chemistry, 25*, 97–105.

Kaats, G. R., Keith, S. C., Keith, P. L., Leckie, R. B., Perricone, N. V., & Preuss, H. G. (2011). A combination of l-arabinose and chromium lowers circulating glucose and insulin levels after an acute oral sucrose challenge. *Nutrition Journal, 10*, 42–42.

Katina, K., Juvonen, R., Laitila, A., Flander, L., Nordlund, E., Kariluoto, S., … Poutanen, K. (2012). Fermented wheat bran as a functional ingredient in baking. *Cereal Chemistry, 89*, 126–134.

Kulp, K., & Bechtel, W. G. (1963). Effect of water-insoluble pentosan fraction of wheat endosperm on the quality of wheat bread. *Cereal Chemistry, 40*, 493–503.

Katapodis, P., Vardakou, M., Kalogeris, E., Kekos, D., Macris, B.J., Christakopouloset. P. (2003). Enzymic production of a feruloylated oligosaccharide with antioxidant activity from wheat flour arabinoxylan. European Journal of Nutrition, 42, 55–60

Lapierre, C., Pollet, B., Christine, M., Ralet, M., & Saulnier, L. (2001). The phenolic fraction of maize bran: Evidence for lignin-heteroxylan association. *Phytochemistry, 57*, 765–772.

Li, L., Ma, S., Fan, L., Zhang, C., Pu, X., Zheng, X., & Wang, X. (2016). The influence of ultrasonic modification on arabinoxylans properties obtained from wheat bran. *International Journal of Food Science and Technology, 51*, 2338–2344.

Li, W., Zhang, S., & Smith, C. (2015). The molecular structure features-immune stimulatory activity of arabinoxylans derived from the pentosan faction of wheat flour. *Journal of Cereal Science, 62*, 81–86.

Maeda, T., & Morita, N. (2003). Flour quality and pentosan prepared by polishing wheat grain on bread making. *Food Research International, 36*, 603–610.

Mansberger, A., D'amico, S., Novalin, S., Schmidt, J., Tomoskozi, S., Berghofer, E., & Schoenlechner, R. (2014). Pentosan extraction from rye bran on pilot scale for application in gluten-free products. *Food Hydrocolloids, 35*, 606–612.

Medcalf, D. G., D'Appolonia, B. L., & Giles, K. A. (1968). Comparison of chemical composition and properties between hard red spring and durum wheat endosperm pentosans. *Cereal Chemistry, 45*, 539–549.

Mendez-Encinas, M. A., Carvajal-Millan, E., Rascon-Chu, A., Astiazaran-Garcia, H. F., & Valencia-Rivera, D. E. (2018). Ferulated arabinoxylans and their gels: Functional properties and potential application as antioxidant and anticancer agent. *Oxidative Medicine and Cellular Longevity, 2018*, 1–22. https://doi.org/10.1155/2018/2314759

Mendis, M., Leclerc, E., & Simsek, S. (2016). Arabinoxylans, gut microbiota and immunity. *Carbohydrate Polymers, 139*, 159–166.

Michniewicz, J., Biliaderis, C. G., & Bushuk, W. (1990). Water-insoluble pentosans of wheat: Composition and some physical properties. *Cereal Chemistry, 67*, 434–439.

Michniewicz, J., Biliaderis, C. G., & Bushuk, W. (1991). Effect of added pentosans on some physical and technological characteristics of dough and gluten. *Cereal Chemistry, 68*, 252–258.

Morales-Ortega, A., Carvajal-Millan, E., & Brown-Bojorquez, F. (2014). Entrapment of probiotics in water extractable arabinoxylan gels: Rheological and microstructural characterization. *Molecules, 19*, 3628–3637.
Ng, A., Greenshields, R. N., & Waldron, K. W. (1997). Oxidative cross-linking of corn bran hemicellulose: Formation of ferulic acid dehydrodimers. *Carbohydrate Research, 303*, 459–462.
Noaman, E., Badr el-Din, N. K., Bibars, M. A., AbouMossallam, A. A., & Ghoneum, M. (2008). Antioxidant potential by arabinoxylan rice bran, MGN-3/biobran, represents a mechanism for its oncostatic effect against murine solid Ehrlich carcinoma. *Cancer Letters, 268*, 348–359.
Nino-Medina, G., Carvajal-Millan, E., Lizardi, J., Rascon, A., Márquez-Escalante, J.A., Gardea, A.A., Martínez-López, A.L., Guerrero, V. (2009). Maize processing waste water arabinoxylans: Gelling capability and cross-linking content. Food Chemistry, 115, 1286–1290.
Nilsson, M., Andersson, R., Andersson, R. E., Autio, K., & Aman, P. (2000). Heterogeneity in a water-extractable rye arabinoxylan with a low degree of disubstitution. Carbohydrate polymers, 41, 397–405.
Onipe, O. O., Jideani, A. I. O., & Beswa, D. (2015). Composition and functionality of wheat bran and its application in some cereal food products. *International Journal of Food Science and Technology, 50*, 2509–2518.
Oscarsson, M., Andersson, R., Salomonsson, A. C., & Aman, P. (1996). Chemical composition of barley samples focusing on dietary fibre components. *Journal of Cereal Science, 24*, 161–170.
Paiva, L. B. D., Goldbeck, R., Santos, W. D. D., & Squina, F. M. (2013). Ferulic acid and derivatives: Molecules with potential application in the pharmaceutical field. *Brazilian Journal of Pharmaceutical Sciences, 49*(3), 395.
Pitkanen, L., Tuomainen, P., Virkki, L., & Tenkanen, M. (2011). Molecular characterization and solution properties of enzymatically tailored arabinoxylans. *International Journal of Biological Macromolecules, 49*, 963–969.
Praznik, W., Buksa, K., Nowotna, A., Gambus, H., Ziobro, R., & Gumul, D. (2009). Improved method of isolation of water soluble pentosans from rye wholemeal. In *36th International Conference of SSCHE*, vol. 182, Slovakia.
Rattan, O., Izydorczyk, M. S., & Biliaderis, C. G. (1994). Structure and rheological behaviour of arabinoxylans from Canadian bread wheat flours. *LWT - Food Science and Technology, 27*, 550–555.
Revanappa, S. B., Nandini, C. D., & Salimath, P. V. (2010). Structural characterization of pentosans from hemicellulose B of wheat varieties with varying chapatimaking quality. *Food Chemistry, 11*, 27–33.
Riviere, A., Selak, M., Lantin, D., Leroy, F., & De Vuyst, L. (2016). Bifidobacteria and butyrate-producing colon bacteria: Importance and strategies for their stimulation in the human gut. *Frontiers in Microbiology, 7*, 979.
Rose, D. J., & Inglett, G. E. (2010). Production of feruloylatedarabinoxylo-oligosaccharides from maize (*Zea mays*) bran by microwave-assisted autohydrolysis. *Food Chemistry, 119*, 1613–1618.
Rouau, X. (1993). Investigation into the effects of an enzyme preparation for baking on wheat flour dough pentosans. *Journal of Cereal Science, 18*, 145–157.
Roubroeks, J. P., Andersson, R., & Aman, P. (2000). Structural features of (1→3)(1→4)-β-D-glucan and arabinoxylan fractions isolated from rye bran. *Carbohydrate Polymers, 42*, 3–11.
Saeed, F., Pasha, I., Anjum, F. M., & Sultan, M. T. (2011). Arabinoxylans and arabinogalactans: A comprehensive treatise. *Critical Reviews in Food Science and Nutrition, 51*, 467–476.
Sapirstein, H. D., Wang, M., & Beta, T. (2013). Effects of debranning on the distribution of pentosans and relationships to phenolic content and antioxidant activity of wheat pearling fractions. *LWT - Food Science and Technology, 50*, 336–342.
Sarker, D. K., Wilde, P. J., & Clark, D. C. (1998). Enhancement of protein foam stability by formation of wheat arabinoxylan-protein crosslinks. *Cereal Chemistry, 75*, 493–499.
Saulnier, L., Sado, P. E., Branlard, G., Charmet, G., & Guillon, F. (2007). Wheatarabinoxylans: Exploiting variation in amount and composition to develop enhanced varieties. *Journal of Cereal Science, 46*, 261–281.

Sciarini, L. S., Maldonado, F., Ribotta, P. D., Perez, G. T., & Leon, A. E. (2009). Chemical composition and functional properties of Gleditsia triacanthos gum. *Food Hydrocolloids, 23*, 306–313.
Shelton, D. R., & D'Appolonia, B. L. (1985). Carbohydrate functionality in the baking process. *Cereal Foods World, 30*, 437–442.
Shyama, P. R. R., & Muralikrishna, G. (2004). Non-starch polysaccharide–phenolic acid complexes from native and germinated cereals and millet. *Food Chemistry, 84*, 527–531.
Sila, A., Bayar, N., Ghazala, I., Bougatef, A., Ellouz-Ghorbel, R., & Ellouz- Chaabouni, S. (2014). Water-soluble polysaccharides from agro-industrial byproducts: Functional and biological properties. *International Journal of Biological Macromolecules, 69*, 236–243.
Snelders, J., Olaerts, H., Dornez, E., Van de Wiele, T., Aura, A. M., Vanhaecke, L., Delcour, J. A., Courtin, C. M. (2014). Structural features and feruloylation modulate the fermentability and evolution of antioxidant properties of arabinoxylan oligosaccharides during in vitro fermentation by human gut derived microbiota. Journal of Functional Foods, 10, 1–12
Sun, J. X., Sun, X. F., Sun, R. C., & Su, Y. Q. (2004). Fractional extraction and structural characterization of sugarcane bagasse hemicelluloses. *Carbohydrate Polymers, 56*, 195–204.
Sun, Y. L., Cui, S. W., Gu, X. H., & Zhang, J. M. (2011). Isolation and structural characterization of water unextractable arabinoxylans from Chinese black-grained wheat bran. *Carbohydrate Polymers, 85*, 615–621.
Szwajgier, D., & Jakubczyk, A. (2010). Biotransformation of ferulic acid by Lactobacillus acidophilus K1 and selected Bifidobacterium strains. *ACTA Scientiarum Polonorum Technologia Alimentaria, 9*, 45–59.
Tran, A. H., & Nordin, P. (1977). Extraction of pentosans from wheat flour with trichloroacetic acid solutions. *The Star, 29*, 153–156.
Vansteenkiste, E., Babot, C., Rouau, X., & Micard, V. (2004). Oxidative gelation of feruloylated arabinoxylan as affected by protein. Influence on protein enzymatic hydrolysis. *Food Hydrocolloids, 18*(4), 557–564.
Vardakou, M., NuenoPalop, C., Christakopoulos, P., Faulds, C. B., Gasson, M. A., Narbad, A. (2008). Evaluation of the prebiotic properties of wheat arabinoxylan fraction and induction of hydrolase activity in gut microflora. International Journal of Food Microbiology, 123, 166–170
Vinkx, C. J., Van Nieuwenhove, C. G., & Delcour, J. A. (1991). Physicochemical and functional properties of rye nonstarch polysaccharides. III. Oxidative gelation of a fraction containing water-soluble pentosans and proteins. *Cereal Chemistry, 68*, 617–622.
Voragen, A. G. I., Schols, H. A., Marijs, J., Rombouts, F. M., & Angelino, S. A. G. F. (1987). Nonstarch polysaccharides from barley: Structural features and breakdown during malting. *Journal of the Institute of Brewing, 93*, 202–208.
Yacoubi, N., Van Immerseel, F., Ducatelle, R., Rhayat, L., Bonnin, E., & Saulnier, L. (2016). Water-soluble fractions obtained by enzymatic treatment of wheat grains promote short chain fatty acids production by broiler cecal microbiota. *Animal Feed Science and Technology, 218*(1), 110–119.
Zeleznak, K. J., & Hoseney, R. C. (1986). The role of water in the retrogradation of wheat starch gels and bread crumb. *Cereal Chemistry, 63*, 407–411.
Zhong, X., Karen, L., Walker, Z., Muir, J. G., Mascara, T., & O'Dea, K. (2000). Arabinoxylan fiber, a byproduct of wheat flour processing, reduces the postprandial glucose response in normoglycemic subjects. *The American Journal of Clinical Nutrition, 71*(5), 1123–1128.

Dietary Gums

Prakhar Chatur, Umar Shah, Asir Gani, Mudasir Ahmad, Adil Gani, and Zakir Khan

Introduction

The diverse properties, applications and safety of natural gums make their widespread use in food and pharmaceutical industries that include their safe oral consumption in the form of food additives (thickeners, emulsifiers, sweeteners etc.), as binders and drug release modified gums (drug carriers) in pharmaceutical dosage forms. Natural gums are polysaccharides obtained usually as plant exudates. They contain various sugars other than glucose and having significant quantities of oxidized groups in adjunct to their normal polyhydroxy format. The polysaccharide gums are considered to be one of the best abundant industrial raw materials due to their sustainability, biodegradability and safety. The phenomenon of uncontrolled rates of hydration, pH dependent solubility, thickening, drop in viscosity on storage, and the possibility of microbial contamination are some problems associated with the use of natural gums. Furthermore, interaction of materials with mucosal layer of the gastrointestinal tract, nasal pathway or airway is known as mucoadhesion. Many polysaccharides do possess mucoadhesive properties and their use for specific oral drug delivery has been identified by many investigators by modifying them (Rana et al. 2011). The use of modified gums for microencapsulation has increased in the nutraceutical industry. The structures of polysaccharides or dietary gums are modified/derivatized by various methods that involves attachment of wide variety of

P. Chatur · U. Shah
School of Molecular and Life Sciences, Curtin University, Perth, WA, Australia

M. Ahmad · A. Gani · A. Gani (✉)
Department of Food Science and Technology, University of Kashmir,
Srinagar, Jammu and Kashmir, India

Z. Khan
Islamic University of Science and Technology, Awantipora, Jammu and Kashmir, India

A. Gani, B. A. Ashwar (eds.), *Food biopolymers: Structural, functional and nutraceutical properties*, https://doi.org/10.1007/978-3-030-27061-2_8

functional groups to the natural gums. This may include the introduction of hydrophobic, acidic, basic, or other functionality into polysaccharide structures that can alter the properties of materials based on these substances. For example, attachment of carboxyl groups, carboxymethyl groups, polyacrylamide groups, phosphate groups etc., has all been extensively investigated for such purposes. The introduction of new functional groups changed the charges, aggregation state of molecular chains, hydrophilic–hydrophobic capability, complexing capacity, stimuli-response ability, and rheological behavior of gums, and so the application domain of gums was greatly extended. But, the derivatization of gums can only improve the properties to a finite degree because the number of introduced functional groups is less and the molecular weight of gums cannot be increased by the simple modification with small molecules. Graft polymerization is anticipated to be a quite promising technique for modifying the properties of a polymer, and the modification of natural polymer materials by graft copolymerization. It offers the opportunity to tailor their physical and chemical properties, functionalize biopolymers to impart desirable properties onto them, and combine the advantages of both natural and synthetic polymers (Battaerd and Tregear 1967 and Kalia et al. 2011).

Classification of Gums

Gum Arabic

Being one of oldest known natural plant gums-gum acacia or gum arabic is a complex mixture of arabinogalactan oligosaccharides, polysaccharides and glycoprotein. It is obtained from the branches of *Acacia Senegal* (L) Willdenow or close species of *Acacia* (leguminosae family). This branched chain, complex polysaccharide exists in nature as a neutral or slightly acidic (D-glucuronic) salt of complex polysaccharides containing calcium, magnesium and potassium ions. Its backbone constitutes of 1,3-linked β-D-galactopyranosyl units and side chains contain two to five 1,3- linked β-D-galactopyranosyl units. Both the main and the side chains contain units of α-L-arabinofuranosyl, α-L-rhamnopyranosyl, β-D-glucopyranosyl and 4-O-methyl-β-D-glucopyranosyl. This gum is available worldwide and is mainly produced in the tropical and subtropical regions of Africa, India, Australia, Central America and the Republic of Sudan with Sudan being the largest producer. The *Acacia senegal* and *A. laetia* are two main species providing this gum and about 80% of commercial production comes from *A. sengal*. It is important to note that the chemical composition of this gum may vary according to the source and age, climatic conditions (soil, environment, and rainfall), method of extraction, age of the tree, and consequently the molecular weight of the gum which varies from 250,000 to 1,000,000. Gum arabic is colorless, odorless, tasteless and is highly soluble in water. It is widely used as a stabilizer, emulsifier, flavoring agent, thickener or surface finishing agent and also retards sugar crystallization. At higher concentrations (above 40%) gum arabic is highly viscous (forms gel like mass) and exhibits a pseudoplastic behavior, whereas at lower concentrations it exhibits a Newtonian behavior.

Gum Tragacanth

This complex heterogeneous polysaccharide occurs naturally as a slightly acidic, calcium, magnesium or potassium salt with a molar mass of approximately 8.4×10^5 g/mol. It is obtained from the stems and branches of *Astragalus gummifer Labillardiere* and other Asiatic species of *Astragalus*. These plants are mostly grown in Turkey, Iran, Iraq, Syria, Lebanon, Afghanistan, Pakistan and in some parts of Russia. The two main components of tragacanth are the water swelling component and the water soluble component. Most times this gum contains small amounts of Uronic acid (3%). The high molar mass tragacanthic acid fraction has a rod like molecular shape and the main chain is formed by $(1 \rightarrow 4)$ linked D-galactose residues with side chains of D-xylose units attached to the main chain by $(1 \rightarrow 3)$ linkages. The neutral water soluble tragacanthin is highly branched arabinogalactan with a spherical molecular shape and is composed of $(1 \rightarrow 6)$ and $(1 \rightarrow 3)$ linked D-galactose with attached chains of $(1 \rightarrow 2)$, $(1 \rightarrow 3)$ and $(1 \rightarrow 5)$ linked L-arabinose. Some of the important physiochemical characteristics of tragacanth gum are acid resistance, resistance to degradation at low pH and bifunctional emulsifying ability. At lower concentrations, it exhibits pseudoplastic behavior. An excellent formulation was seen with an average size of 22 nm and 100% microbial reduction after 12 h stirring, when tragacanth was used as an encapsulating agent.

Pectin

Pectin is a very complex, non-random structural hetero polysaccharide comprising of homogalacturonan, rhamnogalacturonan-I and rhamnogalacturonan-II. The average molecular weight falls in the range of 50,000–180,000 Da. The esterification of the carboxyl group with methanol, classifies pectin into two groups high methoxyl pectin (HM) which has 50–80% esterified carboxyl group and low methoxyl pectin (LM) which has 25–50% esterified carboxyl group. The first reported attempt to study pectin was by Henri Braconnot in 1825. The esterification degree, molecular weight and extraction conditions determines stability, gelling, and textural properties of pectin and thus it finds ample use in food formulation. The type of pectin, concentration of pectin, modification of hydroxyl methoxy pectin, and temperature determine gelling process and consequently gelling activity. Pectin is soluble in water and insoluble in most organic solvents. It exhibits Newtonian behavior at lower concentration and at higher concentration exhibits pseudoplastic behavior.

Guar Gum

Guar gum is a galactomannan also known as locust bean gum, tara gum, carob bean gum. It is a nonionic, biodegradable, water-soluble, and low-cost natural polysaccharide obtained from seeds of *Cyamopsis tetragonoloba* (Carob tree). It is composed of linear chains of 1 attached β-d-(1 → 4) mannopyranosyl unit and short-side branched alfa-D-galactopyranosyl units attached at (1–6) linkages. Locust bean gum is one of the most widely used gums for controlled and targeted delivery in colon and is used in medicinal industry as a binder, disintegrator, hydrophilic matrix and an encapsulating agent. Guar gum is soluble in cold water and remains stable in the pH range of 5–7, but forms a gel like structure in hot water. The solution of this gum exhibits pseudo plastic behavior, and finds use as a thickener, emulsifier and retarding agent in food industry. Guar gum has been used as an encapsulating agent in various formulations as it remains unaffected in the upper digestive tract but gets easily degraded by microflora in the colon. The highly viscous nature of guar gum limits its use in encapsulation and hence various modifications can be employed including physical, chemical, enzymatic or a combination of hydrocolloids to increase its functionality. Guar gum can be modified by derivatization, grafting, network formation and by the use of acids, enzymes, hydrothermal degradation, microwave mediated free radical degradation, and irradiation in particular ultrasonication to improve its properties for diverse biomedical applications. Guar gum nanoparticles have been prepared by nano-precipitation and crosslinking methods. Different factors responsible for the nanoparticle formation are the molecular mass of the galactomannan, solvent surfactant, crosslinking and agitation.

Gum Karaya

Gum karaya also known as *Sterculia* gum is a complex polysaccharide obtained as a calcium and magnesium salt with a molecular mass of up to 1.6×10^7 g/mol. It is obtained from the stems and branches of *Sterculia urens Roxb* and other species of it or from the species of *Cochlospermum*. It is mainly grown in the Asian continent, particularly in India. The backbone of the karaya gum consists of α-d-galacturonic acid, α-l-rhamnose residues and side chains are attached by (1 → 2) or (1 → 3) linkage of β-d-galactose to the main chain. The color of this gum is white to grayish white and has a 10% solubility in room temperature water and 30% soluble in hot water.

Microbial Gums

Carrageenan Gum

Carrageenan is the generic name for high molecular weight sulphated polysaccharide of D-galactose and 3,6-anhydro-D-galactose or a linear structure consisting of D-galactose units alternatively linked by α- (1–3) and α- (1–4) bonds and isolated from red weeds belonging to the class Rhodophyceae, especially *Chondruscrispus*, *Euchema* spp, *Gigartinastellata* and *Iridaea* spp. The polymer chains comprise alternating (1 → 3)-linked β-d-galactopyranosyl and (1 → 4)-linked α-d-galactopyranosyl units. Some of the (1 → 3) linked units occur as the 2- and 4-sulfates, whereas the (1 → 4) linked units occur as the 2- and 6- sulfates, the 2,6-disulfates, the 3,6-anhydride, and the 3,6-andydride-2-sulfate. Carrageenan has been broadly divided into three types based on difference in the number of sulphate groups (kappa (κ), iota (ι) and lambda (λ)). Carrageenan safety in terms of use in food systems has been approved jointly by FAO/WHO. Carrageenan and its derivatives have been used in various pharmaceutical formulations for use as a mucoadhesive material e.g. pravastatin was encapsulated using carrageenan as a wall material.

Alginate Gum

Modification of Dietary Gums

This nontoxic, linear heteropolysaccharide consists of anionic copolymer comprising of 1 → 4 linked β-D-mannuronic acid (M residue) and α-L-guluronic acids (G residue), obtained from brown sea algae and some bacterias. The versatility and positive cellular response of this unbranched binary copolymer polysaccharide makes it a promising candidate for the development of hydrogel particles and scaffolds. It exists in three forms as homopolymeric M-blocks (M-M-M). Ion binding sites in alginates are controlled by the cavities formed by diaxially linked G residues proving junction zone in the gel network. This binding zone between the G-block is described as the egg-box model. Therefore, the length of D-manuronic acid polymer is a major structure contributing factor to gel formation and subsequent instantaneous entrapment of probiotic cells in three-dimensional lattice of alginates. Other factors known to affect gel formation of alginates includes concentration of the alginate and $CaCl_2$ and gel hardening time. These anionic copolymer vary widely in composition and sequence and are arranged in a pattern of blocks along the chain. The average molecular weight of alginate is in the range of 200,000–300,000 Da however, molecular weight and polymerization degree depends on the enzymatic activity at the time of extraction/production, isolation process and biotechnological procedure. The average weight of the alginate is in the range 4×10^4–5×10^5 g/mol.

At lower pH, concentrations and in presence of divalent cations (calcium and zinc), alginates form strong heat stable solution of high viscosity and exhibit Newtonian behavior whereas solution exhibits pseudoplastic behavior at higher concentration. In external gelation method, alginate microparticles are fabricated by the process of water-in-oil emulsion and maintained by surfactants like Tween® 80 and usually calcium chloride in a solution form is added to the emulsion. The internal gelation is very rarely used for microencapsulation purposes in which calcium carbonate is used and is fabricated by the process of water-in-oil emulsion followed by addition of organic acids like acetic acid. Calcium and carbonic acid are formed, when it penetrates into water phase. The major drawbacks associated with alginates are leaching of encapsulating agent during preparation and rapid dissolution in intestinal pH or in the presence of sodium ions and in most of the cases concentration range of 0.5–4% are used and recommended for nano delivery. Besides this, there are various downsides associated with alginate micro-particles such as liable to be influenced to acidic conditions and also splitting and loss of mechanical stability in these conditions, which worsen in the presence of monovalent ions or chelating agents. The highly porous nature consequently results in fast diffusion of moisture and other fluids through beads. To overcome the problem associated with mechanical stability of alginate microspheres, alginate in combination with other natural and artificial hydrocolloids are used and also ionically cross-linked by addition of divalent cations in aqueous solution. Properties like mild gel setting conditions, biocompatibility, biodegradability, pH sensitiveness, low cost and mucoadhesive property of alginates make it suitable for use as an effective matrices for the entrapment and/or delivery of variety of biological agents.

Gellan Gum

Gellan gum which is produced by fermentation using *Sphingomonas paucimobilis* is a high molecular weight anionic linear polysaccharide. This polymer is a tetra saccharide repeating unit of D-glucose, L-rhamnose and D-glucuronate in a molar ratio of 2:1:1 and its molar mass is in the range of 5×10^5 g/mol. The 3-linked glucose units are substituted with glyceryl at O(2) and with acetyl at O(6). It can, therefore, be designated as →4)-l-rhamnopyranosyl-a-(1 → 3)-d-glucopyranosyl-b-(1 → 4)-d-glucuronopyranosyl-b-(1 → 4)-d-glucopyranosyl-b-(1→). The gellan gum chemically is composed of 60% glucose, 20% glucuronic acid and 20% rhamnose and has broadly been divided into three types on the basis of amount of acetyl groups i.e. native gellan gum, deacetylated gellan gum and clarified gellan gum. Gellan gum has been accepted as a food additive by the USA in 1992 proceeded by EU approval as E 418 and followed by Canada, Australia and South Africa.

Xanthan Gum

Xanthan gum is a fermentation derived natural, high molar mass anionic polyelectrolyte from *Xanthomonascampestris.* It consists of 1 → 4 linked β-D-glucose units having a trisaccharide side chain attached to alternate D-glucosyl residues. The molecular weight of this gum varies from 1.5×10^5 g/mol. The solution of gum exhibits pseudoplasticbehaviour. Xanthan gum has been widely been used in targeted and controlled drug delivery in the form of nanoparticles, liposomes, niosomes, microspheres, hydrogels, dendrimers and nanofibers. The main problems associated with unmodified form of xanthan gum are slow dissolution and substantial swelling and hence modification is required to overcome these disadvantage for improved target release. The carboxymethylation and grafting (free radical, microwave irradiation, chemoenzyme) are the main chemical modification done to improve physicochemical properties. Xanthan gum has shown various positive physicochemical changes when used in combination with other hydrocolloids particularly with guar gum.

Dextrin Gum

Dextrin is a linear, water soluble, biodegradable polysaccharide of glucan composed mostly of varying glucose units linked by α (1 → 6) glucosidic bonds and small percentage of α (1 → 3) linkages of bacterial origin. The nature of the attached group in modified dextrin (15–100%) or degree of substitution determines its properties in term of solubility and can form self-organized nano scale particles. The muco adhesive emulsion of cyclo-dextrin was prepared to load antibacterial, antiviral, contraceptive and antifungal agents and was also used as drug solubility enhancer.

Animal Origin

Being the second most abundant natural polymer after cellulose, chitosan is obtained from chitin from shrimp, carp and lobster shells. Chitosan is linear, cationic, biocompatible, non-toxic, and biodegradable polymer consisting of randomly distributed β (1 → 4) linked d-glucosamine and N-acetyl-d-glucosamine and is obtained by alkaline deacetylation of chitin and is an introductory polysaccharide containing a considerable percentage of nitrogen. It is a cationic polymer studied as an excipient in mucoadhesive dosage forms and controlled delivery formulations, attributing to its gelling and adhesive properties. Chitosan has been recommended safe by various regulatory agencies in Japan, USA, Italy, Portugal and England for use as a dietary accessory and food additive. As a polymer, chitosan has a special status due to the presence of nitrogen, cationicity, capacity to form polyelectrolyte complexes, pH sensitivity, bioadhesive ability, solubility, absorbability, controllable biodegradability and mucoadhesive properties consequently allowing the polymer to become water soluble after formation of carboxylate salts such as formate, acetate, lactate, malate, citrate, glyoxylate, pyruvate, glycolate and ascorbate. All the above properties satisfy its need for use in sustained and controlled release including non- viral vector for DNA gene and drug delivery. Native chitosan was replaced by some chemical and biological modifications to enhance susceptibility to degradation by nucleases of their cargo, improve cellular membrane permeability, solubility at physiological pH and consequently increasing colloidal stability, specificity and enabling formation of polyelectrolte complexes in gene delivery. Chemical modifications have been a strong tool to control interaction of chitosan with drugs for enhanced load capability, upgrade bulk properties and to outfitter release profile of nanoparticles. The esterification degree, acetyl-glucosamine, source and isolation technique determine the composition and molar mass. Chitosan is insoluble at higher pH and soluble at easily acidic pH. It has been used as an antioxidant, antimicrobial agent in various formulation and for target delivery of bioactives. This positively charged gum has miscellaneous biotic activities e.g. antitumor activity, immune enhancing effect, antibacterial, antifungal properties and used for delivery of anticancer drugs, antibacterial drugs, antifungal drugs, anti-inflammatory drugs, protein/peptides as well as for DNA/gene delivery. Chitosan capsules are not very often used as they do not increase the viability of probiotic cells but chitosan is mostly used as a coating or a shell. Various modification techniques (physical and chemical) or a combination of chitosan with other hydrocolloids have been employed to extend its biopolymer functional properties. The covalent crosslinking has been widely used for the preparation of chitosan nanoparticles from early times. The intermolecular crosslinking of chitosan nanoparticles was done by the use of water soluble condensation agents of carbodiimide, natural dicarboxylic acid and tricarboxylic acids (Succinic acid, mailic acid, tartaric acid and citric acid). In this case, production of chitosan nanoparticles was due to reaction between carboxylic group of natural acids and amino groups of chitosan. The nanoparticle produced by reaction between carboxylic group of natural acids and pendant amino groups of chitosan group were stable in aqueous media at low pH, neutral and mild alkaline conditions. The toxicity of glutraaldehyde on cell viability, limits its use as a crosslinking agent in the field of drug delivery. Tripolyphosphate (TPP) has been widely used as polyanion crosslinker in various studies and was first used for synthesis of chitosan nanoparticles. The combination of chitosan and tripolyphosphate (TPP) form a gel by ionic interaction between positively charged amino groups of chitosan and negatively charged counter ions of TPP and has widely been used for targeted and controlled release of bioactives and drugs. Chitosan and its derivatives have been widely used as ion removal, fiber forming, in cosmetics, for gene and drug delivery. The production of nanoparticle chitosan reveals biological effects like higher antitumor activity and is because of membrane disruption and apoptosis inducing activity for cancer cells.

The various modification methods and their effect on physiochemical properties of various gums are discussed below:

Methods for Modification of Gums

Carboxymethylation It occurs through Williamson's ether synthesis reaction involving covalent attachment of $-CH_2COOH$ or –COONa moieties at sixth position of hexose replacing –OH moieties and transforms a nonionic gum into an anionic gum. The process of modification is used to increase the aqueous solubility, hydrophilicity and solution clarity of gums. The general reaction involved in carboxylation reaction is shown in Fig. 1. These carboxymethylated aqueous gum solutions were characterized by non-Newtonian pseudoplastic behavior (Rana et al. 2011).

Carboxymethylation of tamarind kernel powder with respect to the concentration of sodium hydroxide, monochloroacetic acid, solvent ratio, reaction time, and reaction temperature was analysed and it was found that the paste quality and microbial resistance of Caboxymethylated tamarind kernel powder was better than that of the native gum (Goyal et al. 2007).

Guar gum is an interesting polymer for preparation of hydrophilic matrix tablets because of its high water swellability, nontoxicity, and low cost (Bhardwaj et al. 2000). Carboxymethylated Guar gum microbeads cross linked with Ba^{2+} and Ca^{2+} ions were found to have potential for gastrointestinal drug delivery. The Ba^{2+} cross-linked products were able to protect drug release under gastric pH conditions while Ca^{2+} ion cross-linked products released the encapsulated drug when exposed to pH 7.4, i.e. intestinal pH (Thimma and Tammishetti 2001).However the safety concerns

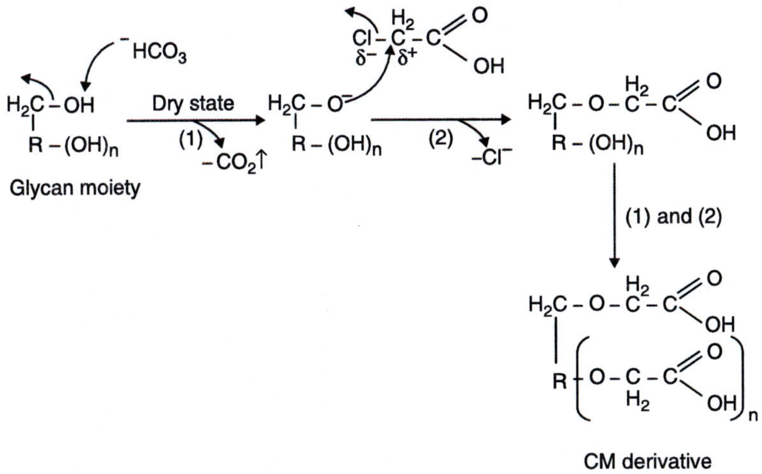

Fig. 1 Carboxymethylation (CM) of gums. (Source: Rana et al. 2011)

with Ba^{2+} were not reported. The cross linking with trivalent ions were found to possess greater that 75% of drug retention at much lower concentrations as compared to divalent metal ions (Ba^{2+}, Ca^{2+}, Cu^{2+} and Cd^{2+}). This may be due to the high valency of trivalent ion (Thimma and Tammishetti 2001). Microencapsulation of Papain with carboxymethylated flamboyant seed gum was found to have a better papain release rate under simulated Intestinal pH conditions than Carboxymethylated Guar gum suggesting it could be an effective delivery system in anti-inflammatory or preventative treatment of gastric or duodenal ulcers (Betancur-Ancona et al. 2011).

Cross-Linking of Gums

The hydrophilicity of natural gums allows them to swell in cold water or other dissolution media. This allows the possibility of entrapped drug leakage prior to the arrival of drug at the site of absorption. This enormous swelling of the gums can be reduced by cross linking. The properties of cross linked gum derivatives depends mainly on their cross linking density, ratio of number of moles of cross linking agent to the moles of polymer repeating units. The type of interactions forming depends upon the nature of the cross linking agents in the network. The cross linked gums or derivatives are of three types: (1) Hybrid polymer network formed by the reaction between a structural unit of a gum or its derivatives and the structural unit of another polymeric chain of another type. (2) Semi or fully interpenetrating polymer network contains a non-reacting polymer added to the gum or it's derivate before cross linking leading to the formation of cross linked gum in which non reacting polymer is entrapped (3) ionically cross linked gums may be formed in the presence of negatively charged entities which form bridges between the positively charged polymeric chains. Various crosslinking agents are used for cross linking of different gums.

Gaur gum was crosslinked with epichlorohydin as the crosslinking agent, sodium hydroxide as the catalyst and ethanol as the solvent and it was found that the cross linked guar gum had improved alkali resistance, acid resistance, retrogradation and viscosity stability, but its freeze-thaw stability, swelling power and solubility decreased (Hongbo et al. 2012). The reaction that may be involved in cross linking of guar gum with epichlorohydrin is shown in Fig. 2.

Glutaraldehyde has been used extensively for cross-linking polymers containing hydroxyl groups. The increase in the concentration of cross linking agent glutaraldehyde results in increase in cross linking density. Guar gum has been cross linked with Glutaraldehyde as shown in Fig. 3.This resulted in a biodegradable hydrogel with reduced enormous swelling as compared to natural guar gum that has the

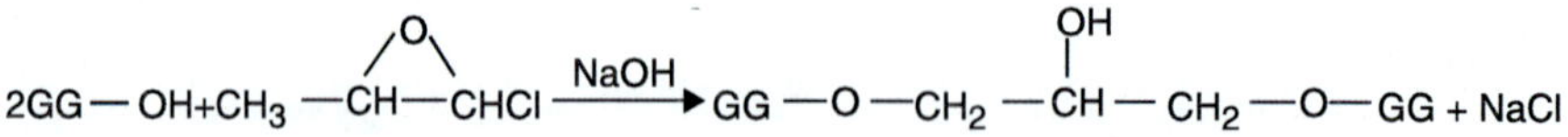

Fig. 2 Cross linking of guar gum with epichlorohydrin source (Hongbo et al. 2012)

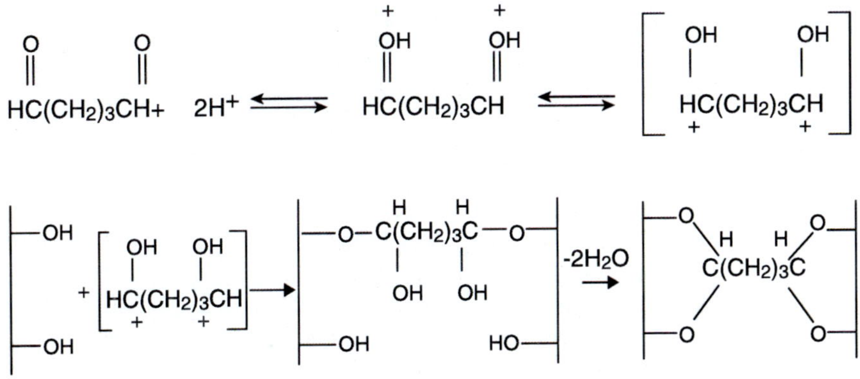

Fig. 3 Cross linking of gum with (A) glutaraldehyde (Kabir et al. 1998)

ability to retain poorly soluble drugs such as indomethacin and budesonide in artificial gastrointestinal fluids (Kabir et al. 1998).

Xanthan gum was modified by treatment with mannitol and the Modified xanthan gum obtained was found to be biodegradable, directly compressible and exhibited desirable swelling dynamics to be used as a hydrophilic excipient for rapidly disintegrating tablets Roxithromycin. The modified xanthan gum was found to retain drugs under in-vitro conditions for about 12 months (Sharma and Pathak 2013).

Ultrasonic Modification of Gums

Ultrasonic technology has also been used for the modification of polysaccharides and certain gums such as pectin, chitin, starch, carrageenan and guar gum. The ultrasound is capable of causing physical and chemical transformations through acoustic cavitation (Mohod and Gogate 2011) but require fluid for cavitation to be generated.

Generally, sonication breaks polymer chains at the center, which structurally is the weakest point. Molecules with high molecular weight and long chain length are more often broken at the center, than shorter and stiffer chains. Furthermore, the linear polymer chains are more easily sonolysed compared to branched ones. Therefore the initial molecular weight of the polymers play an important role in the overall degradation rate. As polymers are broken down, a point is reached where the chains becomes so short and stiff that further degradation is not possible and a lower molecular weight limit is attained (Zhang et al. 2013). The only way to overcome the lower limit is to increase the energy density of the sound field. It is worth noting that longer-chained molecules lead to higher solution viscosity which also affects the rate of cavitation (Sotanaphun et al. 2012).

The degradation of carrageenan by sonication was investigated and it was found that the rate of degradation increased with increased intensity, time, and reduced pH and concentration (Zhou and Ma 2006). The degradation was found more in

K-carrageenan as compared to ʎ-carrageenan which may be due to the presence of additional sulphate groups in ʎ-carrageenan. The sonication breaks the Carrageenan into its oligomers that can be applied as plant growth promoter, antiviral, antitumor, antioxidants, anticoagulant, antithrombotic and hydrogels for burns dressings. These functions are linked to oligomers with specific molecular weight range (Abad 2010).

Sonication was also found to reduce the molecular weight of guar gum. The reduction of Guar Gum (GG) molecular weight for an appropriate modulation of its flow and gelling properties could ultimately be useful in pharmaceutical industries for innovative drug formulations in cosmetics and food industry (Ogutu et al. 2015).

The ultrasonic degradation of CMC gum was studied under different operating parameters such as time of irradiation, and solution concentration has been investigated. The experimental results show that the viscosity of polymer solution decreased with an increase in the ultrasonic irradiation time. It has also been observed that major extent of degradation takes place in the initial period of irradiation time.

Modification of Gums by Grafting

The grafting reaction usually occurs on the –OH groups of gum by a radical polymerization reaction process, in the presence of thermal initiator or redox initiators. The radicals were generated from the decomposition of thermal initiator may strip down the H atom of –OH groups and initiate the macromolecular chains of gums to generate macroradicals, and these radical reactive sites may initiate the vinyl groups of monomers to process the chain propagation. Figure 4 gives the typical grafting mechanism of vinyl monomers onto the gum backbone. First initiators generate radicals which strip down the hydrogen atoms of the –OH groups on gum chains to form macro-radicals and after addition of vinyl monomers, the active radical sites on gum chains may initiate vinyl groups of the monomers to process chain propagation.

The graft copolymerization of gum can introduce new polymer chains and functional groups and increases the molecular weight of gums with the functional groups forming a crosslink or cyclic structure that increases the thermal resistance of gums. Grafting of Poly (methyl methacrylate) onto xyloglucan improved thermal stability that was investigated by Mishra and Malhotra (2012) Gum-g-Copolymer. Samui et al. (2007) showed improvement of degradation-resistance properties of acacia gum by grafting ethyl methacrylate onto its molecular chains. Viscosity of grafted gums also increases as compared to the natural ones due to the change in the molecular weight and charge and therefore can be used as thickening agents at low concentrations.

The grafted gum polymers have strong complexing capability, and therefore showed great prospect in the adsorption of toxic heavy metals. Guar gum grafted

Fig. 4 The common mechanism for the graft copolymerization of gums (source: Wang and Wang 2013)

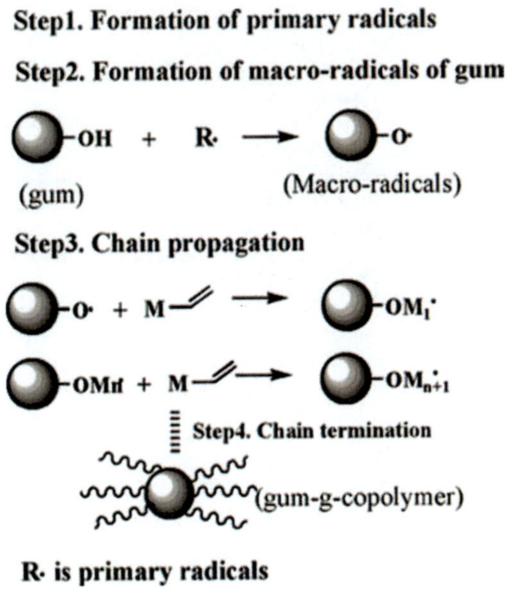

with Poly (N-vinyl-2-pyrrolidone) polymer prepared by Sharma and Lalita for the sorption of Fe^{2+} and Cr^{6+} and was found that the adsorption amount increased with increasing percent of grafting (Sharma and Lalita 2011). (Singh et al. 2007) also proved the Cassia grandis seed gum-graft-poly (methyl methacrylate) has stronger combination action with the Pb (II) ions than the gum. The basic model of adsorption of Pb (II) ions by grafted gum polymers is shown in Fig. 5.

The increasing global resource and environmental crisis continuously propels the development of eco-friendly materials. The naturally occurring raw materials have recently raised a research upsurge owing to their unique "green" characteristics that results in the exploration of new natural polymers. These polymers were modified by various methods to utilize and derive new eco-friendly materials with desirable applicability. Gums are one of the most rapidly developed natural polymers by virtue of its abundance, biodegradability, processability, and structural diversity, and thus play an irreplaceable role in the modern world. The derivatization or modification of gums may tailor the properties of gums by introducing additional functional groups, and deepening the application domains. Furthermore, gums are well tolerated by the human body because they are easily degraded to monosaccharides by colonic bacteria. However, the highly swellable nature of their putative form often restricts their use for delivering drugs to distal parts of the gastrointestinal tract. The derivatization or modification of these gums improves their use for the preparation of various dosage forms with modified drug release profiles.

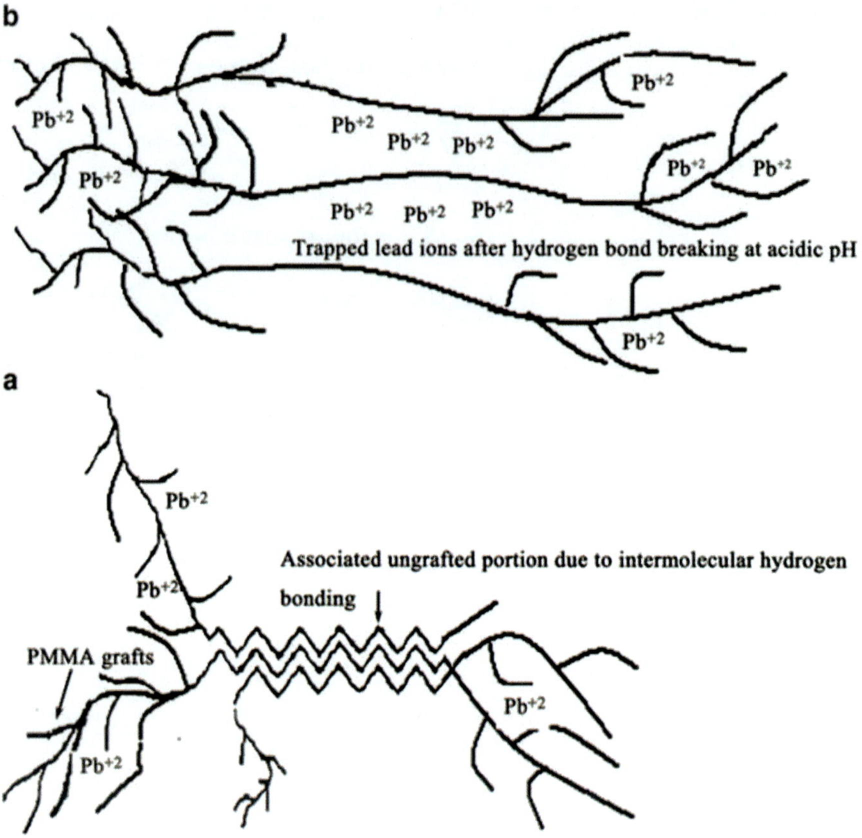

Fig. 5 Model for interaction between the Pb (II) and graft copolymer: (**a**) at normal pH, (**b**) athighly acidic pH. Source: (Singh et al. 2007)

Recent Advances in Beta Glucan Application

Introduction

β-Glucans are polysaccharides of *D*-glucose monomers linked through *β*-glycosidic bonds. It is a kind of dietary fiber and is found in a variety of natural sources such as yeasts, mushrooms, bacteria, algae, barley and oat (Zhu et al. 2015). *β*-Glucan shows a broad spectrum of biological activities which include anti-tumor, immunomodulating, anti-aging and anti-inflammatory properties. *β*-Glucans have attracted attention over the years because of their physical and chemical properties. *β*-Glucan from different sources and with different molecular weights have different biological activities (Du and Xu 2014). *β*-Glucan from baker's yeast consists of *β*-(1 → 3) and (1 → 6) linkages, other *β*-glucans, derived from cereals, are polysaccharides of glucose residues with *β*-(1 → 3) and *β*-(1 → 4) linkages whereas cellulose is a (1 → 4) *β*-glucan; curdlan is a (1 → 3) *β*-glucan; and*β*-glucan from lichen may contain either (1 → 3)(1 → 4) or (1 → 3)(1 → 6)-*β*-glucans. Fungal *β*-glucan has shown effectiveness as an immune system booster and an anti-tumor substance. *B*-Glucan from cereals help in lowering cholesterol and blood glucose levels (Zhu et al. 2015).

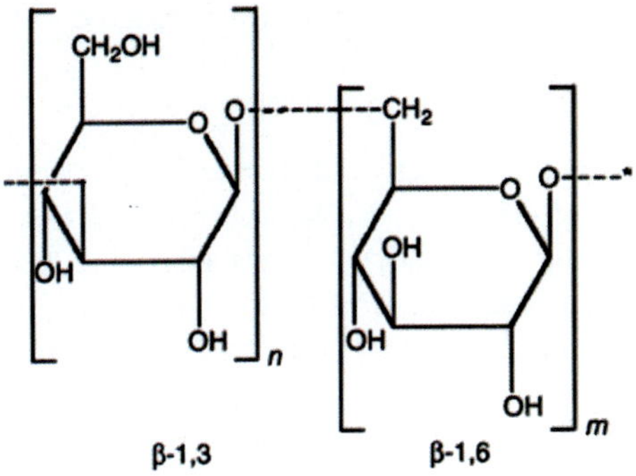

Cereal Beta glucan structure

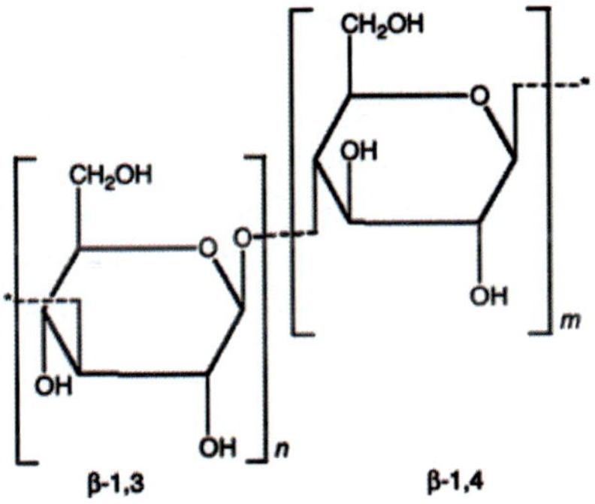

Yeast Beta glucan structure

Cereal Sources of β-glucan and its content.

Sources	β-glucan content (%)
Barley	5–10
Barley mutant (HiAmi × Cheri) × Cheri	7–12
Beans	2.4–3.5
Canary seeds	1.1–2.3
Corn/maize	0.1–1.3
Flax	0.3–0.7
Hull-less barley	5–10
Lentils	0.4–1.1
Millet	0.5–1.0
Oat	3–7
Peas	0.3–0.7

Sources	β-glucan content (%)
Rice	0.4–0.9
Rye	0.7–2.4
Spelt	0.6–1.1
Spring wheat	0.4–1.4
Winter wheat	3.7–11

Source: Ahmad et al. (2012)

The effectiveness of *β*-glucan against various diseases and disorders, and several applications reported in various number of scientific works observed that Beta-glucan consumption can reduce onset of colorectal cancer (Dongowski et al. 2002), increased stool bulk and provide assistance against constipation, reduction in glycaemic index (Cavallero et al. 2002; Jenkins et al. 2002; Granfeldt et al. 2008), lowering of the postprandial blood glucose levels and a rise in insulin levels (Hallfrisch et al. 2003; Li et al. 2003; Jenkins et al. 2002), prevention of insulin resistance (Brennan and Cleary 2007; Hlebowicz et al. 2008), reduction in serum cholesterol levels (Delaney et al. 2003; Kang et al. 2003; Kerckhoffs et al. 2003; Li et al. 2003; Yang et al. 2003; Smith et al. 2004), prevention of coronary heart disease (Jinshui et al. 2002), production of short-chain fatty acids (Wisker et al. 2000), prevention of hepatic damage by reducing taxol-induced hepatic damage (Karaduman et al. 2010), and promotion of the growth of beneficial gut microflora (Crittenden et al. 2002). *β*-glucan has been shown to prevent insulin resistance and can be used for diabetic patients (Östman et al. 2006; Brennan and Cleary 2007; Hlebowicz et al. 2008). The beta glucan containing diet promoted hepatic insulin by decreasing serine phosphorylation (Choi et al. 2010). Granfeldt et al. (2008) suggested intake of 4 g oat *β*-glucans to achieve significant decrease in glucose and insulin responses in healthy subjects thus favoring the diabetic patients.

Application of β-Glucans in Food Products

β-glucan has suitable functional properties such as thickening, stabilizing, emulsification, and gelation. These properties determine the suitability of *β*-glucan to be incorporated in soups, sauces, beverages, and in other food products (Dawkins and Nnanna 1995; Burkus and Temelli 2000). Barley *β*-glucan addition makes a beverage an excellent source of soluble dietary fiber. Its properties enable it to be incorporated alternatively in traditional beverage thickeners as replacement for gum arabic, alginates, pectin, xanthan gum, and carboxymethyl-cellulose. There is evidence that β-glucan from barley and mushroom sources has great potential to enrich snack products that have exhibited better physical and textural characteristics as compared to a control recipe. Such β-glucan-containing products may increase the possibility of controlling glycemic responses to a suitable condition (Brennan et al. 2013). Owing to greater health benefits and the potential of using it as a vital ingredient in the food industry. The US FDA supported the usage of β-glucan as a functional ingredient in food items and endorsed a dose of 3 g daily to provide health benefits.

Beta Glucan Applications as Food Additive

There are supporting reasons for the use of some sorts of Beta-glucans for supplementation of food and nutraceuticals. Beta-glucans act in food as free radical scavengers (prevent lipid peroxidation) and induce postprandial production of interferon by blood leukocytes. Beta-glucans from mushrooms such as *Agaricus-bisporus*, *Auriculariaauricula*, *Flammulinavelutipes*, *Ganodermalucidum*, and *Lentinusedodes* show antioxidant activity and free radical scavenging activity for example,, and have been recommended for food applications. Beta-glucans from the mushroom *L. edodes*have been successfully used as a partial replacement of wheat flour in baked foods and noodles, and in which they improve batter viscosity and elasticity. A reduction of glycemic index was also reported in beta glucan enriched breakfast bars (Jenkins et al. 2002) and beta glucan containing bread (Cavallero et al. 2002).

Beta-Glucan Application in Bakery Products

In bakery preparations these are used as stabilizers, thickening agents, and foaming agents and to promote water retention. These properties give nutraceutical properties to bakery products *p*. Normal breads prepared with ordinary wheat flour contain less than 2.5% of dietary fiber, which is insufficient to provide the daily requirement of dietary fiber if this product is used as a staple. In a research study, Brennan and Cleary (2007) explored the way an increase the nutritional status of bread using dietary fiber can be achieved. They suggested the use of barley β-glucan to improve the dietary fiber content of bread. The addition of β-glucan in bread at a level of 2.5–5% also reduced the release of reducing sugars in an in vitro experiment. A strong correlation between glycemic index and content of β-glucan in bread was observed. Incorporation of oat flour into wheat flour that can increase β-glucan content by a value of 2.5% ensuring the daily consumption of β-glucan recommended by the FDA, provided that the person consumes at least two slices of this bread (Flander et al. 2007). Addition of β-glucan into bread sometimes causes a reduction in loaf volume with a firm crumb structure. This reduction in loaf volume is attributed to dilution of gluten content that reduces its property to hold gases during the fermentation process. Binding of more water in the dough also causes less availability of water to gluten.

β-Glucan Applications in Meat Products

Meat and its derivatives are considered functional foods, but still there is a possibility to add more functional ingredients into them. When mushroom extract was added into meat products, it provided protection against bacteria and oxidation reactions. This was actually attributed to the presence of β-glucan and other dietary fibers along with several natural antioxidants. This addition of mushroom extract also resulted in stabilization of meat color in the final products. Storage duration may be doubled or even tripled by this technique (Aida et al. 2009). Some other functions of β-glucan and several other forms of dietary fibers in processed meat include: as binding agent, as fillers, as extender ingredient, and as fat replacers, as well as to add a synergistic effect with nutrients, to control pH, to improve emulsion stability, to improve water holding capacity, and to expand its sensory parameters (Talukder 2015).

β-Glucan and Dairy Products

Fermented dairy products can be used develop effective delivery systems for β-glucans from various sources. Addition of β-glucans into milk or milk products or into dairy-based derived protein gels during the process of fermentation that produces acids and oxidative products of monosaccharides results in production of glucono-δ-lactone. This substance may modify textural, physiochemical, chemical, and sensory properties of milk products due to phase separation between these polysaccharides and milk proteins, thus altering the gelation process through retardation of interaction between proteins and carbohydrates. Higher temperatures during fermentation also favor acidification in these products, yielding a shorter gelation time. Exopolysaccharides from *Lactobacillus kefiranofaciens* are rich in dietary fiber, including β-glucan, and also possess similar capacities to apply in dairy products (Ahmed et al. 2013).

β-Glucan and Pasta Products

In the modern era, consumers are demanding more products having health implications. These health benefits may be added in commonly consumed foods like pasta products. When β-glucan from an unusual source, *Amaranthusmantegazzianus*, was added in wheat flour, it improved the pasta characteristics and positively influenced the dietary fiber content and the glycemic index, along with an improvement in carbohydrate and protein digestibility. Beneficial effects can be achieved if this source is added at a 30% level (Martinez et al. 2016). The incorporation of barley β-glucan into pasta has documented health benefits such as lowering of the glycaemic index. Beyond its physiological benefits, it also requires suitable changes in processing methods as well to cause improvement in color, flavor, taste, and texture (Izydorczyk and Dexter 2008). Barley β-glucan, when incorporated into wheat flour, imparts increased strength to dough, causes required changes in viscoelastic properties of the dough, and makes the dough high resistance to mixing (Izydorczyk et al. 2001).

While having important functional properties, β-glucan can also be very useful as a nutraceutical supplement. Being versatile, natural and widely available, β-glucan can be used in a variety of food products imparting health-promoting properties to staple foods, thus meeting the demands of today's health conscious consumer.

Conclusion

Enormous advances have been made in studying the mechanisms governing the physiological effects of dietary gums. While the road ahead may seem long, it looks very promising. Dietary gums have gained high importance due to their great structural versatility, high water-affinity and easy processing, which facilitates its wide spread application in food and pharmaceutical industries. Further, polysaccharide gums are very promising ingredients and play a key role in designing of food with high satiating capacity. Ample studies have been conducted on the determining the benefits of dietary interventions with soluble gum rich foods. In the last decade alone, numerous efforts have been made to use gums in encapsulation technology as new instrumental imaging techniques are available, making it possible to understand their action in vivo. However, more research is needed to develop gastrointestinal models simulating different gut areas to allow evaluation of the intestinal absorption of food ingredients and validate them in vitro in order to extrapolate the results to real, in vivo situations.

References

Abad, L. V. (2010). *Radiolysis studies of kappa carrageenan for bio based materials development.* (Thesis). Department of Nuclear Engineering and Management, University of Tokyo.

Ahmad, A., Munir, B., Muhammad, A., Shaukat, B., Muhammad, A., & Tahira, T. (2012). Perspective of β-glucan as functional ingredient for food industry. *Journal of Nutrition & Food Sciences, 2*(2), 133–139.

Ahmed, Z., Wang, Y., Anjum, N., Ahmad, A., & Khan, S. T. (2013). Characterization of exopolysaccharide produced by Lactobacillus kefiranofaciens ZW3 isolated from Tibet kefir—Part II. *Food Hydrocolloids, 30*(1), 343–350.

Aida, F. M. N. A., Shuhaimi, M., Yazid, M., & Maaruf, A. G. (2009). Mushroom as a potential source of prebiotics: A review. *Trends in Food Science and Technology, 20*(11–12), 567–575.

Battaerd, H. A., & Tregear, G. W. (1967). *Graft copolymers* (Vol. 16). New York, NY: Interscience Publishers.

Betancur-Ancona, D., Pacheco-Aguirre, J., Castellanos-Ruelas, A., & Chel-Guerrero, L. (2011). Microencapsulation of papain using carboxymethylated flamboyant (Delonixregia) seed gum. *Innovative Food Science and Emerging Technologies, 12*(1), 67–72.

Bhardwaj, T. R., Kanwar, M., Lal, R., & Gupta, A. (2000). Natural gums and modified natural gums as sustained-release carriers. *Drug Development and Industrial Pharmacy, 26*(10), 1025–1038.

Brennan, C. S., & Cleary, L. J. (2007). Utilisation of glucagel in the β-glucan enrichment of breads: A physicochemical and nutritional evaluation. *Food Research International, 40*(2), 291–296.
Brennan, M. A., Derbyshire, E., Tiwari, B. K., & Brennan, C. S. (2013). Integration of β-glucan fibre rich fractions from barley and mushrooms to form healthy extruded snacks. *Plant Foods for Human Nutrition, 68*(1), 78–82.
Burkus, Z., & Temelli, F. (2000). Stabilization of emulsions and foams using barley beta-glucan. *Food Research International, 33*, 27–33.
Cavallero, A., Empilli, S., Brighenti, F., & Stanca, A. (2002). High (1→3,1→4)- β-glucan barley fractions in bread making and their effects on human glycemic response. *Journal of Cereal Science, 36*, 59–66.
Choi, J. S., Kim, H., Ho Jung, M., Hong, S., & Song, J. (2010). Consumption of barley β-glucan ameliorates fatty liver and insulin resistance in mice fed a high-fat diet. *Molecular Nutrition & Food Research, 54*, 1004.
Crittenden, R., Karppinen, S., Ojanen, S., Tenkanen, M., Fagerstrom, R., Matto, J., … Poutanen, K. (2002). In vitro fermentation of cereal dietary carbohydrates by probiotic and intestinal bacteria. *Journal of the Science of Food and Agriculture, 82*, 781–789.
Dawkins, N. L., & Nnanna, I. A. (1995). Studies on oat gum (1→3, 1→4)-β-d-glucan: Composition, molecular weight estimation and rheological properties. *Food Hydrocolloids, 9*(1), 1–7.
Delaney, B., Nicolosi, R. J., Wilson, T. A., Carlson, T., Frazer, S., Zheng, G. H., … Knutson, N. (2003). Beta-glucan fractions from barley and oats are similarly antiatherogenic in hypercholesterolemic Syrian golden hamsters. *Journal of Nutrition, 133*, 468–475.
Dongowski, G., Huth, M., Gebhardt, E., & Flamme, W. (2002). Dietary fiber rich barley products beneficially affect the intestinal tract of rats. *Journal of Nutrition, 132*, 3704–3714.
Du, B., & Xu, B. J. (2014). Oxygen radical absorbance capacity (ORAC) and ferric reducing antioxidant power (FRAP) of β-glucans from different sources with various molecular weight. *Bioactive Carbohydrates and Dietary Fibre, 3*, 11–16.
Flander, L., Salmenkallio-Marttila, M., Suortti, T., & Autio, K. (2007). Optimization of ingredients and baking process for improved wholemeal oat bread quality. *LWT - Food Science and Technology, 40*, 860–870.
Goyal, P., Kumar, V., & Sharma, P. (2007). Carboxymethylation of tamarind kernel powder. *Carbohydrate Polymers, 69*(2), 251–255.
Granfeldt, Y., Nyberg, L., & Bjöarck, I. (2008). Muesli with 4 g oat β-glucans lowers glucose and insulin responses after a bread meal in healthy subjects. *European Journal of Clinical Nutrition, 62*, 600–607.
Hallfrisch, J., Schofield, D. J., & Behall, K. M. (2003). Physiological responses of men and women to barley and oat extracts (NutrimX). II. Comparison of glucose and insulin responses. *Cereal Chemistry, 80*, 80–83.
Hlebowicz, J., Darwiche, G., Björgell, O., & Alméf, L. (2008). Effect of muesli with 4 g Oat β-glucan on postprandial blood glucose, gastric emptying and satiety in healthy subjects: A randomized crossover trial. *Journal of the American College of Nutrition, 27*(4), 470–475.
Hongbo, T., Yanping, L., Min, S., & Xiguang, W. (2012). Preparation and property of crosslinking guar gum. *Polymer Journal, 44*(3), 211.
Izydorczyk, M., & Dexter, J. (2008). Barley β-glucans and arabinoxylans: Molecular structure, physicochemical properties, and uses in food products—A review. *Food Research International, 41*(9), 850–868.
Izydorczyk, M., Hussain, A., & MacGregor, A. (2001). Effect of barley and barley components on rheological properties of wheat dough. *Journal of Cereal Science, 34*(3), 251–260.
Jenkins, A. L., Jenkins, D. J. A., Zdravkovic, U., Würsch, P., & Vuksan, V. (2002). Depression of the glycemic index by high levels of β-glucan fiber in two functional foods tested in type-2 diabetes. *European Journal of Clinical Nutrition, 56*, 622–628.
Jinshui, W., Cristina, M. R., & De-Barbera, C. B. (2002). Effect of the addition of different fibers on wheat dough performance and bread quality. *Food Chemistry, 79*, 221–226.

Kabir, G. I., Yagen, B., Penhasi, A., & Rubinstein, A. (1998). Low swelling, crosslinked guar and its potential use as colon-specific drug carrier. *Pharmaceutical Research, 15*(7), 1019–1025.

Kalia, S., Kaith, B. S., & Kaur, I. (Eds.). (2011). *Cellulose fibers: Bio-and nano-polymer composites: Green chemistry and technology*. Berlin: Springer Science & Business Media.

Kang, S. A., Jang, K. H., Hong, K., Choi, W. A., & Lee, I. Y. (2003). Effects of dietary beta-glucan on serum lipids and leptin levels in the diet-induced obese rats. *FASEB Journal, 17*, A1141.

Karaduman, D., Eren, B., & Keles, O. N. (2010). The protective effect of beta-1,3-D-glucan on taxol-induced hepatotoxicity: A histopathological and stereological study. *Drug and Chemical Toxicology, 33*(1), 8–16.

Kerckhoffs, D. A. J. M., Hornstra, G., & Mensink, R. P. (2003). Cholesterol lowering effect of β-glucan from oat bran in mildly hypercholersterolemic subjects may decrease when β-glucan is incorporated into bread and cookies. *American Journal of Clinical Nutrition, 78*, 221–227.

Li, J., Kaneko, T., Qin, L. Q., Wang, J., & Wang, Y. (2003). Effects of barley intake on glucose tolerance, lipid metabolism, and bowel function in women. *Nutrition, 19*, 11–12.

Martinez, C. S., Ribotta, P. D., León, A. E., & Yildiz, F. (2016). Influence of the addition of Amaranthus mantegazzianus flour on the nutritional and health properties of pasta. *Cogent Food & Agriculture, 2*(1), 1–12.

Mishra, A., & Malhotra, A. V. (2012). Graft copolymers of xyloglucan and methyl methacrylate. *Carbohydrate Polymers, 87*(3), 1899–1904.

Mohod, A. V., & Gogate, P. R. (2011). Ultrasonic degradation of polymers: Effect of operating parameters and intensification using additives for carboxymethyl cellulose (CMC) and polyvinyl alcohol (PVA). *Ultrasonics Sonochemistry, 18*(3), 727–734.

Ogutu, F. O., Mu, T. H., Elahi, R., Zhang, M., & Sun, H. N. (2015). Ultrasonic modification of selected polysaccharides-review. *Journal of Food Processing & Technology, 6*(5), 1.

Östman, E., Rossi, E., Larsson, H., Brighenti, F., & Bjorck, I. (2006). Glucose and insulin response in healthy men to barley bread with different levels of (1→3) (1→4)-β-glucans; predictions using fluidity measurements of in vitro enzyme digests. *Journal of Cereal Science, 43*, 230–235.

Rana, V., Rai, P., Tiwary, A. K., Singh, R. S., Kennedy, J. F., & Knill, C. J. (2011). Modified gums: Approaches and applications in drug delivery. *Carbohydrate Polymers, 83*(3), 1031–1047.

Samui, S., Ghosh, A. K., Ali, M. A., & Chowdhury, P. (2007). Synthesis, characterization and kinetic studies of PEMA grafted acacia gum. *Indian Journal of Chemical Technology, 14*(2), 126.

Sharma, V., & Pathak, K. (2013). Modified xanthan gum as hydrophilic disintegrating excipient for rapidly disintegrating tablets of roxithromycin. *Indian Journal of Pharmaceutical Education and Research, 47*(4), 79–87.

Singh, V., Tiwari, S., Sharma, A. K., & Sanghi, R. (2007). Removal of lead from aqueous solutions using Cassia grandis seed gum-graft-poly (methylmethacrylate). *Journal of Colloid and Interface Science, 316*(2), 224–232.

Smith, K. N., Queenan, K., Thomas, W., Fulcher, G., & Slavin, J. (2004). Cholesterol-lowering effect of barley beta-glucan in hypercholesterolemic subjects. *FASEB Journal, 18*, 149–152.

Sotanaphun, U., Chaidedgumjorn, A., Kitcharoen, N., Satiraphan, M., Asavapichayont, P., & Sriamornsak, P. (2012). Preparation of pectin from fruit peel of Citrus maxima. *Science, Engineering and Health Studies, 6*(1), 42–48.

Talukder, S. (2015). Effect of dietary fiber on properties and acceptance of meat products: A review. *Critical Reviews in Food Science and Nutrition, 55*(7), 1005–1011.

Thimma, R. T., & Tammishetti, S. (2001). Barium chloride crosslinked carboxymethyl guar gum beads for gastrointestinal drug delivery. *Journal of Applied Polymer Science, 82*(12), 3084–3090.

Wang, A., & Wang, W. (2013). Gum-g-copolymers: Synthesis, properties, and applications. In *Polysaccharide based graft copolymers* (pp. 149–203). Berlin, Germany: Springer.

Wisker, E., Daniel, M., Rave, G., & Feldeim, W. (2000). Short chain fatty acids produced in vitro from fiber residues obtained from mixed diets containing different breads and in human feces during ingestion of diets. *British Journal of Nutrition, 84*, 31–37.

Yang, J. L., Kim, Y. H., Lee, H. S., Lee, M. S., & Moon, Y. K. (2003). Barley beta-glucan lowers serum cholesterol based on the up-regulation of cholesterol 7-alpha-hydroxylase activity and mRNA abundance in cholesterol fed rats. *Journal of Nutritional Science and Vitaminology, 49*, 381–387.

Zhang, L., Ye, X., Ding, T., Sun, X., Xu, Y., & Liu, D. (2013). Ultrasound effects on the degradation kinetics, structure and rheological properties of apple pectin. *Ultrasonics Sonochemistry, 20*(1), 222–231.

Zhou, C., & Ma, H. (2006). Ultrasonic degradation of polysaccharide from a red algae (Porphyra yezoensis). *Journal of Agricultural and Food Chemistry, 54*(6), 2223–2228.

Zhu, F. M., Du, B., Bian, Z. X., & Xu, B. J. (2015). Beta-glucans from edible and medicinal mushrooms: Characteristics, physicochemical and biological activities. *Journal of Food Composition and Analysis, 41*, 165–173.

Part III
Proteins: Structure, Functions and Applications

Food Biopolymers: Structural, Functional, and Nutraceutical Properties: Food Proteins: An Overview

Nisar Ahmad Mir, Mamta Bharadwaj, Basharat Yousuf, Khalid Gul, Charanjit Singh Riar, and Sukhcharan Singh

Introduction

Proteins are considered as the most abundant organic molecules in living systems and are usually required in large quantities because they are the building blocks of the body therefore a way more diverse in structure and function than any other class of macromolecules. All proteins are made up of one or more chains of amino acids which share a general structure consisting of a central carbon atom, also known as alpha (α) carbon, bonded to an amino group (NH_2), a carboxylic group (COOH), and a hydrogen atom (Fig. 1). Each amino acid has another atom or group of atoms bonded to the central atom known as R group which determines the identity of the amino acid. For instance, if R group is a hydrogen atom, then the amino acid is glycine and if it is methyl (CH_3) the amino acid is alanine. Food quality is usually determined by its nutritional and functional properties which primarily depend upon the type of proteins present in the food material. Nutritional properties of proteins include biological value, protein efficiency ratio; protein digestibility corrected amino acid score, nutritional index, and corrected amino acid score (Mir et al. 2018). These nutritional properties depend upon the type of amino acids and their bioavailability or in other words properties affecting the body after passage of food

N. A. Mir · M. Bharadwaj · C. S. Riar · S. Singh
Department of Food Engineering and Technology, SLIET, Sangrur, Punjab, India

B. Yousuf
Department of Post-Harvest Engineering and Technology, Faculty of Agricultural Sciences, Aligarh Muslim University, Aligarh, India

K. Gul (✉)
Department of Food Process Engineering, National Institute of Technology, Rourkela, Odisha, India

Department of Food Science and Engineering, School of Agriculture and Biology, Shanghai Jiao Tong University, Shanghai, China

A. Gani, B. A. Ashwar (eds.), *Food biopolymers: Structural, functional and nutraceutical properties*, https://doi.org/10.1007/978-3-030-27061-2_9

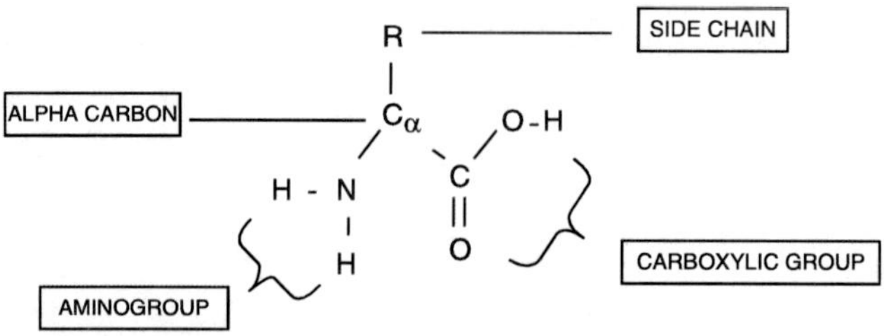

Fig. 1 General structure of proteins

into the alimentary canal. In addition to this the effectiveness of use of proteins largely depends upon the functional characteristics which can be tailored to meet the needs of food products and manufacturers. The functional attributes of the food proteins depend primarily on their molecular size, charge distribution and three dimensional structures. Among functional properties the most important ones include hydration, water/oil combination, gelling, emulsification, foam formation and rheological behaviors. Therefore, proteins in the form of isolates or concentrates can be used in a wide variety of bakery, meat, dairy and other food products in order to impart desirable sensorial and textural characteristics and to produce some novel functionalities.

Protein Structure

Protein structure sets the foundation for its interaction with other molecules in the body and therefore determines its function. Protein molecules are usually single, unbranched chains of amino acids and monomers. In general protein molecules are made up of 20 different amino acids (Fig. 2). The amino acid side chains in a peptide can become modified and in turn extend the functional repertoire of amino acids to more than hundred different amino acids. In addition to this the amino acid sequence of proteins determine its three dimensional structure (confirmation) which in turn determines the functionality of the protein.

In general a protein structure is divided into four different types depending upon the sequence of amino acids present in the structure. These four levels of structure determines the shape of proteins, the four types of structure are:

1. Primary structure

   The primary structure of proteins consists of a sequence of amino acids linked together by peptide bonds and includes any disulfide bonds. It is also characterized by genetic code and post translational covalent modifications and it ultimately determines the physicochemical and functional properties of food. However the secondary, tertiary and quaternary structure determines the biological functions of proteins.
2. Secondary structure

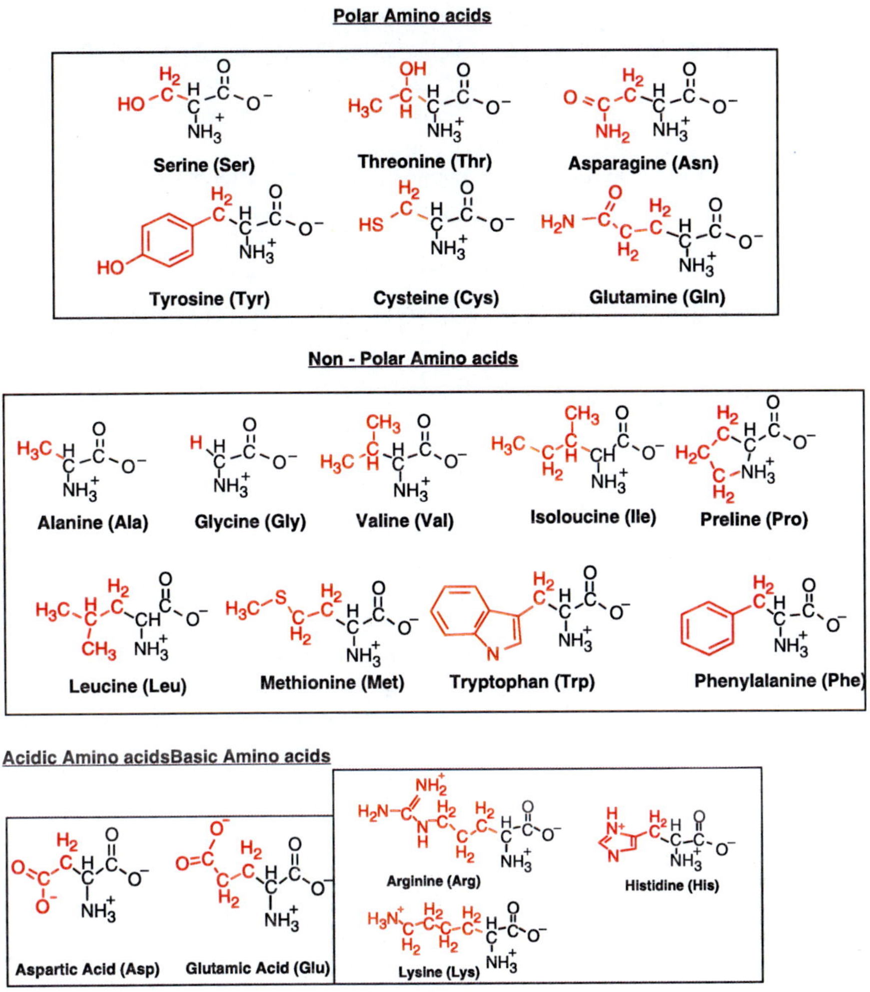

Fig. 2 Structure of twenty different amino acids

The secondary structure comprises of localized parts of a polypeptide chain (e.g. the α-helix or β-sheet with hydrogen bonds as their backbone. These special structures comprise the secondary structure of protein molecules. Secondary structure is usually due to aperiodic and periodic structures. Random coil is an example of aperiodic structure, whereas, helical and extended are examples of periodic structure. Stability of these structures are provided by decrease in local free energy by the rotation of φ and ψ angles, local non-covalent interactions between amino acid side chains, as well as hydrogen bonding between CO and N-H groups (Nelson and Cox 2013; Voet et al. 2013).

(a). *Alpha-helix proteins*: The secondary structure of some proteins is essentially all alpha-helices. Myoglobin is a good example of this type. Damodaran

(2008) proposed that α-helical structure is associated with the amphiphilic nature due to the presence of helix surface being made up of hydrophobic (non polar residues) as well as hydrophyllic (polar) residues. The nonpolar residues may also position themselves in the hydrophobic interior of the protein molecule. The α-helixis sometimes also termed the 3.6_{13}-helix, due to the 13 backbone atoms in the hydrogen bonded loop in the structure of this molecule. It should be noted that out of 20 amino acids, the cyclic imino acid proline cannot form α- helices which is because of its ring structure the N-Cα bond is unable to twist and the φ angle remains fixed at 70° due to this reason it is also known as 'α helix breaker' which is because of its ring structure. It should also be noted that of the 20 amino acids commonly found in foods, only proline is likely to adopt the cis-isomer (Damodaran 2008; Nelson and Cox 2013).

(b). *Beta-sheet proteins*: Some proteins are largely formed from beta—sheets and include some antibodies and T-cell receptors. There are two types of β-pleated sheet structures, parallel β-sheet or antiparallel β-sheet depending on the direction of the polypeptide strands. In parallel β-sheet the N → C strands run parallel, whereas in the antiparallel β-sheet N → C strands run in opposite direction. In the antiparallelβ-sheet structure hydrogen bonds form a straight line which provides additional stability to the structure making antiparallel β-sheet more stable than its parallel β-sheet counterpart. In the parallel β-sheet structure hydrogen bonds are formed at an angle, thus the stability of the hydrogen bonds, and therefore the stability of the structure are reduced. Also, generally, β-sheet is more stable than the α-helix structure; therefore, proteins with large segments of β-sheet structures are likely to be more heat stable or have higher denaturation temperatures (Nelson and Cox 2013; Voet et al. 2013).

(c). *Alpha-helix and beta sheet proteins*: Some proteins contain both alpha helix and beta sheets. Hexokinase is an example of alpha/beta structure

3. Tertiary structure
Tertiary structure of the protein includes the total three dimensional arrangement of the polypeptide chain comprising of hydrophobic interactions, hydrogen bonds (non-covalent bonds in general) and sulfur-bridges. It also includes simple dimmers to homo-oligomers and complexes with defined or variable numbers of subunits. The tertiary structure of proteins depends largely on the sequence of amino acids in a polypeptide and is stabilized through hydrogen bonding between CO and NH groups e.g. heamoglobin.
4. Quaternary structure
It is the association of two or more polypeptides into a multi-subunit complex. Heamoglobin is one example of quaternary structure another example is collagen which is a widespread connective tissue protein and consists of three polypeptide chains. It has been observed that proteins having molecular weight greater than 100 kDa are likely to have more than one polypeptide and more likely to have quaternary structures. Many food proteins having quaternary structures include

cereal and soy proteins, with different polypeptide subunits, bovine milk β-lactoglobulin which is interesting because it is a monomer of 18 kDa at pH >8 and its structure forms a dimer at pH 5–8 and an octamer at pH 3–5 (Damodaran 2008).

All the four structures of protein molecules are shown in the Fig. 3

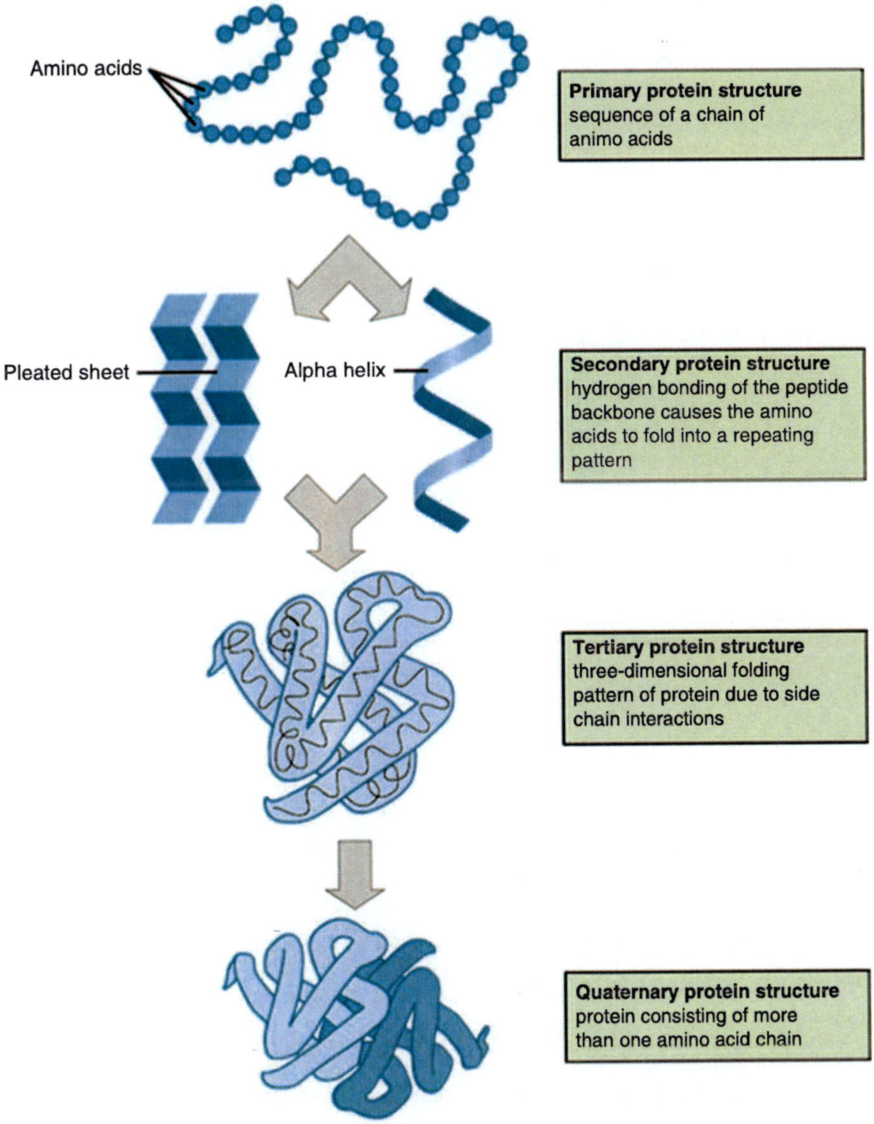

Fig. 3 Primary, secondary, tertiary and quaternary structure of protein molecules

Favorable Interactions in Protein Molecules

The native structure of protein has low energy than the denatured state of the protein. The denaturation of protein is therefore the consequence of breaking labile (non-covalent) bonds that maintain this lower energy of native state. Several non-covalent forces that are responsible for stabilizing the structure of protein are the van der Waals interactions, hydrogen bonding and hydrophobic effect. However these forces are opposed by major destabilizing force, which is associated with conformational entropy loss upon protein folding. Other forces, such as electrostatic interactions, can be either favorable or unfavorable, depending on the context. Nevertheless, the backbone of protein structure is stabilized by covalent bonds (disulfide bonds), however non covalent interactions are required to maintain secondary, tertiary and quaternary structure of proteins molecules.

Non-Covalent Bonds

Hydrophobic Interactions

Regarding hydrophobic interactions they are not attractive in nature but result from the inability of water to form hydrogen bonds with certain side chains. These interactions are the main forces that drive protein folding and are hence important in determining the native structure of protein. Thermodynamically they are unfavorable interactions of protein molecules with water, thus minimize their association with water. Hydropathies are used to describe the hydrophobic and hydrophilic tendencies of each amino acid residue, greater the hydropathy of an amino acid residue, the more likely it will orient orbury itself to the interior of the protein molecule.

Electrostatic Interactions

Electrostatic interactions like van der Waals forces may be attractive or repulsive in nature resulting from induced dipole which is due to the polarization of electron cloud between neutral atoms in protein molecules. However, these forces are relatively weak, the strength of these forces decrease rapidly with increase in distance.

Hydrogen bonds are formed by sharing of a proton between donor and acceptor groups. The strength of hydrogen bond is 2–5 kcal/mol and the ideal distance is 2.8–3 Å. Usually these bonds involve the interaction of hydrogen atom, which is covalently attached to an electronegative atom such as O, N and S, with a second electronegative atom. The most common types of hydrogen bonding include bonding between N–H and C=H groups in α-helix and β-sheet structure of protein.

Ionic interactions are actually salt bridges between ionizable groups of a protein that has negative and positive charge. Damodaran (2008), Li-Chan (2012) proposed that electrostatic interactions between oppositely charged ion pairs are strong, and certainly have an influence on protein folding patterns, therefore they contribute little to the stability of a protein since these charged groups can also interact with water.

Covalent Bonds

Disulfide bonds (S–S) are the only covalent cross linkages found in protein molecules and are formed between sulfhydryl (thiol) groups of two cysteine molecules in the presence of oxidizing environment. Disulfide bonds can be inter or intramolecular and help in the stabilization of folded protein structure.

In general the stability of protein structure is the result of covalent and non-covalent interactions. Table 1 presents the energy of the forces which are involved in the stability of protein structure.

Classification of Proteins

All proteins are remarkably similar in structure because they contain amino acids. As of now little is known about their structure so classification based on this criterion is not completely possible. However various criteria's are used for the classification of proteins which are as under:

a. Classification based on the source of protein molecule
    - Animal proteins usually derived from animal sources like meat, milk, egg and fish and usually are higher in quality because they contain all the essential amino acids.
    - Plant proteins also known as low quality proteins since they contain low content (limiting amount) of one or more of the essential amino acids.
b. Classification based on the shape of protein molecule

Table 1 Adapted and modified from Li-Chan (2012)

S. No.	Type of molecular forces involved	Energy (kJ/mol)
1	Covalent bonds	330–380
2	Electrostatic interactions	42–84
3	Hydrogen bonds	8–40
4	Hydrophobic interactions	4–12
5	Van der Waals	1–9

- Globular or Corpuscular proteins (e.g. Cytochrome C, Blood proteins, Enzymes, nutrient proteins)
- Fibrous or Fibrillar proteins and they can be further classified as collagen, elastin, keratin and fibrion

c. Classification based on composition and solubility

- Simple proteins or holoproteins, they can be further classified mainly on the basis of their solubility like protamines and histones, albumins, globulins, glutelins, prolamines, scleroproteins and albuminoids.
- Conjugated or complex proteins or heteroproteins can be further classified based on the nature of the prosthetic group present. The various divisions are metalloproteins, chromoproteins, glycoproteins, phosphoproteins, lipoproteins and nucleoproteins. (Instead of metalloproteins, chromoproteins etc., the terms metalloproteids, chromoproteids etc., are sometimes used.)
- Derived proteins which are derivatives of proteins resulting from the action of heat, enzymes or chemical agents. They are further classified as primary derived proteins and secondary derived proteins. Primary derived proteins include proteans, metaproteans or infra proteins and coagulated proteins. Secondary derived proteins include proteoses, peptones and polypeptides.

d. Classification based on biological function

- Depending upon their physical and chemical structure and location inside the cell, different proteins perform various functions. Because of their diverse nature proteins may be catgorised under following groups which is based on the metabolic functions they perform and include enzymatic proteins, structural proteins, transport or carrier proteins, nutrient and storage proteins, contractile or motile proteins, defense proteins, regulatory proteins and toxic proteins.

In addition to this Osborne (1924) classified proteins into four groups on the basis of their extraction and solubility in water (albumins), dilute saline (globulins), alcohol hater mixtures (prolamins), and dilute acid or alkali (glutelins). The major seed storage proteins include albumins, globulins, and prolamins. According to this definition, albumins are soluble in water, but globulins are insoluble in water and soluble in dilute salt solutions. Prolamins are alcohol soluble and glutelins are alkali soluble proteins. Albumins and globulins are referred as soluble proteins (Salunkhe et al. 1992). The general classification of proteins is shown in Fig. 4.

Main Storage Proteins of Cereals

It is important to pay attention towards the nature of protein before employing them in any process. Usually they fall into two classes: the prolamins, which are present in cereals and a predominance of globulins with some albumins in pseudo-cereals

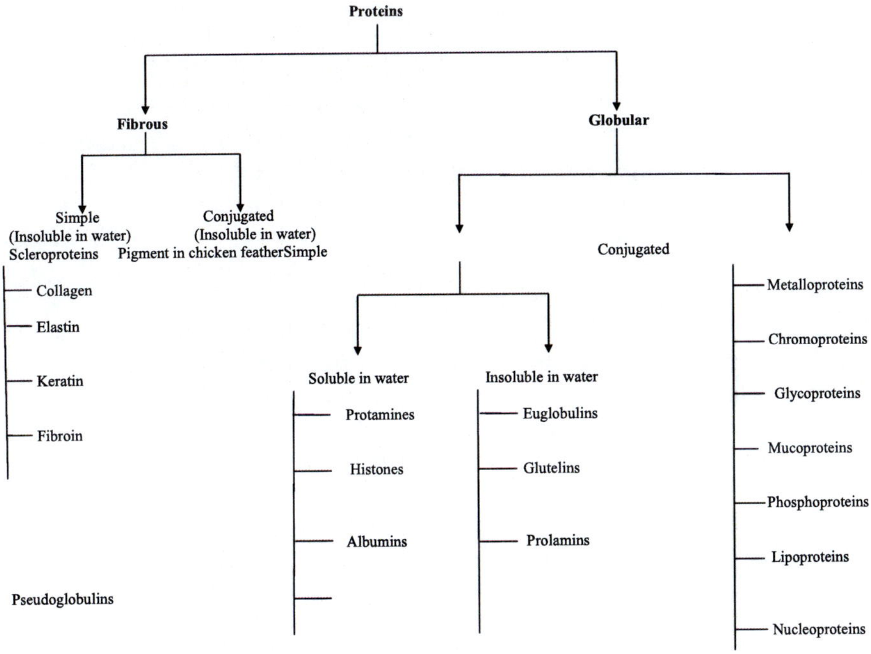

Fig. 4 General classification of proteins (adapted and modified from Voet et al. 2013)

(Shewry et al. 2002). The storage proteins belong to different groups and have significantly different structures and properties. Globulins the major storage proteins present in pseudo-cereals may be classified into groups depending on their sedimentation coefficients (which reflect their molecular masses). Typical 11S globulins are hexameric with a molecular weight of the order of 250–400 k. The subunits consist of two chains which are acidic and basic and linked by single disulphide bond. 7S globulins are trimeric and have molecular weights in the order of 150–190 k and have no disulphide bonds. Regarding albumins, they have a molecular weight of 8–15 k and contain a small and large subunit, which are linked by two disulphide bonds. Taylor et al. (2016) reported that the major storage proteins of pseudo-cereals are similar to the legume proteins, they contain 2S albumin and 11S globulin storage proteins, with 7S globulin present in buckwheat and amaranth. Tandang-Silvas et al. (2012) found that globulins from pseudo-cereals have predominantly β-sheet structure with β-barrel confirmation and therefore they are associated in the formation of good quality doughs. It has also been found that the 11S type globulins of rice oats and pseudo-cereals polymerize by disulphide bonding. Whilst the composition and structure of these storage proteins share some similarities with glutenin, but there are some important differences in terms of aminoacid composition, sequence and secondary, tertiary and quaternary structure.

Functional Properties of Proteins

The functional properties arise from a number of physical and chemical properties and affect the behavior of proteins in food systems during processing, cooking, storage and consumption. In addition to this they are also influenced by other factors such as pH, temperature, radiation or the presence of ions in foods. The functional properties of proteins play an important role as they determine the applications of particular type of protein in different systems. As food systems are usually complex therefore selecting specific type of protein for a particular application will depend upon its functionality. The functional properties depend upon the type of amino acids and functional groups present in the particular amino acid and also depend upon the interactions which are responsible for stabilizing the native structure of protein molecule. The favorable interactions in protein molecules may be covalent, hydrophobic, electrostatic, hydrogen bonds and ionic interactions. These interactions also determine the type of functionality of a protein in different food systems.

In general, several factors affect the functional properties of food proteins, namely intrinsic factors such as amino acid sequence and composition, secondary and tertiary structures, hydrophilic/hydrophobic character of the protein surface, net charge and charge distribution and molecular rigidity/flexibility of the protein and extrinsic factors such as pH, ionic strength, temperature and interactions with other food components (Zhu and Damodaran 1994). It is important to note that processing of foods may lead to structural modification of the native structure of the protein reversibly (unfolding) or irreversibly (denaturation) depending upon the processing conditions and technologies applied. Food, chemical and pharmaceutical industries rely upon these functional properties of proteins with the aim of improving the stability of the formulations or developing novel foods. Some functional of proteins which are important from the technological point of view are discussed below.

Solubility

Solubility is one of the most important properties of proteins since other functional properties like emulsion activity, emulsion stability, water binding capacity, oil binding capacity, foam capacity and foam stability are directly related to solubility (Stefanović et al. 2017). Some researchers have even concluded that solubility is the prerequisite for other functional properties. Solubility is also considered as the most important applicable scale for denaturation and aggregation thus it is a good indicator of protein function. A number of factors which play a predominant role in solubility are amino acid composition and number of hydrophilic groups present in the particular amino acid and the pH. Protein surface has a net charge that depends on the number and identity of the charged amino acids, and also depends upon the pH. For example at a specific pH the positive and negative charges will be balanced and the net charge will be zero this pH is called the iso-electric point. Most of the food proteins have iso-electric pH ranging from 3.5 to 4.5. A protein molecule has lowest solubility at its iso-electric point so if there is a charge at the protein surface, the protein prefers to interact with water, rather than with other protein molecules, thus this charge makes it more soluble. The surfaces of proteins are occupied by amino acid residues that interact with water these amino acids are referred to as hydrophilic amino acids and include arginine, lysine, aspartic acid, and glutamic acid. At pH 7 the side chains of these amino acids carry charges positive for arginine and lysine, negative for aspartic acid and glutamic acid. As the pH increases, lysine and arginine begin to lose their positive charge, and at pHs greater than 12 they are mainly neutral. On the other hand when pH decreases, aspartic acid and glutamic acid begin to lose their negative charge and it has been found that at pH less than 4 they are mainly neutral.

Gel Forming Properties

It is an important functional attribute of proteins which is related to food processing. Many foods are in the form of gels and the main structural building element in such type of foods is protein. In addition to proteins, pectin, starches and gums are also associated in the formation of strong gels. The process of gelation is a basic fundamental for various types of food systems like milk gels, comminuted meat and fish products, other meat products, cake fillings, fruit jellies, bread dough's and others. It is the main criterion which is frequently used to evaluate quality of proteins. Many quality characteristics like adhesiveness, gumminess, juiciness and other textural properties are directly related to the gelling properties of proteins. Visco-elastic properties of many foods are also related to gelation properties of food proteins. The gelation properties also affect other functional properties of proteins like water binding capacity, oil binding capacity, emulsifying capacity, emulsion stability etc. It also plays a major role in stabilizing various types of emulsions and foams. Protein gels can be formed by employing different type of approaches like heating, enzymatic process, heating in combination with salts etc. Whey protein gels can be obtained by heating which proceeds through a series of transitions like denaturation of native proteins, aggregation of unfolded molecules, strand formation from aggregates or association of strands into a network. It should be noted that aggregates are formed in the presence of salts which results in the formation of strong gels. Likewise for soy proteins gelation process is obtained by heating soya bean flour or milk followed by addition of salt (e.g. Ca^{++}, or Mg^{++}) to from a gel (Cayot and Lorient 1997; Jong et al. 2009). In case of milk proteins casein molecules are strongly hydrophobic and thus micelles are hold together by hydrophobic bonds or salt bridges. Gels can be obtained by enzymatic hydrolysis of k-casein obtained from rennet CMP (caseinomacropeptide) and thus causes the micelles to aggregate resulting in rennet gelation. For egg proteins both albumen and yolk of liquid eggs have the capacity to form gels upon heating. Gel formation is a two-step process of denaturation followed by aggregation of denatured proteins (Montejano et al. 1984; Woodward and Cotterill 1986).

Emulsifying Properties

The amphiphillic nature of protein molecules is utilized to stabilize different types of emulsions. The stability of the emulsions by utilizing proteins comes from the fact that protein molecules concentrate at the oil and water interface; with lipophilic portion in the non-polar phase (oil) and the hydrophobic portion in the polar (water) phase (Wilde 2000). The stability is maximum when proteins form a solid visco-elastic structure which results in absorption, unfolding and formation of strong interactions. These interactions are well correlated with emulsion stability. The different method by which unfolding of proteins can be obtained are thermal, enzymatic, radiation and ultrasonic treatment. Some researchers have used combination of pH and heat treatment in order to change the native structure of protein molecule to impart desirable functional properties. Moreover, the unfolding of proteins at interfaces is influenced by the structure in solution, like flexible proteins will unfold quickly and rapidly and hence lower the interfacial tension (Kinsella and Whitehead 1989; Mitchell 1986), whereas globular proteins unfold more slowly as they have more intramolecular bonds stabilizing their structure (Wilde 2000). The unfolded proteins tend to form stronger intermolecular interactions and stabilize against coalescence very effectively (Mitchell 1986). Therefore, changing the structure of proteins by various means has been used as a tool for improving protein functionality, probably by inducing a change in adsorbed conformation. Emulsifying properties of proteins are important in many food systems. The important emulsifying properties of proteins include emulsion activity and emulsion stability. Emulsion activity is defined as the ability of a protein to form an emulsion by adsorbing oil at the oil-water-interface. On the other hand emulsion stability is the ability to stabilize emulsion without forming coalescence and flocculation over a period of time. These properties are important as they determine their ability to act as emulsifiers in various foods such as soup, sauce, confectionary product, and dairy products Karaca et al. (2011). These properties are greatly affected by molecular size, surface hydrophobicity, net charge, steric hindrance and molecular flexibility. Apart from this it has been observed that hydrophobic patches present on the surface of protein molecules are important for protein adsorption at the water oil interface during the formation of emulsion (Timilsena et al. 2016).

Film Forming Properties of Proteins

Biodegradable films developed from hydrocolloid materials are gaining tremendous interest due to their excellent mechanical, and comparable barrier properties. Proteins are well known for their film forming properties because they are far better than the films developed from polysaccharides and lipids. As protein molecules have unique structure due to the presence of 20 different monomers which are responsible for providing a wide range of functional properties, especially a high intermolecular binding potential. In addition to this the films formed from these hydrocolloid materials are usually biodegradable and safe packaging materials thus reducing the pressures on landfill from plastic solid wastes. A large number of animal protein sources like milk proteins, collagen, gelatin, keratin, and myofibrillar protein are readily available for the development of biodegradable films. However due to rising global economic problems and consumer demands originated from health concerns, religious limitations and increasing trend of vegetarianism has recently arisen an interest in the usage of functional plant based proteins as alternative to animal proteins in the food industry for the development of biodegradable films (Dormont 2002; Alonso et al. 2006; Karim and Bhat 2009). Among plant sources the commonly used proteins sources are corn zein, wheat gluten, soy protein, amaranth protein, sunflower, chestnut proteins. Pseudo cereal proteins also are gaining popularity for the development of biodegradable films. The film forming ability is also associated with some desirable functional properties,such as barrier properties (i.e., water vapor permeability), mechanical properties (i.e., tensile strength, elongation, deformability, and elastic modulus) as well as microstructural properties (i.e., dough and fiber formation and texturizing capability) (Wihodo and Moraru 2013). These functional properties are crucial on improving the quality of food products, especially extending the shelf life of processed fruits and vegetables coated with the films. Some researchers have used polysaccharides in combination with proteins for the preparation of edible films however the applicability is limited due to their high water vapour permeability which is due to their hydrophilic nature. To improve the water-barrier properties of hydrocolloid-based films, lipid compounds are frequently incorporated into these structures causing a decrease in the WVP values at the expense of a reduction in the tensile strength and elasticity of the composite films (Morillon et al. 2002; Vargas et al. 2009). In addition to this the good film forming properties of plant proteins as compared to animal proteins makes them potential candidate materials for developing edible films which would serve as an alternative to plastic packaging materials (Bräuer et al. 2007).

In recent years bioactive films and coatings developed from proteins have received increasing attention. Nowadays, packaging plays a decisive role in the improvement of the shelf life of food products and new packaging materials derived from renewable sources are being developed (Lin and Zhao 2007). The potential of edible films to control gas transfer and to improve food quality, has received increasing attention from researchers and industry, possibly due to their numerous advantages over non-biodegradable plastic packaging films (Srinivasa et al. 2007). Edible film or coating can be defined as a thin, continuous layer of edible material formed or placed on or between foods or food components and poses no health hazard to consumers (Bravin et al. 2006). In addition to this edible films or coatings can also serve as a carrier of bioactive compounds, thus enhancing the functional properties of the food product by conferring number of health benefits. Most frequently used bioactive agents in edible films include lysozyme, oregano extract, chitosan, essential oils of clove, garlic and origanum, lactic acid (LA) and propionic acid (PRO), chitooligosaccharides and natamycin (NA) as antimicrobial agents. Incorporation of bioactive compounds to these films improves the functional properties such as water vapour permeability as well as antimicrobial and antioxidant properties

(Garcia et al. 2000; Oussalah et al. 2004; Seydim and Sarikus 2006).This can serve as a novel technique for packaging of many foods. Encapsulation can be a better approach for incorporation of bioactive compounds in edible films. Encapsulation of bioactive compounds into nano-vesicles may promote a number of beneficial effects by protecting them against degradation and undesirable interactions, and increasing their stability, apparent solubility and efficiency (Sozer and Kokini 2009; Brandelli et al. 2017). Besides, the amount of encapsulated bioactive required for a specific effect is often much less than the amount required when non-encapsulated (Reza Mozafari et al. 2008). Liposomes have been used as an interesting platform to deliver bioactive compounds, such as antimicrobials, antioxidants, vitamins in food systems (Fathi et al. 2012).

Protein Modification

The proteins have important functional properties from the technological point of view as they are amphiphillic in nature and the ability to from interfacial films also helps in stabilizing different food systems like emulsions and foam type foods. The stabilizing effect of food proteins is because of their large molecular weight which is associated with bulkier structure as compared with low molecular weight emulsifiers. After employing them in a particular food system they diffuse slowly to the oil water interface through the continuous phase. At the interface protein molecules undergoes surface denaturation and rearrange themselves in order to align their hydrophilic and hydrophobic groups in the oil and aqueous phase respectively thus ultimately results in the decrease in overall interfacial tension and free energy of the system (McClements 2004; Caetano da Silva Lannes and Natali Miquelim 2013).

However, native structure of protein is devoid of desirable functional traits and other important properties which are required in different food systems like creams, comminuted meat products, confectionary and dairy products. In order to increase the functionality of proteins several approaches are used which may be chemical, enzymatic and physical or a combination of these methods. As far protein functionality is concerned usually physical methods of protein modification are employed to achieve desirable functional properties as they are also safer than chemical methods. Moreover they are also considered as efficient methods in comparison with enzymatic approaches because enzymatic methods used for protein modification are usually time consuming. Some of the most commonly used physical methods for protein modification include radiation treatment (Electron beam, gamma and ultraviolet radiations), pulsed electric field, heat treatment and ultrasonic treatment. Modification of food proteins is usually carried out to alter the microstructure and physical performance of the biopolymers used for food, medical and industrial applications. Among the functional properties, the important ones which act as a target for modification include texture, flavor, color, solubility, foam stability, whippability and digestibility (Ball 1987; Hoogenkamp 2001). Table 2 lists different

Table 2 Different approaches used for increasing protein functionality

Approach	Type/modifying agent used	Reference
Physical	Pulsed electric field Heat treatment Ultrasonic treatment Elevated pressure treatment Electron beam irradiation Gamma irradiation Pulsed ultraviolet light Ozone processing	Fernandez-Diaz et al. (2000) Lam and Nickerson (2015) Resendiz-Vazquez et al. (2017) Guyon et al. (2018) Wang et al. (2017) Hassan et al. (2018) Meinlschmidt et al. (2016) Segat et al. (2014)
Chemical	Malondialdehyde modification Addition of succinyl groups PEGylation Acetylation Covalent modification by EGCG (−)-epigallocatechin-3-gallate Oxidative modification Hydrogen peroxide pH modification	Wang et al. (2018) Wan et al. (2018) He et al. (2018) Wang and Arntfield (2016) Jia et al. (2016) Duan et al. (2018) Sutariya and Patel (2017) Romani et al. (2018)
Enzymatic	Transglutaminase and thermolysin modification Corolase PP Alcalase Pepsin and trypsin treatment Corolase PP and flavourzyme hydrolysis Combined effect of alcalase, flavourzyme, neutrase, protamex, pepsin and trypsin	Damodaran and Li (2017) Guan et al. (2018) Ghribi et al. (2015) Ma et al. (2018) Connolly et al. (2014) Tang et al. (2009)

types of approaches which are responsible for increasing the functionality of proteins. It should be kept in mind that the type of technique chosen should not affect the other properties of proteins as they are usually sensitive to temperature and other processing conditions so well balance of processing parameters should be applied to obtain the desirable functional traits.

Conclusion and Future Directions

It is evident from the above discussion that proteins are the essential ingredients of our diet due to their tremendous applications, whether it may be nutritional or functional. The nutritional properties of proteins like essential amino acid index (EAAI), protein efficiency ratio (PER), biological value (BV) and amino acid score are the indicators of proteins overall quality so finding the best protein to be used in supplements and baby foods directly depends upon these properties. In addition to this, the structure of proteins also confers some important functional properties to various food systems like bakery, creams, and comminuted meat products. Moreover there are some novel approaches by which proteins functional properties can be significantly improved for obtaining technological and functional attributes. Modifying approaches like ultrasound and radiation can be used as a green technology for imparting desirable traits. Polymeric materials like proteins can also be used for the encapsulation of antioxidants, minerals, fatty acids, probiotics and other bioactive compounds but these materials are associated with early and uncontrolled release due to their porous nature. Modification is a remedy for this problem as modification of food proteins is associated with the reduction in pore size which in turn increases the efficiency of these polymeric materials. Moreover correct dosage and desirable process parameters are to be taken into consideration which is key factor for increasing the overall efficacy of the process.

References

Alonso, A., Beunza, J. J., Bes-Rastrollo, M., Pajares, R. M., & Martínez-González, M. Á. (2006). Vegetable protein and fiber from cereal are inversely associated with the risk of hypertension in a Spanish cohort. *Archives of Medical Research, 37*(6), 778–786.

Ball, H. R., Jr. (1987). Functional properties of chemically modified egg white proteins. *Journal of the American Oil Chemists' Society, 64*(12), 1718–1725.

Brandelli, A., Lopes, N. A., & Boelter, J. F. (2017). Food applications of nanostructured antimicrobials. In *Food preservation* (pp. 35–74). London, UK: Elsevier.

Bräuer, S., Meister, F., Gottlöber, R.-P., & Nechwatal, A. (2007). Preparation and thermoplastic processing of modified plant proteins. *Macromolecular Materials and Engineering, 292*(2), 176–183.

Bravin, B., Peressini, D., & Sensidoni, A. (2006). Development and application of polysaccharide–lipid edible coating to extend shelf-life of dry bakery products. *Journal of Food Engineering, 76*(3), 280–290.

Caetano da Silva Lannes, S., & Natali Miquelim, J. (2013). Interfacial behavior of food proteins. *Current Nutrition & Food Science, 9*(1), 10e14. https://doi.org/10.2174/1573401113804810914

Cayot, P., & Lorient, D. (1997). Surface-function relationships of whey proteins. In S. Damodaran & A. Paraf (Eds.), *Food proteins and their applications* (pp. 225–256). New York, NY: Marcel Dekker, Inc..

Connolly, A., Piggott, C. O., & FitzGerald, R. J. (2014). Technofunctional properties of a brewers' spent grain protein-enriched isolate and its associated enzymatic hydrolysates. *LWT - Food Science and Technology, 59*(2), 1061–1067.

Damodaran, S. (2008). Amino acids, peptides and proteins. In S. Damodaran, K. Parkin, & O. R. Fennema (Eds.), *Fennema's food chemistry*. Boca Raton, FL: CRC Press.

Damodaran, S., & Li, Y. (2017). A two-step enzymatic modification method to reduce immunoreactivity of milk proteins. *Food Chemistry, 237*, 724–732.

Dormont, D. (2002). Prion diseases: Pathogenesis and public health concerns. *FEBS Letters, 529*(1), 17–21.

Duan, X., Li, M., Shao, J., Chen, H., Xu, X., Jin, Z., & Liu, X. (2018). Effect of oxidative modification on structural and foaming properties of egg white protein. *Food Hydrocolloids, 75*, 223–228.

Fathi, M., Mozafari, M. R., & Mohebbi, M. (2012). Nanoencapsulation of food ingredients using lipid based delivery systems. *Trends in Food Science & Technology, 23*(1), 13–27.

Fernandez-Diaz, M. D., Barsotti, L., Dumay, E., & Cheftel, J. C. (2000). Effects of pulsed electric fields on ovalbumin solutions and dialyzed egg white. *Journal of Agricultural and Food Chemistry, 48*, 2332–2339.

Garcia, M. A., Martino, M. N., & Zaritzky, N. E. (2000). Lipid addition to improve barrier properties of edible starch-based films and coatings. *Journal of Food Science, 65*(6), 941–944.

Ghribi, A. M., Gafsi, I. M., Sila, A., Blecker, C., Danthine, S., Attia, H., … Besbes, S. (2015). Effects of enzymatic hydrolysis on conformational and functional properties of chickpea protein isolate. *Food Chemistry, 187*, 322–330.

Guan, H., Diao, X., Jiang, F., Han, J., & Kong, B. (2018). The enzymatic hydrolysis of soy protein isolate by Corolase PP under high hydrostatic pressure and its effect on bioactivity and characteristics of hydrolysates. *Food Chemistry, 245*, 89–96.

Guyon, C., Le Vessel, V., Meynier, A., & de Lamballerie, M. (2018). Modifications of protein-related compounds of beef minced meat treated by high pressure. *Meat Science, 142*, 32.

Hassan, A. B., Mahmoud, N. S., Elmamoun, K., Adiamo, O. Q., & Ahmed, I. A. M. (2018). Effects of gamma irradiation on the protein characteristics and functional properties of sesame (Sesamum indicum L.) seeds. *Radiation Physics and Chemistry, 144*, 85–91.

He, Q., Sun, X., He, S., Wang, T., Zhao, J., Yang, L., & Sun, H. (2018). PEGylation of black kidney bean (Phaseolus vulgaris L.) protein isolate with potential functironal properties. *Colloids and Surfaces B: Biointerfaces, 164*, 89–97.

Hoogenkamp, H. W. (2001). *Soy protein and meat formulations*. Oxford: CABI Publishing.

Jia, Z., Zheng, M., Tao, F., Chen, W., Huang, G., & Jiang, J. (2016). Effect of covalent modification by (−)-epigallocatechin-3-gallate on physicochemical and functional properties of whey protein isolate. *LWT - Food Science and Technology, 66*, 305–310.

Jong, S. H., Klok, J., & Velde, F. V. (2009). The mechanism behind microstructure formation in mixed whey protein–polysaccharide cold-set gels. *Food Hydrocolloids, 23*, 755–764.

Karaca, A. C., Low, N., & Nickerson, M. (2011). Emulsifying properties of canola and flaxseed protein isolates produced by isoelectric precipitation and salt extraction. *Food Research International, 44*(9), 2991–2998.

Karim, A. A., & Bhat, R. (2009). Fish gelatin: Properties, challenges, and prospects as an alternative to mammalian gelatins. *Food Hydrocolloids, 23*(3), 563–576.

Kinsella, J. E., & Whitehead, D. M. (1989). Proteins in whey: Chemical, physical, and functional properties. In J. E. Kinsella (Ed.), *Advances in food and nutrition research* (Vol. 33, pp. 343–438). San Diego, CA: Academic Press.

Lam, R. S., & Nickerson, M. T. (2015). The effect of pH and temperature pre-treatments on the physicochemical and emulsifying properties of whey protein isolate. *LWT - Food Science and Technology, 60*(1), 427–434.

Li-Chan, E. C. Y. (2012). Proteins: Basic concepts. In Y. H. Hui (Ed.), *Food chemistry: Principles and applications*. West Sacramento, CA: Science Technology Systems.

Lin, D., & Zhao, Y. (2007). Innovations in the development and application of edible coatings for fresh and minimally processed fruits and vegetables. *Comprehensive Reviews in Food Science and Food Safety, 6*(3), 60–75.

Ma, S., Wang, C., & Guo, M. (2018). Changes in structure and antioxidant activity of β-lactoglobulin by ultrasound and enzymatic treatment. *Ultrasonics Sonochemistry, 43*, 227–236.

McClements, D. J. (2004). Protein-stabilized emulsions. *Current Opinion in Colloid & Interface Science, 9*(5), 305–313. https://doi.org/10.1016/j.cocis.2004.09.003

Meinlschmidt, P., Ueberham, E., Lehmann, J., Reineke, K., Schlüter, O., Schweiggert-Weisz, U., & Eisner, P. (2016). The effects of pulsed ultraviolet light, cold atmospheric pressure plasma, and gamma-irradiation on the immunoreactivity of soy protein isolate. *Innovative Food Science & Emerging Technologies, 38*, 374–383.

Mir, N. A., Riar, C. S., & Singh, S. (2018). Nutritional constituents of pseudo cereals and their potential use in food systems: A review. *Trends in Food Science & Technology, 75*, 170–180.

Mitchell, J. R. (1986). Foaming and emulsifying properties of proteins. In B. J. Hudson (Ed.), *Developments in food proteins* (Vol. 4, pp. 291–338). London, UK: Elsevier.

Montejano, J. G., Hamann, D. D., Ball, H. R., Jr., & Lanier, T. C. (1984). Thermally induced gelation of native and modified egg white-rheological changes during processing; final strengths and microstructures. *Journal of Food Science, 49*, 1249–1257.

Morillon, V., Debeaufort, F., Blond, G., Capelle, M., & Voilley, A. (2002). Factors affecting the moisture permeability of lipid-based edible films: A review. *Critical Reviews in Food Science and Nutrition, 42*(1), 67–89.

Nelson, D. L., & Cox, M. M. (2013). *Lehninger principles of biochemistry*. New York, NY: Worth Publishers.

Osborne, T. B. (1924). *The vegetable proteins* (p. 154). London, UK: Longmans.

Oussalah, M., Caillet, S., Salmiéri, S., Saucier, L., & Lacroix, M. (2004). Antimicrobial and antioxidant effects of milk protein-based film containing essential oils for the preservation of whole beef muscle. *Journal of Agricultural and Food Chemistry, 52*(18), 5598–5605.

Resendiz-Vazquez, J. A., Ulloa, J. A., Urías-Silvas, J. E., Bautista-Rosales, P. U., Ramírez-Ramírez, J. C., Rosas-Ulloa, P., & González-Torres, L. (2017). Effect of high-intensity ultrasound on the technofunctional properties and structure of jackfruit (Artocarpus heterophyllus) seed protein isolate. *Ultrasonics Sonochemistry, 37*, 436–444.

Reza Mozafari, M., Johnson, C., Hatziantoniou, S., & Demetzos, C. (2008). Nanoliposomes and their applications in food nanotechnology. *Journal of Liposome Research, 18*(4), 309–327.

Romani, V. P., Machado, A. V., Olsen, B. D., & Martins, V. G. (2018). Effects of pH modification in proteins from fish (Whitemouth croaker) and their application in food packaging films. *Food Hydrocolloids, 74*, 307–314.

Salunkhe, D. K., Chavan, J. K., Adsule, R. N., & Kadam, S. S. (1992). Sunflower. In *World oilseeds: Chemistry, technology and utilization* (pp. 97–139). New York, NY: Van Nostrand Reinhold Publishing.

Segat, A., Misra, N. N., Fabbro, A., Buchini, F., Lippe, G., Cullen, P. J., & Innocente, N. (2014). Effects of ozone processing on chemical, structural and functional properties of whey protein isolate. *Food Research International, 66*, 365–372.

Seydim, A. C., & Sarikus, G. (2006). Antimicrobial activity of whey protein based edible films incorporated with oregano, rosemary and garlic essential oils. *Food Research International, 39*(5), 639–644.

Shewry, P. R., Halford, N. G., Belton, P. S., & Tatham, A. S. (2002). The structure and properties of gluten: An elastic protein from wheat grain. *Philos. Trans. R. Soc. Lond. Ser. B Biol. Sci., 357*, 133–142.

Sozer, N., & Kokini, J. L. (2009). Nanotechnology and its applications in the food sector. *Trends in Biotechnology, 27*(2), 82–89.

Srinivasa, P. C., Ramesh, M. N., & Tharanathan, R. N. (2007). Effect of plasticizers and fatty acids on mechanical and permeability characteristics of chitosan films. *Food Hydrocolloids, 21*(7), 1113–1122.

Stefanović, A. B., Jovanović, J. R., Dojčinović, M. B., Lević, S. M., Nedović, V. A., Bugarski, B. M., & Knežević-Jugović, Z. D. (2017). Effect of the controlled high-intensity ultrasound on improving functionality and structural changes of egg white proteins. *Food and Bioprocess Technology, 10*(7), 1.

Sutariya, S., & Patel, H. (2017). Effect of hydrogen peroxide on improving the heat stability of whey protein isolate solutions. *Food Chemistry, 223*, 114–120.

Tandang-Silvas, M. R., Cabanos, C. S., Carrazco Peña, L. D., De La Rosa, A. P. B., Osuna-Castro, J. A., Utsumi, S., … Maruyama, N. (2012). Crystal structure of a major seed storage protein, 11S proglobulin, from Amaranthus hypochondriacus: Insight into its physico-chemical properties. *Food Chemistry, 135*, 819–826.

Tang, C. H., Wang, X. S., & Yang, X. Q. (2009). Enzymatic hydrolysis of hemp (Cannabis sativa L.) protein isolate by various proteases and antioxidant properties of the resulting hydrolysates. *Food Chemistry, 114*(4), 1484–1490.

Taylor, J. R. N., Taylor, J., Campanella, O. H., & Hamaker, B. R. (2016). Functionality of the storage proteins in gluten-free cereals and pseudocereals in dough systems. *Journal of Cereal Science, 67*, 22–34.

Timilsena, Y. P., Adhikari, R., Barrow, C. J., & Adhikari, B. (2016). Physicochemical and functional properties of protein isolate produced from Australian chia seeds. *Food Chemistry, 212*, 648–656.

Vargas, M., Albors, A., Chiralt, A., & González-Martínez, C. (2009). Characterization of chitosan–oleic acid composite films. *Food Hydrocolloids, 23*(2), 536–547.

Voet, D., Voet, J. G., & Pratt, C. W. (2013). *Fundamentals of biochemistry, life at the molecular level* (4th ed.). Hoboken, NJ: John Wiley & Sons.

Wan, Y., Liu, J., & Guo, S. (2018). Effects of succinylation on the structure and thermal aggregation of soy protein isolate. *Food Chemistry, 245*, 542–550.

Wang, J., Zhao, M., Qiu, C., & Sun, W. (2018). Effect of malondialdehyde modification on the binding of aroma compounds to soy protein isolates. *Food Research International, 105*, 150–158.

Wang, K., & Arntfield, S. D. (2016). Modification of interactions between selected volatile flavour compounds and salt-extracted pea protein isolates using chemical and enzymatic approaches. *Food Hydrocolloids, 61*, 567–577.

Wang, L., Ding, Y., Zhang, X., Li, Y., Wang, R., Luo, X., & Chen, Z. (2017). Effect of electron beam on the functional properties and structure of defatted wheat germ proteins. *Journal of Food Engineering, 202*, 9–17.

Wihodo, M., & Moraru, C. I. (2013). Physical and chemical methods used to enhance the structure and mechanical properties of protein films: A review. *Journal of Food Engineering, 114*(3), 292–302.

Wilde, P. J. (2000). Interfaces: Their role in foam and emulsion behaviour. *Current Opinion in Colloid & Interface Science, 5*, 176–181.

Woodward, S. A., & Cotterill, O. J. (1986). Texture and microstructure of heat-formed egg white gels. *Journal of Food Science, 51*, 333–339.

Zhu, H., & Damodaran, S. (1994). Heat-induced conformational changes in whey protein isolate and its relation to foaming properties. *Journal of Agricultural and Food Chemistry, 42*(4), 846–855.

Bioactive Peptides Derived from Different Sources

Sajid Maqsood, Sabika Jafar, Mehvesh Mushtaq, and Priti Mudgil

Introduction

Bioactive peptides are defined as fractions of proteins, which have a specific sequence that has a positive physiological effect and may influence health (Kitts and Weiler 2003). The structure of bioactive peptides consists of amino acids linked together by covalent (or also known as peptide) bonds, and an overall lower molecular weight compared to proteins. Proteins are building blocks of life which play a critical role in the metabolic functions of the human body. Bioactive peptides derived from proteins may thus impact human health by displaying multifunctional activities and can be classified as antioxidative, antimicrobial, antihypertensive, opioid, immunomodulatory, and anti-thrombotic (Sanchez and Vazquez 2017).

Once released from the parent protein, the activity of a bioactive peptide is dependent on two important parameters (1) amino acid sequence, and (2) amino acid composition. Generally, the bioactive peptides are 2–20 amino acids long, and contains hydrophobic amino acids as well as lysine, arginine or proline groups (Moller et al. 2008; Sanchez and Vazquez 2017). It is reported that bioactive peptides show resistance to gastrointestinal digestion (Kitts and Weiler 2003), which is an important property that is validated to prove the efficacy of the bioactive peptide.

The sources of bioactive peptides can largely be classified on the basis of animal origin (milk, meat and egg), plant origin (cereals, pulses, fruit and fruit seeds) and

S. Maqsood (✉) · S. Jafar · P. Mudgil
Food Science Department, College of Food and Agriculture, United Arab Emirates University, Al-Ain, 15551, United Arab Emirates
e-mail: sajid.m@uaeu.ac.ae

M. Mushtaq
Department of Food Science and Technology, University of Kashmir, Srinagar, Jammu and Kashmir, India

A. Gani, B. A. Ashwar (eds.), *Food biopolymers: Structural, functional and nutraceutical properties*, https://doi.org/10.1007/978-3-030-27061-2_10

marine (fish, lobsters, prawns, etc.) origin. These sources have been extensively studied for their potent bioactivity except plant sources, which are gaining momentum for past few years (Korhonen 2009; El-Salam and El-Shibiny 2013; Mohanty et al. 2016; Park and Nam 2015).

Potential Sources of Bioactive Peptides

Milk and Milk Products

Milk has 3.5% protein, which has been exploited to generate peptides, both from whole milk protein as well as from its constituents (i.e. casein and whey). Whey protein in milk has inherent biological activities like immunoglobulins, enzymes, mineral-binding attributed to its makeup of β-lactoglobulin, α-lactalbumin and some minor proteins. This makes milk a complex and unique food matrix which is why more research is being published that contributes to the growing knowledge of biologically active peptides from milk and lead to their possible incorporation in functional foods, nutraceuticals, cosmetics, and pharmaceuticals (Panchaud et al. 2012; Mohanty et al. 2016).

Apart from this, it's reported that naturally occurring protease enzymes like milk plasmin degrade proteins and liberate bioactive peptides either during food processing or storage (Meisel and FitzGerald 2003). Microbes in the GI tract can also produce biologically active peptides (Saito 2004). Bovine milk proteins and their peptides have been isolated and characterized to show several biological activities (Sanchez and Vazquez 2017). Similarly, milk proteins of non-bovine sources like camel, goat, sheep, buffalo and yak have also been used to obtain bioactive peptides (El-Salam and El-Shibiny 2013). Although human milk is the most studied source of bioactive peptides, milk from animals is also of research interest due to its resemblance to human milk. This makes animal milk a possible alternative for infant consumption (Atanasova and Ivanova 2010; Bidasolo et al. 2012)

Milk products fermented with many types of bacteria are also capable of generating bioactive peptides. For example, hypertensive subjects were administered with *Lactobacillus helveticus* fermented milk containing two biologically active tripeptides Val-Pro-Pro and Ile-Pro-Pro. A reduction in blood pressure was observed (Nakamura et al. 1995). Furthermore, milk proteins have been hydrolyzed to obtain β-casomorphins and caseino macro-peptide-derived peptides (from casein), and lactorphins (from whey) which demonstrated ACE-inhibitory, opioid and antimicrobial activities (Sipola et al. 2002; Jakala and Vapaatalo 2010; Boutrou et al. 2015). Finally, bovine and non-bovine colostrum studies are also documented to release peptides with bioactivities (Park and Nam 2015).

Plants

Proteins derived from plants are emerging as a significant food ingredient to be exploited for the improvement of modern foods in the domain of processing technology, nutrition and health benefits (Duranti 2006). Among the plant sources, legumes and cereals are considered as one of the main sources of bioactive peptides because of their wider distribution in the market and as one of the main sources of carbohydrates and amino acid in the diet. However, spices, alfalfa, seeds, pseudocereals and other common edible plants have also been investigated for their potential to be used as a source of bioactive peptides (Rizzello et al. 2016).

As far as plant-derived bioactive peptides are concerned, soy, potato, wheat germ and corn enzymatic hydrolysates have been reported to have potent antioxidant activities (Mine et al. 2010). Recently, peptides in hydrolysates of cocoa seeds were shown to have antioxidant activities. However, the activity was evaluated in-vitro only and peptides responsible for the activity were not identified. The antihypertensive peptides have been reported mostly in soy and its derivative products (Lee and Hur 2017). However, other sources including wheat, rice, maize, wakame, sunflower, sesame, beans, broccoli, wheat and potatoes have now also been considered as the potent sources of bioactive peptides (Dhaval et al. 2016; Visvanathan et al. 2016). Dipeptides with validated antihypertensive activity were recently quantified and produced from different plant sources like rice, pea, wheat and soy and were related with those of dairy proteins.

Opioid peptides from plant sources have been reported as well. The hydrolysates of zein, hordein, gliadin, gluten and ribulose 1,5-biphosphate oxygenase/carboxylase have been reported to possess peptides with opioid activity, for example rubiscolin 6 (YPLDLF), a peptide obtained from spinach has shown ability to stimulate food intake. Also peptides derived from α-gliadin (glaidinomorphin 7, YPQPQPF), wheat gluten (GYYPT), soy β-conglycinin (soymorphin 5, YPFVV) and soybeans (LPYPR & PGP) have recently been reported to exhibit opioid activities (Stefanucci et al. 2018; Hettiarachchy 2012).

Reports of recent mineral-binding bioactive peptides in plant sources are limited (Xie et al. 2015). Protein hydrolysates of soybean have been demonstrated to have mineral binding capacity however identification of these peptides was not performed (Lv et al. 2008). There are less reported plant peptides with hypocholesterolemic activities, however, soy is the most investigated plant source and the activity of peptides was owed to gene expression alterations (Ramdath et al. 2017; Lovati et al. 2000; Li et al. 2007).

Agro-Industrial Wastes

In the context of technologies pertaining to renewable resources and biosustainable developments, waste and by-products provide a relatively cost effective source for the production of bioactive peptides. Thus would result in reduction of both the waste amount and the disposal cost, while producing value-added nutritional products. The by-products that have been used for such a purpose include soybean meal, rapeseed meal (Xie et al. 2015) and residue of olive oil production (Esteve et al. 2015). In particular food processing by-products and wastes have been considered for the generation of peptides with antioxidant and antihypertensive properties (Banerjee et al. 2017; Capriotti et al. 2012). Cherry seeds were also used for the production of antioxidant and ACE inhibitory peptides by enzymatic hydrolysis of seed proteins (Vasquez-Villanueva et al. 2016).

Production of bioactive peptides from by-products and waste does not profoundly vary for the production from food plants. Most recent examples of bioactive peptides generation from agro-wastes include date seeds (Ambigaipalan and Shahidi 2015), cherry seeds, peach seeds (Vasquez-Villanueva et al. 2016), peanut meal (White et al. 2014), cauliflower waste (Zenezini Chiozzi et al. 2016) and rice bran (Wang et al. 2017; Yan et al. 2015).

Formation of Bioactive Peptides

There are three main ways through which bioactive peptides from parent proteins can be released: (1) enzymatic degradation by digestive enzymes (2) proteolysis by microbial or plant enzymes or (3) fermentation with proteolytic starter cultures. Sometimes a combination of processes like digestive enzymes and fermentations is employed to produce milk products (Muro Urista et al. 2011). Another less common method to produce bioactive peptides is *via* chemical hydrolysis which will be discussed below.

Gastrointestinal (GI) Digestion

The simulated gastrointestinal digestion (SGID) involves mimicking the digestion conditions to determine the stability of potent bioactive peptides produced during the transit (Jakubczyk et al. 2017). The first mechanism is a normal digestive process catalyzed by gastric enzyme like pepsin and pancreatic enzymes like trypsin and chymotrypsin. Most food proteins undergo complete degradation while passing through the stomach and small intestine: (1) hydrochloric acid (HCl) in the stomach converts pepsinogen to pepsin which carry out the protein digestion (2) trypsin and chymotrypsin hydrolyze proteins to oligopeptides in the lumen of small intestine, (3) oligopeptides further broken down to di-, tri- peptides or single amino acids and are ready for absorption (Chung et al. 1979). By the use of this method antioxidant and antihypertensive peptides were generated in past (Lee and Hur 2017; Jamdar et al. 2017; Li et al. 2007). Vilcacundo et al. (2017) has even reported generation of antidiabetic peptides during simulated gastrointestinal digestion of quinoa proteins.

Bioactive peptides can also be generated following the ingestion process either by digestive (gastric and intestinal) enzymes or by microbial enzymes present in the gastro-intestinal tract. Therefore, the proteins during their passage through the digestive tract become the target of enzymes, which are normally present in the gastrointestinal tract. A wide range of enzymes with different specificity may act throughout the whole gastrointestinal tract and under conditions, which may differ depending on various physiological factors (Antalis et al. 2007). In this case, the digestion process is far less easy to control and monitor. Therefore, in vitro simulated gastrointestinal digestion (SGID) is commonly employed to investigate the stability of milk proteins and check the effect on the stability of bioactive peptides upon their transit through SGID. SGID takes into account key factors affecting the digestion process such as the rates of gastric emptying, which are used to determine the period of incubation with gastric proteases, and changes in gastric pH value which determine to a large extent the enzyme activity (De Noni 2008).

GI digestion of casein and/or whey proteins by pepsin, trypsin and chymotrypsin has been reported to released several bioactive peptides (Meisel and FitzGerald 2003; Gobbetti et al. 2002, 2004). The GI digestion requires the generated peptide to be intact at the time of absorption or action in the intestine. Since most dietary proteins fully breakdown while passing through the small intestine, it means that the protein or peptide must be somewhat resistant to proteolysis if an effect is to be observed (Rutherfurd-Markwick 2012). There are examples of proteins (immunoglobulins and lactoferrin) which showcase this property of partial resistance (Roos et al. 1995; Drescher et al. 1999; Moller et al. 2008). Similarly, caseins have also shown to block the active site of proteolytic enzymes thereby shielding certain peptides from digestion (Rutherfurd-Markwick 2012).

Chemical Hydrolysis

This method involves cleavage of peptide bonds with either alkali or acid solutions. Chemical hydrolysis is not only difficult to control but can be damaging to certain amino acid groups like serine and threonine (Rutherfurd-Markwick 2012), destroy protein substrate, and form toxic substances like lysino-alanine (Clemente 2000). Thus, the limitations of this process have significantly reduced its application in generating peptides with biological activities.

Enzymatic Hydrolysis (In Vitro)

The most significant approach for the production of bioactive peptides is the hydrolysis of parent proteins by using multiple proteases or using a single protease. The enzymatic hydrolysis is reported to have profound advantages over fermentation approaches. Unlike fermentation enzymatic hydrolysis is usually rapid and controllable. Also, the optimization of the hydrolysis factors like pH, composition of buffer, temperature and time of reaction results in reproducible profiles of molecular weight and composition of peptides (Clemente 2000; Rizzello et al. 2016). Thus enzymatic hydrolysis may be considered as one of the most suited methods for the generation of target bioactive peptides. Traditionally at laboratory scale, the protein hydrolysis is carried in solution format, but other methods are also available like enzyme immobilization approach that allows the reduction in production cost and scale up of the process (Piovesana et al. 2018).

The enzymes like pepsin, trypsin, alcalase, chymotrypsin, etc., are being used in the enzymatic hydrolysis. These enzymes can mimic gastrointestinal digestion effects on the food peptides (Tavano 2013). Enzymatic hydrolysis process is also much more valuable and milder compared to chemical treatment in generating bioactive peptides. The enzymes can be site unspecific or specific, since enzymes are not equivalent to one another. Particularly, the site-specific enzymes like trypsin allows peptide identification using the established proteomics technologies, as the space for search is strongly reduced when the site of cleavage is known (Piovesana et al. 2018). On the contrary, the use of site-specific enzymes in large-scale productions is not mostly affordable due to high costs. Thus site-unspecific enzymes like alcalase can be used. However, the specificity of the information obtained is lost and thus complicating the finishing step of data management. In recent past, number of studies has been carried out in this direction and the peptidiomics based on modern approaches has been found of great help in identification of peptides in such cases (Rizzello et al. 2016).

ACE-inhibitory peptides from enzymatic digestion have been extensively studied (Hernendez-Ledesma et al. 2007; Murray and FitzGerald 2007; Otte et al. 2007a, b). Studies where trypsin was used to digest β-lactoglobulin found whey protein hydrolysates to show strong antihypertensive activity (da Costa et al. 2007; Ferreira et al. 2007; Roufik et al. 2006). Other proteolytic enzymes such as alcalase, subtilisin, and thermolysin are used in conjunction with pepsin and trypsin to generate peptides with known biological activities (Agyei and Danquah 2011). Enzymes from fungal and bacterial sources has also been utilized to achieve the same (Mohanty et al. 2016).

Liberation of peptides with a wide range of actions can thus be released using different enzymes. Although the bioactivity of peptides has been extensively reported in literature, the mechanism of action is not well understood. Few studies hypothesize it as a structure-activity relationship, while others suggest that the enzymes can be chosen to get the desired fragment and effect (Tavano 2013). Enzymatic hydrolysis of peptides or parent proteins may be performed either in a continuous mode or in batches (Madureira et al. 2010). Continuous process of hydrolysis is being preferred more nowadays to compensate for the high cost of batch processes (Yadav et al. 2015).

Microbial Fermentation (In Vitro)

Biotechnological interventions based on the exploitation of microorganisms for the production of bioactive peptides have been used for past many decades (Rizzello et al. 2016). Fermentation is mostly done for the production of certain foods for example production of natto and tempeh prepared from soybean fermentation were reported to exhibit antioxidant activity because of the presence of bioactive peptides (Vallabha and Tiku 2014; Babini et al. 2017). The amount and the type of bioactive peptides vary with type of microbial cultures used. Recently, the investigation carried to assess the effect of fermentation using *Lactobacillus plantarum* on release of bioactive peptides during gastrointestinal digestion revealed that time and temperature of fermentation affects release of bioactive peptides during gastrointestinal digestion in case of soybeans (Jakubczyk et al. 2017; Singh and Vij 2017).

Since many dairy starter cultures are proteolytic in nature, release of bioactive peptides can be expected during production of fermented dairy products (Korhonen 2009). These cultures consist of cell wall-bound peptidases and proteinases like dipeptidases, tripeptidases, amino-peptidases, and endopeptidases (Griffiths and Tellez 2013). Microbial proteolysis may generate many potent bioactive peptides (Muro Urista et al. 2011). Lactic acid bacteria (LAB) are the most common starter cultures used to hydrolyze milk proteins, especially caseins, due to their highly proteolytic nature (Hernandez-Ledesma et al. 2011; Szwajkowska et al. 2011). For instance, milk hydrolysates with high ACE-inhibitory activity were obtained from LAB fermented milk and subsequent hydrolysis with a microbial enzyme (Chen et al. 2007). Fuglsang et al. (2003) also evaluated the ability of more than 25 wild-type strains of LAB to produce fermented milk with antihypertensive activity. A different study reported casein fragments having potential antihypertensive property when nine different microbial proteolytic enzymes were used (Mizuno et al. 2004). Peptides from yogurt bacteria, probiotic bacteria, and cheese starter bacteria have been studied for their ability to have potential health effects (Gomez-Ruiz et al. 2002; Fuglsang et al. 2003; Gobbetti et al. 2004; Donkor et al. 2007). Apart from bacteria, yeast species like *Kluyveromyces marxianus* and *Saccharomyces cerevisiae* are reported to produce peptides from goat whey proteins (Didelot et al. 2006; Hamme et al. 2009).

Characterization and Analysis of Bioactive Peptides

To identify food-derived peptides in food matrices, blood and other biological samples, potential methods are being evolved. The sequence of peptides is conventionally identified by Edman degradation or enzyme-linked immunoassay (ELISA). However, these techniques require significant fractionation work before the peptides can be identified, rendering the methodology quite complex. To isolate specific peptides from hydrolysates is difficult because of hundreds of peptides present in any given hydrolysate. Also, some peptides are very much alike to each other in terms of physiochemical properties like mass, charge, solubility etc., which makes the separation even more challenging (Capriotti et al. 2016). Additionally, food matrices are intricate too, which combined with the protein hydrolysate makes the identification, characterization and quantification a difficult task (Rutherfurd-Markwick 2012). Nevertheless, advanced analytical techniques such as mass spectrometry (MS), Nuclear Magnetic Resonance (NMR), High Performance Liquid Chromatography (HPLC), Capillary Electrophoresis (CE) are available to carry out complex food matrices studies. Sometimes, a combination of these techniques is used when analyzing multiple components (Bernal et al. 2011). Also, the type of technique employed will differ according to the target compound in the said matrix. For instance, physico-chemical properties like size, polarity, volatility etc. will strongly affect the sample preparation, separation mechanism and technique (HPLC, CE, GC) as well as the type of detector used (UV, Fluorescence Detector, MS, etc.). Advanced analytical techniques also help in the better understanding of bioactivity and bioavailability of the food derived peptides (Bernal et al. 2011). Table 1 provides examples of the types of techniques used to analyze bioactive peptides.

Identification of Sequence by Edman Degradation

Edman degradation is of one of the first and frequent methods used for peptide identification before high-throughput proteomics technologies were introduced. The process required an extensive purification before sequencing, thus the process was laborious and time consuming. The automated instrumentation process was also limited to identification of 50 residues only (Rizzello et al. 2016). Although the Edman degradation process is an outdated approach, however, in recent studies the use of this method has been reported (Jamdar et al. 2017; Furuta et al. 2016).

Capillary Electrophoresis (CE)

For the quantitative determination of peptides, CE is an attractive method that is being used due to its low consumption of sample, reagent and time as well as its ability to separate small to large sized peptides (Herrero et al. 2008; Acunha et al.

Table 1 Examples of different analytical techniques employed to identify and characterize bioactive peptides

Source	Technique	Potential bioactivity	References
Fermented milk	ESI/MS	ACE-inhibitory	Quirós et al. (2007)
Bovine casein peptic and chymosin hydrolysis	ESI/MS	Antibacterial	McCann et al. (2005, 2006)
Lactoferrin hydrolysis by pepsin	ESI/MS	Antihypertensive	Ruiz-Gimenez et al. (2012)
Whey protein concentrate hydrolysis by trypsin	ESI/MS	DPP-IV inhibitory	Silveira et al. (2013)
Fermented milk	MALDI/TOF	Antihypertensive	Pihlanto et al. (2010)
Lactoferrin hydrolysis by pepsin	MALDI/TOF	Antimicrobial	Chan and Li-Chan (2007)
β-casein peptide	RP-HPLC/MS	ACE-inhibitory	Quirós et al. (2009)
Whey protein concentrate hydrolysis by thermolysin	RP-HPLC-MS/MS	Antioxidant	Del Mar Contreras et al. (2011)
Peptides in infant milk formula	CE-TOF-MS	Several activities including antihypertensive, antioxidant, antimicrobial etc.	Catala-Clariana et al. (2013)
Peptides in alcalase hydrolysated pollen	Nano LC–MS/MS and RP-HPLC	Antihypertensive and antioxidant	Maqsoudlou et al. (2019)
Fermented cucumber	MALDESI mass spectrometry and LC-QQQ-MS	Antihypertensive	Fideler et al. (2019)
Soy seeds and soy milk	Nano-LC–MS/MS	ACE Inhibitor, Antioxidative	Capriotti et al. (2015)

2016). Since milk proteins are considered one of the most important sources of bioactive peptides, identification of such peptides has become vital. One such study where bioactive peptides were identified using CE-MS was in commercial hypoallergenic infant milk formulas prepared from bovine milk protein hydrolysates (Catala-Clariana et al. 2013). To measure molecular mass, high resolution TOF-MS was used. A characteristic electrophoretic profile was obtained for each infant milk formula studied. Also, the bioactive peptides identified were mostly found to be ACE inhibitors, but other peptide sequence pertaining to different bioactivities such as antioxidant, antimicrobial, antithrombotic, or hypocholesterolemic were also observed.

Chromatographic Techniques

Classical techniques like PAGE, sodium dodecyl sulfate PAGE (SDS-PAGE) and 2D-PAGE have been used previously, although they are less accurate in identifying biomolecules compared to CE or HPLC coupled to mass spectrometry (Bernal et al. 2011). Based on the peptide physiochemical properties, different chromatographic purification techniques can be used. For example, to retain hydrophilic peptides, hydrophilic interaction chromatography (HILIC) is a better option. Size exclusion chromatography (SEC) is often used for routine and validation analyses because of its reproducibility and speed but coupling this technique with MS is a challenge (Capriotti et al. 2016). Separation of large number of peptides in a sample is performed by HPLC because of its efficiency, versatility and automation capability (Hernandez-Ledesma et al. 2013). To analyze milk-derived bioactive peptides, reversed-phase HPLC (RP-HPLC) is commonly employed to separate small peptides (Recio and Lopez-Fandino 2010); this technique is more effective because most bioactive peptides are small in size (Roberts et al. 1999). In RP-HPLC, retention time depends on the amino acid composition of small peptides while for larger peptides, retention time is influenced by molecular weight and conformational effects. Because RP-HPLC profiles of milk protein hydrolysates cover thousands of peptides of different chemical and genetic variants, statistical tools help researchers obtain only relevant information from such large data (Coker et al. 2005). HPLC technique is often combined with conventional UV and fluorescence detectors. Sometimes, it's necessary to use more than one HPLC step i.e. a combination of techniques to achieve separation (Sandra et al. 2009). An example of a study where RP-HPLC was used to separate and quantify genetic variants of casein and whey protein was done on water buffalo milk. They reported separation of all major protein fractions with improved resolution, and in a shorter time (Bonfatti et al. 2013). Juan et al. (2009) also separated and quantified six major bovine proteins in milk using RP-HPLC.

Mass Spectrometry (MS)

In recent past, MS and proteomics technologies has been considered as gold standard for peptide analysis. This method allows peptide identification in a mixture, therefore, bypassing the cumbersome Edman's degradation based method and does not either require a single pure peptide (Rizzello et al. 2016). The last decade has seen a notable development of the MS technique, which is widely being accepted as a tool in identifying bioactive peptides (Picariello et al. 2012). MS has been helpful in determining molecular mass, protein sequences, protein conformations and detection of new genetic variants (del Mar Contreras et al. 2008). It has also been combined with traditional identification methods like Edman degradation sequencing. Also other techniques have been used in combination with MS for peptide identification (Agrawal et al. 2016, 2017). For example, MALDI MS is frequently used without any upstream fractionation or any real separation. In certain cases, HPLC coupled with MS/MS or high-resolution MS has been used for the identification.

In some cases, HPLC coupled with MS/MS or high-resolution MS is used, and sequences are provided for the peptides in selected fractions or in the whole hydrolysate. It has also been combined with traditional identification methods, like in the study of Pihlanto et al. (2010) where ACE-inhibitory peptides in fermented milk were identified using matrix-assisted laser desorption ionization-time of flight (MALDI-TOF) and Edman degradation sequencing. MS equipment has advantages like sensitive and accurate mass information, with a short-term analysis (D'siva and Mine 2010).

Electrospray (ESI), a type of ionization source, has displaced previous ionization methodology like fast atom bombardment (del Mar Contreras et al. 2008; Sanchez-Rivera et al. 2014). Also, MS coupled with on-line HPLC has allowed discovery of antihypertensive peptides from peptic digestion of lactoferrin (Ruiz-Gimenez et al. 2012). This study used ESI as ionization source and ion trap mass analyzer (IT). Another implementation of ESI was in characterization of bioactive peptides in infant milk formulas where two mass analyzers (ESI-IT and ESI-TOF) were used (Catala-Clariana et al. 2010). MALDI is another type of ionization source that is employed in identification of longer peptides and has some advantages over ESI: higher sensitivity, lower susceptibility to impurities and single charged ions. However, MALDI does not permit on-line chromatographic coupling like ESI does (Mamone et al. 2009). Table 1 shows some examples where MALDI presented with TOF (MALDI-TOF) was used to identify peptides, like antimicrobial activity of peptic hydrolysate from lactoferrin (Chan and Li-Chan 2007).

Processing of Bioactive Peptides

Industrial scale production of bioactive peptides in sufficient amounts for human trials involves various purification and concentration steps. This can lead to changes in the structure of peptides as well as alter how the peptides interact with surrounding matrix, which will consequently influence the bioavailability and activity of peptides. Moreover, the products used in experiment to generate peptides may have the presence of unknown compounds that could affect peptide activity and may give rise to side effects (Bougle and Bouhallab 2017). Lastly, it is reported that storage can shorten the shelf-life of peptides and modify their bioavailability (Rao et al. 2012). Thus, care must be taken when industrial process of producing bioactive peptides is carried out.

Safety of Bioactive Peptides

The safety of food protein-derived bioactive peptides has been of little concern because typically, the body would hydrolyze food proteins into peptides (Wang et al. 2005). Phelan et al. (2009) have reviewed the safety of milk-derived peptides. Also, administration of casein hydrolysates having antihypertensive peptides in rats for 4 weeks showed no detrimental effects (Anadon et al. 2010). Prehypertensive and hypertensive human subjects too have displayed no adverse effects on serum and urine parameters when given bioactive peptides-rich milk protein hydrolysates (Boelsma and Kloek 2010; Germino et al. 2010). Although these studies suggest that bioactive peptides from food proteins are generally safe, more rigorous toxicology studies in humans must be carried out before making any health claims. Also, the production conditions like protein source, enzyme used must be of accepted food grade (GRAS) to avoid any negative effects on the peptide safety and quality (Bougle and Bouhallab 2017).

Future Trends

Bioactive peptides also including the ones derived from vegetable sources are not commercially available in the market except for few products (Carrasco-Castilla et al. 2012). Though few advances have been made in the production technologies of the bioactive peptides, the production at industrial scale is similar to the laboratory-scale because the processing steps including hydrolysis, separation and purification are limited because of the lack of suitable technologies. The downstream processing of bioactive peptides involves high energy-consuming procedures that may include drying, disruption of cellular matrices and extraction of bioactive peptides. Also different technologies are to be studied with respect to cost and production relation (Hayes and Tiwari 2015). Recent reports regarding the use of novel processing technology suggest its limited utilization that depends on number of factors including: (1) affordable optimization technologies for the protein/peptide rich samples; (2) high cost and lack of sustainability associated with pretreatment and extraction process of proteins; (3) expensive isolation and downstream processing of potent peptide/protein isolates, hydrolysates and co-products; (4) safety and health legislation (Hayes and Tiwari 2015). Bioactive peptides are potent candidates for the new era of pharmaceutical products because of the increased concern of side effects and the increased demand for natural products (Danquah and Agyei 2012; Lemes et al. 2016). For last several decades, the production of bioactive peptides was mostly limited to animal sources of proteins, however, from last decade focus has been shifted to plant sources which may pave the ways for the production of cost effective bioactive peptides since the source of production is cheap.

Conclusion

Bioactive peptides derived from food sources is an area that is rapidly growing and interesting researchers to discover new peptides with potential physiological benefits. Currently, enzymatic hydrolysis with proteolytic enzymes and microbial fermentation of the protein source are the two empirical approaches used to obtain bioactive peptides. Food proteins are a good source of peptides that can show potential health effects. With advancements made in techniques to identify, characterize and quantify these bioactive peptides, more information is being obtained which can be used to further study the bioactive peptides in the in vivo systems. This will help validate the bioavailability, efficacy and safety of the peptides and their subsequent usage in development of functional foods and nutraceuticals.

References

Acunha, T., Ibáñez, C., García-Cañas, V., Simó, C., & Cifuentes, A. (2016). Recent advances in the application of capillary electromigration methods for food analysis and Foodomics. *Electrophoresis, 37*(1), 111–141.

Agrawal, H., Joshi, R., & Gupta, M. (2016). Isolation, purification and characterization of antioxidative peptide of pearl millet (*Pennisetum glaucum*) protein hydrolysate. *Food Chemistry, 204*, 365–372.

Agrawal, H., Joshi, R., & Gupta, M. (2017). Isolation and characterization of enzymatic hydrolysed peptides with antioxidant activities from green tender sorghum. *LWT - Food Science and Technology, 84*, 608–616.

Agyei, D., & Danquah, M. K. (2011). Industrial-scale manufacturing of pharmaceutical-grade bioactive peptides. *Biotechnology Advances, 29*(3), 272–277.

Ambigaipalan, P., & Shahidi, F. (2015). Date seed flour and hydrolysates affect physicochemical properties of muffin. *Food Bioscience, 12*, 54–60.

Anadon, A., Martínez, M. A., Ares, I., Ramos, E., Martinez-Larranaga, M. R., Contreras, M. M., & Recio, I. (2010). Acute and repeated dose (4 weeks) oral toxicity studies of two antihypertensive peptides, RYLGY and AYFYPEL, that correspond to fragments (90–94) and (143–149) from αs1-casein. *Food and Chemical Toxicology, 48*(7), 836–1845.

Antalis, T. M., Shea-Donohue, T., Vogel, S. N., Sears, C., & Fasano, A. (2007). Mechanisms of disease: Protease functions in intestinal mucosal pathobiology. *National Clinical Practice and Gastroenetrology and Heapatology, 4*, 393–402.

Atanasova, J., & Ivanova, I. (2010). Antibacterial peptides from goat and sheep milk proteins. *Biotechnology & Biotechnological Equipment, 24*(2), 1799–1803.

Babini, E., Tagliazucchi, D., Martini, S., Dei Piu, L., & Gianotti, A. (2017). LC-ESI-QTOF-MS identification of novel antioxidant peptides obtained by enzymatic and microbial hydrolysis of vegetable proteins. *Food Chemistry, 228*, 186–196.

Banerjee, J., Singh, R., Vijayaraghavan, R., MacFarlane, D., Patti, A. F., & Arora, A. (2017). Bioactives from fruit processing wastes: Green approaches to valuable chemicals. *Food Chemistry, 225*, 10–22.

Bernal, J., Mendiola, J., Ibáñez, E., & Cifuentes, A. (2011). Advanced analysis of nutraceuticals. *Journal of Pharmaceutical and Biomedical Analysis, 55*(4), 758–774.

Bidasolo, I. B., Ramos, M., & Gomez-Ruiz, J. A. (2012). *In vitro* simulated gastrointestinal digestion of donkeys' milk. Peptide characterization by high performance liquid chromatography–tandem mass spectrometry. *International Dairy Journal, 24*(2), 146–152.

Boelsma, E., & Kloek, J. (2010). IPP-rich milk protein hydrolysate lowers blood pressure in subjects with stage 1 hypertension, a randomized controlled trial. *Nutrition Journal, 9*(1), 52.

Bonfatti, V., Giantin, M., Rostellato, R., Dacasto, M., & Carnier, P. (2013). Separation and quantification of water buffalo milk protein fractions and genetic variants by RP-HPLC. *Food Chemistry, 136*(2), 364–367.

Bougle, D., & Bouhallab, S. (2017). Dietary bioactive peptides: Human studies. *Critical Reviews in Food Science and Nutrition, 57*(2), 335–343.

Boutrou, R., Henry, G., & Sanchez-Rivera, L. (2015). On the trail of milk bioactive peptides in human and animal intestinal tracts during digestion: A review. *Dairy Science & Technology, 95*, 815–829.

Capriotti, A. L., Caruso, G., Cavaliere, C., Piovesana, S., Samperi, R., & Lagana, A. (2012). Comparison of three different enrichment strategies for serum low molecular weight protein identification using shotgun proteomics approach. *Analytica Chimica Acta, 740*, 58–65.

Capriotti, A. L., Caruso, G., Cavaliere, C., Samperi, R., Ventura, S., Chiozzi, R. Z., & Lagana, A. (2015). Identification of potential bioactive peptides generated by simulated gastrointestinal digestion of soybean seeds and soy milk proteins. *Journal of Food Composition and Analysis, 44*, 205–213.

Capriotti, A. L., Cavaliere, C., Piovesana, S., Samperi, R., & Lagan, A. (2016). Recent trends in the analysis of bioactive peptides in milk and dairy products. *Analytical and Bioanalytical Chemistry, 408*, 2677–2685.

Carrasco-Castilla, J., Hernandez-Alvarez, A. J., Jimenez-Martínez, C., Jacinto-Hernandez, C., Alaiz, M., Giron-Calle, J., … Daavila-Ortiz, G. (2012). Antioxidant and metal chelating activities of peptide fractions from phaseolin and bean protein hydrolysates. *Food Chemistry, 135*, 1789–1795.

Catala-Clariana, S., Benavente, F., Giménez, E., Barbosa, J., & Sanz-Nebot, V. (2010). Identification of bioactive peptides in hypoallergenic infant milk formulas by capillary electrophoresis–mass spectrometry. *Analytica Chimica Acta, 683*(1), 119–125.

Catala-Clariana, S., Benavente, F., Giménez, E., Barbosa, J., & Sanz-Nebot, V. (2013). Identification of bioactive peptides in hypoallergenic infant milk formulas by CE-TOF-MS assisted by semiempirical model of electromigration behavior. *Electrophoresis, 34*(13), 1886–1894.

Chan, J. C. K., & Li-Chan, E. C. Y. (2007). Production of lactoferricin and other cationic peptides from food grade bovine lactoferrin with various iron saturation levels. *Journal of Agricultural and Food Chemistry, 55*, 493–501.

Chen, G. W., Tsai, J. S., & Pan, B. S. (2007). Purification of angiotensin I-converting enzyme inhibitory peptides and antihypertensive effect of milk produced by protease-facilitated lactic fermentation. *International Dairy Journal, 17*(6), 641–647.

Chung, Y. C., Kim, Y. S., Shadchehr, A., Garrido, A., Macgregor, I. L., & Sleisenger, M. H. (1979). Protein digestion and absorption in human small intestine. *Gastroenterology, 76*(6), 1415–1421.

Clemente, A. (2000). Enzymatic protein hydrolysates in human nutrition. *Trends in Food Science & Technology, 11*(7), 254–262.

Coker, C. J., Crawford, R. A., Johnston, K. A., Singh, H., & Creamer, L. K. (2005). Towards the classification of cheese variety and maturity on the basis of statistical analysis of proteolysis data—a review. *International Dairy Journal, 15*(6-9), 631–643.

da Costa, E. L., da Rocha Gontijo, J. A., & Netto, F. M. (2007). Effect of heat and enzymatic treatment on the antihypertensive activity of whey protein hydrolysates. *International Dairy Journal, 17*(6), 632–640.

del Mar Contreras, M., Hernandez-Ledesma, B., Amigo, L., Martín-Alvarez, P. J., & Recio, I. (2011). Production of antioxidant hydrolyzates from a whey protein concentrate with thermolysin: Optimization by response surface methodology. *LWT - Food Science and Technology, 44*(1), 9–15.

del Mar Contreras, M., Lpez-Expsito, I., Hernndez-Ledesma, B., Ramos, M., & Recio, I. (2008). Application of mass spectrometry to the characterization and quantification of food-derived bioactive peptides. *Journal of AOAC International, 91*(4), 981–994.

D'siva, I., & Mine, Y. (2010). Peptidomics for bioactive peptide analysis. In Y. Mine, E. Li-Chan, & B. Jiang (Eds.), *Bioactive proteins and peptides as functional foods and nutraceuticals* (pp. 307–324). Ames: Wiley-Blackwell.

Danquah, M. K., & Agyei, D. (2012). Pharmaceutical applications of bioactive peptides. *OA Biotechnology, 1*, 5.

De Noni, I. (2008). Release of β-casomorphins 5 and 7 during simulated gastrointestinal digestion of bovine β-casein variants and milk-based infant formulas. *Food Chemistry, 110*, 897–903.

Dhaval, A., Yadav, N., & Purwar, S. (2016). Potential applications of food derived bioactive peptides in management of health. *International Journal of Peptide Research and Therapeutics, 22*, 377–398.

Didelot, S., Bordenave-Juchereau, S., Rosenfeld, E., Piot, J. M., & Sannier, F. (2006). Peptides released from acid goat whey by a yeast-*Lactobacillus* association isolated from cheese microflora. *Journal of Dairy Research, 73*(2), 163–170.

Donkor, O. N., Henriksson, A., Vasiljevic, T., & Shah, N. P. (2007). Proteolytic activity of dairy lactic acid bacteria and probiotics as determinant of growth and *in vitro* angiotensin-converting enzyme inhibitory activity in fermented milk. *Le Lait, 87*(1), 21–38.

Drescher, K., Roos, N., Pfeuffer, M., Seyfert, H. M., Schrezenmeir, J., & Hagemeister, H. (1999). Recovery of 15N-lactoferrin is higher than that of 15N-casein in the small intestine of suckling, but not adult miniature pigs. *The Journal of Nutrition, 129*(5), 1026–1030.

Duranti, M. (2006). Grain legume proteins and nutraceutical properties. *Fitoterapia, 77*, 67–82.

El-Salam, M. H. A., & El-Shibiny, S. (2013). Bioactive peptides of buffalo, camel, goat, sheep, mare, and yak milks and milk products. *Food Reviews International, 29*, 1–23.

Esteve, C., Marina, M. L., & García, M. C. (2015). Novel strategy for the revalorization of olive (*Olea europaea*) residues based on the extraction of bioactive peptides. *Food Chemistry, 167*, 272–280.

Ferreira, I. M. P. L. V. O., Pinho, O., Mota, M. V., Tavares, P., Pereira, A., Goncalves, M. P., & Teixeira, J. A. (2007). Preparation of ingredients containing an ACE-inhibitory peptide by tryptic hydrolysis of whey protein concentrates. *International Dairy Journal, 17*(5), 481–487.

Fideler, J., Johanningsmeier, S. D., Ekelöf, M., & Muddiman, D. C. (2019). Discovery and quantification of bioactive peptides in fermented cucumber by direct analysis IR-MALDESI mass spectrometry and LC-QQQ-MS. *Food Chemistry, 271*, 715–723.

Fuglsang, A., Rattray, F. P., Nilsson, D., & Nyborg, N. C. (2003). Lactic acid bacteria: Inhibition of angiotensin converting enzyme *in vitro* and *in vivo*. *Antonie Van Leeuwenhoek, 83*(1), 27–34.

Furuta, T., Miyabe, Y., Yasui, H., Kinoshita, Y., & Kishimura, H. (2016). Angiotensin I converting enzyme inhibitory peptides derived from phycobiliproteins of dulse *Palmaria palmata*. *Marine Drugs, 14*(2), 32.

Germino, F. W., Neutel, J., Nonaka, M., & Hendler, S. S. (2010). The impact of lactotripeptides on blood pressure response in stage 1 and stage 2 hypertensives. *Journal of Clinical Hypertension, 12*(3), 153–159.

Gobbetti, M., Minervini, F., & Rizzello, C. G. (2004). Angiotensin I-converting-enzyme-inhibitory and antimicrobial bioactive peptides. *International Journal of Dairy Technology, 57*(2-3), 173–188.

Gobbetti, M., Stepaniak, L., De Angelis, M., Corsetti, A., & Di Cagno, R. (2002). Latent bioactive peptides in milk proteins: Proteolytic activation and significance in dairy processing. *Critical Reviews in Food Science and Nutrition, 42*(3), 223–239.

Gomez-Ruiz, J. A., Ramos, M., & Recio, I. (2002). Angiotensin-converting enzyme-inhibitory peptides in Manchego cheeses manufactured with different starter cultures. *International Dairy Journal, 12*(8), 697–706.

Griffiths, M. W., & Tellez, A. M. (2013). *Lactobacillus helveticus*: The proteolytic system. *Frontiers in Microbiology, 4*, 30.

Hamme, V., Sannier, F., Piot, J. M., Didelot, S., & Bordenave-Juchereau, S. (2009). Crude goat whey fermentation by *Kluyveromyces marxianus* and *Lactobacillus rhamnosus*: Contribution to proteolysis and ACE inhibitory activity. *Journal of Dairy Research, 76*(2), 152–157.

Hayes, M., & Tiwari, B. K. (2015). Bioactive carbohydrates and peptides in foods: An overview of sources, downstream processing steps and associated bioactivities. *International Journal of Molecular Sciences, 16*, 22485–22508.

Hernandez-Ledesma, B., del Mar Contreras, M., & Recio, I. (2011). Antihypertensive peptides: Production, bioavailability and incorporation into foods. *Advances in Colloid and Interface Science, 165*(1), 23–35.

Hernandez-Ledesma, B., Martinez-Maqueda, D., Miralles, B., Amigo, L., & Gomez-Ruiz, J. A. (2013). Peptides. In M. L. Nollet & F. Toldra (Eds.), *Food analysis by HPLC* (pp. 69–95). Boca Raton: CRC Press.

Hernendez-Ledesma, B., Miguel, M., Amigo, L., Aleixandre, M. A., & Recio, I. (2007). Effect of simulated gastrointestinal digestion on the antihypertensive properties of synthetic β-lactoglobulin peptide sequences. *Journal of Dairy Research, 74*(3), 336–339.

Herrero, M., Ibañez, E., & Cifuentes, A. (2008). Capillary electrophoresis-electrospray-mass spectrometry in peptide analysis and peptidomics. *Electrophoresis, 29*(10), 2148–2160.

Hettiarachchy, N. S. (2012). *Bioactive food proteins and peptides: Applications in human health.* Boca Raton: CRC Press.

Jakala, P., & Vapaatalo, H. (2010). Antihypertensive peptides from milk proteins. *Pharmaceuticals, 3*, 251–272.

Jakubczyk, A., Karas, M., Ziotek, U., & Szymanowska, U. (2017). Identification of potential inhibitory peptides of enzymes involved in the metabolic syndrome obtained by simulated gastrointestinal digestion of fermented bean (*Phaseolus vulgaris* L.) seeds. *Food Research International, 100*, 489–496.

Jamdar, S. N., Deshpande, R., & Marathe, S. A. (2017). Effect of processing conditions and *in vitro* protein digestion on bioactive potentials of commonly consumed legumes. *Food Bioscience, 20*, 1–11.

Juan, W., Zhang, Q. H., Zhi-Hua, W., & Hong-Mei, L. I. (2009). Determination of major bovine milk proteins by reversed phase high performance liquid chromatography. *Chinese Journal of Analytical Chemistry, 37*(11), 1667–1670.

Kitts, D. D., & Weiler, K. (2003). Bioactive proteins and peptides from food sources. Applications of bioprocesses used in isolation and recovery. *Current Pharmaceutical Design, 9*, 1309–1323.

Korhonen, H. (2009). Milk-derived bioactive peptides: From science to applications. *Journal of Functional Foods, 1*, 177–187.

Lee, S. Y., & Hur, S. J. (2017). Antihypertensive peptides from animal products, marine organisms, and plants. *Food Chemistry, 228*, 506–517.

Lemes, A. C., Sala, L., Ores, J. D. C., Braga, A. R. C., Egea, M. B., & Fernandes, K. F. (2016). A review of the latest advances in encrypted bioactive peptides from protein-rich waste. *International Journal of Molecular Science, 17*, 950.

Li, G. H., Qu, M. R., Wan, J. Z., & You, J. M. (2007). Antihypertensive effect of rice protein hydrolysate with *in vitro* angiotensin I-converting enzyme inhibitory activity in spontaneously hypertensive rats. *Asia Pacific Journal of Clinical Nutrition, 16*(Suppl 1), 275–280.

Lovati, M. R., Manzoni, C., Gianazza, E., Arnoldi, A., Kurowska, E., & Carroll, K. K. (2000). Soy protein peptides regulate cholesterol homeostasis in Hep G2 cells. *Journal of Nutrition, 130*, 2543–2549.

Lv, Y., Bao, X. L., Yang, B. C., Ren, C. G., & Guo, S. T. (2008). Effect of soluble soybean protein hydrolysate-calcium complexes on calcium uptake by Caco-2 cells. *Journal of Food Science, 73*, H168. https://doi.org/10.1111/j.1750-3841.2008.00873.x

Madureira, A. R., Tavares, T., Gomes, A. M., Pintado, M. E., & Malcata, F. X. (2010). Invited review: physiological properties of bioactive peptides obtained from whey proteins. *Journal of Dairy Science, 93*(2), 437–455.

Mamone, G., Picariello, G., Caira, S., Addeo, F., & Ferranti, P. (2009). Analysis of food proteins and peptides by mass spectrometry-based techniques. *Journal of Chromatography. A, 1216*, 7130–7142.

Maqsoudlou, A., Mahoonak, A. S., Mora, L., Mohebodini, H., Toldrá, F., & Ghorbani, M. (2019). Peptide identification in alcalase hydrolysated pollen and comparison of its bioactivity with royal jelly. *Food Research International, 116*, 905–915.

Meisel, H., & FitzGerald, R. J. (2003). Biofunctional peptides from milk proteins: Mineral binding and cytomodulatory effects. *Current Pharmaceutical Design, 9*, 1289–1295.

McCann, K., Shiell, B., Michalski, W., Lee, A., Wan, J., Roginski, H., & Coventry, M. (2005). Isolation and characterisation of antibacterial peptides derived from the f (164–207) region of bovine αS2-casein. *International Dairy Journal, 15*(2), 133–143.

McCann, K., Shiell, B., Michalski, W., Lee, A., Wan, J., Roginski, H., & Coventry, M. (2006). Isolation and characterisation of a novel antibacterial peptide from bovine αS1-casein. *International Dairy Journal, 16*(4), 316–323.

Mine, Y., Li-Chan, E., & Jiang, B. (Eds.). (2010). *Bioactive proteins and peptides as functional foods and nutraceuticals*. Oxford: Wiley-Blackwell.

Mizuno, S., Nishimura, S., Matsuura, K., Gotou, T., & Yamamoto, N. (2004). Release of short and proline-rich antihypertensive peptides from casein hydrolysate with an *Aspergillus oryzae* protease. *Journal of Dairy Science, 87*(10), 3183–3188.

Mohanty, D., Jena, R., Choudhury, P. K., Pattnaik, R., Mohapatra, S., & Saini, M. R. (2016). Milk derived antimicrobial bioactive peptides: A review. *International Journal of Food Properties, 19*, 837–846.

Moller, N. P., Scholz-Ahrens, K. E., Roos, N., & Schrezenmeir, J. (2008). Bioactive peptides and proteins from foods: Indication for health effects. *European Journal of Nutrition, 47*, 171–182.

Muro Urista, C., Álvarez Fernández, R., Riera Rodriguez, F., Arana Cuenca, A., & Tellez Jurado, A. (2011). Production and functionality of active peptides from milk. *Food Science and Technology International, 17*(4), 293–317.

Murray, B. A., & FitzGerald, R. J. (2007). Angiotensin converting enzyme inhibitory peptides derived from food proteins: Biochemistry, bioactivity and production. *Current Pharmaceutical Design, 13*(8), 773–791.

Nakamura, Y., Yamamoto, N., Sakai, K., Okubo, A., Yamazaki, S., & Takano, T. (1995). Purification and characterization of angiotensin I-converting enzyme inhibitors from sour milk. *Journal of Dairy Science, 78*(4), 777–783.

Otte, J., Shalaby, S. M., Zakora, M., & Nielsen, M. S. (2007b). Fractionation and identification of ACE-inhibitory peptides from α-lactalbumin and β-casein produced by thermolysin-catalysed hydrolysis. *International Dairy Journal, 17*(12), 1460–1472.

Otte, J., Shalaby, S. M., Zakora, M., Pripp, A. H., & El-Shabrawy, S. A. (2007a). Angiotensin-converting enzyme inhibitory activity of milk protein hydrolysates: Effect of substrate, enzyme and time of hydrolysis. *International Dairy Journal, 17*(5), 488–503.

Panchaud, A., Affolter, M., & Kussmann, M. (2012). Mass spectrometry for nutritional peptidomics: How to analyze food bioactives and their health effects. *Journal of Proteomics, 75*(12), 3546–3559.

Park, Y. W., & Nam, M. S. (2015). Bioactive peptides in milk and dairy products: A review. *Korean Journal for Food Science of Animal Resources, 35*, 831–840.

Phelan, M., Aherne, A., FitzGerald, R. J., & O'Brien, N. M. (2009). Casein-derived bioactive peptides: biological effects, industrial uses, safety aspects and regulatory status. *International Dairy Journal, 19*(11), 643–654.

Picariello, G., Mamone, G., Addeo, F., & Ferranti, P. (2012). Novel mass spectrometry-based applications of the 'omic' sciences in food technology and biotechnology. *Food Technology and Biotechnology, 50*(3), 286.

Pihlanto, A., Virtanen, T., & Korhonen, H. (2010). Angiotensin I converting enzyme (ACE) inhibitory activity and antihypertensive effect of fermented milk. *International Dairy Journal, 20*(1), 3–10.

Piovesana, S., Capriotti, A. L., Cavaliere, C., La Barbera, G., Montone, C. M., Chiozzi, R. Z., & Lagana, A. (2018). Recent trends and analytical challenges in plant bioactive peptide separation, identification and validation. *Analytical and Bioanalytical Chemistry, 410*, 3425–3444.

Quirós, A., Ramos, M., Muguerza, B., Delgado, M. A., Miguel, M., Aleixandre, A., & Recio, I. (2007). Identification of novel antihypertensive peptides in milk fermented with *Enterococcus faecalis*. *International Dairy Journal, 17*(1), 33–41.

Quirós, A., del Mar Contreras, M., Ramos, M., Amigo, L., & Recio, I. (2009). Stability to gastrointestinal enzymes and structure–activity relationship of β-casein-peptides with antihypertensive properties. *Peptides,30*(10), 1848–1853.

Ramdath, D., Padhi, E., Sarfaraz, S., Renwick, S., & Duncan, A. (2017). Beyond the cholesterol-lowering effect of soy protein: A review of the effects of dietary soy and its constituents on risk factors for cardiovascular disease. *Nutrients, 9*, 324.

Rao, Q., Rocca-Smith, J. R., Schoenfuss, T. C., & Labuza, T. P. (2012). Accelerated shelf-life testing of quality loss for a commercial hydrolysed hen egg white powder. *Food Chemistry, 135*(2), 464–472.

Recio, I., & Lopez-Fandino, R. (2010). Peptides. In L. Nollet & F. Toldrá (Eds.), *Handbook of dairy foods analysis* (pp. 33–77). Boca Raton: Taylor & Francis.

Rizzello, C. G., Tagliazucchi, D., Babini, E., Sefora Rutella, G., Taneyo Saa, D. L., & Gianotti, A. (2016). Bioactive peptides from vegetable food matrices: Research trends and novel biotechnologies for synthesis and recovery. *Journal of Functional Foods, 27*, 549–569.

Roberts, P. R., Burney, J. D., Black, K. W., & Zaloga, G. P. (1999). Effect of chain length on absorption of biologically active peptides from the gastrointestinal tract. *Digestion, 60*, 332–337.

Roos, N., Mahe, S., Benamouzig, R., Sick, H., Rautureau, J., & Tome, D. (1995). 15N-labeled immunoglobulins from bovine colostrum are partially resistant to digestion in human intestine. *The Journal of Nutrition, 125*(5), 1238–1244.

Roufik, S., Gauthier, S. F., & Turgeon, S. L. (2006). *In vitro* digestibility of bioactive peptides derived from bovine β-lactoglobulin. *International Dairy Journal, 16*(4), 294–302.

Ruiz-Gimenez, P., Salom, J. B., Marcos, J. F., Valles, S., Martínez-Maqueda, D., & Recio, I. (2012). Antihypertensive effect of a bovine lactoferrin pepsin hydrolysate: Identification of novel active peptides. *Food Chemistry, 131*, 266–273.

Rutherfurd-Markwick, K. J. (2012). Food proteins as a source of bioactive peptides with diverse functions. *British Journal of Nutrition, 108*(S2), S149–S157.

Saito, T. (2004). Selection of useful probiotic lactic acid bacteria from the *Lactobacillus acidophilus* group and their applications to functional foods. *Animal Science Journal, 75*(1), 1–13.

Sanchez, A., & Vazquez, A. (2017). Bioactive peptides: A review. *Food Quality and Safety, 1*(1), 29–46.

Sanchez-Rivera, L., Martínez-Maqueda, D., Cruz-Huerta, E., Miralles, B., & Recio, I. (2014). Peptidomics for discovery, bioavailability and monitoring of dairy bioactive peptides. *Food Research International, 63*, 170–181.

Sandra, K., Moshir, M., D'hondt, F., Tuytten, R., Verleysen, K., & Kas, K. (2009). Highly efficient peptide separations in proteomics. Part 2: Bi- and multidimensional liquid-based separation techniques. *Journal of Chromatography B, 877*, 1019–1039.

Silveira, S. T., Martínez-Maqueda, D., Recio, I., & Hernández-Ledesma, B. (2013). Dipeptidyl peptidase-IV inhibitory peptides generated by tryptic hydrolysis of a whey protein concentrate rich in β-lactoglobulin. *Food Chemistry, 141*(2), 1072–1077.

Singh, B. P., & Vij, S. (2017). Growth and bioactive peptides production potential of *Lactobacillus plantarum* strain C2 in soy milk: A LC-MS/MS based revelation for peptides biofunctionality. *LWT, 86*, 293–301.

Sipola, M., Finckenberg, P., Vapaatalo, H., Pihlanto-Leppala, A., Korhonen, H., Korpela, R., & Nurminen, M. L. (2002). Alpha-lactorphin and beta-lactorphin improve arterial function in spontaneously hypertensive rats. *Life Sciences, 71*, 1245–1253.

Stefanucci, A., Mollica, A., Macedonio, G., Zengin, G., Ahmed, A. A., & Novellino, E. (2018). Exogenous opioid peptides derived from food proteins and their possible uses as dietary supplements: A critical review. *Food Reviews International, 34*, 70–86.

Szwajkowska, M., Wolanciuk, A., Barłowska, J., Krol, J., & Litwińczuk, Z. (2011). Bovine milk proteins as the source of bioactive peptides influencing the consumers' immune system—a review. *Animal Science Papers and Reports, 29*(4), 269–280.

Tavano, O. L. (2013). Protein hydrolysis using proteases: An important tool for food biotechnology. *Journal of Molecular Catalysis B: Enzymatic, 90*, 1–11.

Vallabha, V. S., & Tiku, P. K. (2014). Antihypertensive peptides derived from soy protein by fermentation. *International Journal of Peptide Research and Therapeutics, 20*, 161–168.

Vasquez-Villanueva, R., Marina, M. L., & Garcia, M. C. (2016). Identification by hydrophilic interaction and reversed-phase liquid chromatography-tandem mass spectrometry of peptides with anti-oxidant capacity in food residues. *Journal of Chromatography A, 1428*, 185–192.

Vilcacundo, R., Martínez-Villaluenga, C., & Hernández-Ledesma, B. (2017). Release of dipeptidyl peptidase IV, α-amylase and α-glucosidase inhibitory peptides from quinoa (*Chenopodium quinoa* Willd.) during *in vitro* simulated gastrointestinal digestion. *Journal of Functional Foods, 35*, 531–539.

Visvanathan, R., Jayathilake, C., Chaminda Jayawardana, B., & Liyanage, R. (2016). Health-beneficial properties of potato and compounds of interest. *Journal of Science and Food Agriculture, 96*, 4850–4860.

Wang, W., Mejia, D., & Gonzalez, E. (2005). A new frontier in soy bioactive peptides that may prevent age-related chronic diseases. *Comprehensive Reviews in Food Science and Food Safety, 4*(4), 63–78.

Wang, X., Chen, H., Fu, X., Li, S., & Wei, J. (2017). A novel antioxidant and ACE inhibitory peptide from rice bran protein: Biochemical characterization and molecular docking study. *LWT - Food Science and Technology, 75*, 93–99.

White, B. L., Sanders, T. H., & Davis, J. P. (2014). Potential ACE-inhibitory activity and nanoLC-MS/MS sequencing of peptides derived from aflatoxin contaminated peanut meal. *LWT - Food Science and Technology, 56*, 537–542.

Xie, N., Huang, J., Li, B., Cheng, J., Wang, Z., & Yin, J. (2015). Affinity purification and characterisation of zinc chelating peptides from rapeseed protein hydrolysates: Possible contribution of characteristic amino acid residues. *Food Chemistry, 173*, 210–217.

Yadav, J. S. S., Yan, S., Pilli, S., Kumar, L., Tyagi, R. D., & Surampalli, R. Y. (2015). Cheese whey: A potential resource to transform into bioprotein, functional/nutritional proteins and bioactive peptides. *Biotechnology Advances, 33*(6), 756–774.

Yan, Q. J., Huang, L. H., Sun, Q., Jiang, Z. Q., & Wu, X. (2015). Isolation, identification and synthesis of four novel antioxidant peptides from rice residue protein hydrolyzed by multiple proteases. *Food Chemistry, 179*, 290–295.

Zenezini Chiozzi, R., Capriotti, A. L., Cavaliere, C., La Barbera, G., Piovesana, S., & Laganà, A. (2016). Identification of three novel angiotensin-converting enzyme inhibitory peptides derived from cauliflower by-products by multidimensional liquid chromatography and bioinformatics. *Journal of Functional Foods, 27*, 262–273.

Nutraceutical Properties of Bioactive Peptides

Sajid Maqsood, Sabika Jafar, and Priti Mudgil

Introduction

Diet is one of the major environmental factors that is recognized to impact our health and the emergence of certain diseases. During gastrointestinal digestion, the major nutrient proteins consumed undergoes breakdown and releases several peptides and amino acids. Some of these peptides may act as hormones or neurotransmitters due to their similar characteristics to endogenous peptides. These exogenous food-derived peptides can act on receptors in the body and exert agonistic or antagonistic activities. One example of such peptides is opioid peptides (Teschemacher 2003). The bioactivity of food-derived peptides depends on their chain length, molecular charge, hydrophobicity and side-chain bulkiness of the constituent amino acid residues (Pripp et al. 2005). Normally, the peptide activity against disease targets is considered as lower than synthetic drugs, but the benefits of dietary bioactive peptides like low health cost, safety, and additional nutritional benefits makes these peptides seem attractive (Udenigwe and Aluko 2012).

The last two decades have seen a tremendous amount of literature published regarding identification of peptides with different biological activity. Majority of the studies have been performed in vitro, but animal models and clinical trial studies are also available depending on which activity is being investigated (Hernández-Ledesma et al. 2014). The nutraceutical effects of the bioactive peptides are diverse and widely reported (Korhonen 2009; Moller et al. 2008). Consumption of food-derived bioactive peptides with multifunctional effects have the potential to positively impact overall health of individuals by preventing disease. This chapter will focus on milk derived peptides and a few of their biological activities.

S. Maqsood (✉) · S. Jafar · P. Mudgil
Department of Food Science, College of Food and Agriculture, United Arab Emirates University, Al Ain, United Arab Emirates
e-mail: sajid.m@uaeu.ac.ae

A. Gani, B. A. Ashwar (eds.), *Food biopolymers: Structural, functional and nutraceutical properties*, https://doi.org/10.1007/978-3-030-27061-2_11

Prevention of Cancer

Cancer is a leading cause of death around the globe. There are several environmental and lifestyle factors which play a role in the development of cancer like tobacco, alcohol consumption, sun exposure etc. (Anand et al. 2008; Hernández-Ledesma et al. 2014). Dietary patterns and food consumption is also reported to be associated with many types of cancer. On the other hand, several studies (in vitro and in vivo) have demonstrated the potential of food compound to lower the risk of cancer (De Kok et al. 2008).

Peptides that are able to decrease or inhibit cell proliferation are called antitumor peptides (Blanco-Míguez et al. 2016). These peptides are mostly reported in in vitro studies with cell-culture models. A few clinical trials on animal models and humans have also shown potential anticancer effects (Zhang et al. 2014). However, clear results of anticancer peptides are still insufficient. It is suggested that the low intrinsic toxicity and high tissue-penetration and permeability make bioactive peptides a good anticancer agent (Cicero et al. 2017). Bioactive peptides have also exhibited their cytotoxic activity in cancer cell lines by either acting as cytotoxic agents themselves or acting as carriers of cytotoxic agents and radioisotopes that may target cancer cells (Thundimadathil 2012). They are not genotoxic and may affect specific molecular pathways that participate in carcinogenesis (Blanco-Míguez et al. 2016). Generally, the main mechanisms of action shown by bioactive peptides are antioxidant activity, induction of apoptosis, inhibition of tumor angiogenesis, inhibition of cell migration, cytotoxicity and inhibition of cell proliferation (Schweizer 2009; Tyagi et al. 2014). Table 1 summarizes the different types of bioactive peptides that show anticancer effects.

Colon and mammary tumorigenesis in murine model studies have demonstrated whey proteins to be better than other dietary proteins in terms of suppressing tumor development (Sasaki and Kume 2007). This is due to whey protein containing high concentrations of cysteine and glutamylcyst(e)ine dipeptides. These peptides act as substrates for synthesis of a unique cellular antioxidant glutathione, that removes reactive oxygen species and detoxifies carcinogens (Sasaki and Kume 2007). Intact whey proteins have comparably tight structure and are less approachable to mutagens, while caseins are more accessible to mutagens because of their loose micellar structure (Bosselaers et al. 1994). Moreover, in silico studies indicate that amino acids like lysine, alanine, glycine, and serine predominate various positions in anticancer peptides obtained from whey proteins (Sah et al. 2015).

Amongst the whey proteins explored, lactoferrin remains the most extensively studied, followed by β-lactoglobulin, α-lactalbumin, and serum albumin (Parodi 2007). A minor component of whey proteins, lactoferrin inhibits intestinal tumors by mechanisms of apoptosis, inhibiting angiogenesis, and regulating carcinogen metabolizing enzymes (Pepe et al. 2013). Zhang et al. (2015) reported the growth inhibition of MCF-7 breast cancer cells by bovine lactoferrin through apoptotic pathway. A different in vivo study on bovine lactoferrin in mouse model showed improvement in chemotherapeutic effects of tamoxifen in basal-like breast cancer

Table 1 Examples of bioactive peptides showing anticancer effects

Source	Cell line/animal model	Effects/mechanisms of action	References
Bovine lactoferrin	Stomach cancer cell line SGC-7901	Antitumor effect through inhibition of Akt activation and apoptosis	Xu et al. (2010)
Bovine lactoferrin	Human myeloid leukemia cells (HL-60)	Inhibition of cell proliferation through induction of apoptosis	Roy et al. (2002)
Bovine lactoferricin	Jurkat T leukemia cells	Selective apoptosis through generation of reactive oxygen species	Mader et al. (2005)
Lactoferricin B	Raji and Ramos human B-lymphoma cells	Cytotoxic activity through induced apoptosis	Furlong et al. (2010)
Lactoferricin B from peptic digestion of bovine lactoferrin	Human MYCN-amplified and non-MYCN amplified neuroblastoma cell lines	Cytotoxic activity through disruption of cytoplasmic and mitochondrial membrane	Eliassen et al. (2006)
Lactobacillus helveticus 1315 fermented bovine skim milk	Colon cancer HT-29 cell line	Significant growth inhibition through apoptosis induction and cytoplasmic membrane disruption	Elfahri et al. (2016)
Casomorphin peptides (from bovine α- and β-casein)	Human breast cancer cell line T47D	Antiproliferative action through interaction with opioid and somatostatin receptors	Hatzoglou et al. (1996)
Casomorphin peptides from casein	Intestinal tumor HT-29 and AZ-97 cells	Apoptosis induction	Perego et al. (2012)
Cationic INKKI peptide from bovine β-casein	B16F10 melanoma tumor-bearing mice	Antitumor effect	Azevedo et al. (2012)

(Sun et al. 2012). Other mechanisms through which bovine lactoferrin exhibits antitumor activity include delayed angiogenesis and reduced endothelial cell proliferation (Shimamura et al. 2004; Hoskin and Ramamoorthy 2008).

Lactoferricin obtained by peptic digestion of bovine lactoferrin has also shown to be potent against many cancer cell lines including breast, colon, leukemia, and ovarian cancer without any toxic effects of harming normal cells (Furlong et al. 2010). Lactoferricin's activity is attributed to its strong cationic character that helps in interaction with negatively charged cancer cells, and the subsequent weakening of cancer cell membranes (Hoskin and Ramamoorthy 2008). Other mechanisms by which lactoferricin shows its anticancer activity is by inducing apoptosis, modulating gene expression, and preventing angiogenesis (De Mejia and Dia 2010). The cytotoxic effect of bovine lactoferricin fragment obtained during hydrolysis of bovine lactoferrin has also been reported in several types of rat and human cancer cell lines in vivo (Yoo et al. 1997; Eliassen et al. 2002, 2006; Mader et al. 2005).

Other components of whey protein too have shown anti-carcinogenic effects, for instance, antiproliferative effects of α-lactalbumin in colon adenocarcinoma cell lines (Sternhagen and Allen 2001; Sah et al. 2015), protective effects of β-lactoglobulin against cancer development in animal models (Goldman et al. 1990; Chatterton et al. 2006), and inhibitory effects of bovine serum albumin in MCF-7 human breast cancer cell growth (Laursen et al. 1989) and in a Chinese hamster epithelial cell line (Bosselaers et al. 1994).

Various peptides purified from bovine casein proteins have shown antitumor activity, for instance: (1) β-casein peptides inhibiting proliferation of human ovarian cancer cells partially (Wang et al. 2013) and displaying cytotoxic activity toward B16F1O melanoma cells (Azevedo et al. 2012), (2) α-casein peptides inhibiting growth of T47D human breast cancer cells (Kampa et al. 1996) and inducing necrosis of leukemic B- and T-cells (Otani and Suzuki 2003), and (3) κ-casein peptides displaying cytotoxicity toward mammalian cells (Matin and Otani 2002).

Another group of bioactive peptides generated during digestion of caseins are caseinophosphopeptides (CPPs), containing high number of phosphorylated sites which can bind and solubilize calcium (Berrocal et al. 1989). This characteristic helps CPPs to mediate the growth and apoptosis of cells, thereby demonstrating their antitumor activities against intestinal cancerous cells (Perego et al. 2012; Sah et al. 2015).

Thus, several studies have displayed the anti-tumoral and cytotoxic activity of milk derived bioactive peptides against different cancer cell lines. However, clinical data is still lacking, and the anticancer activity of such peptides must be thoroughly investigated in controlled trials before drawing any conclusions (Cicero et al. 2017)

Hypoglycemic Effects

Diabetes mellitus is a widely prevalent disease which results in high blood sugar levels and is known to cause complications like hypertension and cardiovascular disease. Impaired insulin secretion and insulin resistance characterizes type 2 diabetes, whereas type 1 diabetes is characterized by body's inability to secrete insulin (Brandelli et al. 2015; Shori 2015).

Management of type 2 diabetes is mainly focused on increasing insulin availability, either by direct delivery of insulin or through agents that have specific roles (e.g. improving insulin sensitivity, enhancing insulin secretion, increasing glucose excretion) (DeFronzo et al. 2014). Upon ingesting food, incretin hormones like glucagon-like peptide 1 (GLP-1) and glucose-dependent insulinotropic polypeptide (GIP) are secreted which modulate gut motility, helps in nutrient absorption, and stimulate insulin secretion (Drucker and Nauck 2006). However, a serine exopeptidase, dipeptidyl-peptidase IV (DPP-IV) inhibits the GLP-1 and GIP hormones making treatment of type 2 diabetes complicated. Hence, DPP-IV inhibitors and GLP-1 analogues resistant to DPP-IV action are currently the therapies available to manage type 2 diabetes (Fadini and Avogaro 2011).

Influence on glycaemia by bioactive peptides may be exerted through different mechanisms including incretin secretion, insulinotropic activity and regulation of

DPP-IV secretion (Cicero et al. 2016; Horner et al. 2016). Milk-derived caseins and whey proteins are reported to have a stimulating effect on secretion of insulin in obese, healthy, pre-diabetic and type-2 diabetes subjects (Power et al. 2009; Pal and Ellis 2010; Pal et al. 2010; Jonker et al. 2011; Hoefle et al. 2015). Bovine milk caseins like β-casein are very good sources of DPP-IV inhibitors (Patil et al. 2015). The insulinotropic effect of whey proteins was more pronounced compared to caseins. Also, protein hydrolysates of whey showed faster effect than intact proteins consumption (Horner et al. 2016). Intact whey protein has also shown enhanced GIP (Hall et al. 2003; Frid et al. 2005; Nilsson et al. 2007) and GLP-1 (Ma et al. 2009; Akhavan et al. 2014) response. Whey protein supplementation has demonstrated insulin release and postprandial glucose control in clinical trials of healthy and type 2 diabetic subjects (Sousa et al. 2012). It is hypothesized that whey proteins release peptides while passing through gastrointestinal tract and help regulate the glycaemic response (Akhavan et al. 2010). A study on obese mice also showed an improvement in insulin secretion is response to glucose and lower blood glucose levels when orally administered with whey protein hydrolysates (Gaudel et al. 2013). Moreover, recent studies have shown DPP-IV inhibitory effects of bovine casein and whey protein derived peptides (Gunnarsson et al. 2006; Tulipano et al. 2011; Nongonierma and Fitzgerald 2013). Peptides derived from *β*-lactoglobulin are efficient DPP-IV inhibitors: peptide with sequence VAGTWY was able to demonstrate hypoglycaemic effects in mice in the oral glucose tolerance test (Uchida et al. 2011). An *in-silico* study also showed both casein and whey proteins as precursors of DPP-IV inhibitory peptides (Lacroix and Li-Chan 2012).

Other important enzymes that play a role in insulin modulation are α-glucosidase and α- amylase. Hence, the inhibition of these enzymes may lead to delayed carbohydrate digestion and glucose absorption may be helpful in managing diabetes (Bharatham et al. 2008; Vankudre et al. 2015). Milk fermentation with *L. bulgaricus* and *L. acidophilus* showed increased α-glucosidase inhibition (Apostolidis et al. 2007). In a different study, peptic digestion of whey proteins displayed high α-glucosidase inhibition (Lacroix and Li-Chan 2013). Thus, milk derived bioactive peptides may help in glycemic management strategy, although further work is needed to fully determine the bioavailability of the peptides in humans.

Antihypertensive Effects

One of the major risk factors for cardiovascular disease is elevated blood pressure (Erdmann et al. 2008). The angiotensin-I converting enzyme (ACE) helps regulate blood pressure by catalyzing the conversion of angiotensin I to angiotensin II (a potent vasoconstrictor) and breakdown of bradykinin (a vasodilator) (Sánchez and Vázquez 2017). Since angiotensin II elevates the blood pressure levels, inhibiting ACE is a way to control blood pressure (Korhonen and Pihlanto 2007). Inhibitors of ACE are currently available in the markets to treat hypertnesio like capotril, alacepril, lisinopril and enalapril. However, their synthetic nature causes unwanted side effects like dry cough, loss of taste, skin rashes, proteinuria and blood dyscrasias (Atkinson and Robertson 1979). Thus, the search for natural ACE inhibitors is growing which can be safely consumed, with little to no side effects, and treat hypertension.

Bioactive peptides with antihypertensive and ACE inhibitory activity have been mainly isolated from enzymatic digestion of bovine and human caseins or fermented foods (Yamamoto et al. 1999; Mohanty et al. 2016). Studies that have investigated ACE inhibitors suggest that the tripeptide sequence of substrate at *C*-terminal greatly affects the binding to ACE. Hence, a substrate or competitive inhibitor with hydrophobic amino acid residues, be it branched or aromatic side chains, at the three *C*-terminal position has greater affinity for ACE (Gobbetti et al. 2002; Sánchez and Vázquez 2017). Antihypertensive action of bioactive peptides can be due to competitive and/or non-competitive inhibition of ACE. Competitive inhibition is competition of peptides with ACE substrate for the enzyme active sites (Sato et al. 2002). Non-competitive (Leu-Trp, Ile-Tyr) and uncompetitive (Ile-Trp, Phe-Tyr) peptides are characterized by peptides binding to other enzyme sites which changes enzyme conformation and decreases activity. Thus, the interaction between peptides and ACE is greatly influenced by a single amino acid substitution as well as isomers (Sato et al. 2002). For this, structure-activity studies are vital to understand more (Udenigwe and Aluko 2012). Other possible mechanisms of action include increase in activity of some vasodilating agents like endothelial NOS, inhibition of renin (by mixed-type inhibition mode), induction of vasodilation and reducing the activity of sympathetic system (Aluko 2015). A few examples of ACE-inhibitory peptides from bovine milk and its proteins are presented in Table 2. The tripeptides Val-Pro-Pro (VPP) and Ile-Pro-Pro (IPP) derived from fermented milk are recognized to be potent ACE inhibitors (Pihlanto et al. 2010). IPP has also been identified and quantified from digested camel milk (Tagliazucchi et al. 2016). Both IPP and VPP have been reported to lower blood pressure in mildly hypertensive patients (Seppo et al. 2003) and rats (Sipola et al. 2002), when administered with fermented milk products. Similarly, Mizuno et al. (2004) identified casein fragments IPP and VPP from *Aspergillus oryzae* fermented milk and suggested that IPP consumption lowers blood pressure when compared to placebo. Peptide from β-lactoglobulin f(142–145)

Table 2 ACE-inhibitory activity of bioactive peptides derived from bovine milk and its proteins

Source	Peptide sequence	Reference
ß-casein, αs1-casein	Tyr-Pro-Phe-Pro, Ala-Val-Pro-Tyr-Pro--Gln-Arg, Thr-Thr-Met-Pro-Leu-Trp	Gobbetti et al. (2002)
ß-casein, κ-casein	Val-Pro-Pro, Ile-Pro-Pro	Seppo et al. (2003)
β-lactoglobulin	Ala-Leu-Pro-Met	Murakami et al. (2004)
Lactobacillus helveticus (LBK16H) fermented milk	Val-Pro-Pro, Ile-Pro-Pro	Sipola et al. (2002)
Whey protein tryptic hydrolysate	Ala-Leu-Pro-Met-His-Ile-Arg	Ferreira et al. (2007)

was isolated and orally given to spontaneously hypertensive rates which showed a significant decrease in their blood pressure levels (Murakami et al. 2004).

Because pharmacological drugs have strong side effects in hypertensive patients, ACE inhibitor peptides from foods may provide an alternative and natural option to regulate blood pressure (Mohanty et al. 2016). However, to prove the efficacy of these peptides in vivo still remains a challenge. This is because peptides need to reach the target organ in their active and intact form once orally ingested. For this purpose, several studies have been carried out to evaluate the resistance of small antihypertensive peptides to gastrointestinal digestion in cell line models (Quiros et al. 2008; Contreras et al. 2012; Picariello et al. 2013). Another important factor to consider is the association between in vitro and in vivo experiments. Various studies has reported that some peptides that showed potent antihypertensive activity in vitro failed to exhibit the same effects in animal models. For instance, no hypotensive effect of αs1-casein f(23–27) was observed in anesthetized rats in contrast to significant ACE inhibitory activity in vitro (Maruyama et al. 1987); presumably due to deterioration to inactive fragments during oral ingestion (FitzGerald et al. 2004). This suggests that although valuable information from in vitro studies is obtained, only in vivo studies can encompass important parameters like nature, mechanism and bioavailability and reliably assess the hypotensive effects of a given peptide (FitzGerald et al. 2004; Vermeirssen et al. 2004). Several fermented milk products (yogurt, drinks, hydrolyzed milk) are available in the markets that contain the ACE inhibitory peptides such as IPP, VPP, RY and TTMPLW. Specifically, casein hydrolysate (Casein DP, Kanebo Ltd., Japan), whey protein hydrolysate (Biozate, Davisco, US), and C12 peptide (DMV, the Netherlands) have been developed for consumption (FitzGerald et al. 2004). Valio Company in Finland, for instance, has developed two fermented dairy products with IPP and VPP peptides (FitzGerald et al. 2004; Gobbetti et al. 2004), while a lacto tri-peptide is an alternative to drugs available to maintain blood pressure (Jäkälä and Vapaatalo 2010).

A few epidemiological studies have also shown the association between milk consumption and hypertension, where individuals with low to non-milk consumption had a higher incidence of hypertension (McCarron et al. 1984; Garcia-Palmieri et al. 1984). Clinical trials have also shown decreased blood pressure by ACE inhibitory peptides from fermented milk (Seppo et al. 2003). Lactotripeptides from casein hydrolysate were reported to modulate pulse wave velocity in mildly hypertensive subjects (Cicero et al. 2011, 2016). Thus, such peptides may be a safe and natural alternative to synthetic drugs, although their potency is not the same as the pharmacological ones (Sharma et al. 2011).

Opioid Properties

Milk-derived opioid peptides are those which have structurally similar sequences to endogenous peptides (Table 3). They are opioid receptor ligands hidden within bovine and human β-casein and released enzymatically in vitro (Brantl 1984). Other places where opioid peptides can be found are the nervous, endocrine and immune systems of mammals (Mohanty et al. 2016). These peptides may exert an agonist or antagonist effect when they interact with receptors in the human body (Fernández-Tomé et al. 2016).

Opioid agonist peptides which are endogenously produced can modulate the function and growth of cells in the central nervous system (Calvo et al. 2000). In contrast, exogenous opioid agonist peptides like β-casomorphins interact with opiate receptors present in intestinal epithelium and show similar pharmacological properties to morphine and naloxone like modulating social behavior, inducing analgesia, stimulating insulin and somatostatin secretion and exerting an antidiarrheal effect (Meisel and Schlimme 1990). β-casomorphin studies in rats suggest that they may play a significant role in regulating dietary fat intake, with β-casomorphin 1–7 suppressing high carbohydrate diet intake and stimulating high fat diet intake in satiated rats (Lin et al. 1998). Moreover, depressive effects on the central respiratory system by bovine β-casomorphins have been reported in rats and rabbits (Hedner and Hedner 1987). Other important examples of opioid agonist peptides are serorphin from bovine serum albumin and lactorphins from bovine whey protein.

Opioid antagonists have also been identified from bovine κ-casein. Bovine casoxins A and B are reported to have low antagonist potency (Meisel 1998), whereas casoxin C is a potent opioid antagonist peptide of κ-casein, with greatest biological potency (Xu 1998). These peptides are known to suppress enkephalin's agonist activity and have affinity for the μ- and κ-type of opioid receptors (Clare and Swaisgood 2000). Both endo- and exogenously produced opioid peptides share a common feature—tyrosine at the *N*-terminus and phenylalanine or tyrosine in the third or fourth position—which allows for a better fit within the active site of the opioid receptor (Clare and Swaisgood 2000). Other protein sources of opioid peptides include rice, gluten or soy from plant proteins (Yoshikawa 2015).

Opioid-like sequences are present in bovine whey protein's primary structure, namely α-lactalbumin f50–53 and β-lactoglobulin f102–105 (Antila et al. 1991). These tetrapeptides are known as α- and β-lactorphins and can be released via proteolytic enzymes in vitro (Nagpal et al. 2011). The biological functions of both lactorphins vary vastly even though structurally, they differ from each other by only one amino acid (Rutherfurd-Markwick 2012). In vitro, α-lactorphin shows an

Table 3 Examples of bovine milk-derived opioid peptides

Protein substrate	Bioactive compound	Bioactivity	Reference
β-casein (region 60–70)	β-casomorphins	Opioid agonist	Teschemacher (2003)
Bovine κ-casein f(25–34)	Casoxin C	Opioid antagonist	Meisel and FitzGerald (2000)
α-lactalbumin, β-lactoglobulin	α- and β-lactorphins	Opioid agonist	Antila et al. (1991)
Bovine serum albumin (f399–404)	Serorphin	Opioid agonist	Meisel and FitzGerald (2000)

inhibitory effect on guinea pig ileum contractions, while β-lactorphin exhibits a non-opioid stimulatory effect. Both peptides display very low affinity for opioid receptors and their actions are induced via the μ-receptor (Pihlanto-Leppälä 2000).

Other Bioactive Properties Exerted by Bioactive Peptides

Maintenance of the physiological functioning of cells and body organs is done by an important protein called calmodulin (CaM) (Chung et al. 2000). This protein is a calcium-binding protein with molecular weight of 16.7 kDa and consisting of 148 amino acid residues (Sánchez and Vázquez 2017). CaM helps in cell proliferation, vasodilation, smooth muscle contraction neurotransmission, and calcium-dependent cell division (Rasmussen and Means 1987; Cho et al. 1998). Abnormally high levels of CaM can be damaging to the human body and may lead to the progression of cancer (Sánchez and Vázquez 2017). Thus, CaM-binding natural compounds may help in prevention and mitigation of diseases that are induced by increased activity of CaM-dependent enzyme (Martínez-Luis et al. 2007). The natural CaM-binding peptides generally have cationic and hydrophobic amino acid residues at the CaM-binding sites (O'Neil and DeGrado 1990). Because of these features, research has led to identification of food-protein derived cationic peptides that can bind to negatively charged CaM, and inhibit CaM-dependent enzymes. For example, Kizawa et al. (1995) and Kizawa (1997) reported CaM-binding peptides from pepsin hydrolysis of bovine casein which inhibited CaM-dependent phosphodiesterase (CaMPDE) activation.

Another interesting property of bioactive peptides is anti-inflammation. Bovine caseinomacropeptide (CMP) has shown activity against intestinal inflammation and is the most extensively studied peptide. Several animal models have established the anti-inflammatory effect of CMP at the gastrointestinal level (Sánchez and Vázquez 2017). For example, mice with colilitis and ileitis induced by trinitrobenzene sulphonic acid (TNBS) exhibited a decrease in colonic damage and lower interleukin 1 and trefoil factor 3 levels. Although mechanism was unknown, inhibition of activation of immune cells was consistent with previous literature (Daddaoua et al. 2005). Other mechanisms hypothesized by de Medina et al. (2010) include activation of macrophages and increased differentiation of regulatory T cells by CMP. Apart from CMP, hydrolysates from *β*-casein have also shown anti-inflammatory effects in a murine TNBS-induced colitis model (Espeche Turbay et al. 2012). Hydrolysates from whey protein and bovine β-casein are reported to have anti-inflammatory effects via decreased NF-κB activity in cell-culture models (Altmann et al. 2016; Ma et al. 2016). A different study on glycomacropeptide (GMP) derived from milk resulted in decreased allergen-specific IgE concentrations and caused a suppression of inflammatory cell recruitment in asthmatic rat model. It was concluded that oral treatment with GMP prevented airway inflammation and inhibited the development of allergic asthma (Roldán et al. 2016).

Antimicrobial activity of bioactive peptides is displayed by various mechanisms such as prebiotic effect, peptides that are able to block attachment or invasion of pathogen microorganism, or peptides that could induce a killing or inhibiting effect on the growth of microorganism (Hernández-Ledesma et al. 2014). Milk-derived antimicrobial peptides contain highly positive charged sequence. Thus, the net positive charge may help in binding of the peptides to negatively charged bacterial membranes. Also, the amphiphilic nature of peptides aids in bacterial membrane disruption (Hernández-Ledesma et al. 2014). Milk derived antimicrobial activity is attributed to factors like immunoglobulins, lysozyme, lactoferrin and lactoperoxidase inherently present in milk proteins (Jabbari et al. 2012). Thus, dairy proteins act as precursors for antimicrobial peptides which can boost an organism's natural defense system against pathogens (Pellegrini 2003).

The fragment 17–41 of lactoferrin (also known as lactoferricin) has shown potent antimicrobial activity due to its attachment to the lipid A part of bacterial lipopolysaccharides, along with increased membrane permeability (Cicero et al. 2017). Enzymatic hydrolysates of lactoferrin has displayed more antimicrobial activity than that of parent protein (Tomita et al. 1991). also, peptides purified from bovine lactoferrin has shown high antibacterial activity against Gram-positive and Gram-negative bacteria (Wakabayashi et al. 2003). Whey proteins like β-lactoglobulin and α-lactalbumin are said to have broad spectrum antimicrobial potential (Sultan et al. 2017). Tryptic hydrolysate of whey protein generated two anionic peptides that exhibited growth inhibition activity against *Listeria monocytogenes* and *Staphylococcus aureus* (Demers Mathieu et al. 2013). Casein proteins too are a good source of antimicrobial peptides (Expósito and Recio 2006). Thus, peptides from milk proteins could be used as potential antimicrobial agents in food industries.

Conclusions

There are peptide fragments encrypted within the structure of food proteins that show biological functions upon release. The bioactivities of the peptides are wide ranging; antioxidative, antimicrobial, antihypertensive, anticancer, anti-diabetic, opioid, mineral binding and anti-inflammatory. Thus, bioactive peptides could potentially be used to reduce risk and prevent chronic diseases like diabetes mellitus, hypertension, obesity and cancer. It is speculated that with more in vivo studies designed around long-term randomized-control clinical trials will further elucidate the mechanism by which bioactive peptides showcase their physiological effects. Their potential application in the food industry is an exciting and novel prospect which will contribute to improving human health.

References

Akhavan, T., Luhovyy, B. L., Brown, P. H., Cho, C. E., & Anderson, G. H. (2010). Effect of premeal consumption of whey protein and its hydrolysate on food intake and postmeal glycemia and insulin responses in young adults. *American Journal of Clinical Nutrition, 91*, 966–975.

Akhavan, T., Luhovyy, B. L., Panahi, S., Kubant, R., Brown, P. H., & Anderson, G. H. (2014). Mechanism of action of pre-meal consumption of whey protein on glycemic control in young adults. *The Journal of Nutritional Biochemistry, 25*(1), 36–43.

Altmann, K., Wutkowski, A., Klempt, M., Clawin-Rädecker, I., Meisel, H., & Lorenzen, P. C. (2016). Generation and identification of anti-inflammatory peptides from bovine β-casein using enzyme preparations from cod and hog. *Journal of the Science of Food and Agriculture, 96*(3), 868–877.

Aluko, R. E. (2015). Antihypertensive peptides from food proteins. *Annual Review of Food Science and Technology, 6*, 235–262.

Anand, P., Kunnumakara, A. B., Sundaram, C., Harikumar, K. B., Tharakan, S. T., Lai, O. S., . . . Aggarwal, B. B. (2008). Cancer is a preventable disease that requires major lifestyle changes. *Pharmaceutical Research, 25*(9), 2097–2116.

Antila, P., Paakkari, I., Järvinen, A., Mattila, M. J., Laukkanen, M., Pihlanto-Leppälä, A., ... Hellman, J. (1991). Opioid peptides derived from in-vitro proteolysis of bovine whey proteins. *International Dairy Journal, 1*(4), 215–229.

Apostolidis, E., Kwon, Y. I., Ghaedian, R., & Shetty, K. (2007). Fermentation of milk and soymilk by *Lactobacillus bulgaricus* and *Lactobacillus acidophilus* enhances functionality for potential dietary management of hyperglycemia and hypertension. *Food Biotechnology, 21*(3), 217–236.

Atkinson, A. B., & Robertson, J. I. S. (1979) Captopril in the treatment of clinical hypertension and cardiac failure. *The Lancet 314*(8147), 836–839.

Azevedo, R. A., Ferreira, A. K., Auada, A. V. V., Pasqualoto, K., Marques-Porto, R., Maria, D. A., & Lebrun, I. (2012). Antitumor effect of cationic INKKI peptide from bovine β-casein on melanoma B16F10. *Journal of Cancer Therapy, 3*, 237–244.

Berrocal, R., Chanton, S., Juillerat, M. A., Favillare, B., Scherz, J. C., & Jost, R. (1989). Tryptic phosphopeptides from whole casein. II. Physicochemical properties related to the solubilization of calcium. *Journal of Dairy Research, 56*(03), 335–341.

Bharatham, K., Bharatham, N., Park, K. H., & Lee, K. W. (2008). Binding mode analyses and pharmacophore model development for sulfonamide chalcone derivatives, a new class of α-glucosidase inhibitors. *Journal of Molecular Graphics and Modelling, 26*(8), 1202–1212.

Blanco-Míguez, A., Gutiérrez-Jácome, A., Pérez-Pérez, M., Pérez-Rodríguez, G., Catalán-García, S., Fdez-Riverola, F., ... Sánchez, B. (2016). From amino acid sequence to bioactivity: The biomedical potential of antitumor peptides. *Protein Science, 25*(6), 1084–1095.

Bosselaers, I. E. M., Caessens, P. W. J. R., Van Boekel, M. A. J. S., & Alink, G. M. (1994). Differential effects of milk proteins, BSA and soy protein on 4NQO-or MNNG-induced SCEs in V79 cells. *Food and Chemical Toxicology, 32*(10), 905–909.

Brandelli, A., Daroit, D., & Corrêa, A. (2015). Whey as a source of peptides with remarkable biological activities. *Food Research International, 73*, 149–161.

Brantl, V. (1984). Novel opioid peptides derived from human β-casein: Human β-casomorphins. *European Journal of Pharmacology, 106*(1), 213–214.

Calvo, C. F., Cesselin, F., Gelman, M., & Glowinski, J. (2000). Identification of an opioid peptide secreted by rat embryonic mixed brain cells as a promoter of macrophage migration. *European Journal of Neuroscience, 12*(8), 2676–2684.

Chatterton, D. E., Smithers, G., Roupas, P., & Brodkorb, A. (2006). Bioactivity of β-lactoglobulin and α-lactalbumin—Technological implications for processing. *International Dairy Journal, 16*(11), 1229–1240.

Cho, M. J., Vaghy, P. L., Kondo, R., Lee, S. H., Davis, J. P., Rehl, R., ... Johnson, J. D. (1998). Reciprocal regulation of mammalian nitric oxide synthase and calcineurin by plant calmodulin isoforms. *Biochemistry, 37*(45), 15593–15597.

Chung, W. S., Lee, S. H., Kim, J. C., Do Heo, W., Kim, M. C., Park, C. Y., . . . Harper, J. F. (2000). Identification of a calmodulin-regulated soybean Ca2+-ATPase (SCA1) that is located in the plasma membrane. *The Plant Cell, 12*(8), 1393–1407.

Cicero, A. F., Colletti, A., Rosticci, M., Cagnati, M., Urso, R., Giovannini, M., ... D'Addato, S. (2016). Effect of lactotripeptides (isoleucine–proline–proline/valine–proline–proline) on blood pressure and arterial stiffness changes in subjects with suboptimal blood pressure control and metabolic syndrome: A double-blind, randomized, crossover clinical trial. *Metabolic Syndrome and Related Disorders, 14*(3), 161–166.

Cicero, A. F., Fogacci, F., & Colletti, A. (2017). Potential role of bioactive peptides in prevention and treatment of chronic diseases: A narrative review. *British Journal of Pharmacology, 174*(11), 1378–1394.

Cicero, A. F., Rosticci, M., Gerocarni, B., Bacchelli, S., Veronesi, M., Strocchi, E., & Borghi, C. (2011). Lactotripeptides effect on office and 24-h ambulatory blood pressure, blood pressure stress response, pulse wave velocity and cardiac output in patients with high-normal blood pressure or first-degree hypertension: A randomized double-blind clinical trial. *Hypertension Research, 34*(9), 1035.

Clare, D. A., & Swaisgood, H. E. (2000). Bioactive milk peptides: A prospectus. *Journal of Dairy Science, 83*(6), 1187–1195.

Contreras, M. M., Sancho, M. A., Recio, I., & Mills, C. (2012). Absorption of casein antihypertensive peptides through an in vitro model of intestinal epithelium. *Food Digestion, 3*, 16–24.

Daddaoua, A., Puerta, V., Zarzuelo, A., Suárez, M. D., de Medina, F. S., & Martínez-Augustin, O. (2005). Bovine glycomacropeptide is anti-inflammatory in rats with hapten-induced colitis. *The Journal of Nutrition, 135*(5), 1164–1170.

De Kok, T. M., van Breda, S. G., & Manson, M. M. (2008). Mechanisms of combined action of different chemopreventive dietary compounds. *European Journal of Nutrition, 47*, 51–59.

de Medina, F. S., Daddaoua, A., Requena, P., Capitán-Cañadas, F., Zarzuelo, A., Suárez, M. D., & Martínez-Augustin, O. (2010). New insights into the immunological effects of food bioactive peptides in animal models of intestinal inflammation. *Proceedings of the Nutrition Society, 69*(3), 454–462.

De Mejia, E. G., & Dia, V. P. (2010). The role of nutraceutical proteins and peptides in apoptosis, angiogenesis, and metastasis of cancer cells. *Cancer Metastasis Review, 29*, 511–528.

DeFronzo, R. A., Triplitt, C. L., Abdul-Ghani, M., & Cersosimo, E. (2014). Novel agents for the treatment of type 2 diabetes. *Diabetes Spectrum, 27*(2), 100–112.

Demers Mathieu, V., Gauthier, S., Britten, M., Fliss, I., Robitaille, G., & Jean, J. (2013). Antibacterial activity of peptides extracted from tryptic hydrolysate of whey protein by nanofiltration. *International Dairy Journal, 28*, 94–101.

Drucker, D. J., & Nauck, M. A. (2006). The incretin system: Glucagon-like peptide-1 receptor agonists and dipeptidyl peptidase-4 inhibitors in type 2 diabetes. *Lancet, 368*, 1696–1705.

Elfahri, K. R., Vasiljevic, T., Yeager, T., & Donkor, O. N. (2016). Anti-colon cancer and antioxidant activities of bovine skim milk fermented by selected Lactobacillus helveticus strains. *Journal of Dairy Science, 99*(1), 31–40.

Eliassen, L., Berge, G., Sveinbjørnsson, B., Svendsen, J., Vorland, L. H., & Rekdal, O. (2002). Evidence for a direct antitumor mechanism of action of bovine lactoferricin. *Anticancer Research, 22*, 2703–2710.

Eliassen, L. T., Berge, G., Leknessund, A., Wikman, M., Lindin, I., Løkke, C., ... Flægstad, T. (2006). The antimicrobial peptide, lactoferricin B, is cytotoxic to neuroblastoma cells in vitro and inhibits xenograft growth in vivo. *International Journal of Cancer, 119*(3), 493–500.

Erdmann, K., Cheung, B. W. Y., & Schröder, H. (2008). The possible roles of food derived bioactive peptides in reducing the risk of cardiovascular disease. *Journal of Nutritional Biochemistry, 19*, 643–654.

Espeche Turbay, M. B., de Moreno de LeBlanc, A., Perdigón, G., Savoy de Giori, G., & Hebert, E. M. (2012). β-Casein hydrolysate generated by the cell envelope associated proteinase of

Lactobacillus delbrueckii ssp. lactis CRL 581 protects against trinitrobenzene sulfonic acid-induced colitis in mice. *Journal of Dairy Science, 95*, 1108–1118.

Expósito, I. L., & Recio, I. (2006). Antibacterial activity of peptides and folding variants from milk proteins. *International Dairy Journal, 16*(11), 1294–1305.

Fadini, G. P., & Avogaro, A. (2011). Cardiovascular effects of DPP-4 inhibition: Beyond GLP-1. *Vascular Pharmacology, 55*, 10–16.

Fernández-Tomé, S., Martínez-Maqueda, D., Girón, R., Goicoechea, C., Miralles, B., & Recio, I. (2016). Novel peptides derived from αs1-casein with opioid activity and mucin stimulatory effect on HT29-MTX cells. *Journal of Functional Foods, 25*, 466–476.

Ferreira, I. M. P. L. V. O., Pinho, O., Mota, M. V., Tavares, P., Pereira, A., Goncalves, M. P., … Teixeira, J. A. (2007). Preparation of ingredients containing an ACE-inhibitory peptide by tryptic hydrolysis of whey protein concentrates. *International Dairy Journal, 17*(5), 481–487.

FitzGerald, R. J., Murray, B. A., & Walsh, D. J. (2004). Hypotensive peptides from milk proteins. *The Journal of Nutrition, 134*(4), 980S–988S.

Frid, A. H., Nilsson, M., Holst, J. J., & Björck, I. M. (2005). Effect of whey on blood glucose and insulin responses to composite breakfast and lunch meals in type 2 diabetic subjects. *The American Journal of Clinical Nutrition, 82*(1), 69–75.

Furlong, S. J., Mader, J. S., & Hoskin, D. W. (2010). Bovine lactoferricin induces caspase-independent apoptosis in human B-lymphoma cells and extends the survival of immune-deficient mice bearing B-lymphoma xenografts. *Experimental and Molecular Pathology, 88*, 371–375.

Garcia-Palmieri, M. R., Costas, R., Cruz-Vidal, M., Sorlie, P. D., Tillotson, J., & Havlik, R. J. (1984). Milk consumption, calcium intake, and decreased hypertension in Puerto Rico. Puerto Rico Heart Health Program study. *Hypertension, 6*(3), 322–328.

Gaudel, C., Nongonierma, A. B., Maher, S., Flynn, S., Krause, M., Murray, B. A., … Newsholme, P. (2013). A whey protein hydrolysate promotes insulinotropic activity in a clonal pancreatic β-cell line and enhances glycemic function in ob/ob mice. *Journal of Nutrition, 143*(7), 1109–1114.

Gobbetti, M., Minervini, F., & Rizzello, C. G. (2004). Angiotensin i-converting-enzyme-inhibitory and antimicrobial bioactive peptides. *International Journal of Dairy Technology, 57*(2-3), 173–188.

Gobbetti, M., Stepaniak, L., De Angelis, M., Corsetti, A., & Di Cagno, R. (2002). Latent bioactive peptides in milk proteins: Proteolytic activation and significance in dairy processing. *Critical Reviews in Food Science and Nutrition, 42*(3), 223–239.

Goldman, A. S., Goldblum, R. M., & Hanson, L. Å. (1990). Anti-inflammatory systems in human milk. In *Antioxidant nutrients and immune functions* (pp. 69–76). Boston, MA: Springer.

Gunnarsson, P. T., Winzell, M. S. r., Deacon, C. F., Larsen, M. O., Jelic, K., Carr, R. D., & Ahrén, B. (2006). Glucose-induced incretin hormone release and inactivation are differently modulated by oral fat and protein in mice. *Endocrinology, 147*(7), 3173–3180.

Hall, W. L., Millward, D. J., Long, S. J., & Morgan, L. M. (2003). Casein and whey exert different effects on plasma amino acid profiles, gastrointestinal hormone secretion and appetite. *British Journal of Nutrition, 89*(2), 239–248.

Hatzoglou, A., Bakogeorgou, E., Hatzoglou, C., Martin, P. M., & Castanas, E. (1996). Antiproliferative and receptor binding properties of α-and β-casomorphins in the T47D human breast cancer cell line. *European Journal of Pharmacology, 310*(2-3), 217–223.

Hedner, J., & Hedner, T. (1987). β-Casomorphins induce apnea and irregular breathing in adult rats and newborn rabbits. *Life Sciences, 41*(20), 2303–2312.

Hernández-Ledesma, B., García-Nebot, M. J., Fernández-Tomé, S., Amigo, L., & Recio, I. (2014). Dairy protein hydrolysates: Peptides for health benefits. *International Dairy Journal, 38*(2), 82–100.

Hoefle, A. S., Bangert, A. M., Stamfort, A., Gedrich, K., Rist, M. J., Lee, Y. M., … Daniel, H. (2015). Metabolic responses of healthy or prediabetic adults to bovine whey protein and sodium caseinate do not differ1–3. *The Journal of Nutrition, 145*(3), 467–475.

Horner, K., Drummond, E., & Brennan, L. (2016). Bioavailability of milk protein-derived bioactive peptides: A glycaemic management perspective. *Nutrition Research Reviews, 29*(1), 91–101.
Hoskin, D. W., & Ramamoorthy, A. (2008). Studies on anticancer activities of antimicrobial peptides. *Biochimica et Biophysica Acta, Biomembranes, 1778*(2), 357–375.
Jabbari, S., Hasani, R., Kafilzadeh, F., & Janfeshan, S. (2012). Antimicrobial peptides from milk proteins: A prospectus. *Annals of Biological Research, 3*, 5313–5318.
Jäkälä, P., & Vapaatalo, H. (2010). Antihypertensive peptides from milk proteins. *Pharmaceuticals, 3*, 251–272.
Jonker, J. T., Wijngaarden, M. A., Kloek, J., Groeneveld, Y., Gerhardt, C., Brand, R., ... Smit, J. W. A. (2011). Effects of low doses of casein hydrolysate on post-challenge glucose and insulin levels. *European Journal of Internal Medicine, 22*(3), 245–248.
Kampa, M., Loukas, S., Hatzoglou, A., Martin, P., Martin, P. M., & Castanas, E. (1996). Identification of a novel opioid peptide (Tyr-Val-Pro-Phe-Pro) derived from human αS1 casein (αS1-casomorphin, and αS1-casomorphin amide). *Biochemical Journal, 319*(3), 903–908.
Kizawa, K. (1997). Calmodulin binding peptide comprising α-casein exorphin sequence. *Journal of Agricultural and Food Chemistry, 45*(5), 1579–1581.
Kizawa, K., Naganuma, K., & Murakami, U. (1995). Calmodulin-binding peptides isolated from α-casein peptone. *Journal of Dairy Research, 62*(4), 587–592.
Korhonen, H. (2009). Milk-derived bioactive peptides: From science to applications. *Journal of Functional Foods, 1*, 177–187.
Korhonen, H., & Pihlanto, A. (2007). Technological options for the production of health-promoting proteins and peptides derived from milk and colostrum. *Current Pharmaceutical Design, 13*(8), 829–843.
Lacroix, I. M., & Li-Chan, E. C. (2013). Inhibition of dipeptidyl peptidase (DPP)-IV and α-glucosidase activities by pepsin-treated whey proteins. *Journal of Agricultural and Food Chemistry, 61*(31), 7500–7506.
Lacroix, I. M. E., & Li-Chan, E. C. Y. (2012). Evaluation of the potential of dietary proteins as precursors of dipeptidyl peptidase (DPP)-IV inhibitors by an in silico approach. *Journal of Functional Foods, 4*, 403–422.
Laursen, I., Briand, P., & Lykkesfeldt, A. E. (1989). Serum albumin as a modulator on growth of the human breast cancer cell line, MCF-7. *Anticancer Research, 10*(2A), 343–351.
Lin, L., Umahara, M., York, D. A., & Bray, G. A. (1998). β-Casomorphins stimulate and enterostatin inhibits the intake of dietary fat in rats. *Peptides, 19*(2), 325–331.
Ma, J., Stevens, J. E., Cukier, K., Maddox, A. F., Wishart, J. M., Jones, K. L., ... Rayner, C. K. (2009). Effects of a protein preload on gastric emptying, glycemia, and gut hormones after a carbohydrate meal in diet-controlled type 2 diabetes. *Diabetes Care, 32*(9), 1600–1602.
Ma, Y., Liu, J., Shi, H., & Yu, L. L. (2016). Isolation and characterization of anti-inflammatory peptides derived from whey protein. *Journal of Dairy Science, 99*(9), 6902–6912.
Mader, J. S., Salsman, J., Conrad, D. M., & Hoskin, D. W. (2005). Bovine lactoferricin selectively induces apoptosis in human leukemia and carcinoma cell lines. *Molecular Cancer Therapeutics, 4*(4), 612–624.
Martínez-Luis, S., Pérez-Vásquez, A., & Mata, R. (2007). Natural products with calmodulin inhibitor properties. *Phytochemistry, 68*(14), 1882–1903.
Maruyama, S., Mitachi, H., Tanaka, H., Tomizuka, N., & Suzuki, H. (1987). Studies on the active site and antihypertensive activity of angiotensin I-converting enzyme inhibitors derived from casein. *Agricultural and Biological Chemistry, 51*(6), 1581–1586.
Matin, M. A., & Otani, H. (2002). Cytotoxic and antibacterial activities of chemically synthesized κ-casecidin and its partial peptide fragments. *Journal of Dairy Research, 69*(02), 329–334.
McCarron, D. A., Morris, C. D., Henry, H. J., & Stanton, J. L. (1984). Blood pressure and nutrient intake in the United States. *Science, 224*(4656), 1392–1398.
Meisel, H. (1998). Overview on milk protein-derived peptides. *International Dairy Journal, 8*(5-6), 363–373.

Meisel, H., & FitzGerald, R. J. (2000). Opioid peptides encrypted in intact milk protein sequences. *British Journal of Nutrition, 84*(S1), 27–31.

Meisel, H., & Schlimme, E. (1990). Milk proteins: Precursors of bioactive peptides. *Trends in Food Science & Technology, 1*, 41–43.

Mizuno, S., Nishimura, S., Matsuura, K., Gotou, T., & Yamamoto, N. (2004). Release of short and proline-rich antihypertensive peptides from casein hydrolysate with an Aspergillus oryzae protease. *Journal of Dairy Science, 87*(10), 3183–3188.

Mohanty, D., Jena, R., Choudhury, P. K., Pattnaik, R., Mohapatra, S., & Saini, M. R. (2016). Milk derived antimicrobial bioactive peptides: A review. *International Journal of Food Properties, 19*, 837–846.

Moller, N. P., Scholz-Ahrens, K. E., Roos, N., & Schrezenmeir, J. (2008). Bioactive peptides and proteins from foods: Indication for health effects. *European Journal of Nutrition, 47*, 171–182.

Murakami, M., Tonouchi, H., Takahashi, R., Kitazawa, H., Kawai, Y., Negishi, H., & Saito, T. (2004). Structural analysis of a new anti-hypertensive peptide (β-lactosin B) isolated from a commercial whey product. *Journal of Dairy Science, 87*(7), 1967–1974.

Nagpal, R., Behare, P., Rana, R., Kumar, A., Kumar, M., Arora, S., … Yadav, H. (2011). Bioactive peptides derived from milk proteins and their health beneficial potentials: An update. *Food & Function, 2*(1), 18–27.

Nilsson, M., Holst, J. J., & Björck, I. M. (2007). Metabolic effects of amino acid mixtures and whey protein in healthy subjects: Studies using glucose-equivalent drinks. *The American Journal of Clinical Nutrition, 85*(4), 996–1004.

Nongonierma, A. B., & Fitzgerald, R. J. (2013). Dipeptidyl peptidase IV inhibitory and antioxidative properties of milk protein-derived dipeptides and hydrolysates. *Peptides, 39*, 157–163.

O'Neil, K. T., & DeGrado, W. F. (1990). How calmodulin binds its targets: Sequence independent recognition of amphiphilic α-helices. *Trends in Biochemical Sciences, 15*(2), 59–64.

Otani, H., & Suzuki, H. (2003). Isolation and characterization of cytotoxic small peptides, α-casecidins, from bovine αs1-casein digested with bovine trypsin. *Animal Science Journal, 74*(5), 427–435.

Pal, S., & Ellis, V. (2010). The acute effects of four protein meals on insulin, glucose, appetite and energy intake in lean men. *British Journal of Nutrition, 104*(8), 1241–1248.

Pal, S., Ellis, V., & Dhaliwal, S. (2010). Effects of whey protein isolate on body composition, lipids, insulin and glucose in overweight and obese individuals. *British Journal of Nutrition, 104*(5), 716–723.

Parodi, P. W. (2007). A role for milk proteins and their peptides in cancer prevention. *Current Pharmaceutical Design, 13*(8), 813–828.

Patil, P., Mandal, S., Tomar, S. K., & Anand, S. (2015). Food protein-derived bioactive peptides in management of type 2 diabetes. *European Journal of Nutrition, 54*(6), 863–880.

Pellegrini, A. (2003). Antimicrobial peptides from food proteins. *Current Pharmaceutical Design, 9*(16), 1225–1238.

Pepe, G., Tenore, G. C., Mastrocinque, R., Stusio, P., & Campiglia, P. (2013). Potential anticarcinogenic peptides from bovine milk. *Journal of Amino Acids, 2013*, 1.

Perego, S., Cosentino, S., Fiorilli, A., Tettamanti, G., & Ferraretto, A. (2012). Casein phosphopeptides modulate proliferation and apoptosis in HT-29 cell line through their interaction with voltage-operated L-type calcium channels. *The Journal of Nutritional Biochemistry, 23*(7), 808–816.

Picariello, G., Iacomino, G., Mamone, G., Ferranti, P., Fierro, O., Gianfrani, C., . . . Addeo, F. (2013). Transport across Caco-2 monolayers of peptides arising from in vitro digestion of bovine milk proteins. *Food Chemistry, 139*(1–4), 203–212.

Pihlanto, A., Virtanen, T., & Korhonen, H. (2010). Angiotensin I converting enzyme (ACE) inhibitory activity and antihypertensive effect of fermented milk. *International Dairy Journal, 20*(1), 3–10.

Pihlanto-Leppälä, A. (2000). Bioactive peptides derived from bovine whey proteins: Opioid and ace-inhibitory peptides. *Trends in Food Science & Technology, 11*(9-10), 347–356.

Power, O., Hallihan, A., & Jakeman, P. (2009). Human insulinotropic response to oral ingestion of native and hydrolysed whey protein. *Amino Acids, 37*(2), 333–339.

Pripp, A. H., Isaksson, T., Stepaniak, L., Sørhaug, T., & Ardö, Y. (2005). Quantitative structure activity relationship modelling of peptides and proteins as a tool in food science. *Trends in Food Science & Technology, 16*(11), 484–494.

Quiros, A., Davalos, A., Lasuncion, M. A., Ramos, M., & Recio, I. (2008). Bioavailability of the antihypertensive peptide LHLPLP: Transepithelial flux of HLPLP. *International Dairy Journal, 18*, 279–286.

Rasmussen, C. D., & Means, A. R. (1987). Calmodulin is involved in regulation of cell proliferation. *The EMBO Journal, 6*, 3961–3968.

Roldán, N. R., Jiménez, M., Cervantes-García, D., Marín, E., & Salinas, E. (2016). Glycomacropeptide administration attenuates airway inflammation and remodeling associated to allergic asthma in rat. *Inflammation Research, 65*(4), 273–283.

Roy, M. K., Kuwabara, Y., Hara, K., Watanabe, Y., & Tamai, Y. (2002). Peptides from the N-terminal end of bovine lactoferrin induce apoptosis in human leukemic (HL-60) cells. *Journal of Dairy Science, 85*(9), 2065–2074.

Rutherfurd-Markwick, K. J. (2012). Food proteins as a source of bioactive peptides with diverse functions. *British Journal of Nutrition, 108*(S2), S149–S157.

Sah, B. N. P., Vasiljevic, T., Mckechnie, S., & Donkor, O. N. (2015). Identification of anticancer peptides from bovine milk proteins and their potential roles in management of cancer: A critical review. *Comprehensive Reviews in Food Science and Food Safety, 14*(2), 123–138.

Sánchez, A., & Vázquez, A. (2017). Bioactive peptides: A review. *Food Quality and Safety, 1*(1), 29–46.

Sasaki, H., & Kume, H. (2007). Nutritional and physiological effects of peptides from whey milk whey proteins/peptides, natural beneficial modulators of inflammation. *Bulletin-International Dairy Federation, 417*, 43.

Sato, M., Hosokawa, T., Yamaguchi, T., … Nakano, T. (2002). Angiotensin I-converting enzyme inhibitory peptides derived from Wakame (Undaria pinnatifida) and their antihypertensive effect in spontaneously hypertensive rats. *Journal of Agricultural and Food Chemistry, 50*(21), 6245–6252.

Schweizer, F. (2009). Cationic amphiphilic peptides with cancer-selective toxicity. *European Journal of Pharmacology, 625*(1-3), 190–194.

Seppo, L., Jauhiainen, T., Poussa, T., & Korpela, R. (2003). A fermented milk high in bioactive peptides has a blood pressure–lowering effect in hypertensive subjects. *The American Journal of Clinical Nutrition, 77*(2), 326–330.

Sharma, S., Singh, R., & Rana, S. (2011). Bioactive peptides: A review. *International Journal Bioautomation, 15*(4), 223–250.

Shimamura, M., Yamamoto, Y., Ashino, H., Oikawa, T., Hazato, T., Tsuda, H., & Iigo, M. (2004). Bovine lactoferrin inhibits tumor-induced angiogenesis. *International Journal of Cancer, 111*(1), 111–116.

Shori, A. B. (2015). Camel milk as a potential therapy for controlling diabetes and its complications: A review of in vivo studies. *Journal of Food and Drug Analysis, 23*(4), 609–618.

Sipola, M., Finckenberg, P., Korpela, R., Vapaatalo, H., & Nurminen, M. L. (2002). Effect of long-term intake of milk products on blood pressure in hypertensive rats. *Journal of Dairy Research, 69*(1), 103–111.

Sousa, G. T., Lira, F. S., Rosa, J. C., de Oliveira, E. P., Oyama, L. M., Santos, R. V., & Pimentel, G. D. (2012). Dietary whey protein lessens several risk factors for metabolic diseases: a review. *Lipids in Health and Disease, 11*(1), 67.

Sternhagen, L. G., & Allen, J. C. (2001). Growth rates of a human colon adenocarcinoma cell line are regulated by the milk protein alpha-lactalbumin. In *Bioactive components of human milk* (pp. 115–120). Boston, MA: Springer.

Sun, X., Jiang, R., Przepiorski, A., Reddy, S., Palmano, K. P., & Krissansen, G. W. (2012). "Iron-saturated" bovine lactoferrin improves the chemotherapeutic effects of tamoxifen in the treatment of basal-like breast cancer in mice. *BMC Cancer, 12*(1), 591.

Sultan, S., Huma, N., Butt, M. S., Aleem, M., & Abbas, M. (2017). Therapeutic potential of dairy bioactive peptides: A contemporary perspective. *Critical Reviews in Food Science and Nutrition, 58*(1), 105–115.

Tagliazucchi, D., Shamsia, S., & Conte, A. (2016). Release of angiotensin converting enzyme-inhibitory peptides during in vitro gastro-intestinal digestion of camel milk. *International Dairy Journal, 56*, 119–128.

Teschemacher, H. (2003). Opioid receptor ligands derived from food proteins. *Current Pharmaceutical Design, 9*(16), 1331–1344.

Thundimadathil, J. (2012). Cancer treatment using peptides: Current therapies and future prospects. *Journal of Amino Acids, 2012*, 1.

Tomita, M., Bellamy, W., Takase, M., Yamauchi, K., Wakabayashi, H., & Kawase, K. (1991). Potent antibacterial peptides generated by pepsin digestion of bovine lactoferrin. *Journal of Dairy Science, 74*(12), 4137–4142.

Tulipano, G., Sibilia, V., Caroli, A. M., & Cocchi, D. (2011). Whey proteins as source of dipeptidyl dipeptidase IV (dipeptidyl peptidase-4) inhibitors. *Peptides, 32*, 835–838.

Tyagi, A., Tuknait, A., Anand, P., Gupta, S., Sharma, M., Mathur, D., … Raghava, G. P. (2014). CancerPPD: A database of anticancer peptides and proteins. *Nucleic Acids Research, 43*(D1), D837–D843.

Uchida, M., Ohshiba, Y., & Mogami, O. (2011). Novel dipeptidyl peptidase-4-inhibiting peptide derived from b-lactoglobulin. *Journal of Pharmacological Sciences, 117*, 63–66.

Udenigwe, C. C., & Aluko, R. E. (2012). Food protein-derived bioactive peptides: production, processing, and potential health benefits. *Journal of Food Science, 77*(1), R11–R24.

Vankudre, M., Balpande, A., & Athale, M. (2015). Comparative analysis of α-amylase inhibition and antioxidant activity of whey from cow and buffalo milk fermented with Lactobacillus species. *Bioscience Biotechnology Research Communications, 8*(1), 25–28.

Vermeirssen, V., Van Camp, J., & Verstraete, W. (2004). Bioavailability of angiotensin I converting enzyme inhibitory peptides. *British Journal of Nutrition, 92*(3), 357–366.

Wakabayashi, H., Takase, M., & Tomita, M. (2003). Lactoferricin derived from milk protein lactoferrin. *Current Pharmaceutical Design, 9*(16), 1277–1287.

Wang, W., Gu, F., Wei, C., Tang, Y., Zheng, X., Ren, M., & Qin, Y. (2013). PGPIPN, a therapeutic hexapeptide, suppressed human ovarian cancer growth by targeting BCL2. *PLoS One, 8*(4), e60701.

Xu, R. J. (1998). Bioactive peptides in milk and their biological and health implications. *Food Reviews International, 14*(1), 1–16.

Xu, X. X., Jiang, H. R., Li, H. B., Zhang, T. N., Zhou, Q., & Liu, N. (2010). Apoptosis of stomach cancer cell SGC-7901 and regulation of Akt signaling way induced by bovine lactoferrin. *Journal of Dairy Science, 93*(6), 2344–2350.

Yamamoto, N., Maeno, M., & Takano, T. (1999). Purification and characterization of an antihypertensive peptide from a yogurt-like product fermented by Lactobacillus helveticus CPN4. *Journal of Dairy Science, 82*(7), 1388–1393.

Yoo, Y. C., Watanabe, R., Koike, Y., Mitobe, M., Shimazaki, K. I., Watanabe, S., & Azuma, I. (1997). Apoptosis in human leukemic cells induced by lactoferricin, a bovine milk protein-derived peptide: Involvement of reactive oxygen species. *Biochemical and Biophysical Research Communications, 237*(3), 624–628.

Yoshikawa, M. (2015). Bioactive peptides derived from natural proteins with respect to diversity of their receptors and physiological effects. *Peptides, 72*, 208–225.

Zhang, Y., Lima, C. F., & Rodrigues, L. R. (2014). Anticancer effects of lactoferrin: Underlying mechanisms and future trends in cancer therapy. *Nutrition Reviews, 72*(12), 763–773.

Zhang, Y., Lima, C. F., & Rodrigues, L. R. (2015). In vitro evaluation of bovine lactoferrin potential as an anticancer agent. *International Dairy Journal, 40*, 6–15.

Recent Advances in Analysis of Food Proteins

Mehnaza Manzoor, Jagmohan Singh, Aratrika Ray, and Adil Gani

Introduction

Proteins are the biological macromolecules composed of amino acids linked by peptide bonds and differ from each other in size, molecular structure, charge, physiochemical properties. Depending upon their amino acid composition, they are extremely functionally versatile posing unique functional properties such as solubility, crystallization ability and hydrophobicity that result in their varied nutritional profile, bioavailability, and digestibility essential for human growth and maintenance. In addition quantitative analysis of protein content of major agricultural commodities such as cereal grains, legumes, flour, oilseeds, milk, and livestock feeds determines their market value (Krotz et al. 2008; Wiles et al. 1998). Therefore, being such important constituents it is crucial to have an adequate knowledge for analyzing and understanding proteins important for accurate nutritional labelling, quality control, functional property investigation, safety and economic implications in food industry (Chang 2010).

Protein analysis has been a major area of development in food science for analysis of nutritionally important proteins chiefly in terms of determining the quality and authenticity of food products by exploring the sequence of amino acids and their conformation. However, is not a straightforward procedure and is influenced by the complexity of foods in terms of composition, structural or spatial organisation in addition to external environmental and processing conditions leading to

M. Manzoor · J. Singh
Department of Food Science and Technology, Sher-e-Kashmir University of Agricultural Sciences and Technology, Jammu, India

A. Ray
Department of Food Engineering and Technology, Institute of Chemical Technology Mumbai, Mumbai, India

A. Gani (✉)
Department of Food Science and Technology, University of Kashmir, Srinagar, India

A. Gani, B. A. Ashwar (eds.), *Food biopolymers: Structural, functional and nutraceutical properties*, https://doi.org/10.1007/978-3-030-27061-2_12

underestimation of protein content (Li-Chan and Lacroix 2018). Certain unknown proteins from food sources are prone to showing allergic reaction but because of their extensive use, their detection and analysis are important. More recently, a number of newer protein sources like recombinant therapeutic proteins and peptides are being incorporated in the market and their analysis and estimation is necessary due to the potential complexity of product degradation during pre-formulation and formulation studies that compromise protein integrity, leading to a potentially harmful, unstable product. To address this issue, new techniques such as capillary electrophoresis (CE) based separation techniques has become a popular choice for the separation and analysis of therapeutic proteins and peptides (Creamer et al. 2014).

A wide array of different analytical methods has been developed throughout the years for analysis of proteins. However, recent developments in protein analyses have been mainly in innovative instrumental modification simulated by rapid development in molecular biology that deals with minute quantity of protein preferably in the range of milligram to microgram and with high activity. These methods either improve existing ones or develop new approaches to old methods in order to generate information more quickly, accurately and economically. These methods are highly sensitive even detecting as low as nano mole (10^{-9} mole) amount of protein. Some of the oldest and significant methods for protein analysis are mentioned in Table 1.

The aim of this chapter is to present an overview of recent and most cited methods of food protein analysis such as spectrophotometric, immunobiological, chemical and chromatographic methods, their techniques and usefulness with an aim to

Table 1 Chronological order of food protein analysis techniques

Technique/eponym	Date
Dumas method	1831
Nesseler reagent	1843
Biuret method	1849
Bethelot method (alkali-phenol reagent)	1859
Kjeldahl method	1883
Folin-Ciocalteu	1927
Dye-binding	1944
Lowry method	1951
Direct alkaline distillation	1960
NIR (near infrared reflectance)	1960
Modified Berthelot reaction	1971
Modified Lowry method	1975
Bradford method (Commassic blue dye-binding method)	1976
BCA(Bicinchoninic acid)method	1985
3-(4-Carboxy benzyl) quinoline-2-carboxaldehyde	1997

Source: Owusu-Apenten (2002), You et al. (1997), Simonian (2004), Nobel and Bailey (2009)

inform users how to select the most appropriate analysis technique for specific proteins or mixture thereof.

Spectrophotometric Methods

Almost all biological macromolecules are characterised by their optical properties via use of spectroscopic measurements. The technique employs electromagnetic radiation (200–800 nm) to interact with matter relative to a standard, or using an assigned extinction coefficient. The functional groups or regions, such as basic groups, aromatic groups, peptide bonds or aggregated proteins within the protein absorb light in the ultraviolet or visible range of the electromagnetic spectrum (200–800 nm). This absorbance is read by spectrophotometer and used for calculating protein concentration by comparing it with reference standard or known absorptivity for specific protein (Maehre et al. 2018). Depending on properties of electromagnetic radiation and its interaction with matter, different spectroscopic methods are used for analysis of proteins in sample. A wide range of spectrophotometric methods are available for accurate determination of protein concentration essential for quantitative biochemical, biophysical, molecular, and structural biology studies.

UV Visible Light Spectroscopy

These techniques are most frequently used for analytical and research work. In this type of spectroscopic technique proteins can be quantified by their intrinsic chromophores such as peptide bonds, aromatic amino acids (tyrosine, tryptophan), and certain prosthetic groups and coenzymes (e.g. porphyrine groups such as in haem). In UV-Vis region absorption of light by chromophores arises due to electronic transitions within atoms or molecules from a bonding or non bonding to an anti-bonding/ orbital. The amount of excitation energy depends on energy difference between ground and exited state.

Among the following possible transitions (σ-π^*, π-π^*, π-σ^*, σ-σ^*), conjugated π systems exhibit π-π^* transition absorb in region between 200 and 800 nm while other type require light in the vacuum UV region. Such molecules find application in colorimetric methods. For a given concentration, greater the degree of conjugation in molecule smaller will be difference in energy between the ground and excited state and hence greater will be intensity of absorbance. UV Visible spectroscopy can be divided into two classes, absorbance and fluorescence, based on sample- radiation interactions.

UV Visible Absorption Spectroscopy

This method is based on accurate measurement of fraction of an incident light beam absorbed by sample protein in solution. Absorbance measured at 280 nm (A_{280}) and at 205 nm (A_{205}) can be used to quantitate total protein in crude lysates and purified or partially purified protein. Both of these methods are simple and can be completed quickly (Simonian 2004). Although the A_{280} method is most commonly used, the A_{205} method can detect lower concentrations of protein and can quantitate dilute protein samples, but is more susceptible to interference from solvents and biological buffer components than the A_{280} method.

Protein Estimation by Near UV Absorbance; A_{280} Method (Range 20–3000 μg)

This method is based on the measuring absorbance of UV light at 280 nm by the aromatic amino acids, tryptophan, tyrosine, and by cystine, disulfide-bonded cysteine residues, in protein solutions to calculated protein concentration (Simonian 2004). The assay is suitable for protein concentration ranging between 20 μg/mL and 3000 μg/mL assuming that protein is pure containing no nonprotein component such as nucleotide cofactors, haem, or iron sulfur centres (Nobel and Bailey 2009). Protein concentration can be quantified using Beer-Lambert's law if molar absorptivity at 280 nm (a_{280}) is known or by comparing with standard curve of known standard protein solution. Using Beer-Lamberts law protein concentration (Pc) in sample is given as:

$$P_c\left(\text{mg / mL}\right) = \frac{A_{280}}{a_{280} * b} \tag{1}$$

In this equation Pc is sample concentration in mg/ml, a_{280} is molar absorptivity with unit of ml/mg cm, and b is path length in cm.

Protein Estimation by Far UV Absorbance; A_{205} Method (Range 1–100 μg)

This method is based on absorption of photons by peptide bonds at wavelength below 210 nm with a broad absorption peak allowing measurements at longer wavelength. It can be used to quantitate dilute protein solutions or for short path length applications, such as in column chromatography, or analyzing peptides (Nobel and Bailey 2009). However, the absorptivity for a given protein at 205 nm is several-fold greater than that at 280 nm (Scopes 1974; Stoscheck 1990).The disadvantage of this method is the interference from some buffers and other components that absorb at 205 nm (Stoscheck 1990). Sample concentration (Pc) can be calculated using following equation:

$$P_c\left(\text{mg}/\text{mL}\right)=\frac{A_{205}}{31*b} \quad (2)$$

In this equation, the absorptivity value, 31, has units of ml/mg cm and b is the path length in cm.

Fluorescence Emission Method

Fluorescence emission is a phenomenon that uses spectroflurometer or a filter flurometer to measure radiation energy emitted by fluorescent molecule, fluorophore such as aromatic or highly unsaturated organic compounds, after subsequent absorption of light in UV, Visible, or near infrared region. It is basically a two step method that first involves excitation of electrons from singlet ground state to one of excited states by absorption of light followed by a vibrational relaxation or internal conversion from an upper excited state to a lowest excited state, without any radiation. Finally, the fluorescence occurs, typically 10–8 s after excitation when electrons return to the ground state. Emitting light has energy equal to the difference between energies having maximum absorbance at 280 nm of ground and excited states. The intensity of emitted light is directly proportional to probability of the transition from the electronic excited to the ground state and concentration can be calculated from standard curve based on fluorescent emission of standard protein solution.

Fluorescence Properties of Aromatic Amino Acids

This assay can be used for quantification of protein solutions with concentrations of 5–50 μg/mL (Simonian 2004). Fluorescence properties of aromatic amino acids are demonstrated in Table 2.

Apart from tryptophan having maximum absorption maximum at 280 nm, there are several other amino acids whose absorption maxima fall under the UV range. Table 3 demonstrates the wavelengths of absorption maxima and corresponding molar absorptivity (ε) for the amino acids with appreciable absorbance in the UV range (Simonian 2004).

The major disadvantages of these methods can be summed up as follows:

- Since proteins from different sources vary in their proportion of aromatic amino acids, so too do their molar absorptivity coefficient for individual proteins.

Table 2 Fluorescence properties of aromatic amino acids (pH 7 and 25 °C)

Amino acid	Excitation wavelength	Emission wavelength	Quantum yield
Phenylalanine	260 nm	283 nm	0.04
Tryptophan	285 nm	360 nm	0.20
Tyrosine	275 nm	310 nm	0.21

Source: Hawkins and Honigs (1987), Fasman (1989)

Table 3 Absorption maxima and molar absorptivity (ε) of amino acid (pH 7.1)

Amino acid	Wavelength maxima (nm)	ε × 10–3 (l/mol cm)
Cysteine	250	0.3
Histidine	211	5.9
Phenylalanine	188	60.0
	206	9.3
	257	0.2
Tryptophan	219	47.0
	279	5.6
Tyrosine	193	48.0
	222	8.0
	275	1.4

Source: Freifelder (1982), Fasman (1989)

Table 4 Concentration limits of interfering reagents for A_{205} and A_{280} protein assays

Reagent	A_{205}	A_{280}
Ammonium sulphate	9% (w/v)	>50% (w/v)
Brij 35	1% (v/v)	1% (v/v)
DTT[a]	0.1 mM	3 mM
EDTA[a]	0.2 mM	30 mM
Glycerol	5% (v/v)	40% (v/v)
KCl[a]	50 mM	100 mM
2-ME[a]	<10 mM	10 mM
NaCl[a]	0.6 M	>1 M
NaOH[a]	25 mM	>1 M
Phosphate buffer	50 mM	1 M
SDS[a]	0.10% (w/v)	0.10% (w/v)
Sucrose	0.5 M	2 M
Tris buffer	40 mM	0.5 M
Triton X-100	<0.01% (v/v)	0.02% (v/v)
TCA[a]	<1% (w/v)	10% (w/v)
Urea	<0.1 M	>1 M

Source: Stoscheck (1990)

[a]Abbreviations: *DTT* dithiothreitol, *EDTA* ethylenediaminetetraacetic acid, *2-ME* 2-mercaptoethanol, *SDS* sodium dodecyl sulfate, *TCA* trichloroacetic acid

- These spectrophotometric measurements, however, is subjected to interference by presence of many components such as solvents and biological buffers that also absorb strongly at this wavelength. Table 4 represents the concentration limits of interfering reagents for A_{205} and A_{280} protein assays.
- absorbance values >2.0 should not be used for sample proteins measured by the A_{280} or A_{205} method as stray light can affect the linearity of absorbance versus concentration.
- In addition to these, nucleic acids also have substantial absorbance at 260 nm and can interfere with A_{280} quantification of protein in crude samples. However,

knowing the absorption ratio of A_{280} to A_{260} for pure protein, concentration can be calculated.
- For known absorptivity, these assays require <30 min depending on the number of samples. However, these assays usually require 1 h for intrinsic fluorescence quantization or quantization using standards or reference protein.

Circular Dichroism Spectroscopy

Circular Dichroism (CD) is the form of chiroptical spectroscopy, spectroscopic techniques that measures differential absorption of left and right circularly polarized light by a chiral molecule, particularly biologically important molecules such as proteins, carbohydrates, nucleic acids, lipids, and therapeutic drugs (Siligardi and Hussain 2017).

When asymmetric molecules interact with beam of polarised light, they may differently absorb right and left handed circularly polarized light (hence termed circular dichroism), which trace out circles to different extents and changes the plane of polarisation with different indices of refraction for the two sinusoidal waves. The resultant wave rotates and traces out an ellipse and is said to be elliptically polarized (Greenfield 2007).

CD is thus defined as the difference in absorption (ΔE) for the left and the right (CPL) circularly polarized light ($\Delta\varepsilon = \varepsilon_L - \varepsilon_R$, where ε_L and ε_R is absorptions of the left and the right helical rays, respectively).It is reported either in units of ΔE or in degrees ellipticity, which is defined as the angle whose tangent is the ratio of the minor to the major axis of the ellipse. $[\theta]$, the molar ellipticity in deg.cm^2/dmol = 3298ΔE. Circular dichroism spectra are measured using a circular dichroism spectrometer, such as the Chirascan, which is a highly specialised derivative of an ordinary absorption spectrometer. CD spectrometers measure alternately the absorption of L- and R-CPL, usually at a frequency of 50 kHz, that can be used to calculate the circular dichroism signal.

In proteins CD is useful in characterizing and estimating protein folding in terms of the secondary structure content of proteins, such as, α-helical and β-pleated sheets, as a function of asymmetric environment such as concentration, pH, temperature, aqueous buffer composition, ionic strength, denaturants, and UV irradiation, age, and ligand interactions whether an expressed, purified protein is folded, or if its stability or conformation behaviour is affected by mutation (Greenfield 2007). The conformational studies in peptides (like proteins) can include:

- Monitoring conformational changes (e.g. monomer-oligomer, substrate binding, denaturation, etc.), and
- Estimation of secondary structural content (e.g. under certain conditions this particular peptide is 25% helical (Robb et al. 2006)

CD spectra in far UV (180–250 nm) can be used to monitor, characterize and predict different protein secondary structural types: alpha helix, parallel and anti-parallel beta sheet, and turns while measurement in the near-UV region (250–330 nm) provides the detailed fingerprint of tertiary structure of aromatic amino acid residues (Tryptophan, Tyrosine, and Phenylanaline) and dihedral angles of disulfide bonds can be used for quality control (Siligardi and Hussain 2017). Protein molecules exhibit absorption in the ultraviolet region of the spectrum by peptide bonds (amid chromophores), side chains in proteins, and any prosthetic groups (Johnson 1992). For example, CD spectra of α-helical proteins have negative band near 222 nm due to the strong hydrogen-bonding environment of this conformation. A second transition at 190 nm is split into a negative band near 208 nm and a positive band near 192 nm. Both bands are reduced in intensity in short helices. Antiparallel β-pleated sheets show a negative band at 216 nm, a positive band between 195 and 200 nm, and a negative band near 175 nm. However, the position and magnitude of these bands is variable, resulting in less accurate predictions for β-structure than for a-helices by CD (Pelton and McLean 2000). Collagens are unique class of proteins having triple helical structure with conformation of each strand resembling that of poly-L-proline in an extanded helical conformation where all bonds are trans to each other (poly-L-proline II). CD spectra associated with some of the common protein secondary structures are shown below (Fig. 1):

Mass Spectrometry (MS)

MS is a unique spectroscopy technique based on ionization to produce gas phase protonated ions by an electrostatic field (mass analyzer) and their subsequent separation in a mass spectrometer according to their mass-to-charge ratio (*m*/*z*). The charged ions generated by ionisation are then counted on the electrode and presented in dependence on their m/z ratio. This technique provides information about molecular weights and chemical structures of the peptides, proteins, carbohydrates, oligonucleotides, natural products, and drug metabolites (Biemann 2014).

Improvements in instrumentation, advances in on- line separation techniques and in data processing have contributed to great expansion of MS in food-related analysis. Post era 1981 saw the rise of so-called “soft ionization” techniques such as fast atom bombardment technique (FAB) followed by the electrospray ionization (ESI) and matrix-assisted laser desorption/ ionization (MALDI) techniques that made analyses by mass spectrometry possible for biomolecules, with high accuracy (0.01%).The common feature of these ionization techniques is ionization without fragmentation, accurate mass determination, broad applicability and picomole to femtomole sensitivity. A brief discussion on FAB-, ESI-and and MALDI-MS is being given below (Leonil et al. 2000).

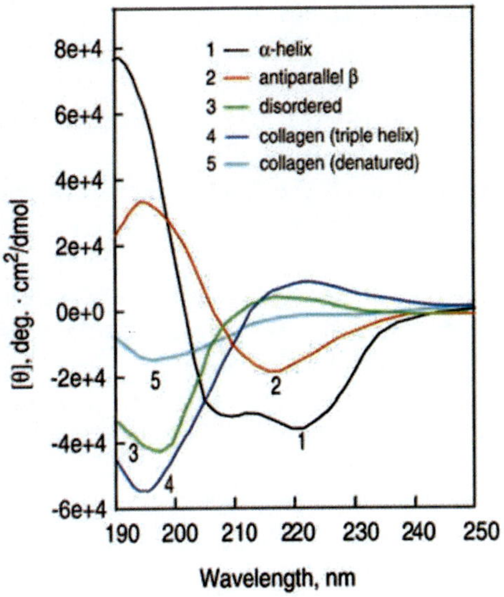

Fig. 1 CD spectra of proteins and polypeptides with representative secondary structures. (**a**) CD spectra of poly-L-lysine at pH 11.1 in the 1 (black) α-helical and 2 (red) antiparallel β-sheet conformations and at pH 5.7 in the 3 (green) extended conformations and placental collagen in its 4 (blue) native triple-helical and 5 (cyan) denatured forms. (**b**) CD spectra of representative proteins with varying conformations: 1 (black) sperm whale myoglobin; 2 (green) chicken heart lactate dehydrogenase; 3 (red) bovine α-chymotrypsin and 4 (cyan) human Bence Jones protein REI light chain, which is a human immunoglobulin light chain of κ type (Johnson 1992)

Modes of Mass Spectroscopy

Fast Atom Bombardment Mass Spectrometry (FAB-MS)

This is basically a desorption technique that significantly contributes to determination of peptides and proteins with masses larger than 10,000 on magnetic sector instruments (Barber and Green 1987).This method is based on ionization of atoms or ions (typically Xe or Cs at 8–40 keV) by bombardment with an energetic beam.

FAB has been successfully coupled to liquid chromatography for analysis of peptides (Leonil et al. 2000). A related technique, Continuous Flow-FAB, although, having certain persistent advantages like decreasing the ion suppression effect with hydrophilic peptides, and the matrix-derived background peaks, however, FAB-MS and these related techniques (CF-FAB and (stainless steel frit) frit-FAB) contribute some drawbacks such as liability to generate good signal intensity with large molecular mass proteins (greater than about 15,000) (Nguyen et al. 1995). To overcome drawbacks, an attractive alternative technique has been developed called the ESI-MS, which is more easily interfaced to liquid chromatography (Leonil et al. 2000).

Electro-Spray Ionization Mass Spectrometry (ESI-MS)

This is a "soft ionisation"technique that enables the analysis of large intact (underivatised) biomolecules, such as proteins and DNA. It involves atomisation of solvent containing analyte into a spray of charged liquid droplets at an atmospheric pressure by application of strong electric field to it. The charged liquid droplets produced are then transmitted into the vacuum chamber of mass spectrometer which vaporizes the solvent. The droplet size decreases and eventually charged analyte (free of solvent) remains. Theory based research on ESI-MS has led to the development of nano electrospray, in which the electrostatic force establishes flow fate and is thus independent of the solvent delivery (Wilm and Mann 1996).

The electrospray process results in multiple-protonation of proteins and peptides, and finds extensive application for accurate molecular weight determination of thermally unstable, high molecular mass macromolecules such as proteins, oligonucleotides, and synthetic polymers (Ho et al. 2003). It has been significantly used for mass determination of milk proteins, detection of modifications in proteins caused by insertion, deletion or modification of amino acids, identification of genetic variants and post-translational modifications like deamidation, oxidation, glycosylation, phos- post-translational modifications, genetic polymorphorylation, sulfation (Alomirah et al. 2000)

Since, ESI requires constant delivery of liquids; it can be easily coupled online with liquid based separation systems such as HPLC chromatography for accurate mass determination of proteins. It has been shown that liquid chromatography techniques coupled with tandem mass spectrometry (MS–MS) enhanced the capacity of electrospray ionization to elucidate the structure of protein by sequence assignment from peptide mapping (Leonil et al. 2000).

Other advances in the field of MS analysis of biomolecules include the use of collision-cell CID of proteins in ESI-MS–MS experiments for elucidation purposes. Liquid chromatography–tandem mass spectrometry with ESI ionization was successfully applied for the separation and identification of low molecular-mass peptides released during milk sterilization (Careri et al. 2002)

Matrix-Assisted Laser Desorption Mass Spectrometry (MALD-MS)

Though a very old technique introduced in 1988 by Karas and Hillenkamp, this laser desorption mass spectrometry technique is an important tool for complex protein analysis which is valuable even in the present era. The procedure involves mixing of the sample with excess of matrix material usually organic acid of low molecular mass in solvent. The matrix/analyte mixture is irradiated by a laser such as nitrogen laser of wavelength 337 nm, nowadays, by UV laser which causes desorption and ionisation of both matrix and analyte and their subsequent ejection to vapour phase (Leonil et al. 2000). The analyte ions are finally measured on mass spectrometer. MALDI-MS has proven useful for peptide mapping/MS and there are several advantages in using these techniques. One crucial step of MALDI-MS is the choice of matrix which is important for good MALDI-MS spectra. From aspect of peptide analysis, following matrices can be considered most effective which includes cinnamic acid derivatives (α-cyano-4-hydroxycinnamic) and benzoic acid derivatives (2,5-dihydroxybenzoic and that for protein includes sinapinic acid (trans-3,5-dimethoxy 4-hydroxycinnamic acid) (Beavis et al. 1992; Beavis and Chait 1989).

An advance of this technique is MALDI-TOF, MALDI coupled with a time-of-flight (TOF) mass spectrometer. This conjugation is used due to pulsed nature of MALDI which makes the process much convenient. This technique allows the simultaneous measurement of masses from 1000 to 100,000 without chromatographic purification in the low picomol range within few minutes. Thus, not only intact proteins, but also protein hydrolyzates can be analyzed. The technique has been proved very helpful in evaluating authenticity of milk by detecting presence of even small amount of cow milk in either raw ewe and buffalo milk samples and the addition of powdered milk to samples of fresh raw milk, adulteration was evaluated from protein patterns coming from the most abundant whey proteins, alpha-lactalbumin and beta-lactoglobulin, used as molecular markers (Cozzolino et al. 2001)

Chromatographic Methods of Food Protein Analysis

In the post-genomic era some of the greatest challenges faced by scientist were to identify and quantify all expressed protein components in cells, tissues, and organisms. In rescue of this situation chromatography has found one important position in protein analysis scenario. Proteins can be purified based on characteristics such as size and shape, total charge, hydrophobic groups present on the surface, and binding capacity with the stationary phase. Four separation techniques based on molecular characteristics and interaction type use mechanism of ion exchange, surface adsorption, partition, and size exclusion. Other chromatography techniques are based on the stationary bed, including column, thin layer, and paper chromatography.

Column Chromatography

Column Chromatography is one of the most frequently used methods of protein purification (Coskun 2016). The technique is performed on chromatography columns filled with a variety of column matrices (mostly derivatives of agarose) using different characteristics of proteins for separation such as size, charge, affinity toward substrates or inhibitors, hydrophobicity, lectin, or antibody binding. In column chromatography stationary phase suitably moistened with mobile phase is either coated onto discrete small particles (the matrix) and packed into the column or applied as a thin film to the inside wall of the metal or glass column. The extract material or sample to be separated, is then applied and mobile phase (eluent), is poured into column over the sample either by use of pumping system or applied gas pressure. As mobile phase flows through the column, the analytes separate on the basis of their affinity towards the stationary and mobile phase. Those with greater affinity to mobile phase travels fast and reach the bottom early while analytes having higher affinity to stationary phase travel slow and reach bottom late and finally emerge individually in eluate as it leaves the column (Fig. 2).

Column Chromatography has been used for a long time to characterize, separate, and quantitate cereal protein fractions. In particular, gel permeation (GP) chromatography and reversed-phase (RP) chromatography have been widely used for separating according to different molecular weights and hydrophobicities (Wieser 2008). The detection and quantization of proteins eluted from the column is carried out by UV absorbance in the range of 200–220 nm. At these wavelengths, the absorbance units are highly correlated with protein quantity (Wieser 1998). The detection limit is around 1–2 μg protein. One advance in column chromatography is use of GP-HPLC on Superdex 200 HR for quantification of both gliadins and total gluten in a series of wheat starches (Wieser and Antes 2002) and it was found that 23 starch samples had gliadin contents between 15 and 574 mg/kg (Wieser and Seilmeier 2003).

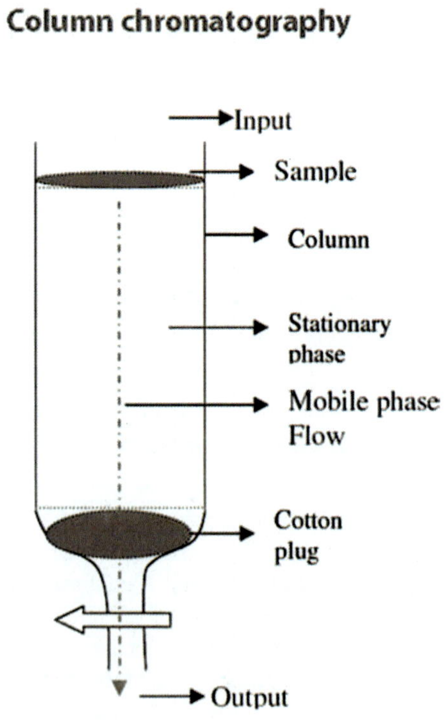

Fig. 2 Column chromatopragpy

Ion-Exchange Chromatography

The next important technique of protein separation, identification and purification is the ion-exchange chromatography. It is one of the powerful and perhaps most frequently used chromatographic technique for purification and separation of proteins, polypeptides, nucleic acids, poly nucleotides, and other charged bio-molecules. This technique is based on ionic interaction of solid support material (matrix) with oppositely charged protein in a solution. Proteins are separated from the column either by changing pH, concentration of ion salts or ionic strength of the buffer solution. This technique can be subdivided into two types namely cation and anion exchangers that possess negatively charged groups attracting positively charged cations and positively charged groups that will attract negatively charged anions respectively. Cation exchangers are also referred to as acidic ion exchangers because their negative charges result from the ionisation of acidic groups and anion exchangers are also called as basic ion exchangers, as positive charges generally result from the association of protons with basic groups (Wilson 2010).

Principle and Method

Proteins are complex ampholytes having both positive and negative charges. Positive charges are usually provided by free N-terminal amine below pH 8 and by arginines, lysines and histidines, depending of the pH of the surrounding buffer. Negative charges are principally provided by aspartate and glutamate residues and the C-terminal carboxyl group. Virtually all these residues are ionized above pH 6. At higher pH values (>8) cysteines may become ionized too. At a certain pH point, called isoelectric point (pI) of protein, the net charge is zero (Righetti and Caravaggio 1976) and it depends on the proportions of ionizable amino acid residues in its structure. In an ion exchange chromatography separation a protein at a pH above its pI will bind to a positively charged medium or anion exchanger and, at a pH below its pI it will bind to a negatively charged medium or cation exchanger (Jadaun et al. 2017).

Anion exchanger matrices such as quaternary amines or diethylaminoethyl groups coupled via a linker to a cellulose matrix are used for binding of anionic proteins. For the binding of cationic proteins, sulfopropyl or carboxymethyl groups can be used. The bound proteins of interest are eluted by changing the pH of the eluting buffer or increasing ionic strength by sodium chloride gradient to break the interaction of protein with charged matrix.

Affinity Chromatography

When it comes to separation and analysis of proteins with respect to their specific and natural properties, affinity chromatography has found wide application. In 1968, the technique was initially used for purification of enzymes and has since been extended for purification of other bio molecules such as receptor proteins, immunoglobulin's, glycoprotein, nucleotides, nucleic acids and even to whole cells and cell fragments (Wilson and Walker 2010).

Unlike most other forms of chromatography, this chromatographic technique is most specific as it does not rely on the than physico-chemical properties of analytes. Rather, it is based on strong specific biological ligand-protein interaction (Narayanan 1994). These interactions not only depend on general properties such as isoelectric point or hydrophobicity, but on selective properties such as the interactions between antigen and antibody, enzyme and substrate analogue, nucleic acid and binding proteins as well as hormone and receptors (Fanali et al. 2017).

This technique thus finds wide application in preserving immunological and biological activity of isolated proteins and antibodies. Theoretically it is capable of giving very high purification, even from complex mixtures, in just one simple process step.

Principle and Method

The technique involves reversible binding of protein of interest to a specific ligand coupled to column matrices. In general, immobilization of ligand occurs in three steps:

- Activation of matrix for its reaction with the functional group of the ligand.
- covalently coupling of ligand through some chemical reaction
- Finally, blocking of residual unreacted groups by an excess of a suitable low molecular weight substance such as ethanolamine.

This provides a higher degree of certainty that all binding will be between the ligand and the sample. Since only the intended protein is adsorbed from the extract passing through the column. It is retained in column by making a complex with ligand, other substances will be washed away. The interaction of ligand-protein can be reversed, and bound proteins can leave the column either specifically using a competitive ligand, or non-specifically by decreasing the pH of the buffer, increasing ionic strength or polarity to elute the target protein in bioactive form. This technique is able to concentrate dilute amounts of the expensive molecule, purification of theraupeutic products and stabilize the protein when adsorbed onto a ligand for which it has a natural affinity.

Another important feature of affinity chromatography is that proteins often can be separated from denatured or functionally different forms, since the technique relies on functional properties. One form of affinity chromatography uses Protein A and Protein G as ligand which are highly specific for immunoglobulin's, immune complexes and monoclonal antibodies. Binding to Protein A and Protein G is usually optimum in physiological buffers at a high pH of 8- and elution requires low pH 2.5–3.0 (Narayanan 1994).

Immunobiological Methods

The analysis of intact proteins by chromatographic methods is still challenging, especially when biological functions as antigenicity of proteins or peptides are in the focus. Immunobiological methods are analytical techniques based on the specific and high affinity binding of antibodies (immunoglobulins) with particular target antigens, the substance to be determined (proteins or peptides) (Wieser 2008). Immunobiological methods are useful in detecting electrophoretic fraction of proteins, detecting heat treatment and bacterial contamination, and provide information about the entirety of antigenic proteins/peptides, e.g., in ELISA assays. In an immunoassay, antigens and antibodies are used either as target molecules or capture molecules. In other words, a particular antigen can be used as capture molecule to trap its specific antibody in complex sample, or a specific antibody can be used as capture molecule to trap the target antigen in a sample. The binding affinity between antibody and antigen is one of the strongest non-covalent interactions and an important factor in determining the sensitivity of an immunoassay.

Various immunological methods include Western blotting, immuno-electrophorosis, immuno-precipitation, as well as more advanced combination methods, e.g., immuno-precipitation with MS detection have been developed based on specific antibody-antigen affinity and more. The basis of every immunobiological method is the detection and measurement of the primary antigen–antibody interaction to indicate the presence of particular proteins in a sample. In its simplest form, antibody capture of antigen can involve a simple precipitation and be detected visually (Peggy Hsieh 2010).

Western Blotting

Western blotting also known as protein blotting or immunoblotting is a laboratory based method that combines two techniques: polyacrylamide gel electrophoresis (PAGE) and immunoassay where complex protein mixtures are first electrophoretic separated on the based on their molecular mass followed by subsequent transfer from sodium dodecyl sulfate polyacrylamide gels and electroblotting them onto nitrocellulose (NC) or polyvinylidene difluoride (PVDF) membrane. The nitrocellulose along with the transferred protein is referred to as a blot. Once transferred onto nitrocellulose, the membranes are washed and incubated with primary and then secondary antibodies, the primary antibody detects the specific protein of interest and secondary antibody recognises the primary antibody (recognizing one) and is conjugated to reporter enzyme usually either horseradish peroxidase (HRP) or alkaline phosphatase that enables detection of coloured electrophoretic band indicating position of protein of interest. In general the technique uses three elements to accomplish this task: (1) separation by size, (2) transfer to a solid support, and (3) marking target protein using a proper primary and secondary antibody for visualization (Mahmood and Yang 2012).

Alkaline phosphatase converts colourless 5-bromo-4-chloro-indolylphosphate (BCIP) substrate into a blue product and horseradish peroxidise oxidises either 3-amino-9-ethylcarbazole into an insoluble brown product, or 4-chloro-lnaphthol into an insoluble blue product using H_2O_2 as a substrate (Wilson and Walker 2010). Other enzymes such as β-galactosidase, glucose oxidase and glucose- 6-phosphate dehydrogenase can also be used. The method thus finds wide application in cell biology and molecular biology and helps to identify specific proteins from a complex mixture of proteins extracted from cells. Due to the high resolution power of the electrophoresis coupled with the specificity of the antibody, the method has been used for the detection of various molecular forms of proteolytic enzymes in whey, such as the various forms of cathepsin D; procathepsin D, pseudocathepsin D, mature single-chained and mature two-chained cathepsin D (Larsen and Petersen 1995). It is thus very powerful for discrimination of different molecular forms of the target protein e.g., to follow activation by proteolytic cleavage such as during purification.

Immunoelectrophoresis

This is a two-stage technique for characterization of antibodies that combines the principle of electrophoresis and immunodiffusion. The proteins of interest are first separated by electrophoresis in a supporting media such as agarose and then immunodiffused with the addition of antiserum in a well that is cut from agarose gel. When an antibody interacts with a specific antigen, a white precipitin band is formed on black background (immunodiffusion). This precipitate band shows the presence of antibody specific to target antigen. Presence of one precipitin band indicates homogeneity of antibody while as antibody heterogeneity is represented by presence of multiple precipitate bands. Likewise absence of any precipitin band indicates no specific antibody against target antigen (Buyukkoroglu and Senel 2018). Mobility of the molecules during electrophoresis is dependent on a number of factors such as net charge of molecules, size & shape of the molecules, pH of buffer & ionic strength of buffer etc. Four types of immune-electrophoresis have been used, such as electro-immunoassay (EIA) or rocket/Laurell rocket electro-immunoassay, classical immune-electrophoresis, immuno-fixation electrophoresis (IFE), and capillary electrophoresis (Levinson 2009).

Nowadays, capillary electrophoresis is the most commonly used analysis technique primarily due to its faster separation times and the use of multi-capillary arrays analysing hundreds of samples per day. Secondly, capillaries used are of micro scale dimensions, only microlitres of reagent are needed for analysis of nanolitres of sample, along with the ability for on-line detection down to femto-mole (10^{-15} moles). The use of micro capillaries result in more heat dissipation because of their large surface area to volume rations thus higher resolution than traditional electrophoresis. It can be used to separate a wide spectrum of biological molecules including amino acids, peptides, proteins, DNA fragments (e.g. synthetic oligonucleotides) and nucleic acids, as well as any number of small organic molecules such as drugs or even metal ions. This technique was used for detection of proteolysis of whey proteins using a coated fused-silica capillary column (Miralles et al. 2003). CE-based separation techniques such as capillary zone electrophoresis (CZE), Capillary gel electrophoresis (CGE), Capillary isoelectric focusing (CIEF), and Capillary isoelectric focusing (CEC) has become a popular choice for characterization of therapeutic proteins and peptides (Creamer et al. 2014) In additional of having the potential for high throughput analysis using capillary arrays, these provide versatile, efficient, and fast analyses of proteins.

Immuno-Precipitation (IP)

This is a technique where a protein antigen is precipitated out of a complex sample mixture using a specific antibody coupled to sedimentable matrix. This technique analyzes already isolated proteins by other biochemical methods such as density gradient dependent sedmentation or gel filtration methods. The antigens isolated by immunoprecipitation are further analysed by western blotting or SDS-PAGE. Antibodies used can be either polyclonal or monoclonal obtained from various animal species. They can bind non covalently to immunoadsorbents such as protein A or G–agarose beads, or can be coupled covalently to a solid-phase matrix (Bonifacino et al. 2001). Immunoprecipitation or co-immunoprecipitation is used to enrich a specific protein or protein complex from a tissue homogenate, cell lysate, or culture supernatant (Kaboord and Perr 2008).

In IP firstly, the antigen (specific protein of interest) is enriched from tissue homogenate, suspension cultures or cell lysates. Enrichment is achieved by binding antigen with a specific antibody that is non-covalently attached to solid phase matrix such as protein A or protein G aragose beads. Subsequently, enriched antigens are incubated with immobilized antibody solid-phase matrix followed by thorough washing. The resultant precipitated complex are denatured and resolved by SDS-PAGE and can further be analysed using number of different methods such as western blot or protein mass spectrometry (Xie et al. 2017).

Mass Spectrometry Based Immunoprecipitation

Immunoprecipitation, eventually in combination with MS detection of precipitates, is an alternative method that provide a sensitive and accurate way of characterising protein complexes and their response to regulatory mechanisms, especially of minor whey proteins or protein complexes when suitable antibodies are available. This has been used for the study of proteins complexing with Lf (Sokolov et al. 2014), and for identification of prolactin-binding protein in human milk (Kline and Clevenger 2001). Usually the majority of proteins identified in IP experiments are non-specific binders. In biological processes, MS- based immunoprecipitation have greater potential for studying protein complexes e.g., in the study of heat-induced protein complexes (Le et al. 2019) (Fig. 3).

Immunostaining

It is a biochemical and antibody based technique to detect a specific protein in a sample. In 1941 Albert Coons described the immuno-histochemical staining of tissue sections. However it is now used in histology, cell biology, and molecular biology that use antibody-based staining methods.

Immunohistochemistry (IHC) is a morphology based technique for detecting proteins or antigens in cells that are part of a tissue section by exploiting the

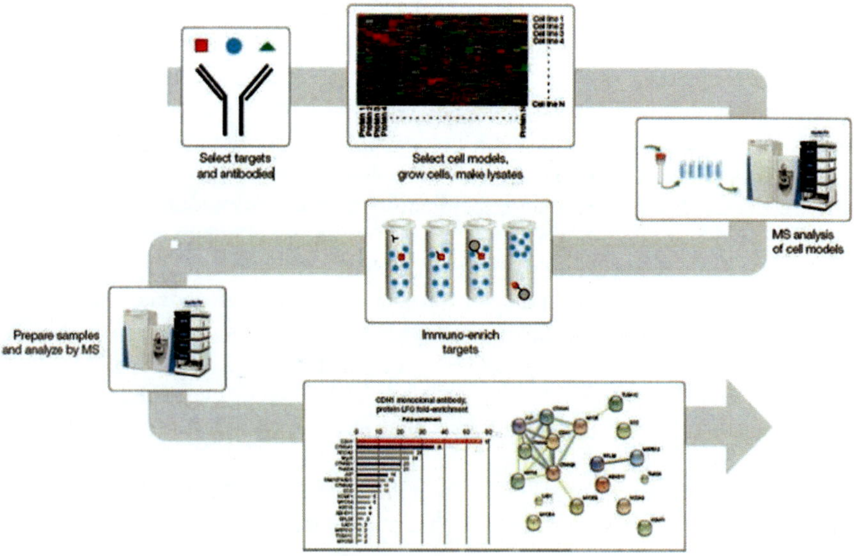

Fig. 3 Antibody verification workflow by IP-MS analysis

principle of antibodies binding specifically to antigens in biological tissues. The antibody-antigen binding can be visualized after immunohistochemical staining by a fluorescent dye, radioactive tracer, colloidal gold particle, enzyme such as Horseradish Peroxidase (HRP) or Alkaline Phosphatase (AP). This technique can be performed on formaldehyde fixed and paraffin-embedded (FFPE) tissues allowing access to archival material. A key advantage of IHC is that it not only allows analysis of the anatomy of the tissue of interest but also visualization of the spatial distribution and expression of specific antigens or cellular components in a variety of tissue sections. Immunohistochemical analysis of soy protein allows the detection of all soy protein with all forms of soy additives (texturates, concentrates, soy flour, isolates) with appropriate epitomes (Pospiech et al. 2009).

SDS-Polyacrylamide Gel Electrophoresis (SDS-PAGE)

Electrophoresis is a phenomenon that describes migration of charged particles towards opposite electrode under the influence of electric field. This technique finds wide application in separating proteins according to their electrophoretic mobility which depends on charge, molecule size and structure of the proteins. Electrophoresis in polyacrylamide gels (PAG) is frequently referred to as gel electrophoresis (PAGE). PAG is a three-dimensional mesh network polymer formed from polymerisation of acrylamide monomer and a cross-linker, bis-acrylamide under the catalyzation of ammonium per-sulphate and base tetra methylenediamine (TEMED). PAG is a versatile supporting matrix because of its stability with little adsorption and electro-osmosis effect provided by its neutrally charged nature. Acrylamide gels with low gel percentage (10–40%) have small pore size. They are used in techniques such as SDS-PAGE to separate proteins according to their size and to identify their relative molecular mass.

Two different buffer systems, continuous and discontinuous, can be used in electrophoresis. In continuous system, only one separating gel is used with same buffer in the tanks and the gel. Whereas in the discontinuous system two-sided gel preparation with different buffers is used (Buyukkoroglu et al. 2018).

To prepare protein samples for SDS-PAGE, they are first boiled in buffer solution containing an anionic detergent (sodium dodecyl sulphate) that denatures the protein secondary and tertiary structures and opens up into a rod-shaped structure as well as gives negative charge to the molecule and a strong reducing agent such as mercaptoethanol and Dithiothreitol (DTT) that could disrupt any disulfide bridge holding protein tertiary structures. The buffer solution along with protein sample at an appropriate pH is loaded onto the top of stacking gel with large pore size to concentrate the protein into narrow bands prior to their entry into resolving gel of smaller pore size. Stacking gel is poured on top of the resolving gel and it is into this gel that the wells are formed. On applying electric field, a voltage gradient is formed between the chloride (high negative charge) and glycerine ions (low negative charge) in the electrode buffer at pH 6.8, which serves to stack the proteins into narrow bands between the ions. Migration into the resolving gel of a different pH 8.8 disrupts this voltage gradient and allows separation of the proteins into discrete bands (Nielsen 2010).

The sample buffer also contains an ionisable tracking dye, (Bromophenol blue) that monitors electrophoretic separation progress, and sucrose or glycerine, which gives the sample solution density allowing the sample to settle easily at the bottom when injected into the loading well. Being a small molecule, the dye migrates ahead of proteins and remains totally un-retarded till it reaches the bottom of gel where current is turned off. On completion of electrophoresis protein band are made visible as blue bands on clear background by treating gel with protein dye (Coomassie Brilliant Blue) followed by de-staining to removes unbound background dye from the gel. Specific antibodies or enzyme stains can be used to detect a protein of interest.

Two-Dimensional Gel Electrophoresis (2-DE)

2-DE technique was initially introduced in 1975 by P. H. O'Farrell and J. Klose (O'Farrell 1975; Klose 1975). It is one of the leading powers in proteomics and study of proteins. This is the most versatile method for fractionating and visualizing advanced protein complex extracted from cells, tissues, or alternative biological samples with an exceptional ability to separate thousands of proteins at once. It is the only currently available method which is capable of separating thousands of protein molecules by two consecutive techniques. In the first step, isoelectric focusing separates proteins according to their charge and in second step; proteins are separated according to their molecular mass in a single gel. The separated protein on the gel with isoelectric focusing is negatively charged by treatment with SDS, and the electrophoresis is performed by inserting the gel horizontally into the SDS-PAGE gel. The separated compounds are visualized by staining with Coomassie stains, silver stains or fluorescent dyes. Following separation by 2-DE, the protein spot detected on gel are cut out, de-stained, digested for further analyses of typtic peptides by peptide fingerprinting using MALDI-TOF or nano-LC-ion trap mass spectrometry or probed by antibodies, then followed by computer- assisted software for image evaluation. This technique finds wide application in detection of post- and co-translational protein modifications, study cell differentiation(Jungblut and Seifert 1990), detection of biomarkers and disease markers, drug discovery, cancer research(Wu et al. 2002), bacterial pathogenesis (Enany et al. 2013), purity checks, micro scale protein purification, and product characterization.

Although 2 DE is capable of high resolution, it has some limitations, including limited molecular mass range, poor separation of highly acidic or basic proteins, and exclusion of the majority of membrane proteins (Kline and Wu 2009). Modern proteomics approaches combine high resolution 2-DE technique with mass spectrometry using soft ionisation such as matrix-assisted laser desorption/ionisation (MALDI) or electrospray ionisation (ESI) followed by time-of-flight (TOF), ion trap or triple quadrupole detection (Ferranti et al. 2007). An alternative modern approach to 2-DE is a multi-dimensional liquid phase based separation techniques such as MudPIT (Multidimensional Protein Identification Technology) pioneered by Yates and colleagues (Washburn et al. 2001).

Chemical Methods

Foods being heterogenic materials are composed of different nutrients, such as carbohydrates, lipids and other micronutrients. There components and interactions between them may result in inaccuracy in protein content measurements. Moreover, results of these measurements are used in other calculations like measurement of enzyme activity, proteins loading of SDS-PAGE gels, such measurements are thus critical. Any error in such measurement results in errors in calculation. Thus methods that provide quicker and accurate assessment of protein concentrations are acceptable. Many chemical methods determining the amount of protein either directly or indirectly in a sample are based on different analytical principles, such as nitrogen content determination, presence of peptide bond and aromatic amino acids, dye binding capacity, light scattering and ultraviolet absorptivity properties of protein molecules.

Direct methods are based on analysis of amino acid residues in the sample while indirect protein determination relies on nitrogen content determination with subsequent conversation to protein using nitrogen-to-protein conversion factor that varies with sample type, usually 6.25 or interference from other chemical substances (Maehre et al. 2018).

Most of these are colorimetric methods, where a portion of the protein solution is reacted with a reagent that produces a coloured product which is then measured spectrophotometrically. The amount of colour relates to the amount of protein present by appropriate calibration. However, none of these methods is absolute.

Direct Protein Determination

Amino Acid Analysis

The direct method for analysis of amino acid is considered the gold standard for protein quantitation as it measures the quantity of each individual amino acid in a protein. The method is composed of three basic steps hydrolysis of peptide bonds, which link amino acids with each other and chromatographic separation of individual amino acids and detection and quantification of separated amino acids. Protein content is calculated after subtraction of molecular mass of water from sum total of individual amino acid residues. Hydrolysis of peptide bond is usually carried out under vacuum using 6 M HCl and heating at 100–165 °C for upto 72 h.

Peptide bonds being stable at neutral pH are hydrolysed by strong acids and bases. However, enzymatic hydrolysis is not so effective. The recent developments in technology are microwave assisted acid hydrolysis that speeds up the hydrolysis and replacement of ion exchange chromatography by reverse phase high performance liquid chromatographic technique (Fountoulakis and Lahm 1998). This is achieved by supplementing the existing chromatographic techniques of HPLC and GC-MS by number of new techniques such as development of new LC column, isotope dilution MS technique, Ultra HPLC combined with MS technique, capillary electrophoresis (Otter 2012).

Indirect Protein Determinations

Kjeldahl Method

This is a general chemical method of indirect total protein determination in sample by direct nitrogen measurement and came into existence in 1883 by a brewer Johan Kjeldahl, originally used for quantifying protein content of beer. The technique measures sample nitrogen as ammonia and subsequently multiplied by nitrogen-to-protein conversion factor (F) to convert the measured nitrogen to crude protein. Frequently it is given a default value of 6.25 or 5.7 (equivalent to 16% and 17.5% nitrogen content respectively). The Kjeldahl method is a three step process including: (1) sample digestion (mineralization), (2) neutralization, and distillation and (3) titration (ammonia determination) (Owusu-Apenten 2002; Sáez-Plaza et al. 2013). This method first uses sulphuric acid that causes dehydration and charring in presence of catalysts and salts converting any organically bound nitrogen in food to ammonium sulphate. The digest is then neutralized with alkali to form ammonia which is then distilled and trapped using 4% of boric acid solution. Ammonium borate ions produced are proportional to amount of nitrogen and are titrated against standardized acid in presence of suitable indicator (methylene blue).

The nitrogen content in sample that weighs m grams, is equivalent to concentration of hydrogen ions (in moles) required to reach the end and can be determined using × M HCl solution for the titration by the equation given below.

$$\%N = \frac{x \text{ moles}}{1000\text{cm}^3} * \frac{(v_s - v_b)\text{cm}^3}{\text{mg}} * \frac{14\text{g}}{\text{moles}} * 100 \tag{3}$$

where v_s and v_b are sample and blank titration volumes respectively, and 14 g is the molecular weight of nitrogen (N). Sample and reagent blank should be analysed at same time.

Though AOAC international recognises this technique as an official protein determination method (Latimer 2016) but it has some drawbacks as analytical step of nitrogen estimation suffer from non-protein nitrogenous compound interferences, the relative contents of which are higher in vegetable proteins than in animal proteins (Imafidon and Sosulski 1990) requiring different conversion factors. Also use of concentrated sulphuric acid at high temperature and catalysts poses a serious threat. This method, however, is modified to measure micro quantities of proteins such as micro and macro Kjeldhal analysis, automated Kjeldhal analysis and colorimetric Kjeldhal analysis of nitrogen (Owusu-Apenten 2002). In spite of introduction of sophisticated instrumental methods e.g., spectrophotometric, potentiometric with ion selective electrode, FIA, ion chromatography, chemiluminiscence, and others, modification of commercial sample digestion by Kjeldahl microwave digestion decreases the digestion time and improves precision and is thus considered as a worthy alternative to overcome the drawbacks of classical Kjeldhal method (Sáez-Plaza et al. 2013).

UV Absorption Method

With the advent of UV-Vis spectrometer, there has been an improvement in the analytical techniques of proteins. This method is based on the principle that most of proteins in food contain fair amount of aromatic amino acids such as tryptophan and tyrosine that have strong absorbance in ultraviolet region of electromagnetic spectrum at 280 nm. The absorbance can be used for measurement of protein concentration using Beer-Lambert's law given as:

$$A = \log\left(I_0 / I\right) = E_m cl \tag{4}$$

where, A = absorbance of solution, I_0 = intensity of light incident on sample cell, I = intensity of light exiting the sample cell, c = molar concentration of absorbing species, l = length of sample cell (cm), E_m = molar absorptivity.

It is thus clear from the Beer-Lambert law that greater the number of molecules capable of absorbing light of a given wavelength, the greater the extent of light absorption.

Since proportions of these aromatic amino acids in proteins vary, extinction coefficients (E_{280}) or molar absorptivity (E_m), which for individual proteins lies in the range 0.4–1.5, must be determined for estimation of protein content. Moreover, presence of nonprotein chromatophores such as nucleic acid contaminations that have 10 times more absorbance at this wavelength (280 nm) obscures the quantitation of protein in crude sample. This interference can be, to some degree, eliminated by measuring absorbance at 260 and 280 nm and then protein concentration (Pc) can be calculated using following equation (Warburg and Christian 1942; Layne 1957):

$$Pc\left(mg / ml\right) = 1.55 * A_{280\,nm} - 0.76 * A_{260\,nm} \tag{5}$$

Since nucleic acids have identical absorbance at 280 and 235 nm, interference by nucleic acids in protein quantification can be neglected and protein concentration can be calculated by following formula (Whitaker and Granum 1980):

$$Pc(mg/mL) = (A_{235} - A_{280})/2.51 \quad (6)$$

Where 2.51 is difference between average extinction coefficients measured at 235 and 280 nm.

Protein estimation with this method is very sensitive detecting protein concentration as low as 10 μg cm^{-3} and unlike other methods, this method is non-destructive i-e sample in cuvette can be reused and can be measured continuously such as in chromatographic column effluents (Wilson and Walker 2010). Nowadays, this method is frequently used for determination of proteins and peptides in chromatographic effluents.

There are number of other protein quantitation methods that can also be conveniently accompanied using spectroscopic methods. The Biuret method, Lowry method and Bradford method are most frequently used and reliable procedures.

Biuret Method and Its Modification

This method is based on purplish-violet color production, resulting from complex formation when cupric ions (Cu^{2+}) interact with peptide bonds under alkaline conditions (Sapan et al. 1999).The intensity of color is proportional to amount of protein present. The typical biuret reagent is mixed with a protein solution and then absorbance is measured at 549 nm after allowing standing for 15–30 minutes. This technique has its own advantage that there is no interference from materials that adsorb at lower wavelengths, and is less sensitive to protein composition but is influenced by protein purity and association state.

Compared with other colorimetric methods of protein determination, this method is, however, somewhat insensitive. A recent modification of biuret reaction is 'a reverse biuret method' that combines biuret reaction with copper (1)-bathocuproine chealate reaction and is based on reduction of the protein-complexed Cu^{+2} to cu^{+} by ascorbic acid, allowing Cu^{+} to form Cu^{+}-bathocuproine complex. Protein concentration is inversely proportional to amount of Cu^{+}-bathocuproine chelate complex formed. The sensitivity of this method is found to be higher than that of the original Lowry and CBBG·250 assay detecting down to 0.5 μg cm^{-3} (Matsushita et al. 1993)

Lowry (Folin-Ciocalteau) Method

The Lowry method developed by Lowry et al. (1951) was the first biochemical assay used for quantitation of total proteins in water or in a mild buffer solution. It is based on blue-purplish color formation when Folin-Ciocalteau phenol reagent together with mixture of copper sulphate and sodium-potassium tartarate reacts with protein solution. The reaction is basically combination of two methods:

1. Biuret method, in which a copper ion (Cu + 2) complexes with peptide bonds of protein and peptide and under appropriate alkaline conditions is reduced to Cu+, which reacts with Folin reagent, and
2. Folin-Ciocalteau reaction, which involves the reduction of phosphomolybdo-tungstate to hetero-polymolybdenum blue by the copper-catalysed oxidation of aromatic amino acids tyrosine and tryptophan (Kolakowski 2005)

This blue color can be quantified by its absorbance at 660 nm. The intensity of blue color is partly dependent on amount of tyrosine and tryptophan amino acids present in protein sample. This technique is more sensitive to low concentrations of proteins than the Biuret method detecting down to 10 μg cm^{-3} of protein. However, interference by an exhaustive number of compounds such as Tris, zwitterionic buffers such as Pipes and Hepes, and EDTA affects the effectiveness of this method. These interferences are eliminated by number of modifications including heating sample before and after treating it with Folin reagent, addition of perhydrate, SDS, chloramines-T and lipid removal by extraction with organic solvents (Kolakowski 2001).

Bradford Method (CB Dye-Binding Assay)

Bradford assay is a colorimetric method that relies on electrostatic interactions between basic amino acid molecules such as arginine, lysine and histidine with Coomassie brilliant blue G-250 (CBB) in an acidic matrix which results in a dye-protein complex with srectral shift from reddish to bluish form of dye (Redmile-Gordon et al. 2013). At low PH free dye is protonated with an absorption spectrum maximum at 465 nm. On binding to protein, a metachromatic response is observed due to formation of bluish dye-protein complex with maximum absorbance at 595 nm. At this wavelength the unprotonated species absorb. This dye protein complex absorbance, however, varies with the type of dye used. For example dye-protein complex is formed only when blue (deprotonated) form of dye binds to protein molecules, or may occur when green (protonated) form of free dye interacts with protein molecules (Atherton et al. 1996). The assay is monitored using spectrophotometer at 595 nm and can be performed in 10 min or less. The greatest advantage of this assay is its ease and high sensitivity, perceived linearity and speed of determination of protein concentration in wide variety of protein samples (Sapan et al. 1999).Moreover, this assay is comparatively resistant to interference from polyphenols, carbohydrates such as sucrose, cations such as sodium potassium present in protein sample except detergents such as sodium dodecyl sulphate.

The two measure formats of this assay are micro-assay and macro-assay that can measure protein concentration between 1 and 20 μg protein cm^{-3} and 20 and 100 μg protein cm^{-3} respectively. The assay can be modified by using perchloric acid (3.15%) or hydrochloric acid in place of perchloric acid as solvent for dye that varies maximum absorbance of dye-protein complex from 595 to 620 nm (Sedmark and Grossberg 1977).

Conclusion

Several techniques for analysis and purification of proteins have been developed and used for more than half a century and each has their own pros and cons. Recent advances in protein analysis have been mainly on innovative instrumentation design to optimize the procedures and to standardize methodology. However to draw a comparison between the latest trends of protein estimation with the available one, it is apt to say that due to the availability of newer proteins like therapeutic proteins, the need for newer sophisticated methods like affinity and chromatographic techniques, that are constantly undergoing development and improvement have risen which provides accuracy to the analysis of these micro molecules because of their high sensitivity and resolution. Hence modern techniques are necessity for their efficient analysis methodologies for sensitive molecules and to advent protein measurement beyond total nitrogen based methods.

Acknowledgments Authors are thankful to Department of Science and Technology Goverment of India for providing INSPIRE Fellowship to Mehnaza Manzoor vide letter no. DST/INSPIRE Fellowship/2017/IF170581.

References

Alomirah, H. F., Alli, I., & Konishi, Y. (2000). Applications of mass spectrometry to food proteins and peptides. *Journal of Chromatography. A, 893*(1), 1–214.

Atherton, B. A., Cunningham, E. L., & Splittgerber, A. G. (1996). A mathematical model for the description of the Coomassie Brilliant Blue protein assay. *Analytical Biochemistry, 233*, 160–168.

Barber, M., & Green, B. N. (1987). The analysis of small proteins in the molecular weight range 10-24 kDa by magnetic sector mass spectrometry. *Rapid Communications in Mass Spectrometry, 1*(5), 80–83.

Beavis, R. C., & Chait, B. T. (1989). Matrix-assisted laser-desorption mass spectrometry using 355 nm radiation. *Rapid Communications in Mass Spectrometry, 3*(12), 436–439.

Beavis, R. C., Chaudhary, T., & Chait, B. T. (1992). α-Cyano-4-hydroxycinnamic acid as a matrix for matrix assisted laser desorption mass spectrometry. *Organic Mass Spectrometry, 27*(2), 156–158.

Biemann, K. (2014). Laying the groundwork for proteomics: Mass spectrometry from 1958 to 1988. *Journal of Proteomics, 107*, 62–70.

Bonifacino, J. S., Dell'Angellica, E. C., & Springer, T. A. (2001). Immunoprecipitation. *Current Protocols in Immunology, 41*, 8.3.1–8.3.28. https://doi.org/10.1002/0471142735.im0803s41

Buyukkoroglu, G., Dora, D. D., Ozdemir, F., & Hizel, C. (2018). Techniques for protein analysis. In D. Barh & V. Azevedo (Eds.), *Omics technologies and bio-engineering: Towards improving quality of life* (Vol. 1, pp. 317–346). London, UK: Academic Press. https://doi.org/10.1016/B978-0-12-804659-3.00015-4

Buyukkoroglu, G., & Senel, B. (2018). Engineering monoclonal antibodies: Production and applications. In D. Barh & V. Azevedo (Eds.), *Omics technologies and bio-engineering: Towards improving quality of life* (Vol. 1, pp. 353–385). London, UK: Academic Press. https://doi.org/10.1016/B978-0-12-804659-3.00016-6

Careri, M., Bianchi, F., & Corradini, C. (2002). Recent advances in the application of mass spectrometry in food-related analysis. *Journal of Chromatography A, 970*, 3–64.
Chang, S. K. C. (2010). Protein analysis. In S. S. Nielsen (Ed.), *Food analysis* (p. 135). Heidelberg, Germany: Springer.
Coskun, O. (2016). Separation techniques: Chromatography. *Northern Clinics of Istanbul, 3*(2), 156–160. https://doi.org/10.14744/nci.2016.3275
Cozzolino, R., Passalacqua, S., Salemi, S., et al. (2001). Identification of adulteration in milk by matrix-assisted laser desorption/ionization time-of-flight mass spectrometry. *Journal of Mass Spectrometry, 36*(9), 1031–1037.
Creamer, J. S., Oborny, N. J., & Lunte, S. M. (2014). Recent advances in the analysis of therapeutic proteins by capillary and microchip electrophoresis. *Analytical Methods, 6*(15), 5427–5449.
Enany, S., Yoshida, Y., Magdeldin, S., et al. (2013). Two dimensional electrophoresis of the exoproteome produced from community acquired methicillin resistant Staphylococcus aureus belonging to clonal complex 80. *Microbiological Research, 11*, 504–511.
Fanali, S., Haddad, P. R., Poole, C., & Riekkola, M. L. (2017). *Liquid chromatography: applications* (2nd ed.). Amsterdam, The Netherlands: Elsevier.
Fasman, G. D. (1989). *Practical handbook of biochemistry and molecular biology*. Boca Raton, FL: CRC Press.
Ferranti, P., Mamone, G., Picariello, G., et al. (2007). Mass spectrometry analysis of gliadins in celiac disease. *Journal of Mass Spectrometry, 42*, 1531–1548.
Fountoulakis, M., & Lahm, H. W. (1998). Hydrolysis and amino acid composition analysis of proteins. *Journal of Chromatography. A, 826*, 109–134.
Freifelder, D. (1982). *Physical biochemistry. Applications to biochemistry and molecular biology*. New York, NY: W. H. Freeman.
Greenfield, N. J. (2007). Using circular dichroism spectra to estimate protein secondary structure. *Nature Protocols, 1*(6), 2876–2890.
Hawkins, B. K., & Honigs, D. E. (1987). A comparison of spectroscopic techniques for protein quantification in aqueous solutions. *American Biotechnology Laboratory, 5*, 26–37.
Ho, C. S., Lam, C. W. K., Chan, M. H. M., Cheung, R. C. K., et al. (2003). Electrospray ionisation mass spectrometry: Principles and clinical applications. *Clinical Biochemist Reviews, 24*(1), 3–12.
Imafidon, G. I., & Sosulski, F. W. (1990). Non-protein nitrogen contents of animal and plant foods. *Journal of Agricultural and Food Chemistry, 38*, 114–118.
Jadaun, G. P. S., Dixit, S., Saklani, V., et al. (2017). HPLC for peptides and proteins: Principles, methods and applications. *Pharmaceutical Methods, 8*(1), 139–144.
Johnson, W. C., Jr. (1992). Analysis of circular dichroism spectra. *Methods in Enzymology, 210*, 426–447.
Jungblut, P. R., & Seifert, R. (1990). Analysis by high-resolution two-dimensional electrophoresis of differentiation-dependent alterations in cytosolic protein pattern of HL-60 leukemic cells. *Journal of Biochemical and Biophysical Methods, 11*, 47–58.
Kaboord, B., & Perr, M. (2008). Isolation of proteins and protein complexes by immunoprecipitation. In A. Posch (Ed.), *Methods in molecular biology. 2D PAGE: Sample preparation and fractionation* (Vol. 424, pp. 349–364). Totowa, NJ: Humana Press.
Kline, J. B., & Clevenger, C. V. (2001). Identification and characterisation of the prolactin binding protein in human serum and milk. *Journal of Biological Chemistry, 278*, 24760–24766.
Kline, K. G., & Wu, C. C. (2009). MudPIT analysis: Application to human heart tissue. *Methods in Molecular Biology, 528*, 281–293.
Klose, J. (1975). Protein mapping by combined isoelectric focusing and electrophoresis of mouse tissues. A novel approach to testing for induced point mutation in mammals. *Humangenetik, 26*, 231–243.
Kolakowski, E. (2001). Protein determination and analysis in food systems. In Z. E. Sikorski (Ed.), *Chemical and functional properties of food proteins* (pp. 57–112). Lancaster, PA: Technomic Publishing Company.

Kolakowski, E. (2005). Analysis of proteins, peptides, and amino acids in foods. In S. Otles (Ed.), *Methods of analysis of food components and additives* (2nd ed., pp. 60–93). Boca Raton, FL: Taylor & Francis.

Krotz, L., Cicerci, E., & Giazzi, G. (2008). Protein determination in cereals and seeds. Food Quality, 15(4), 37–39.

Larsen, L. B., & Petersen, T. E. (1995). Identification of five molecular forms of cathepsin D in bovine milk. In K. Takahashi (Ed.), *Aspartic proteinases: Structure, function, biology, and biomedical implications* (pp. 279–283). New York, NY: Plenum Press.

Latimer, G. W. (2016). *Official methods of analysis of AOAC international*. Gaithersburg, MD: AOAC International.

Layne, E. (1957). Spectrophotometric and turbidimetric methods for measuring proteins. *Methods in Enzymology, 3*, 447–454.

Le, T. T., Zhao, D., & Larsen, L. B. (2019). Analytical methods for measuring or detecting whey proteins. In *Whey proteins* (pp. 155–184). London, UK: Academic Press. https://doi.org/10.1016/B978-0-12-812124-5.00005-9

Leonil, J., Gagnaire, V., Molle, D., et al. (2000). Application of chromatography and mass spectrometry to the characterization of food proteins and derived peptides. *Journal of Chromatography. A, 881*, 1–21.

Levinson, S. S. (2009). *Immunoelectrophoresis*. New York, NY: Wiley. https://doi.org/10.1002/9780470015902.a0001136.pub2

Li-Chan, E. C. Y., & Lacroix, I. M. E. (2018). Properties of proteins in food systems: An introduction. In R. Y. Yada (Ed.), *Proteins in food processing* (2nd ed., pp. 1–25). Duxford, UK: Woodhead Publishing. https://doi.org/10.1016/B978-0-08-100722-8.00002-4

Lowry, O. H., Rosebrough, N. J., Farr, A. L., et al. (1951). Protein measurement with the Folin phenol reagent. *The Journal of Biological Chemistry, 193*(1), 265–275.

Maehre, H. K., Dalheim, L., Edvinsen, G. K., et al. (2018). Protein determination—Method matters. *Food, 7*, 5. https://doi.org/10.3390/foods7010005

Mahmood, T., & Yang, P. C. (2012). Western blot: Technique, theory, and trouble shooting. *North American Journal of Medical Sciences, 4*(9), 429–434.

Matsushita, M., Irino, T., Komoda, T., et al. (1993). Determination of proteins by a reverse biuret method combined with the copper-bathocuproine chelate reaction. *Clinica Chimica Acta, 216*, 103–111.

Miralles, B., Ramos, M., & Amigo, L. (2003). Influence of proteolysis of milk on the whey protein to total protein ratio as determined by capillary electrophoresis. *Journal of Dairy Science, 86*, 2813–2817.

Narayanan, S. R. (1994). Preparative affinity chromatography of proteins. *Journal of Chromatography. A, 658*(2), 237–258.

Nguyen, D. N., Becker, G. W., & Riggin, R. M. (1995). Protein mass spectrometry: Applications to analytical biotechnology. *Journal of Chromatography. A, 705*(1), 21–45.

Nielsen, S. S. (Ed.). (2010). *Food analysis*. New York, NY: Springer.

Nobel, J. E., & Bailey, M. J. A. (2009). Quantitation of proteins. In R. R. Burgess & M. P. Deutscher (Eds.), *Methods in enzymology. Guide to protein purification* (Vol. 463, pp. 73–95). Amsterdam, The Netherlands: Elsevier Inc. https://doi.org/10.1016/S0076-6879(09)63008-1

O'Farrell, P. H. (1975). High resolution two-dimensional electrophoresis of proteins. *The Journal of Biological Chemistry, 250*, 4007–4021.

Otter, D. E. (2012). Standardised methods for amino acid analysis of food. *British Journal of Nutrition, 108*, S230–S237. https://doi.org/10.1017/S0007114512002486

Owusu-Apenten, R. K. (2002). Kjeldahl method, quantitative aminoacid analysis and combustion analysis. In *Food protein analysis. Quantitative effects on processing* (pp. 1–45). New York, NY: Marcel Dekker.

Peggy Hsieh, Y. H. (2010). Immunoassays. In S. S. Nelson (Ed.), *Food analysis* (4rd ed., pp. 303–311). Heidelberg, Germany: Springer.

Pelton, J. T., & McLean, L. R. (2000). Spectroscopic methods for analysis of protein secondary structure. *Analytical Biochemistry, 277*(2), 167–176.

Pospiech, M., Tremlová, B., Renčová, E., et al. (2009). Immunohistochemical detection of soya protein – Optimisation and verification of the method. *Czech Journal of Food Sciences, 27*(1), 11–19.

Redmile-Gordon, M. A., Armenise, E., White, R. P., et al. (2013). A comparison of two colorimetric assays, based upon Lowry and Bradford techniques, to estimate total protein in soil extracts. *Soil Biology and Biochemistry, 67*(100), 166–173.

Righetti, P. G., & Caravaggio, T. (1976). Isoelectric points and molecular weights of proteins: A table. *Journal of Chromatography. A, 127*(1), 1–28.

Robb, F. T., Shukla, H. D., & Clark, D. S. (2006). Heat shock proteins in hyperthermophiles. In *Methods in microbiology* (Vol. 35, pp. 233–252). London, UK: Academic Press. https://doi.org/10.1016/S0580-9517(05)35010-0

Sáez-Plaza, P., Michałowski, T., Navas, M. J., Asuero, A. G., & Wybraniec, S. (2013). An overview of the Kjeldahl method of nitrogen determination. Part II. Sample preparation, working scale, instrumental finish, and quality control. *Critical Reviews in Analytical Chemistry, 43*, 224–272. https://doi.org/10.1080/10408347.2012.751787

Sapan, C. V., Lundblad, R. L., & Price, N. C. (1999). Colorimetric protein assay techniques. *Biotechnology and Applied Biochemistry, 29*(2), 99–108.

Scopes, R. K. (1974). Measurement of protein by spectrophotometry at 205 nm. *Analytical Biochemistry, 59*, 277–282.

Sedmark, J. J., & Grossberg, S. E. (1977). A rapid, sensitive, and versatile assay for protein using Coomassie brilliant blue G 250. *Analytical Biochemistry, 79*, 544–552.

Siligardi, G., & Hussain, R. (2017). Circular dichroism, applications. In J. Lindon, G. Tranter, & D. Koppenaal (Eds.), *Encyclopedia of spectroscopy and spectrometry* (Vol. 1, 3rd ed., pp. 293–298). Oxford, UK: Elsevier.

Simonian, M. H. (2004). Spectrophotometric determination of protein concentration. *Current Protocols in Toxicology*. https://doi.org/10.1002/0471140856.txa03gs21

Sokolov, A. V., Zakahrova, E. T., Kostevich, V. A., et al. (2014). Lactoferrin, myeloperoxidase, and ceruloplasmin: Complementary gearwheels cranking physiological and pathological processes. *Biometals, 27*, 815–828.

Stoscheck, C. M. (1990). Quantitation of protein. *Methods in Enzymology, 182*, 50–68.

Warburg, O., & Christian, W. (1942). Isolation and crystallization of enolase. *Biochemische Zeitschrift, 310*, 384–421.

Washburn, M. P., Wolters, D., & Yates, I. I. I., Jr. (2001). Large-scale analysis of the yeast proteome by multidimensional protein identification technology. *Nature Biotechnology, 19*(3), 242–247.

Whitaker, J. R., & Granum, P. E. (1980). An absolute method for protein determination based on difference in absorbance at 235 and 280 nm. *Analytical Biochemistry, 109*, 156–159.

Wieser, H. (1998). Investigations on the extractability of gluten proteins from wheat bread in comparison with flour. *Zeitschrift für Lebensmitteluntersuchung und-Forschung A, 207*, 128–132.

Wieser, H. (2008). Detection of gluten. In E. K. Arendt & F. D. Bello (Eds.), *Gluten-free cereal products and beverages* (pp. 47–80). Heidelberg, Germany: Academic Press.

Wieser, H., & Antes, S. (2002). Development of a non-immunochemical method for the quantitative determination of gluten in wheat starch. In M. Stern (Ed.), *Proceedings of the 16th meeting of the working group on prolamin analaysis and toxicity* (pp. 19–23). Zwickau, Germany: Verlag Wissenschaftliche Scripten.

Wieser, H., & Seilmeier, W. (2003). Determination of gliadin and gluten in wheat starch by means of alcohol extraction and gel permeation chromatography. In M. Stern (Ed.), *Proceedings of the 17th meeting of the working group on prolamin analaysis and toxicity* (pp. 53–57). Zwickau, Germany: Verlag Wissenschaftliche Scripten.

Wiles, P. G., Gray, I. K., & Kissling, R. C. (1998). Routine analysis of proteins by Kjeldahl and Dumas methods. Review and interlaboratory study using dairy products. *Journal of AOAC International, 81*(3), 620–632.

Wilm, M., & Mann, M. (1996). Analytical properties of the nanoelectrospray ion. *Analytical Chemistry, 68*(1), 1–8. https://doi.org/10.1021/ac9509519

Wilson, K. (2010). Chromatographic techniques. In K. Wilson & J. Walker (Eds.), *Principles and techniques of biochemistry and molecular biology* (pp. 459–461). Cambridge, UK: Cambridge University Press.

Wilson, K., & Walker, J. (2010). *Principles and techniques of practical biochemistry*. Cambridge, UK: Cambridge University Press.

Wu, W., Tang, X., Hu, W., et al. (2002). Identification and validation of metastasis-associated proteins in head and neck cancer cell lines by two-dimensional electrophoresis and mass spectrometry. *Clinical & Experimental Metastasis, 11*, 319–326.

Xie, Y., Kang, R., & Tang, D. (2017). Assessment of posttranslational modifications of ATG proteins. In *Methods in Enzymology* (Vol. 587, p. 171). San Diego, CA.: ISSN 0076-6879: Academic Press. https://doi.org/10.1016/bs.mie.2016.09.057

You, W. W., Haugland, R. P., Ran, D. K., & Haugland, R. P. (1997). 3-(4-Carboxy benzyl) quinoline-2-carboxaldehyde, a reagent with broad dynamic range for the assay of proteins and lipoproteins in solution. *Analytical Biochemistry, 244*(2), 277–282.

Proteins as Enzymes

Sajad A. Rather, F. A. Masoodi, Jahangir A. Rather, Tariq A. Ganaie, Rehana Akhter, and S. M. Wani

Introduction

Living cells have the cell factories operate as a collection of efficient molecular characteristics. The success of these factories depends on the efficiency of a particular class of biomolecules-protein enzymes (Agarwal 2006). Enzymes are the complex protein molecules that catalyze chemical reactions, i.e. transformations from one or more substrates to one or more products (Bugg 2004). An integrated view of protein structure, dynamics and function is emerging, where proteins are considered as dynamically active machines and internal protein motions are closely linked to function such as enzyme catalysis (Agarwal 2006). Enzymes exhibit the physicochemical properties including solubility, electrophoretic properties, electrolytic behaviors and chemical reactivity of proteins (Lee 2006; Bhatia 2018). The sequence of amino acid of an enzyme also called as primary structure of enzyme plays an important role in enzyme function including substrate/cofactor binding or release (Yadav and Tiwari 2015). Thus the degree of biocatalytic activity chiefly depends on the integrity of the enzymes structure as a protein. The complete biochemically active enzyme is composed of a protein part (apoenzyme) with a co-enzyme or a metal ion and is called a holoenzyme. The co-enzyme in the enzyme structure may bind covalently or non-covalently to the apoenzyme. When the co-enzyme is tightly and permanently bound to protein part (apoenzyme) in this case it is known as a prosthetic group.

$$\underset{(\text{Protein})}{\text{Apoenzyme}} + \underset{(\text{Non}-\text{protein})}{\text{Prosthetic group}} = \underset{(\text{Complete Enzyme})}{\text{Holoenzyme}}$$

S. A. Rather · F. A. Masoodi (✉) · J. A. Rather · T. A. Ganaie · R. Akhter · S. M. Wani
Department of Food Science and Technology, University of Kashmir, Srinagar, India

A. Gani, B. A. Ashwar (eds.), *Food biopolymers: Structural, functional and nutraceutical properties*, https://doi.org/10.1007/978-3-030-27061-2_13

An International Commission on enzymes was established by the International Union of Biochemistry [now termed the International Union of Biochemistry and Molecular Biology (IUBMB)] in 1956 to address the problems of enzyme classification and nomenclature based on the overall chemical transformation they catalyze. Enzymes are now named and classified systematically with an EC number to a four level hierarchical description depending on the overall chemical transformation of substrates into products (Cuesta et al. 2015). The EC classification is still made on the basis of the main reaction catalyzed. The EC denotes the six classes of enzymes based on general type of reaction being carried out including (EC-1) oxidoreductases, (EC-2) transferases, (EC-3) hydrolases, (EC-4) lyases, (EC-5) isomerases, and (EC-6) ligases, where EC stands for Enzyme Commission (Kumar et al. 2015).

Enzyme function is intrinsically linked to its structure, determining how it performs substrate binding, catalysis and regulation. The amino acid-based enzymes are globular proteins that range in size from <100 to >2000 amino acid residues. These amino acids can be arranged into polypeptide chains that are folded and bent to form a specific three-dimensional structure (Robinson 2015). Some of the amino acids in enzymes are involved in binding ligands (substrates, intermediates, products, organic cofactors, metal cofactors or allosteric regulators) and some are actively involved in catalysis by interacting with the substrate, intermediate or product of the reaction (Soding et al. 2005). The structures of enzymes can be elucidating by techniques such as spectroscopic methods, X-ray crystallography and more recently, multidimensional NMR methods. The X-ray crystallography has been the most widely used technique for structural characterization of enzymes. The first enzyme to be crystallized and its structure successfully solved was chicken egg lysozyme in 1965. NMR spectroscopy is a powerful tool for elucidating the structure–function relationships of enzymes. It yields detailed information regarding structure of enzyme and the specific ligands which bind to the enzyme. The structure of the ligands at the binding sites of enzymes and the structure of enzyme–ligand complexes can also be obtained, as well as the dynamics of the ligand and the associated structure of the protein binding site (Monasterio 2014). The aim of this chapter is to present and update the existing knowledge about basic principles of enzymes such as proteinaceous nature and substrate binding, detailed description of the enzyme classification and structural characterization.

Proteinaceous Nature of Enzymes and Substrate Binding

All enzymes are proteins made up of amino acids linked together by peptide bonds except small group of RNAase molecules (Bhatia 2018). The structure and reactivity of a protein depends its amino acid sequence, called primary structure, which is genetically determined by the deoxyribonucleotide sequence in the structural gene that codes for it (Illanes 2008). The deoxyribonucleotide sequence is transcribed into a mRNA molecule. The mRNA molecule upon reaching the ribosome of cell is translated into an amino acid sequence and synthesizes a polypeptide chain. The polypeptide chain is finally transformed into a three dimensional structure, called native structure, which is having the biological functionality (Schumacher et al. 1986; Longo and Combes 1999). The secondary three-dimensional structure is the result of interactions of amino acid residues in the primary structure, mainly by hydrogen bonding of the amide groups. For the globular proteins, like enzymes, these interactions dictate a predominantly ribbon-like coiled configuration termed ɣ-helix. The tertiary three-dimensional structure is the result of interactions of amino acid residues located apart in the primary structure that produce a compact and twisted configuration in which the surface is rich in polar amino acid residues, while the inner part is abundant in hydrophobic amino acid residues. This tertiary structure is essential for the biological functionality of the protein. Some proteins have a quaternary three-dimensional structure, which is common in regulatory proteins, that is the result of the interaction of different polypeptide chains constituting subunits that can display identical or different functions within a protein complex (Dixon and Webb 1979; Creighton 1993).

In enzymes, proteins (apoenzyme) can be conjugated or associated with other molecules like, co-enzyme or co-factor or a prosthetic group (Fig. 1). However catalysis always occurs in the protein portion of an enzyme. The co-enzyme in the enzyme structure may bind covalently or noncovalently to the apoenzyme. When the co-enzyme is tightly and permanently bound to protein part (apoenzyme) in this case it is known as a prosthetic group (Yadav and Tiwari 2015). Prosthetic groups may be organic macromolecules, like carbohydrates (glycoproteins), lipids (lipoproteins) and nucleic acids (nucleoproteins), or simple in organic entities, like metalions. Prosthetic groups are tightly bound (usually covalently) to the apoenzyme and do not dissociate during catalysis (Union of Pure and Applied Chemistry

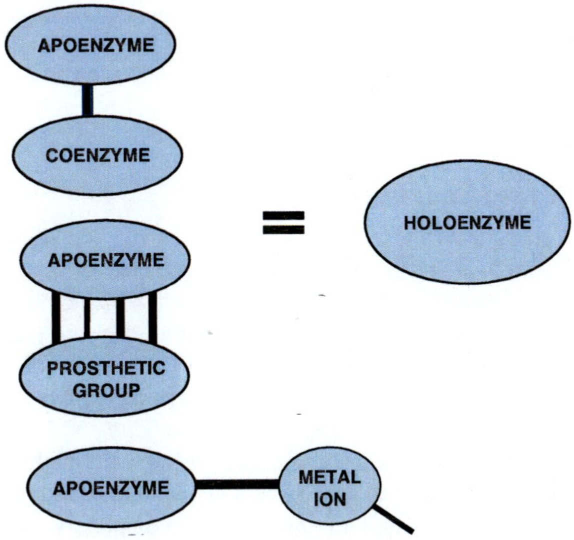

Fig. 1 The components of a holoenzyme

2005–2009; Illanes 2008). Although there are also prosthetic groups that are not cofactors (e.g. retinal in light receptors), only those prosthetic groups that are located in the active site of an enzyme are denoted cofactors. Therefore a prosthetic group is distinguished from a coenzyme in that it stays with the enzyme over many catalytic cycles, possibly until the enzyme is degraded. The coenzyme, on the other hand, binds to the enzyme at the beginning of each catalytic cycle and leaves at the end of it (Union of Pure and Applied Chemistry 2005–2009).

Small portion of the enzyme (active site) is involved in catalysis which is usually formed by very few amino acid residues. In enzymatic reaction substrate binds to the enzyme at the active site and produces changes in the distribution of electrons in its chemical bonds which lead to the reactions that result to the formation of products. The products formed are then released from the enzyme and is ready for the next catalytic cycle. It is the shape and charge properties of the active site of enzyme which enable it to bind to a specific substrate molecule, and demonstrate it specificity in catalytic activity (Whitehurst and van Oort 2009). According to the early lock and key hypothesis proposed by the German chemist Emil Fischer in 1894, the active site has a unique geometric shape that is complementary to the geometric shape of the substrate molecule that fits into it. However this rigid hypothesis hardly explains many experimental evidences of enzyme biocatalysis (Sonkaria et al. 2004). Later on through some techniques such as X ray crystallography, it became clear that enzymes are quite flexible but not rigid structures. In the light of this finding, induced-fit theory was proposed by Daniel Koshland in 1958 according to which the substrate induces a change in the enzyme conformation after binding that may orient the catalytic groups in a way prone for the subsequent reaction. This theory has been extensively used to explain enzyme catalysis (Yousef et al. 2003). Since, it is the active site alone that binds to the substrate. The rest of protein acts to stabilize the active site and provide an appropriate environment for interaction of the site with the substrate molecule (Robinson 2015). According to the transition-state theory, enzyme catalysis is the transition state complementariness, which considers the preferential binding of the transition state rather than the substrate or product (Benkovic and Hammes-Schiffer 2003).

Classification of Enzymes

Classifying enzymes in different groups based on the type of reaction they catalyze is a possible way to gain an understanding of the bonds they create or break. Classification of enzymes is developing constantly and one current issue is that the recommendations for enzyme classification and nomenclature are inappropriate for several enzyme groups (e.g. carbohydrate-active enzymes), especially in case of enzymes with multiple substrate specificity and for isoenzymes. The enzyme classification system is being constantly updated with new enzymes or corrections to existing entries and the details of recommendations for enzyme classification are provided. Because of the growing complexity in the naming of enzymes, the

International Union of Biochemistry [now termed the International Union of Biochemistry and Molecular Biology (IUBMB)] set up the Enzyme Commission (EC) for providing a systematic approach to the naming of enzymes and published first report in 1961. The sixth edition, published in 1992, contained details of nearly 3200 different enzymes, and supplements published annually have now extended this number to over 5000 (Robinson 2015). The E.C. number classification is a four level hierarchical system of an enzyme's overall reaction or function. The E.C. first level corresponds to six classes according to the type of reaction being carried out includes oxidoreductases catalyze oxidation/reduction reactions (EC 1), transferases transfer a chemical group (EC 2), hydrolases perform hydrolysis of chemical bonds (EC 3), lyases also cleave chemical bonds by other means than by oxidation or hydrolysis (EC 4), isomerases catalyze geometric and structural changes between isomers (EC 5), and ligases joins two compounds with associated hydrolysis of a nucleoside triphosphate molecule (EC6). The next two classification levels are sub-class and sub-sub-class (level 2 and level 3) depends on a various criteria such as chemical bond cleaved or formed, the reaction center, the transferred chemical group or the cofactor used for catalysis. The final level (fourth) gives a serial number for each enzyme reaction, substrate specificity. One E.C. number denotes an overall chemical reaction of an enzyme. Thus, several enzymes, which may be non-homologous, may be identified by the same E.C. number if they catalyze the same overall reaction. For example, the enzyme with the trivial name lactate dehydrogenase has the EC number 1.1.1.27, is an oxidoreductase (indicated by the first digit) with the alcohol group of the lactate molecule as the hydrogen donor (second digit) and NAD+ as the hydrogen acceptor (third digit), and is the 27th enzyme to be categorized within this group (fourth digit). The basic E.C. number classification layout of enzymes is described in Table 1.

The EC classification is still made on the basis of main reaction being catalyzed (Cuesta et al. 2015). Nowadays the assignment of EC numbers to enzyme is a common routine in the functional annotation of proteins and protein-coding genes in databases such as UniprotKB (UniProt Consortium 2013) and Ensembl (Kersey et al. 2014) and has been adopted by the widely uses Gene Ontology (GO) (Ashburner et al. 2000). However possible changes between EC classes are observed. There are some preferences such as transferases (EC 2) becoming oxidoreductases (EC 1), hydrolases (EC 3) and lyases (EC 4) (Martınez Cuesta et al. 2014). Exchanges between different EC classes suggest that the chemistry of enzymes is more complex than previously classified with close relationships between enzymes with radically different EC numbers. The substrate specificity of enzyme is represented by the last digit of the EC number, while the first three digits describe the type of the reaction. In case the sequence identity is below 70%, all the four digits of the EC number start to diverge quickly (Rost 2002). This creates an urgent need to choose alternative methods to sub-group enzymes that reflects their function or substrate specificity. The chemistry of related enzyme functions can now be explored using robust computational approaches like EC-BLAST (Rahman et al. 2014). This tool searches and compares reactions on the basis of bond charges, reaction centers, and structures of substrates and products (Cuesta et al. 2015; Rausch et al. 2005).

Table 1 The E.C. classification layout of enzymes according to the IUBMB enzyme nomenclature

Class	Sub-class	sub-sub-class	Reaction type
EC 1: Oxidoreductases			
	EC 1.1		Acting on the CH–OH group of donors
		EC 1.1.1	NAD or NADP as acceptor
		EC 1.1.2	Cytochrome as acceptor
		EC 1.1.3	Oxygen as acceptor
		EC 1.1.4	Disulfide as acceptor
		EC 1.1.5	Quinine or similar compound as acceptor
		EC 1.1.99	Other acceptors
	EC 1.2		Acting on the aldehyde or oxo group of donors
		EC 1.2.1	NAD or NADP as acceptor
		EC 1.2.2	Cytochrome as acceptor
		EC 1.2.3	Oxygen as acceptor
		EC 1.2.4	Disulfide as acceptor
		EC 1.2.7	Iron–sulfur protein as acceptor
		EC 1.2.99	Other acceptors
	EC 1.3		Acting on the CH–CH group of donors
		EC 1.3.1	NAD or NADP as acceptor
		EC 1.3.2	Cytochrome as acceptor
		EC 1.3.3	Oxygen as acceptor
		EC 1.3.5	Quinine or similar compound as acceptor
		EC 1.3.7	Iron–sulfur protein as acceptor
		EC 1.3.99	Other acceptors
	EC 1.4		Acting on the CH–NH_2 group of donor
		EC 1.4.1	NAD or NADP as acceptor
		EC 1.4.2	Cytochrome as acceptor
		EC 1.4.3	Oxygen as acceptor
		EC 1.4.4	Disulfide as acceptor
		EC 1.4.7	Iron–sulfur protein as acceptor
		EC 1.4.99	Other acceptors
	EC 1.5		Acting on the CH–NH group of donors
		EC 1.5.1	NAD or NADP as acceptor
		EC 1.5.3	Oxygen as acceptor
		EC 1.5.4	Disulfide as acceptor
		EC 1.5.5	Quinine or similar compound as acceptor
		EC 1.5.99	Other acceptors
	EC 1.6		Acting on NADH or NADPH
		EC 1.6.1	NAD or NADP as acceptor
		EC 1.6.2	Cytochrome as acceptor
		EC 1.6.4	Disulfide as acceptor
		EC 1.6.5	Quinine or similar compound as acceptor
		EC 1.6.6	Nitrogenous group as acceptor

(continued)

Table 1 (continued)

Class	Sub-class	sub-sub-class	Reaction type
		EC 1.6.8	Flavin as acceptor
		EC 1.6.99	Other acceptors
	EC 1.7		Acting on other nitrogenous compounds as donors
		EC 1.7.2	Cytochrome as acceptor
		EC 1.7.3	Oxygen as acceptor
		EC 1.7.7	Iron–sulfur protein as acceptor
		EC 1.7.99	Other acceptors
	EC 1.8		Acting on a sulfur group of donors
		EC 1.8.1	NAD or NADP as acceptor
		EC 1.8.2	Cytochrome as acceptor
		EC 1.8.3	Oxygen as acceptor
		EC 1.8.4	Disulfide as acceptor
		EC 1.8.5	Quinine or similar compound as acceptor
		EC 1.8.7	Iron–sulfur protein as acceptor
		EC 1.8.99	Other acceptors
	EC 1.9		Acting on a heme group of donors
		EC 1.9.3	Oxygen as acceptor
		EC 1.9.6	Nitrogenous group as acceptor
		EC 1.9.99	Other acceptors
	EC 1.10		Diphenols and related substances as donors
		EC 1.10.1	NAD or NADP as acceptor
		EC 1.10.2	Cytochrome as acceptor
		EC 1.10.3	Oxygen as acceptor
		EC 1.10.99	Other acceptors
	EC 1.11		Acting on a peroxide as acceptor
		EC 1.11.1	NAD or NADP as acceptor
	EC 1.12		Acting on hydrogen as donor
		EC 1.12.1	NAD or NADP as acceptor
		EC 1.12.2	Cytochrome as acceptor
		EC 1.12.99	Other acceptors
	EC 1.13		Acting on single donors with incorporation of molecular oxygen (oxygenases)
		EC 1.13.11	Incorporation of two atoms of oxygen
		EC 1.13.12	Incorporation of one atom of oxygen

(continued)

Table 1 (continued)

Class	Sub-class	sub-sub-class	Reaction type
		EC 1.13.99	Miscellaneous
	EC 1.14		Acting on paired donors, with incorporation or reduction of molecular oxygen
		EC 1.14.11	2-oxoglutarate as one donor, and incorporation of one atom each of oxygen into both donors
		EC 1.14.12	NADH2 or NADPH2 as one donor, and incorporation of two atoms of oxygen into one donor
		EC 1.14.13	NADH2 or NADPH2 as one donor, and incorporation of one atom of oxygen
		EC 1.14.14	Reduced flavin or flavoprotein as one donor, and incorporation of one atom of oxygen
		EC 1.14.15	Reduced iron–sulfur protein as one donor, and incorporation of one atom of oxygen
		EC 1.14.16	Reduced pteridine as one donor, and incorporation of one atom of oxygen
		EC 1.14.17	Ascorbate as one donor, and incorporation of one atom of oxygen
		EC 1.14.18	Another compound as one donor, and incorporation of one atom of oxygen
		EC 1.14.99	Miscellaneous
	EC 1.15		Acting on superoxide radicals as acceptor
	EC 1.16		Oxidising metal ions
		EC 1.16.1	NAD or NADP as acceptor
		EC 1.16.3	Oxygen as acceptor
	EC 1.17		Acting on CH or CH_2 groups
		EC 1.17.1	NAD or NADP as acceptor
		EC 1.17.3	Oxygen as acceptor
		EC 1.17.4	Disulfide as acceptor
		EC 1.17.99	Other acceptors
	EC 1.18		Acting on iron-sulfur proteins as donors
		EC 1.18.1	NAD or NADP as acceptor
		EC 1.18.6	Dinitrogen as acceptor
		EC 1.18.99	H+ as acceptor
	EC 1.19		Acting on reduced flavodoxin as donor
		EC 1.19.6	Dinitrogen as acceptor

(continued)

Table 1 (continued)

Class	Sub-class	sub-sub-class	Reaction type
	EC 1.20		Acting on phosphorus or arsenic in donors
	EC 1.21		Acting on X–H and Y–H to form an X–Y bond
	EC 1.22		Acting on halogen in donors
	EC 1.97		Other oxidoreductases
EC 2: Transferases			
	EC 2.1		Transferring one-carbon groups
		EC 2.1.1	Methyltransferases
		EC 2.1.2	Hydroxymethyl-, formyl- and related transferases
		EC 2.1.3	Carboxyl- and carbamoyltransferases
		EC 2.1.4	Amidinotransferases
	EC 2.2		Transferring aldehyde or ketonic groups
		EC 2.2.1	a single subclass containing the transaldolases
	EC 2.3		Acyltransferases
		EC 2.3.1	Acyltransferases
		EC 2.3.2	Aminoacyltransferases
	EC 2.4		Glycosyltransferases
		EC 2.4.1	Hexosyltransferases
		EC 2.4.2	Pentosyltransferases
		EC 2.4.99	Transferring other glycosyl groups
	EC 2.5		Transferring alkyl or aryl groups, other than methyl groups
		EC 2.5.1	A single subclass that includes a rather mixed group of such enzymes
	EC 2.6		Transferring nitrogenous groups
		EC 2.6.1	Transaminases (aminotransferases)
		EC 2.6.3	Oximinotransferases
		EC 2.6.99	Transferring other nitrogenous groups
	EC 2.7		Transferring phosphorus-containing groups
		EC 2.7.1	Phosphotransferases with an alcohol group as acceptor
		EC 2.7.2;	Phosphotransferases with a carboxyl group as acceptor
		EC 2.7.3	Phosphotransferases with a nitrogenous group as acceptor
		EC 2.7.4	Phosphotransferases with a phosphate group as acceptor
		EC 2.7.6	Diphosphotransferases
		EC 2.7.7	Nucleotidyltransferases
		EC 2.7.8	Transferases for other substituted phosphate groups
		EC 2.7.9	Phosphotransferases with paired acceptors
	EC 2.8		Transferring sulfur-containing groups
		EC 2.8.1	Sulfurtransferases
		EC 2.8.2	Sulfotransferases

(continued)

Table 1 (continued)

Class	Sub-class	sub-sub-class	Reaction type
		EC 2.8.3	CoA-transferases
	EC 2.9		Transferring selenium-containing groups
	EC 2.10		Transferring molybdenum- or tungsten-containing groups
EC 3: Hydrolases			
	EC 3.1		Acting on ester bonds
		EC 3.1.1	Carboxylic ester hydrolases
		EC 3.1.2	Thiolester hydrolases
		EC 3.1.3	Phosphoric monoester hydrolases
		EC 3.1.4	Phosphoric diester hydrolases
		EC 3.1.5	Triphosphoric monoester hydrolases
		EC 3.1.6	Sulfuric ester hydrolases
		EC 3.1.7	Diphosphoric monoester hydrolases
		EC 3.1.8	Phosphoric triester hydrolases
		EC 3.1.11	Exodeoxyribonucleases producing 5′-phosphomonoesters
		EC 3.1.13	Exoribonucleases producing 5′-phosphomonoesters
		EC 3.1.14	Exoribonucleases producing other than 5′-phosphomonoesters
		EC 3.1.15	Exonucleases active with either ribo- or deoxyribonucleic acids and producing 5′-phosphomonoesters
		EC 3.1.16	Exonucleases active with either ribo- or deoxyribonucleic acids and producing other than 5′-phosphomonoesters
		EC 3.1.21	Endodeoxyribonucleases producing 5′-phosphomonoesters
		EC 3.1.22	Endodeoxyribonucleases producing other than 5′-phosphomonoesters
		EC 3.1.25	Site-specific endodeoxyribonucleases specific for altered bases
		EC 3.1.26	Endoribonucleases producing 5′-phosphomonoesters
		EC 3.1.27	Endoribonucleases producing other than 5′-phosphomonoesters
		EC 3.1.30	Endonucleases active with either ribo- or deoxyribonucleic acid and producing 5′-phosphomonoesters
		EC 3.1.31	Endonucleases active with either ribo- or deoxyribonucleic acid and producing other than 5′-phosphomonoesters
	EC 3.2		Glycosidases
		EC 3.2.1	Hydrolysing O-glycosyl compounds
		EC 3.2.2	Hydrolysing N-glycosyl compounds
		EC 3.2.3	Hydrolysing S-glycosyl compounds
	EC 3.3		Acting on ether bonds
		EC 3.3.1	Thioether hydrolases
		EC 3.3.2	Ether hydrolases
	EC 3.4		Acting on peptide bonds (peptidases)
		EC 3.4.11	Aminopeptidases
		EC 3.4.13	Dipeptidases

(continued)

Table 1 (continued)

Class	Sub-class	sub-sub-class	Reaction type
		EC 3.4.14	Dipeptidyl-peptidases and tripeptidyl-peptidases
		EC 3.4.15	Peptidyl-dipeptidases
		EC 3.4.16	Serine-type carboxypeptidases
		EC 3.4.17	Metallocarboxypeptidases
		EC 3.4.18	Cysteine-type carboxypeptidases
		EC 3.4.19	Omega peptidases
		EC 3.4.21	Serine endopeptidases
		EC 3.4.22	Cysteine endopeptidases
		EC 3.4.23	Aspartic endopeptidases
		EC 3.4.24	Metalloendopeptidases
		EC 3.4.99	Endopeptidases of unknown catalytic mechanism
	EC 3.5		Acting on carbon-nitrogen bonds, other than peptide bonds
		EC 3.5.1	In linear amides
		EC 3.5.2	In cyclic amides
		EC 3.5.3	In linear amidines
		EC 3.5.4	In cyclic amidines
		EC 3.5.5	In nitriles
		EC 3.5.99	In other compounds
	EC 3.6		Acting on acid anhydrides
		EC 3.6.1	In phosphorus-containing anhydrides
		EC 3.6.2	In sulfonyl-containing anhydrides
	EC 3.7		Acting on carbon-carbon bonds
		EC 3.7.1	In ketonic substances
	EC 3.8		Acting on halide bonds
		EC 3.8.1	In C-halide compounds
	EC 3.9		Acting on phosphorus-nitrogen bonds
	EC 3.10		Acting on sulfur-nitrogen bonds
	EC 3.11		Acting on carbon-phosphorus bonds
	EC 3.12		Acting on sulfur-sulfur bonds
EC 4: Lyases			
	EC 4.1		Carbon-carbon lyases
		EC 4.1.1	Carboxy-lyases
		EC 4.1.2	Aldehyde-lyases
		EC 4.1.3	Oxo-acid-lyases
		EC 4.1.99	Other carbon–carbon lyases
	EC 4.2		Carbon-oxygen lyases
		EC 4.2.1	Hydro-lyases
		EC 4.2.2	Acting on polysaccharides

(continued)

Table 1 (continued)

Class	Sub-class	sub-sub-class	Reaction type
		EC 4.2.99	Other carbon–oxygen lyase
	EC 4.3		Carbon-nitrogen lyases
		EC 4.3.1	Ammonia-lyases
		EC 4.3.2	Amidine-lyases
		EC 4.3.3	Amine-lyases
		EC 4.3.99	Other carbon–nitrogen-lyases
	EC 4.4		Carbon-sulfur lyases
	EC 4.5		Carbon-halide lyases
	EC 4.6		Phosphorus-oxygen lyases
	EC 4.99		Other lyases
EC 5: Isomerases			
	EC 5.1		Racemases and epimerases
		EC 5.1.1	Acting on amino acids and derivatives
		EC 5.1.2	Acting on hydroxy acids and derivatives
		EC 5.1.3	Acting on carbohydrates and derivatives
		EC 5.1.99	Acting on other compounds
	EC 5.2		cis-trans-Isomerases
	EC 5.3		Intramolecular isomerases
		EC 5.3.1	Interconverting aldoses and ketoses
		EC 5.3.2	Interconverting keto- and enol-groups
		EC 5.3.3	Transposing C=C bonds
		EC 5.3.4	Transposing S–S bonds
		EC 5.3.99	Other intramolecular oxidoreductases
	EC 5.4		Intramolecular transferases (mutases)
		EC 5.4.1	Transferring acyl groups
		EC 5.4.2	Phosphotransferases (phosphomutases)
		EC 5.4.3	Transferring amino groups
		EC 5.4.99	Transferring other groups
	EC 5.5		Intramolecular lyases
	EC 5.99		Other isomerases
EC 6: Ligases			
	EC 6.1		Forming carbon—oxygen bonds
		EC 6.1.1	Ligases forming aminoacyl-tRNA and related compounds
	EC 6.2		Forming carbon—sulfur bonds
		EC 6.2.1	Acid–thiol ligases
	EC 6.3		Forming carbon—nitrogen bonds
		EC 6.3.1	Acid–ammonia (or amine) ligases (amide synthases)
		EC 6.3.2	Acid–amino-acid ligases (peptide synthases)
		EC 6.3.3	Cyclo-ligases

(continued)

Table 1 (continued)

Class	Sub-class	sub-sub-class	Reaction type
		EC 6.3.4	Other carbon–nitrogen ligases
		EC 6.3.5	Carbon–nitrogen ligases with glutamine as amido-N-donor
	EC 6.4		Forming carbon—carbon bonds
	EC 6.5		Forming phosphoric ester bonds
	EC 6.6		Forming nitrogen—metal bonds

Table 2 Enzyme classification attempts based on sequence similarity, structural similarity and protein descriptors

Method	Feature used	Classification accuracy/result	References
BLAST, FASTA	Sequence information	40% of enzyme classes predicted correctly	Shah and Hunter (1997)
BLAST	Sequence information	Found putative analogy of 40.5% for all EC classes	Audit et al. (2007)
Bayesian	Structural information	45% of enzyme classes predicted correctly	Borro et al. (2006)
Support vector machine	Structural properties	60% accuracy in functional annotation of enzymes	Dobson and Doig (2005)
Structure template matching	Structural information	87% accuracy in functional annotation of enzymes	Kristensen et al. (2008)
Nearest neighbor algorithm	Sequence Descriptor: Amino acid composition	95% accuracy to the level of enzyme class	Nasibov and Kandemir-Cavas (2009)
Nearest neighbor algorithm	Domain composition and pseudo amino acid composition	98% accuracy to the level of enzyme class	Cai et al. (2005)
Self-organizing maps	Reaction descriptors	Accuracies up to 92%, 80% and 70% for class, subclass and sub-subclass levels, respectively	Latino et al. (2008)
Support vector machine	Amino Acid Composition and Conjoint triad feature	81–98% accuracy in predicting the first three EC digits	Wang et al. (2011)
Recursive feature elimination technique (RFE)	sequence information	Accuracies up to 97.8%, 87.3%, and 85.6%, for the first, second and third level	Kumar et al. (2015)

For a dataset of functionally known protein sequences belonging to different enzyme groups, group-specific features can be extracted to build models using machine learning algorithms or computational approaches to predict the function of an unknown protein sequence or to assign a group label to it (Juncker et al. 2009; Ong et al. 2007). Table 2 shows the enzyme classification attempts based on sequence similarity, structural similarity and protein descriptors.

Structural Characterization of Enzymes

The proteins in enzyme molecules fold into three-dimensional structures determining how it performs substrate binding, catalysis and regulation. Some of the amino acids are involved in binding ligands (substrates, intermediates, products, organic cofactors, metal cofactors or allosteric regulators) and some are actively involved in catalysis by interacting with the substrate, intermediate or product of the reaction (Soding et al. 2005). Thus the catalytic activity of enzymes depends on the integrity of their native protein conformation. The structures of enzymes can be elucidating by techniques such as spectroscopic methods, x-ray crystallography and more recently, multidimensional NMR methods.

X-ray Crystallography

X-ray crystallography has been the most explored technique for obtaining three-dimensional structures of proteins and in particular enzymes. Knowledge of three-dimensional structures is essential to understand reaction mechanisms at the atomic level (Feiten et al. 2017). One of the pioneers of enzyme crystallography was David Blow (1931–2004); he shared the Wolf Prize in Chemistry in 1987 for this research along with David Phillips (1924–1999), who first successfully solved the structure of chicken egg lysozyme in 1965 (Helliwell 2017). The Wolf Prize 1987 citation stated *"for their contributions to protein X-ray crystallography and to the elucidation of structures of enzymes and their mechanisms of action"*. Its structure was solved to a resolution of 2°A. The diffraction of X-rays caused by a single protein molecule is too weak to be measured (Rhodes 2006). Therefore, protein crystals are used for X-ray structure determination to amplify the signal. A protein crystal contains many copies of the molecule neatly arranged in a highly ordered regular three dimensional array or crystal lattice (Rhodes 2006). The suitability of enzyme crystals for structure determination is based on their ability to interact with X-rays. In the experimental setup (Fig. 2) a narrow beam of monochromatic X-rays of suitable wavelength is directed to the crystal which either traverses straight through the crystal, in between the enzyme molecules, or hit the electron clouds of the atoms in the enzyme molecules. The molecules arranged side-by-side in a periodic way form a lattice from which the waves diffracted to the same directions accumulate and strengthen each other to produce diffraction maxima that can be recorded by sensitive detectors (Petsko and Ringe 2004). Enzyme crystals are almost invariably frozen during the X-ray crystallography achieved by directing a cold stream of nitrogen gas onto the crystal or soaking in a solution called "cryoprotectant" so that, when frozen, vitrified water, rather than crystalline ice, is formed. Freezing makes the crystal tolerant to damage by the radiation and usually allows a higher quality and higher resolution diffraction data, while providing more accurate structural information (Ilari and Savino 2008). Additionally, freezing may sometimes help in

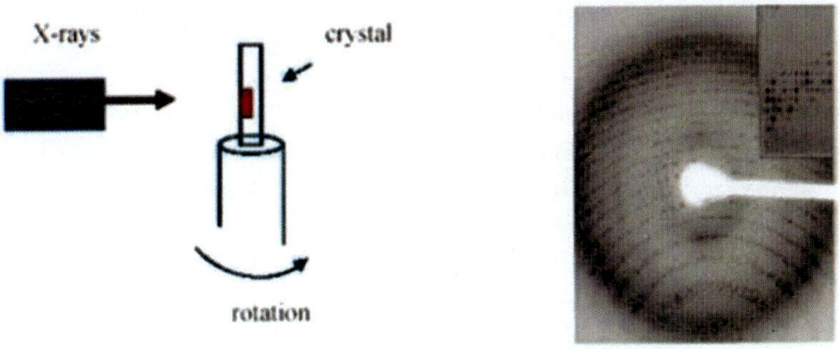

Fig. 2 Structural characterization of enzymes by X-ray crystallography

trapping substrates or other molecules that bind to the enzyme to become part of the structure, which is fundamental for structure-function studies (Rhodes 2000).

Atomic' resolution at ≥1.2°A resolution allows the placement of atoms with fewer geometrical restraints and gives a better picture of the protein structure. Advances in X-ray sources and cryo-crystallography have led to increasing numbers of structures solved at these high resolutions (Kleywegt et al. 1996). The three-dimensional representation of the protein may be displayed in a molecular structure viewer as a model that was created by the crystallographer to be chemically realistic and to match the observed electron density as precisely as possible. The resolution of a crystal structure is measured in angstrom and refers to the minimum distance between two points that can be distinguished. Although there is a large number of quality assessment methods available, resolution is a straightforward and robust parameter to assess the quality of a protein structure model (Kleywegt et al. 2004).

Nuclear Magnetic Resonance (NMR) Spectroscopy

NMR spectroscopy is a powerful tool for elucidating the structure–function relationships of substrates, peptides, proteins and in particular enzymes. It yields detailed information regarding structure of enzyme and the specific ligands which bind to the enzyme. The structure of ligands at the binding sites of enzymes and the structure of enzyme–ligand complexes can also be obtained, as well as the dynamics of the ligand and the associated structure of the protein binding site. The tertiary structures of proteins can now be obtained independently of diffraction data in solution by homo nuclear and hetero nuclear multi-dimensional NMR. In principle one can investigate the magnetic nuclei of each of the atoms within the molecule of the enzyme (^{1}H, ^{13}C, ^{15}N, …) or ligands which bind to the enzyme (^{1}H, ^{19}F, ^{31}P, ^{13}C, …), or of the environment of the active-site (solvent $^{1}H^{2}O$, $^{2}D^{2}O$, ^{23}Na, ^{39}K, ^{35}Cl, …) (Monasterio 2014). Until recently, NMR spectroscopy has yielded structures of protein complexes with small and medium size (~30 to 40 kDa). Major breakthroughs during recent past especially in isotope-labeling techniques, have enabled NMR characterization of large protein systems with molecular weights of hundreds of kDa. This has provided unique insights into the binding, dynamic, and allosteric properties of enzymes (Huang and Kalodimos 2017).

The useful approach to study enzyme structure by protein NMR is the observation of the resonances from histidine. The C-2 and C-5 proton resonances are downfield from the aromatic protons (Markley 1975). The classical use of these properties was with the small enzyme (Mr = 23,500) RNAase (Meadows and Jardetzky 1986) and the large enzyme (Mr = 237,000) pyruvate kinase (Meshitsuka et al. 1981). The C-2 proton resonance is especially sensitive to the ionization state of the imidazole nitrogens, thus the pKa for each individual histidine within the native enzyme can be obtained from titration studies. The binding of a ligand or metal ion to a specific histidine or histidines could result in a change in the magnetic environment (chemical shift) of the resonance and an alteration in the pKa. This application of NMR has been useful in some limited number of enzymes. Enzymes enriched with ^{13}C and ^{15}N have been used to increase the range of chemical shifts of these nuclei in order to enhance spectral dispersion and increases the possibility of resolving more resonances. The detailed structural and dynamic studies of larger proteins have been done with ^{13}C and ^{15}N isotope labels through NMR and nuclear Overhauser effect (Redfield et al. 1989). This type of studies is routine for determining the structure of enzymes and their dynamics using multidimensional NMR (Kevin et al. 1998; Bachovchin 2001). An alternative approach is use of a reporter group such as ^{19}F on the enzyme or on the substrate to obtain information regarding enzyme structure and the effects of ligand binding on the enzyme (Geric 1981; Danielson and Falke 1996). ^{19}F nucleus is 83% as sensitive as ^{1}H, and has a large range of chemical shifts in addition there are no back ground resonances of ^{19}F to cause interference. The ^{19}F reporter groups can be incorporated by different methods. A fluorinated amino acid i.e. fluorotyrosine, fluor-oalanine can be added to growth medium and incorporated into the protein (Sykes and Weiner 1980). The amino acids i.e. tyrosines, alanines containing the ^{19}F are labeled and will exhibit a resonance. The hetero dimer of tubulin, the principal protein of microtubules, fluoro tyrosine can be incorporated to α-subunit on the C-terminal amino acid through the reaction catalyzed by tubulin–tyrosine-ligase (Monasterio et al. 1995). An alternative approach is to covalently label the enzyme at a specific residue with a fluorine-containing reagent like trifluoroacetic anhydride, trifluoroacetyliodide, or 3-bromo-1, 1, 1-trifluoro-propanone. The chemical shift and/or the line width (1/T2) of the ^{19}F label, a "reporter" for a change in the enzyme structure, must reflect ligand binding and/or catalysis. In case ^{19}F resonance is sensitive to conformational changes in the enzyme then site-specific modification of groups at the active site will be reflected by changes in the ^{19}F resonance. The method of using reporter groups can be also be elucidated by using other labels like ^{2}H or ^{13}C labels. However, most other labels are less sensitive than fluorine. A potential strength of using these labels is the incorporation of ^{2}H for ^{1}H or ^{13}C for ^{12}C into the protein will have a very minor, if any, effect on the protein itself. Use of reporter groups yield information regarding the environment of the group. But not the specific structural features of the enzyme, comparative structural changes can be studied by photo-chemically induced nuclear polarization (photo

CIDNP) originating from free radical reactions. This has been developed as a sensitive method to measure structural changes on the surface of proteins (Kaptein 1982; Berliner 1989). Photo-chemically induced nuclear polarization (photo CIDNP) requires a modified spectrometer and a proper light source (laser) to begin to probe surface changes. This technique has the advantage of high sensitivity, and it yields general conformation information (Monasterio 2014).

Conclusions

Enzymes are proteins responsible for catalysis of biochemical reactions. The classification information-rich EC number given by the Enzyme Commission as a simple identifier still persists. However robust approaches to quantitatively compare catalytic reactions or to accurately predict enzyme mechanisms are just beginning to appear. Further combining bond changes and reaction centers with structural information about the substrates, products and mechanisms are needed to capture the essence of enzyme chemistry in a functional classification.

X-ray Crystallography and NMR are most explored technique for structural characterization of proteins and in particular enzymes. Recent technical advances in crystallography, as well as better computational programs have made it much more rapid in solving enzyme crystal structures. Modern NMR spectroscopy techniques make extensive use of isotopically enriched proteins and should prove a powerful approach for structural characterization of proteins in particular enzymes in the future. Further technological advances are needed to establish NMR as the primary tool for obtaining atomic structures of challenging systems with even higher complexity. The accumulating data on enzyme structures—and novel approaches, particularly genome projects and bioinformatics—are expected to increase our understanding of enzyme function and mechanisms in the future.

References

Agarwal, P. K. (2006). Enzymes: An integrated view of structure, dynamics and function. *Microbial Cell Factories, 5*, 2. https://doi.org/10.1186/1475-2859-5-2

Ashburner, M., Ball, C. A., & Sherlock, G. (2000). The Gene Ontology Consortium. Gene ontology: tool for the unification of biology. *Nature Genetics, 25*, 25–29.

Audit, B., Levy, E. D., Gilks, W. R., Goldovsky, L., & Ouzounis, C. A. (2007). CORRIE: Enzyme sequence annotation with confidence estimates. *BMC Bioinformatics, 8*(Suppl 4), S3.

Bachovchin, W. W. (2001). Contributions of NMR spectroscopy to the study of hydrogen bonds in serine protease active sites. *Magnetic Resonance in Chemistry, 39*, 199–213.

Benkovic, S. J., & Hammes-Schiffer, S. (2003). A perspective on enzyme catalysis. *Sci, 301*(5637), 1196–1202.

Berliner, L. J. (1989). Laser chemically induced dynamic nuclear polarization studies in proteins: α-lactalbumin. *Archivos de Biología y Medicina Experimentales, 22*, 123–128.

Bhatia, S. (2018). Enzymes, proteins and bioinformatics. In *Introduction to pharmaceutical biotechnology* (2nd ed.). Bristol, UK: IOP Publishing. https://doi.org/10.1088/978-0-7503-1302-5ch1

Borro, L. C., Oliveira, S. R. M., Yamagishi, M. E. B., Mancini, A. L., Jardine, J. G., Mazoni, I., … Neshich, G. (2006). Predicting enzyme class from protein structure using Bayesian classification. *Genetics and Molecular Research, 5*(1), 193–202.

Bugg, T. D. H. (2004). *Introduction to enzyme and coenzyme chemistry* (2nd ed.). Oxford, UK: Blackwell Publishing Ltd.

Cai, Y. D., Zhou, G. P., & Chou, K. C. (2005). Predicting enzyme family classes by hybridizing gene product composition and pseudo-amino acid composition. *Journal of Theoretical Biology, 234*(1), 145–149.

Creighton, T. E. (1993). *Proteins* (2nd ed.). New York, NY: W. H. Freeman and Company, 507pp.

Cuesta, S. M., Rahman, S. A., Furnham, N., & Thornton, J. M. (2015). The classification and evolution of enzyme function. *Biophysical Journal, 109*, 1082–1086.

Danielson, M. A., & Falke, J. J. (1996). Use of 19F NMR to probe protein structure and conformational changes. *Annual Review of Biophysics and Biomolecular Structure, 25*, 163–195.

Dixon M, Webb EC (1979) Enzymes 3rd ed. Academic Press, New York, NY, 1116pp. Drauz K, Waldmann H (2002) Enzyme catalysis in organic synthesis, vol II. Weinhein, Germany: Wiley VCH, 998pp.

Dobson, P. D., & Doig, A. J. (2005). Predicting enzyme class from protein structure without alignments. *Journal of Molecular Biology, 345*(1), 187–199.

Feiten, M. C., Luccio, M. D., Santos, K. F., de Oliveira, D., & Oliveira, J. V. (2017). X-ray crystallography as a tool to determine three-dimensional structures of commercial enzymes subjected to treatment in pressurized fluids. *Applied Biochemistry and Biotechnology, 182*(2), 429–451.

Geric, J. T. (1981). Fluorine magnetic resonance in biochemistry. In L. J. Berliner & J. Reuben (Eds.), *Biological magnetic resonance* (Vol. 1, pp. 139–203). New York, NY: Plenum Press.

Helliwell, J. R. (2017). New developments in crystallography: exploring its technology, methods and scope in the molecular biosciences. *Bioscience Reports, 37*, BSR20170204. https://doi.org/10.1042/BSR20170204

Huang, C., & Kalodimos, C. G. (2017). Structures of large protein complexes determined by nuclear magnetic resonance spectroscopy. *Annual Review of Biophysics, 46*, 317–336.

Ilari, A., & Savino, C. (2008). Protein structure determination by X-ray crystallography. In J. M. Keith (Ed.), *Bioinformatics, volume I: Data, sequence analysis, and evolution*. Totowa, NJ: Humana Press. https://doi.org/10.1007/978-1-60327-159-2

Illanes, A. (2008). *Enzyme biocatalysis: Principles and applications*. Dordrecht, Netherlands: Springer.

Juncker, A. S., Jensen, L. J., Pierleoni, A., Bernsel, A., Tress, M. L., Bork, P., … Brunak, S. (2009). Sequence-based feature prediction and annotation of proteins. *Genome Biology, 10*(2), 206.

Kaptein, R. (1982). Photo-CIDNP studies. In L. J. Berliner & J. Reuben (Eds.), *Proteins in biological magnetic resonance* (Vol. 4, pp. 145–191). New York, NY: Plenum Press.

Kersey, P. J., Allen, J. E., & Staines, D. M. (2014). Ensembl Genomes 2013: scaling up access to genome-wide data. *Nucleic Acids Research, 42*, D546–D552.

Kevin, H., Lewis, G., & Kay, E. (1998). The use of 2H, 13C, 15N multidimensional NMR to study the structure and dynamics of proteins. *Annual Review of Biophysics and Biomolecular Structure, 27*, 357–406.

Kleywegt, G., Hoier, H., & Jones, T. (1996). A re-evaluation of the crystal structure of chloromuconate cycloisomerase. *Acta Crystallographica. Section D, Biological Crystallography, 52*, 858–863.

Kleywegt, G. J., Harris, M. R., Zou, J. Y., Taylor, T. C., Wahlby, A., & Jones, T. A. (2004). The Uppsala Electron-Density server. *Acta Crystallographica. Section D, Biological Crystallography, 60*, 2240–2249.

Kristensen, D. M., Ward, R. M., Lisewski, A. M., Erdin, S., Chen, B. Y., Fofanov, V. Y., … Lichtarge, O. (2008). Prediction of enzyme function based on 3D templates of evolutionarily important amino acids. *BMC Bioinformatics, 9*, 17.

Kumar, S., Bhola, A., & Tiwari, A. K. (2015). Classification of enzyme functional classes and subclasses using support vector machine. In: *2015 1st International Conference on Futuristic trend in Computational Analysis and Knowledge Management (ABLAZE-2015)*.

Latino, D. A. R. S., Zhang, Q. Y., & Aires-De-Sousa, J. (2008). Genome-scale classification of metabolic reactions and assignment of EC numbers with self-organizing maps. *Bioinformatics, 24*(19), 2236–2244.

Lee, H. C. (2006). Structure and enzymatic functions of human CD38. *Molecular Medicine, 12*, 317–323.

Longo, M., & Combes, D. (1999). Thermostability of modified enzymes: A detailed study. *Journal of Chemical Technology and Biotechnology, 74*, 25–32.

Markley, J. L. (1975). Observation of histidine residues in proteins by means of nuclear magnetic resonance spectroscopy. *Accounts of Chemical Research, 8*, 70–80.

Martınez Cuesta, S., Furnham, N., & Thornton, J. M. (2014). The evolution of enzyme function in the isomerases. *Current Opinion in Structural Biology, 26*, 121–130.

Meadows, D. H., & Jardetzky, O. (1986). Nuclear magnetic resonance studies of the structure and binding sites of enzymes IV. Cytidine 30-monophosphate binding to ribonuclease. *Proceedings of the National Academy of Sciences, 61*, 406–413.

Meshitsuka, S., Smith, G. M., & Mildvan, A. S. (1981). Proton NMR studies of the histidine residues of rabbit muscle pyruvate kinase and of its phosphoenol pyruvate complex. *The Journal of Biological Chemistry, 256*, 4460–4465.

Monasterio, O. (2014). Nomenclature for the applications of nuclear magnetic resonance to the study of enzymes. *Perspectives on Science, 1*, 88–97.

Monasterio, O., Nova, E., Lopez-Brauet, A., & Lagos, R. (1995). Tubulin–tyrosine ligase catalyzes covalent binding of mfluorotyrosine to tubulin. Kinetic and 19F-NMR Studies. *FEBS Letters, 374*, 165–168.

Nasibov, E., & Kandemir-Cavas, C. (2009). Efficiency analysis of KNN and minimum distance-based classifiers in enzyme family prediction. *Computational Biology and Chemistry, 33*(6), 461–464.

Ong, S. A., Lin, H. H., Chen, Y. Z., Li, Z. R., & Cao, Z. (2007). Efficacy of different protein descriptors in predicting protein functional families. *BMC Bioinformatics, 8*, 300.

Petsko, G. A., & Ringe, D. (2004). *Protein structure and function*. London, UK: New Science Press. 195pp.

Rahman, S. A., Cuesta, S. M., & Thornton, J. M. (2014). EC-BLAST: A tool to automatically search and compare enzyme reactions. *Nature Methods, 11*, 171–174.

Rausch, C., Weber, T., Kohlbacher, O., Wohlleben, W., & Huson, D. H. (2005). Specificity prediction of adenylation domains in nonribosomal peptide synthetases (NRPS) using transductive support vector machines (TSVMs). *Nucleic Acids Research, 33*(18), 5799–5808.

Redfield, A. G., Mclntosh, L. P., & Dahlquist, F. W. (1989). Use of 13C and 15N isotope labels for proton nuclear magnetic resonance and nuclear Overhauser effect. Structural and dynamic studies of larger proteins and nucleic acids. *Archivos de Biología y Medicina Experimentales, 22*, 129–138.

Rhodes, G. (2000). *Crystallography made crystal clear*. San Diego, CA: Academic Press, 269pp.

Rhodes, G. (2006). *Crystallography made crystal clear–A guide for users of macromolecular models* (3rd ed.). London, UK: Academic Press Publications.

Robinson, P. K. (2015). Enzymes: Principles and biotechnological applications. *Essays in Biochemistry, 59*, 1–41.

Rost, B. (2002). Enzyme function less conserved than anticipated. *Journal of Molecular Biology, 318*(2), 595–608.

Schumacher, G., Sizmann, D., & Haug, H. (1986). Penicillin acylase from *E. coli*: Unique gene–protein relation. *Nucleic Acids Research, 14*(14), 5713–5727.

Shah, I., & Hunter, L. (1997). Predicting enzyme function from sequence: A systematic appraisal. *Proceedings of the International Conference on Intelligent Systems for Molecular Biology, 5*, 276–283.

Soding, J., Biegert, A., & Lupas, A. (2005). The HHpred interactive server for protein homology detection and structure prediction. *Nucleic Acids Research, 33*, 244–248.

Sonkaria, S., Boucher, G., & Fl'orez-Alvarez, J. (2004). Evidence for 'lock and key character in an anti-phosphonate hydrolytic antibody catalytic site augmented by non-reaction centre recognition: Variation in substrate selectivity between an anti-phosphonate antibody, an anti-phosphate antibody and two hydrolytic enzymes. *The Biochemical Journal, 381*, 125–130.

Sykes, E. D., & Weiner, J. H. (1980). Biosynthesis and 19F NMR characterization of fluoro amino acid containing proteins. In J. S. Cohen (Ed.), *Magnetic resonance in biology* (Vol. 1, pp. 1–196). New York, NY: Wiley.

Union of Pure, I. & Applied Chemistry. (2005–2009). IUPAC compendium of chemical terminology - The gold book. http://goldbook.iupac.org/.

UniProt Consortium. (2013). Update on activities at the Universal Protein Resource (UniProt) in 2013. *Nucleic Acids Research, 41*, D43–D47.

Wang, Y. C., Wang, Y., Yang, Z. X., & Deng, N. Y. (2011). Support vector machine prediction of enzyme function with conjoint triad feature and hierarchical context. *BMC Systems Biology, 5*, S6.

Whitehurst, R. J., & van Oort, M. (2009). *Enzymes in Food Technology* (2nd ed.). Chichester, UK: Wiley-Blackwell.

Yadav, S. K., & Tiwari, A. K. (2015). Classification of enzymes using machine learning based approaches: A review. *Machine Learning and Applications: An International Journal, 2*, 30–49.

Yousef, M. S., Clark, S. A., Pruett, P. K., Somasundaram, T., Ellington, W. R., & Chapman, M. S. (2003). Induced fit in guanidino kinases – Comparison of substrate-free and transition state analog structures of arginine kinase. *Protein Science, 12*, 103–111.

Exogenous Enzymes

Saqib Farooq, Manzoor Ahmad Shah, Tariq Ahmad Ganaie, and Shabir Ahmad Mir

Introduction

Enzymes, also known as biocatalysts are the large biomolecules required for the numerous chemical inter-conversions that sustain life. They accelerate all the metabolic processes in the body and carry out a specific task (Gurung et al. 2013). Certain enzymes are of special interest and are utilized as organic catalysts in numerous processes on an industrial scale. Enzyme-mediated processes are rapidly gaining interest because of reduced process time, intake of low energy input, cost effective, nontoxic and eco-friendly characteristics (Li et al. 2012; Choi et al. 2015).

Enzymes can be obtained from different sources such as plants, animals and microorganisms. Microbial enzymes are known to be superior enzymes, particularly for applications in industries on commercial scales. Many enzymes from microbial sources are already being used in various commercial processes. Selected microorganisms including bacteria, fungi and yeasts have been globally studied for the bio-synthesis of economically viable preparations of various enzymes for commercial applications (Pandey et al. 1999).

Global market for industrial enzymes was estimated about \$4.2 billion in 2014 and expected to develop at a compound annual growth rate of approximately 7% over the period from 2015 to 2020 to reach nearly \$6.2 billion (Markets and Markets Watch 2015). Part of this market is ascribed to enzymes used in large-scale

S. Farooq · T. A. Ganaie (✉)
Department of Food Technology, Islamic University of Science and Technology, Awantipora, Jammu and Kashmir, India

M. A. Shah
Department of Food Science and Technology, Government PG College for Women, Gandhi Nagar, Jammu, Jammu and Kashmir, India

S. A. Mir
Department of Food Science and Technology, Government College for Women, M. A. Road, Srinagar, Jammu and Kashmir, India

A. Gani, B. A. Ashwar (eds.), *Food biopolymers: Structural, functional and nutraceutical properties*, https://doi.org/10.1007/978-3-030-27061-2_14

applications, among them are those used in food and feed applications (Binod et al. 2008). These include enzymes used in baking, beverages and brewing, dairy and meat industry, dietary supplements, as well as fats and oils (Berka and Cherry 2006; Kirk et al. 2002).

Moreover, with the advent of recombinant DNA technology and protein engineering a microbe can be manipulated and cultured in large quantities to meet increased demand (Liu et al. 2013). Associated driving factors that motivate the use of microbial enzymes in industrial applications are increasing demand of consumer goods, need of cost reduction, natural resources depletion, and environmental safety (Choi et al. 2015).

Production of Exogenous Enzymes

Microorganisms are being the most important source of commercial enzymes today. Enzyme manufacturers have optimized microorganisms for the production of enzymes through natural selection and classical breeding techniques (Agarwal and Sahu 2014). A few years later, for the first time, an enzyme (a protease) was produced by fermentation of *Bacillus licheniformis*. In this, way, large-scale production of enzymes became possible, thus facilitating the industrial application of enzymes (Chaudhary et al. 2015). The primary source of industrial enzymes is microorganisms, out of which, 50% originate from fungi and yeast, 35% from bacteria, while the remaining 15% are either of plant or animal origin (Anisa and Girish 2014).

Amylolytic Enzymes

Amylase can be obtained from different species of microorganisms, but for commercial use, α-amylase derived from *Bacillus licheniformis*, *Bacillus stearothermophilus*, and *Bacillus amyloliquefaciens* has number of applications in different industries such as in food, fermentation, textiles and paper industries (Konsoula and Liakopoulou-Kyriakides 2007; Pandey et al. 2000). Fungal enzymes are limited to terrestrial isolates, mostly to *Aspergillus and Penicillium* (Kathiresan and Manivannan 2006). *Aspergillus oryzae* is considered to be the favorable host for the production of commercial enzymes including α-amylase (Jin et al. 1998).

Several bacterial isolates were isolated from Egyptian soil and were capable to grow and produce amylases. Among these isolates, *Baccilus amyloliquefaciens* was found to produce the highest amylases activity. For the production of amylases, nine agro-industrial residues were added as carbon sources to the basal medium. The medium supplemented with potato starchy waste as the sole carbon source enhanced the enzyme activity more than soluble starch as control for α, β and γ amylases activity, as it increased by *B. amyloliquefaciens* about 1.26 and 4 and eightfold, respectively after 48 h at 50 °C using rotary shaker at 150 rpm. *B. amyloliquefaciens* gave the maximum values of α, β and γ amylases activity on medium supplemented with 2% potato starchy waste after 30, 30, and 36 h of fermentation periods at 50 °C using shake flasks technique as a batch culture. These values were 155.2 U mL^{-1}, 1.0 U mL^{-1} and 2.4 U mL^{-1}, respectively. It could be stated that productive medium supplemented with 2% potato starchy waste as a low price substrate could be more favorable than basal medium containing 1% starch for amylases production in submerged fermentation, as it increased α, β and γ amylase activity by 1.98, 7.69 and 12-fold than that produced in basal medium (control), respectively (Abd-Elhalem et al. 2015).

Residues of wheat processing were used as substrate for amylase production. The medium was supplemented (in g L^{-1}) of: peptone 24; urea 1.9; glycerol 1.5; KH_2PO_4 0.6 MgSO4 0.5; $(NH_4)_2SO_4$ 0.25 and distilled water. The initial pH was 5.0, and the medium was inoculated with 1.10^6 spores mL^{-1} of a spore suspension from *Aspergillus oryzae* NRRL 695. Submerged fermentation was carried out in a rotary shaker (150 rpm) at 30 °C for 96 h (Kammoun et al. 2008). In another study Ramachandran et al. (2004) investigated the production of α-amylases under solid-state fermentation (SSF) by *Aspergillus oryzae* using coconut oil cake (a by-product obtained after oil extraction from dried copra) as substrate. It contains starch, soluble sugars, soluble proteins, lipids and trace amounts of nitrogen. They achieved 3388 U gds^{-1} when coconut oil cake was supplemented by 1% of peptone in 72 h of fermentation at 30 °C, carried out with 2 mL spore suspension (6.108 spores mL^{-1}) with the initial moisture content of 66%.

Anto et al. (2006) analyzed starch content of the raw materials and correlated it with the glucoamylase production under solid-state fermentation. Wheat bran (75.6% of starch content) and coarse waste (71.1% of starch content) presented highest glucoamylase production (264 U gds^{-1} and 211.5 U gds^{-1} respectively) compared to the enzyme production with rice powder (55.8%), medium waste (48.6%) and fine waste (34.2%). Higher enzyme production using wheat bran and coarse waste can be correlated with their starch content. The production media contained solid substrate and mineral solution (1:2 m/v), pH 7.0, inoculate with 106 spores mL^{-1} collected from 72 h grown culture of *Aspergillus* sp. HA-2.

Proteolytic Enzymes

Proteases are important industrial enzymes synthesized by different types of microbes like fungi, bacteria and yeasts. Because of their rapid growth, less space requirement for their cultivation, microbes serve as a preferred source of protease enzymes (Anwar and Saleemuddin 2000). Microbial proteases represent one of the largest classes of industrial enzymes, accounting for 40% of the total worldwide sales of enzymes by value Santhi (2014). *Bacillus* sp. is one among the best protease producers. Several *Bacillus* species involved in protease production are *B. sterothermophilus*, *B. cereus*, *B. megaterium*, *B. mojavensis*, and *B. subtilis* (Anwar and Saleemuddin 2000).

The microbial production of protease by *Bacillus Cereus* using cassava waste water was studied by (Santhi 2014) and it was reported that maximum protease production was observed at 48 h with pH 7; increase in production was observed when glucose and peptone were used as carbon and nitrogen sources in the production medium respectively. Further, cassava waste water was also used as an alternative carbon source for enzyme production, showing a maximum of protease production when supplemented with 6% in the production medium.

An alkaline protease produced by a thermophilic bacteria *Bacillus subtilis*DM-04 was studied by Mukherjee et al. (2008) in SSF using potato peel (51.7% of carbon; 2.6% of nitrogen and 19.9% of C/N ratio) achieving 400 U gds^{-1}. The higher alkaline protease (2382 U gds^{-1}) was obtained with a mixture of potato peel and *Imperata cylindrica* grass in the proportion 1:1 in tray-type bioreactor recovered with aluminum foil, incubated at 50 °C during 24 h. Low levels of acid and alkaline proteases were produced during fermentation using residues (cassava peel, corn cob, corn husk, oat husk and sugar cane bagasse) by *Penicillium janthinellum* CRC87M-115 (Oliveira et al. 2006). In another study alkaline protease production under SSF was investigated using isolated alkalophilic *Bacillus sp.* and green gram husk as substrate. Maltose and yeast extract supplementation increased protease production. The fermentation was conducted at 33 °C, pH 9.0, and moisture content (140%) during 60 h. Protease activity reached more than 35,000 U g^{-1}(Prakasham et al. 2006). Futhermore, SSF was conducted with rice bran and *Rhizopus oligosporus* ACM 145F, incubated at 37 °C for 72 h, achieved a maximal production of an acid protease (1.6 U mL^{-1}) at pH 2.0, decreasing its activity in pH values above 5.0 (Ikasari and Mitchell 1996). In addition to this *Pseudomonas aeruginosa* was a strain isolated from the tannery wastewater for its ability to produce alkaline protease (1160–1175 U mL^{-1}) (Kumar et al. 2008).

Cellulases

Cellulases are inducible enzymes synthesized by a large diversity of microorganisms including both fungi andbacteria during their growth on cellulosic materials (Kubicek 1993; Lee and Koo 2001). These microorganisms can be aerobic, anaerobic, mesophilic or thermophilic. Among them, the genera of *Clostridium, Cellulomonas, Thermomonospora, Trichoderma,* and *Aspergillus* are the most extensively studied cellulose producers (Sun and Cheng 2002; Kuhad et al. 1999). Cellulolytic microbes are primarily carbohydrate degraders and are generally unable to use proteins or lipids as energy sources for growth (Sukumaran et al. 2005).

Rodrigues (2011) reported that different concentrations of cellulose, ranging from 20 g·L^{-1} to 60 g·L^{-1}, were assayed as the sole carbon source of the growth medium of *Aspergillus terreus* A-1 and N-Y strains. The activity of cellulases produced by *Aspergillus terreus* A-1 strain had a maximum (13.2 U/mL) at the concentration of 30 g × L^{-1} of cellulose. On the other hand, the maximum production of cellulases by N-Y strain (10.2 U/mL) was obtained on 20 g × L^{-1} of cellulose. For both strains the maximum cellulasic activity occurred between 72 h and 96 h of incubation at 30 °C.

Waseem et al. (2014) reported that CB-2 and CB-3 strains of *Bacillus subtilis* were used for the flask-scale production of cellulase through submerged fermentation. Results revealed the highest cellulolytic activity (CMCase) with 120.321 U/ml and filter paper activity (FPase) with 1.076 U/ml by CB-2 strain followed by CB-3. Optimum temperature and pH of the medium for cellulase production was 37.5 °C, pH and 9 respectively, with 2% untreated cotton stalk as carbon source, yeast as organic nitrogen source and ammonium sulphate as inorganic nitrogen source with 3% inoculum size.

Ariffin et al. (2006) conducted a study to produce cellulase by local isolate *Bacillus pumilus* EB3, using carboxymethyl cellulose (CMC)as substrate. Following that, cellulase produced from *Bacillus pumilus* EB3 was purified using ion exchange chromatography with anion exchanger (HiTrap QXL) for characterization of the cellulase. Cellulase was successfully produced in 2 L stirred tank reactor (STR) with the productivity of 0.53, 3.08 and 1.78 U/L h and the maximum enzyme activity of 0.011, 0.079 and 0.038 U/mL for FPase, CMCase and β-glucosidase, respectively. Purification of cellulase from *Bacillus pumilus* EB3 using ion exchange chromatography showed that 98.7% of total CMCase was recovered.

Xylanases

Xylanase production has been documented in a wide spectrum of microorganisms, including bacteria, actinomycetes, yeasts and filamentous fungi (Nascimento et al. 2003; Bakri et al. 2008). Studying the effect of orange pomace, orange peel, lemon pomace, lemon peel, apple pomace, pear peel, banana peel, melon peel and hazelnut shell on the production of xylanase, using *Trichoderma harzianum* 1073-D3, Seyis and Aksoz (2005) discovered that molasses are able to reduce the time of production in 50% when used as an additional carbon source. The maximum activity has been observed on 2.5% melon peel medium (26.5 U mg^{-1} of protein)incubated at 30 °C for 7 days on a rotary shaker (150 rpm).

Rose and Van Zyl (2008) optimized xylanase production using a recombinant strain of *Aspergillus niger* D15[xyn2]pyrG$^-$. The highest xylanase activities of 226 and 209 U mL^{-1} were produced with 20 and 30% molasses, respectively, at 30 °C, pH 6.5, agitation of 100 rpm and a spore inoculum of 1.106 spores mL^{-1}. In another study Maciel et al. (2009) also produced xylanase with sugarcane bagasse under SSF by *Aspergillus niger* LPB 326. The highest xylanase activity was 2327 U gdm^{-1} using 65% of sugarcane bagasse and 35% of soybean meal supplemented with a mineral salt solution, 85% initial moisture, 106 spores gdm^{-1},at 30 °C for 4 days. Dobrev et al. (2007) also used corn cobs in medium composition for increasing xylanase production by *Aspergillus niger* B03. The optimization process was performed. The fermentation was carrying out in flasks inoculated with 10% inoculums, cultivated at 28 °C for 64 h at 180 rpm shaking. The xylanase activity obtained with the basic nutrient medium was 750.37 U mL^{-1}. After optimization, xylanase activity attained 996.3 U mL^{-1}, which was 33% higher than the activity obtained with the basic medium. The nutrient medium optimized was composed (in g L^{-1}) by (NH4)2HPO4 2.6, urea 0.9, corn cobs 24, wheat bran 14.6 and malt sprout 6.

Paddy husk was used as support to optimize xylanase production by SSF using *Bacillus pumilus*. The medium contained 200 g of paddy husk with 800 mL of liquid fermentation medium (20 g L^{-1} of xylan; 2 g L^{-1} of peptone; 2.5 g L^{-1} of yeast extract and mineral solution at pH 9.0). The highest xylanase activity of 142 U gdm^{-1} was obtained after 6 days of fermentation at 30 °C. The xylanase activity was highest (147.3 U gdm^{-1}) with a 2:9 ratio of paddy husk and liquid fermentation medium on the sixth day at 40 °C achieving 177.5 U gdm^{-1} (Kapilan and Arasaratnam 2011). In another study Bocchini et al. (2005)used *Bacillus circulans* D1 under Submerged Fermentation containing mineral medium and hydrolysated of bagasse with initial sugar concentration of 2.5 g L^{-1}, agitation of 200 rpm, incubated at 45 °C. The authors achieved 8.4 U $mL^{-1}$of xylanase in 24 h of cultivation.

A mutant *Pseudomonas* sp. WLUN024 grown on xylosidic materials, such as hemicellulose, xylan, xylose, and wheat bran was used for xylanase production by Xu et al. (2005). Batch fermentations were carried out on a rotary shaker at 220 rpm, 37 °C for 24 h. After optimization the maximum activity of xylanase reached 1245 U mL^{-1}. The optimized medium consisted of 70 g wheat bran, 8 g $(NH_4)_2$ SO_4 and 4 g L^{-1} K_2HPO_4 and initial pH adjusted to 8.5 before autoclaving.

Meshram et al. (2008) produced xylanase by Submerged Fermentation using *Penicillium janthinellum* NCIM in 50 mL of Mandels–Weber medium, sugarcane bagasse and beef extract, incubated at 28–30 °C at 180 rpm. The best parameters observed were: carbon source 1.63%, nitrogen source 0.16%, pH 4.1, and inoculum 5.5%, achieving a maximum xylanase activity of 28.98 U mL^{-1}. In addition to this Antoine et al. (2010) tested five agro-industrial wastes (soya oil cake, soya meal, wheat bran, whole wheat bran, and pulp beet) for xylanase production. The fungus, *Penicillium canescens* was investigated in SSF. A xylanase production level of 18,895 U g^{-1} in Erlenmeyer flasks and 9300 U g^{-1} in plastic bags were reached after 7 days of incubation with 5 g of soya oil cake crushed (5 mm particle size) supplemented with 20 mL of distilled water, 3% of casein peptone, 4% Na_2HPO_4 $2H_2O$ at 30 °C and 80% of initial moisture.

Pectinases

Many bacteria, fungi and higher plants are known to produce pectinolytic enzyme called pectinase, that breakdown pectin, a polysaccharide substrate that is found in the cell wall of plants. Microbial pectinases account for 25% of the global food enzymes sales (Singh et al. 1999). Almost all the commercial preparations of pectinases are produced from fungal sources. *Aspergillus niger* is the most commonly used fungal species for industrial production of pectinolytic enzymes (Jayani et al. 2005).

Orange peel could be an attractive and promising substrate used in pectinase production by submerged fermentation using *Aspergillus niger*. Results indicate that at pH 2.2, a pectin yield of 15.5% was obtained from orange peels at 70 °C, and of the three pectinolytic fungi isolated from natural environment and induced with pectin from orange peels as a sole carbon source for pectinase production; *Aspergillus niger* produced more pectinase than others. Upon partial purification, a twofold increase in pectinase activity whose pH and temperature optima were 5.0 and 40 °C was obtained (Ezike et al. 2014). Furthermore, Patil and Dayanand (2006) reported pectinase production using deseeded sunflower head under Submerged Fermentation by *Aspergillus niger* DMF 27 and *Aspergillus niger* DMF 45 in SSF. In both fermentations processes the pH of 5.0 and temperature of 34 °C were ideal. Under optimum conditions, maximum production of exo-pectinase was 34.2 U g^{-1} SSF (65% initial moisture) and endo-pectinase was 12.6 U mL^{-1} in Submerged Fermentation.

Pectinases production by *Thermoascus aurantiacus* in SSF was reported by Martins et al. (2002) using wheat bran. The authors reached 43 U g^{-1} of polygalacturonase at the 4th day of fermentation while pectin lyase (11,600 U g^{-1}) was produced in the 14th day. The same authors reported the production using orange bagasse (composed in dry material 11.8% fibre, 6.4% protein, 63% nitrogen, 6.7% ash, 19% total sugar (9% reducing sugar) and 0.1% pectin), reaching 43 U g^{-1} of polygalacturonase after 6 days and 19,320 U g^{-1} of pectin lyase. Solid State Fermentation was conducted with initial moisture of 67%.

Mannanases

Mannanses are enzymes produced mainly from microorganisms but mannanases produced from plants and animals have also been reported. Bacterial mannanases are mostly extracellular and can act in a wide range of pH and temperature, though acidic and neutral mannanases are more common (Dhawan and Kaur 2007). Galactomannan-rich substrate locust bean gum (LBG) has been used widely as an inducer of β-mannanase (Kote et al. 2009; Kim et al. 2011). Other substrates like konjac powder, copra meal and wheat bran have also been practiced for the same purpose, since they offer significant benefit due to their cheaper cost and abundant availability (Zhang et al. 2009; Meenakshi et al. 2010; Chauhan et al. 2012).

Various microbes require different incubation times for maximum β-mannanase production. In case of bacteria, it ranges from 24 h in *Acinetobacter* sp. ST 1-1 (Titapoka et al. 2008) to 96 h in *Bacillus* sp. MG-33 (Meenakshi et al. 2010). In contrast to bacteria, fungi require 3 days in case of *Streptomyces* sp. PG-08-03 (Bhoria et al. 2009) to 11 days in *Aspergillus* ATCC 20114 (Mohamad et al. 2011). The optimum temperature for mannanase production has been reported in the mesophilic range in most of the cases, and it corresponds with the growth temperature of the respective microorganism. In general, bacteria prefer neutral to alkaline pH and fungi acidic pH for best growth and mannanase production (Mabrouk and El Ahwany 2008; Abdeshahian et al. 2010).

The most potent mannanase producer was reported to be *Aspergillus niger,* which produced the highest extracellular mannanase activity (2.90 U/mL), followed by *Aspergillus flavus* (2.54 U/mL) and *Aspergillus ochraceous* (2.16 U/mL). The optimal operating conditions for β-mannanase activity by *Aspergillus niger* arising from this study are as follows: temperature of 30 °C, 6 days incubation period, initial pH 5.0 and inoculum size of 3×10^6 spore/mL (Alsarrani 2011). In another study mannanase production was carried out with *Aspergillus siydowii* grown on 1.0% (w/v) banana stem as the carbon source, producing 1.229 U mL^{-1}, and *Emericella nidulans* grown on 1.0% (w/v) dirty cotton residue producing 0.455 U mL^{-1} of mannanase activity. Fermentation was conducted using as inoculum 108 spores mL^{-1} during 7 days and agitation of 120 rpm (de Siqueira et al. 2010).

Heck et al. (2005) optimized the mannanase production by *Bacillus circulans* in SSF using industrial fibrous soy residue and nutritive solution in a 500 mL cylindrical bioreactor achieving 0.54 U mg^{-1}. Bacteria of *Bacillus amyloliquefaciens* produce the mannanase enzyme and the activity of the mannanase enzyme with a substrate that is a combination of coconut and tofu waste is not much different compared to with a substrate using locust bean gum, with only a 13.34% difference and the optimum pH and temperature are the same. The results showed that the best conditions were a substrate ratio of 80% coconut waste to 20% tofu waste, a 48 h incubation time, a pH of 6.5 and a temperature of 40 °C yielding a mannanase enzymatic activity of 5.13 U mL^{-1}. When locust bean gum was added to the substrate composed of coconut and tofu waste, the best conditions were a dose of 0.6%, a 48 h incubation time, a pH = 6.5 and a temperature of 40 °C yielding a 5.92 U mL^{-1} mannanase enzymatic activity (Zurmiati et al. 2017).

Safety of Exogenous Enzymes

Safety concerns associated with food enzymes in general are possible allergenic, irritative and otherwise toxic properties. Oral toxicity is especially relevant to consumers of food enzymes. The regulation of enzymes internationally is quite varied between countries, with specific country either requiring a full approval process, a notification of enzyme or no requirement thereof. Pre-market approval may depend on whether an enzyme is classified as processing aid or a food additive, though the point of consideration regardless of classification is that the safety of the enzyme must be assured. Challenges to regulators and industry arise from unresolved issues and from lack of harmonization of both legislation and safety evaluation (Agarwal and Sahu 2014).

In occupational contexts enzyme exposure includes mainly to dust or liquid aerosols that are set free while handling enzyme preparation in either manufacturing of the enzyme itself or using enzyme preparations in other industrial contexts. This is generally true for all enzymes regardless of the particular end-use. The particles are deposited on the skin or on the mucous membranes of the respiratory tract. When an enzyme comes into contact with the respiratory tract or the skin, the body's immune system may be stimulated to produce antibodies resulting in respiratory allergy or contact urticaria, respectively. And because skin has a protein structure, enzymes which catalyse breakdown of proteins such as proteases, are potential skin irritants. The individuals who are exposed to a possible antigen (here: the enzyme) for the first time may develop antigen-specific IgG and/or IgE antibodies. The formation of IgG indicates exposure, and IgE antibodies indicate allergic sensitization but not allergic disease. Once an individual has developed an immune response as a result of inhalation or skin contact with the enzyme, re-exposure produces increasingly severe responses becoming dangerous or even fatal (Spök 2006; Chabane et al. 1994). The enzyme manufacturing industry introduced quite extensive measures to diminish and monitor the exposure of workers, including encapsulation of enzymes, using immobilized preparations, avoiding direct contact, introduction of safe working practices and training of workers, replacement of older products by antigenically distinct proteases, and exclusion of potentially pre-disposed and especially sensitive workers from directly working with enzyme preparations (Spök et al. 1998).

The micro-organisms used in the production of enzymes may themselves be sources of hazardous materials and have been the chief focus of attention by the regulatory authorities. Microbial toxins that are active via the oral route may be produced by certain bacteria or certain filamentous fungi (molds). Yeasts, by contrast, are not known to produce such toxins. The safety of the production strain should be the primary consideration in evaluating enzyme safety. The primary issue in evaluating the safety of a production strain is its *toxigenic potential*, specifically the possible synthesis by the production strain of toxins that are active via the oral route. *Pathogenic potential* is not usually an area of concern for consumer safety because enzyme preparations rarely contain viable organisms. Pathogenicity is, however, important to worker safety (Pariza and Johnson 2001).

Applications of Exogenous Enzymes in Different Food Systems

With the advancement of technology, exogenous enzymes with wide range of applications have great utility in food systems which are as under.

Food Industry

Enzymes have always been important to food technology because of their ability to act as catalysts, transforming raw materials into improved food products. Food processing enzymes are used as food additives to modify food properties. In the twentieth century, enzymes began to be isolated from living cells, leading to a large-scale commercial production and with wider application in the food industry. Food processing enzymes are used in starch processing, meat processing, dairy industry, wine industry and in manufacture of pre-digested foods (Chaudhary et al. 2015).

The use of rennet in cheese manufacturing was among the earliest applications of exogenous enzymes in food processing. In recent years, proteinases have found additional applications in dairy technology, for example in acceleration of cheese ripening, modification of functional properties and preparation of dietic products (IDF 1990). Animal rennet (bovine chymosin) is conventionally used as a milk-clotting agent in dairy industry for the manufacture of quality cheeses with good flavour and texture. Rennin acts on the milk protein in two stages, by enzymatic and by non-enzymatic action, resulting in coagulation of milk (Bhoopathy 1994). Microbial coagulants are low-cost substitutes of rennet, since they are easily produced by fermentation. They have higher resistance to heat and stronger proteolytic activity during cheese making and ripening (Jacob et al. 2011). Lipases in dairy industry are used for flavour enhancement in cheese products (e.g. enzyme-modified cheese flavour) and acceleration of cheese ripening and lipolysis of milk/vegetable oil/fat to obtain specific flavours (Jooyandeh et al. 2009).

The supplementation of flour and dough with enzyme improvers (technical enzymes) is a usual practice for flour standardization and also as baking aids. Enzymes are usually added to modify dough rheology, gas retention and crumb softness in bread manufacture, to modify dough rheology in the manufacture of pastry and biscuits, to change product softness in cake making and to reduce acrylamide formation in bakery products (Cauvain and Young 2006). The enzymes most frequently used in breadmaking are the α-amylases from different origins (Penella et al. 2008). Amylases can degrade starch and produce small dextrins for the yeast to act upon. Enzymes such as hemicellulases, xylanases, lipases and oxidases can directly or indirectly improve the strength of the gluten network and so improve the quality of the finished bread. The addition of certain types of pentosanases or xylanases at the correct dosage can improve dough machinability yielding more flexible, easier-to-handle dough. The addition of functional lipases modifies the natural flour lipids so they become better at stabilizing the dough. The addition of lipases has been claimed to retard the rate of staling in baked products (Cauvain and Young 2006; Siswoyo et al. 1999). Lipoxygenases are also employed to improve mixing tolerance and dough handling properties (Cumbee et al. 1997).

Enzymes are processing aids used worldwide for fruit processing, particularly for the production of clear fruit juice and concentrate. Enzymes can increase the yield of solid recovery during pulp washing, facilitate the production of highly concentrated citrus bases, improve essential oil recovery from peel, de bitter juice, clarify lemon juice or increase the worth of waste products (Grassin and Fauquembergue 1996). Pectinases are one of the important upcoming enzymes of the commercial sector especially for fruit juice industry as prerequisites for obtaining well clarified and stable juices with higher yields (Lee et al. 2006; Sandri et al. 2011). Amylases are added together with pectinases at the start of the processing season when apples contain starch. Vegetable juice processing therefore requires more cellulases in addition to pectinases to reduce viscosity sufficiently for juice extraction using a decanter (Chaudhary et al. 2015).

Animal Feed

The use of exogenous enzymes as a feed additive strategy have attracted growing attention and proved to be useful in improving production efficiency of ruminants (Beauchemin et al. 2003). Morgavi et al. (2000) reported that enzymes improved fiber degradation in the rumen by acting synergistically with the rumen microflora, thereby increasing their hydrolytic capacity in the rumen (Beauchemin et al. 2004). Moreover, the use of fibrolytic enzymes in ruminant diets is generally characterized by an increases dry matter (DM)intake, cellulose degradation and/or nutrient digestibility, and consequently increase animal performance (Yang et al. 2000). The supplementation of exogenous fibrolytic enzymes milk yield was improved significantly (41.0 vs. 39.5 kg/cow/day) compared to untreated dairy cows. In addition, the energy corrected milk (40.6 vs. 39.4 kg) and feed efficiency in early lactating dairy cows were improved significantly compared to the control group (Mohamed et al. 2013). In another study Lopuszanska-Rusek and Bilik (2011) observed enhanced milk production with xylanase-esterase supplementation and a tendency of improving dry matter intake and milk production with xylanase and cellulase enzyme supplementation.

Apart from fibrolytic enzymes there is evidence that exogenous proteolytic enzyme could increase the total tract digestibilities of dry matter, organic matter, acid detergent fiber and neutral detergent fiber with larger increases in digestibility of cows though the feeding of proteolytic enzyme unexpectedly decreased feed intake of cows. As a result, milk production was suppressed, nevertheless, dairy efficiency, expressed as milk/dry matter intake, was increased. Supplemented proteolytic enzyme enhanced some milk composition factors such as milk fat and milk lactose percentages but decreased milk protein percentage (Eun and Beauchemin 2005).

In a study conducted by Gado et al. (2009), they found that addition of enzymes increased rumen microbial N synthesis. Intake of dry matter (DM) and organic matter (OM) was positively influenced by supplementation, and digestibility of all nutrients was higher in the total tract of supplemented cows, although the magnitude of the improvement varied among nutrients, with the highest improvement in neutral detergent fiber (aNDFom) and acid detergent fiber(ADFom) (418–584 and 401–532 g/kg respectively) than the other nutrients. Supplementation of enzymes also increased rumen ammonia N and total short chain fatty acid (SCFA) concentrations, and individual SCFA proportions were also altered with an increase in acetate (61.0–64.8 mol/100 mol) before feeding, and acetate and propionate increased 3 h post-feeding (60.0–64.0 and 18.3–20.8 mol/100 mol respectively).

Recent Advances in the Development of Food Enzymes

With the increase in market requirements for food packaging and preservation, the quest for optimum performance of enzymes has given enzyme engineering, particularly enzyme/cell immobilization or cell encapsulation, prime importance in production of biocatalyst with improved properties (Cao et al. 2003; Hamilton 2009).

Immobilization

Immobilization implies associating the enzymes or cells with an insoluble matrix so that it is retained for further economic use, i.e., giving the optimal immobilization yield and having the activity stability in long term (Miladi et al. 2012). Over the last few decades, intensive research in the area of enzyme technology has shown promise, i.e., the immobilization of enzymes (for extracellular enzymes) and cells (for intracellular enzymes). Immobilization enzymes and cells are widely used in the fermentation industry. Also biosensors are designed on the principle of immobilization of enzymes as it is convenient, economical and a time-efficient process of isolation and purification of intracellular enzymes (Mishra et al. 2016). Immobilization can be performed by several methods, namely, entrapment/microencapsulation, binding to a solid carrier, and cross-linking of enzyme aggregates, resulting in carrier-free macromolecules. The latter presents an alternative to carrier-bound enzymes, since these introduce a large portion of non-catalytic material. This can account to about 90% to more than 99% of the total mass of the biocatalysts, resulting in low space-time yields and productivities (Sheldon 2007).

Entrapment/(micro)encapsulation, where the enzyme is contained within a given structure. This can be: a polymer network of an organic polymer or a sol-gel; a membrane device such as a hollow fiber or a microcapsule; or a (reverse) micelle. Apart from the hollow fiber, the whole process of immobilization is performed insitu. The polymeric network is formed in the presence of the enzyme, leading to supports that are often referred to as beads or capsules. Still, the latter term could preferably be used when the core and the boundary layer(s) are made of different materials, namely, alginate and poly-l-lysine. Although direct contact with an adverse environment is prevented, mass transfer limitations may be relevant, enzyme loading is relatively low, and leakage, particularly of smaller enzymes from hydrogels (namely, alginate, gelatin), may occur. This may be minimized by previously cross-linking the enzyme with multifunctional agent (namely, glutaraldehyde) (Brady and Jordaan 2009; de Segura et al. 2003) or by promoting cross-linkage of the matrix after the entrapment (de Assis et al. 2004). The use of LentiKats, a polyvinyl-alcohol-based support in lens-shaped form, has been used for several applications in carbohydrate processing. Among these are the synthesis of oligosaccharides with dextransucrase (de Segura et al. 2003), maltodextrin hydrolysis with glucoamylase (Rebroš et al. 2006), lactose hydrolysis with lactase (Grosová et al.

2008), and production of invert sugar syrup with invertase (Rebroš et al. 2007). Flavourzyme, (a fungal protease/peptidase complex) entrapped in calcium alginate (Anjani et al. 2007), k-carragenan, gellan, and higher melting-fat fraction of milk fat (Kailasapathy and Lam 2005), was effectively used in cheese ripening, in order to speed up the process, while avoiding the problems associated with the use of free enzyme. These include deficient enzyme distribution, reduced yield and poor-quality cheese, partly ascribed to excessive proteolysis and whey contamination (Anjani et al. 2007). Calcium alginate beads were also used to immobilize glucose isomerase (Tumturk et al. 2008) and α-amylase for starch hydrolysis to whey (Rajagopalan and Krishnan 2008).

In a particularly favored technique immobilization of enzymes in liposomes, known as dehydration-rehydration vesicles (DRVs), small (diameters usually below 50 nm) unilamellar vesicles (SUVs) is prepared in distilled water and mixed with an aqueous solution of the enzyme to be encapsulated. The resulting vesicle suspension is then dehydrated under freeze drying or equivalent method. Upon rehydration, the resulting DRVs are multilamellar and larger (from 200 nm to a little above 1000 nm) than the original SUVs, and can capture solute molecules (Walde and Ichikawa 2001;Grosová et al. 2008). Recent work in this particular application has used lactase as enzyme model and has focused on the optimization and characterization of the liposome-based immobilized system (Rodríguez-Nogales and López 2006). Cocktails of enzymes, namely, Flavourzyme, bacterial proteases and Palatase M (a commercial lipase preparation), were immobilized in liposomes and successfully used to speed up cheddar cheese ripening (Kheadr et al. 2003). Encapsulation in lipid vesicles has been proved a mild method, providing high protection against proteolysis.

Binding to a solid carrier, where enzyme-support interaction can be of covalent, ionic, or physical nature. Curiously, the first reported application of enzyme immobilization was of invertase onto activated charcoal (Nelson and Griffin 1916). Another example is the immobilization of pectinase in egg shell for the preparation of low-methoxyl pectin. The immobilized biocatalyst could be reused for 32 times at 30 °C, and it was used in a fluidized-bed reactor, operated at an optimum flow rate of 5 mL h^{-1} and 35 °C (Nighojkar et al. 1995). Other examples are the surface immobilizations of α-amylase on alumina (Reshmi et al. 2006) and in zirconia (Reshmi et al. 2007).

Carrier-free macroparticles, where a bifunctional reagent (namely, glutaraldehyde), is used to cross-link enzyme aggregates (CLEAs) or crystals (CLECs), leading to a biocatalyst displaying highly concentrated enzyme activity, high stability and low production costs(Sheldon 2007; Roy and Abraham 2004). The use of CLEAs is favored given the lower complexity of the process. This approach is recent, as compared with entrapment and binding to a solid carrier, and there are still relatively few examples of its application to enzymes used in the area of food processing (Fernandes 2010) such as the immobilization of lactase for the hydrolysis of lactose, where, under similar operational conditions as for the free enzyme, the CLEA yielded 78% monosaccharides in 12 h as compared to 3.9% of the free form (Gaur et al. 2006).

Conclusion

Commercial-scale enzyme catalysis has been implemented in several industries such as food, chemical and pharmaceutical. Microbial enzymes are the preferred source to plants or animals due to their economic feasibility, high yields, consistency, ease of product modification and optimization, regular supply due to absence of seasonal fluctuations, rapid growth of microbes on inexpensive media, stability, and greater catalytic activity. In comparison with plant and animal enzymes, microbial enzymes can be produced very effectively by different fermentation techniques like solid-state and submerged fermentations. It is also easy to produce microbial enzymes on a large scale. The recombinant DNA technology has further improved production processes and helped to produce enzymes commercially that could not be produced previously. Global market for exogenous/commercial enzymes is expanding rapidly and this has given enzyme engineering particularly enzyme immobilization a prime importance in production of biocatalyst with improved properties.

References

Abd-Elhalem, B. T., El-Sawy, M., Gamal, R. F., & Abou-Taleb, K. A. (2015). Production of amylases from Bacillus amyloliquefaciens under submerged fermentation using some agro-industrial by-products. *Annals of Agricultural Sciences, 60*(2), 193–202.

Abdeshahian, P., Samat, N., Hamid, A. A., & Yusoff, W. M. W. (2010). Utilization of palm kernel cake for production of β-mannanase by Aspergillus niger FTCC 5003 in solid substrate fermentation using an aerated column bioreactor. *Journal of Industrial Microbiology & Biotechnology, 37*(1), 103.

Agarwal, S., & Sahu, S. (2014). Safety and regulatory aspects of food enzymes: An industrial perspective. *International Journal of Interdisciplinary and Multidisciplinary Studies, 1*(6), 253–267.

Alsarrani, A. Q. (2011). Production of Mannan-degrading enzyme by Aspergillus niger. *Journal of Taibah University for Science, 5*(1), 1–6.

Anisa, S. K., & Girish, K. (2014). Pectinolytic activity of Rhizopus sp. and Trichoderma viride. *International Journal of Research in Pure and Applied Microbiology, 4*(2), 28–31.

Anjani, K., Kailasapathy, K., & Phillips, M. (2007). Microencapsulation of enzymes for potential application in acceleration of cheese ripening. *International Dairy Journal, 17*(1), 79–86.

Anto, H., Trivedi, U. B., & Patel, K. C. (2006). Glucoamylase production by solid-state fermentation using rice flake manufacturing waste products as substrate. *Bioresource Technology, 97*(10), 1161–1166.

Antoine, A. A., Jacqueline, D., & Thonart, P. (2010). Xylanase production by Penicillium canescens on soya oil cake in solid-state fermentation. *Applied Biochemistry and Biotechnology, 160*(1), 50–62.

Anwar, A., & Saleemuddin, M. (2000). Alkaline protease from Spilosoma oblique: Potential applications in bio-formulations. *Biotechnology and Applied Biochemistry, 31*(2), 85–89.

Ariffin, H., Abdullah, N., UmiKalsom, M. S., Shirai, Y., & Hassan, M. A. (2006). Production and characterization of cellulase by Bacillus pumilus EB3. *International Journal of Engineering and Technology, 3*(1), 47–53.

Bakri, Y., Manal, A., & Ghassan, A. (2008). Xylanase production by a newly isolated Aspergillus niger SS7 in submerged culture. *Polish Journal of Microbiology, 57*, 248–251.

Beauchemin, K. A., Colombatto, D., Morgavi, D. P., & Yang, W. Z. (2003). Use of fibrolytic enzymes to improve feed utilization by ruminants. *Journal of Animal Science, 81*, E37–E47.

Beauchemin, K. A., Colombatto, D., Morgavi, D. P., Yang, W. Z., & Rode, L. M. (2004). Mode of action of exogenous cell wall degrading enzymes for ruminants. *Canadian Journal of Animal Science, 84*, 13–22.

Berka, R. M., & Cherry, J. R. (2006). Enzyme biotechnology. In C. Ratledge & B. Kristiansen (Eds.), *Basic biotechnology* (3rd ed., pp. 477–498). Cambridge, UK: Cambridge University Press.

Bhoopathy, R. (1994). Enzyme technology in food and health industries. *Indian Food Industry, 13*, 22–31.

Bhoria, P., Singh, G., & Hoondal, G. S. (2009). Optimization of mannanase production from Streptomyces sp. PG-08-03 in submerged fermentation. *BioResources, 4*(3), 1130–1138.

Binod, P., Singhania, R. R., Soccol, C. R., & Pandey, A. (2008). Industrial enzymes. In A. Pandey, C. Larroche, C. R. Soccol, & C. G. Dussap (Eds.), *Advances in fermentation technology* (pp. 291–320). New Delhi, India: Asiatech Publishers.

Bocchini, D. A., Oliveira, O. M. M. F., Gomes, E., & Da Silva, R. (2005). Use of sugarcane bagasse and grass hydrolysates as carbon sources for xylanase production by Bacillus circulans D1 in submerged fermentation. *Process Biochemistry, 40*(12), 3653–3659.

Brady, D., & Jordaan, J. (2009). Advances in enzyme immobilisation. *Biotechnology Letters, 31*(11), 1639–1650.

Cao, L., van Langen, L., & Sheldon, R. A. (2003). Immobilised enzymes: Carrier-bound or carrier-free? *Current Opinion in Biotechnology, 14*(4), 387–394.

Cauvain, S., & Young, L. (2006). Ingredients and their influences. In S. Cauvain & L. Young (Eds.), *Baked products. Science, technology and practice* (pp. 72–98). Oxford, UK: Blackwell Publishing.

Chabane, M. H., Abuaf, N., & Leynadier, F. (1994). Why are some allergens enzymes? *Annales de Biologie Clinique, 52*(6), 425–431.

Chaudhary, S., Sagar, S., Kumar, M., Sengar, R. S., & Tomar, A. (2015). The use of enzymes in food processing: A review. *South Asian Journal of Food Technology and Environment, 1*(4), 2394–5168.

Chauhan, P. S., Puri, N., Sharma, P., & Gupta, N. (2012). Mannanases: Microbial sources, production, properties and potential biotechnological applications. *Applied Microbiology and Biotechnology, 93*(5), 1817–1830.

Choi, J. M., Han, S. S., & Kim, H. S. (2015). Industrial applications of enzyme biocatalysis: Current status and future aspects. *Biotechnology Advances, 33*(7), 1443–1454.

Cumbee, B., Hildebrand, D. F., & Addo, K. (1997). Soybean flour lipoxygenase isozymes effects on wheat flour dough rheological and breadmaking properties. *Journal of Food Science, 62*(2), 281–283.

de Assis, S. A., Ferreira, B. S., Fernandes, P., Guaglianoni, D. G., Cabral, J. M., & Oliveira, O. M. M. F. (2004). Gelatin-immobilized pectin methylesterase for production of low methoxyl pectin. *Food Chemistry, 86*(3), 333–337.

de Segura, A. G., Alcalde, M. J., Plou, F., Remaud-simeon, M., Monsan, P., & Ballesteros, A. (2003). Encapsulation in LentiKats of dextransucrase from Leuconostoc mesenteroides NRRL B-1299, and its effect on product selectivity. *Biocatalysis and Biotransformation, 21*(6), 325–331.

de Siqueira, F. G., de Siqueira, E. G., Jaramillo, P. M. D., Silveira, M. H. L., Andreaus, J., Couto, F. A., … Ferreira Filho, E. X. (2010). The potential of agro-industrial residues for production of holocellulase from filamentous fungi. *International Biodeterioration and Biodegradation, 64*(1), 20–26.

Dhawan, S., & Kaur, J. (2007). Microbial mannanases: An overview of production and applications. *Critical Reviews in Biotechnology, 27*(4), 197–216.

Dobrev, G. T., Pishtiyski, I. G., Stanchev, V. S., & Mircheva, R. (2007). Optimization of nutrient medium containing agricultural wastes for xylanase production by Aspergillus niger B03 using optimal composite experimental design. *Bioresource Technology, 98*(14), 2671–2678.

Eun, J. S., & Beauchemin, K. A. (2005). Effects of a proteolytic feed enzyme on intake, digestion, ruminal fermentation, and milk production. *Journal of Dairy Science, 88*(6), 2140–2153.

Ezike, T. C., Eze, S. O. O., Nsude, C. A., & Chilaka, F. C. (2014). Production of pectinases from Aspergillus niger using submerged fermentation with orange peels as carbon source. *Sylwan, 158*(8), 434–440.

Fernandes, P. (2010). Enzymes in food processing: A condensed overview on strategies for better biocatalysts. *Enzyme Research, 2010*, 1–19.

Gado, H. M., Salem, A. Z. M., Robinson, P. H., & Hassan, M. (2009). Influence of exogenous enzymes on nutrient digestibility, extent of ruminal fermentation as well as milk production and composition in dairy cows. *Animal Feed Science and Technology, 154*(1–2), 36–46.

Gaur, R., Pant, H., Jain, R., & Khare, S. K. (2006). Galacto-oligosaccharide synthesis by immobilized Aspergillus oryzae β-galactosidase. *Food Chemistry, 97*(3), 426–430.

Grassin, C., & Fauquembergue, P. (1996). Fruit juices. In T. Godfrey & S. West (Eds.), *Industrial enzymology* (2nd ed., p. 227). London, UK: Macmillan.

Grosová, Z., Rosenberg, M., Rebroš, M., Šipocz, M., & Sedláčková, B. (2008). Entrapment of β-galactosidase in polyvinylalcohol hydrogel. *Biotechnology Letters, 30*(4), 763–767.

Gurung, N., Ray, S., Bose, S., & Rai, V. (2013). A broader view: Microbial enzymes and their relevance in industries, medicine, and beyond. *BioMed Research International, 2013*, 1–18.

Hamilton, S. (2009). Introduction to a special issue on food and innovation. *Business History Review, 83*, 233–238.

Heck, J. X., de Barros Soares, L. H., & Ayub, M. A. Z. (2005). Optimization of xylanase and mannanase production by Bacillus circulans strain BL53 on solid-state cultivation. *Enzyme and Microbial Technology, 37*(4), 417–423.

IDF, International Dairy Federation. (1990). *Bull, 247*, 24–38.

Ikasari, L., & Mitchell, D. A. (1996). Leaching and characterization of Rhizopus oligosporus acid protease from solid-state fermentation. *Enzyme and Microbial Technology, 19*(3), 171–175.

Jacob, M., Jaros, D., & Rohm, H. (2011). Recent advances in milk clotting enzymes. *International Journal of Dairy Technology, 64*(1), 14–33.

Jayani, R. S., Saxena, S., & Gupta, R. (2005). Microbial pectinolytic enzymes: A review. *Process Biochemistry, 40*(9), 2931–2944.

Jin, B., Van Leeuwen, H. J., Patel, B., & Yu, Q. (1998). Utilisation of starch processing wastewater for production of microbial biomass protein and fungal α-amylase by Aspergillus oryzae. *Bioresource Technology, 66*(3), 201–206.

Jooyandeh, H., Amarjeet, K., & Minhas, K. S. (2009). Lipases in dairy industry: A review. *Journal of Food Science and Technology, 46*(3), 181–189.

Kailasapathy, K., & Lam, S. H. (2005). Application of encapsulated enzymes to accelerate cheese ripening. *International Dairy Journal, 15*(6–9), 929–939.

Kammoun, R., Naili, B., & Bejar, S. (2008). Application of a statistical design to the optimization of parameters and culture medium for α-amylase production by Aspergillus oryzae CBS 819.72 grown on gruel (wheat grinding by-product). *Bioresource Technology, 99*(13), 5602–5609.

Kapilan, R., & Arasaratnam, V. (2011). Paddy husk as support for solid state fermentation to produce xylanase from Bacillus pumilus. *Rice Science, 18*(1), 36–45.

Kathiresan, K., & Manivannan, S. (2006). Amylase production by Penicilliumfellutanum isolated from mangrove rhizosphere soil. *African Journal of Biotechnology, 5*(10), 829–832.

Kheadr, E. E., Vuillemard, J. C., & El-Deeb, S. A. (2003). Impact of liposome-encapsulated enzyme cocktails on cheddar cheese ripening. *Food Research International, 36*(3), 241–252.

Kim, D. Y., Ham, S. J., Lee, H. J., Kim, Y. J., Shin, D. H., Rhee, Y. H., … Park, H. Y. (2011). A highly active endo-1,4-β-mannanase produced by Cellulosimicrobium sp. strain HY-13, a hemicellulolytic bacterium in the gut of Eisenia fetida. *Enzyme and Microbial Technology, 48*, 365–370.

Kirk, O., Borchert, T. V., & Fuglsang, C. C. (2002). Industrial enzyme applications. *Current Opinion in Biotechnology, 13*(4), 345–351.

Konsoula, Z., & Liakopoulou-Kyriakides, M. (2007). Co-production of α-amylase and β galactosidase by Bacillus subtilis in complex organic substrates. *Bioresource Technology, 98*(1), 150–157.

Kote, N. V., Patil, A. G. G., & Mulimani, V. H. (2009). Optimization of the production of thermostable endo-β-1, 4 mannanases from a newly isolated Aspergillus niger gr and Aspergillus flavus gr. *Applied Biochemistry and Biotechnology, 152*(2), 213–223.

Kubicek, C.P. (1993). From cellulose to cellulase inducers: Facts and fiction. In: *Proceedings of the second TRICEL symposium on Trichoderma reesei cellulases and other hydrolytic enzymes* (pp. 181–188). Espoo, Finland: Foundation of Biotechnical and Industrial Fermentation Research.

Kuhad, R. C., Manchanda, M., & Singh, A. (1999). Hydrolytic potential of extracellular enzymes from a mutant strain of Fusarium oxysporum. *Bioprocess Engineering, 20*(2), 133–135.

Kumar, A. G., Swarnalatha, S., Sairam, B., & Sekaran, G. (2008). Production of alkaline protease by Pseudomonas aeruginosa using protein aceous solid waste generated from leather manufacturing industries. *Bioresource Technology, 99*(6), 1939–1944.

Lee, S. M., & Koo, Y. M. (2001). Pilot-scale production of cellulase using Trichoderma reesei Rut C-30 Fed-Batch Mode. *Journal of Microbiology and Biotechnology, 11*(2), 229–233.

Lee, W. C., Yusof, S., Hamid, N. S. A., & Baharin, B. S. (2006). Optimizing conditions for enzymatic clarification of banana juice using response surface methodology (RSM). *Journal of Food Engineering, 73*(1), 55–63.

Li, S., Yang, X., Yang, S., Zhu, M., & Wang, X. (2012). Technology prospecting on enzymes: Application, marketing and engineering. *Computational and Structural Biotechnology Journal, 2*(3), e201209017.

Liu, L., Yang, H., Shin, H. D., Chen, R. R., Li, J., Du, G., & Chen, J. (2013). How to achieve high-level expression of microbial enzymes: Strategies and perspectives. *Bioengineered, 4*(4), 212–223.

Lopuszanska-Rusek, M., & Bilik, K. (2011). Influence of pre-and postpartum supplementation of fibrolytic enzymes and yeast culture, or both, on performance and metabolic status of dairy cows. *Annals of Animal Science, 11*(4), 531–545.

Mabrouk, M. E., & El Ahwany, A. M. (2008). Production of 946-mannanase by Bacillus amylolequifaciens 10A1 cultured on potato peels. *African Journal of Biotechnology, 7*(8), 1123–1128.

Maciel, G. M., de Souza Vandenberghe, L. P., Fendrich, R. C., Della Bianca, B. E., Haminiuk, C. W. I., & Soccol, C. R. (2009). Study of some parameters which affect xylanase production: Strain selection, enzyme extraction optimization, and influence of drying conditions. *Biotechnology and Bioprocess Engineering, 14*(6), 748–755.

Markets and Markets Watch. (2015). Industrial enzymes market by type (carbohydrases, proteases, non-starch polysaccharides and others), application (food and beverage, cleaning agents, animal feed and others), brands and by region—Global trends and forecasts to 2020. www.bccresearch.com. http://www.marketsandmarkets.com/Market-Reports/industrial-enzymes-market-237327836.html. Accessed on 24 Mar 2016.

Martins, E. S., Silva, D., Da Silva, R., & Gomes, E. (2002). Solid state production of thermostable pectinases from thermophilic Thermoascus aurantiacus. *Process Biochemistry, 37*(9), 949–954.

Meenakshi, Singh, G., Bhalla, A., & Hoondal, G. S. (2010). Solid state fermentation and characterization of partially purified thermostable mannanase from Bacillus sp. MG-33. *BioResources, 5*(3), 1689–1701.

Meshram, M., Kulkarni, A., Jayaraman, V. K., Kulkarni, B. D., & Lele, S. S. (2008). Optimal xylanase production using Penicilium janthinellum NCIM 1169: A model based approach. *Biochemical Engineering Journal, 40*(2), 348–356.

Miladi, B., El Marjou, A., Boeuf, G., Bouallagui, H., Dufour, F., Di Martino, P., & Elm'selmi, A. (2012). Oriented immobilization of the tobacco etch virus protease for the cleavage of fusion proteins. *Journal of Biotechnology, 158*(3), 97–103.

Mishra, S. S., Ray, R. C., Rosell, C. M., & Panda, D. (2016). Microbial enzymes in food applications: History of progress. In *Microbial enzyme technology in food applications* (pp. 3–18). Boca Raton, FL: Taylor & Francis Group, CRC Press.

Mohamad, S. N., Ramanan, R. N., Mohamad, R., & Ariff, A. B. (2011). Improved mannan-degrading enzymes' production by Aspergillus niger through medium optimization. *New Biotechnology, 28*(2), 146–152.

Mohamed, D. E. D. A., Borhami, B. E., El-Shazly, K. A., & Sallam, S. M. A. (2013). Effect of dietary supplementation with fibrolytic enzymes on the productive performance of early lactating dairy cows. *Journal of Agricultural Science, 5*(6), 146–155.

Morgavi, D. P., Beauchemin, K. A., Nsereko, V. L., Rode, L. M., Iwaasa, A. D., Yang, W. Z., … Wang, Y. (2000). Synergybetween ruminalfibrolytic enzymes and enzymes from Trichoderma longibrachiatum 1. *Journal of Dairy Science, 83*, 1310–1321.

Mukherjee, A. K., Adhikari, H., & Rai, S. K. (2008). Production of alkaline protease by a thermophilic Bacillus subtilis under solid-state fermentation (SSF) condition using Imperata cylindrica grass and potato peel as low-cost medium: Characterization and application of enzyme in detergent formulation. *Biochemical Engineering Journal, 39*(2), 353–361.

Nascimento, R. P., Coelho, R. R., Marques Alver, S. L., Girio, E. P. S., & Amaral-Collago, M. T. (2003). A novel strain of Streptomyces malaysiensis from Brazilian soil produces high endo-1,4xylanasetitres. *World Journal of Microbiology and Biotechnology, 19*, 879–881.

Nelson, J. M., & Griffin, E. G. (1916). Adsorption of invertase. *Journal of the American Chemical Society, 38*(5), 1109–1115.

Nighojkar, A., Srivastava, S., & Kumar, A. (1995). Production of low methoxyl pectin using immobilized pectinesterase bioreactors. *Journal of Fermentation and Bioengineering, 80*(4), 346–349.

Oliveira, L. A., Porto, A. L., & Tambourgi, E. B. (2006). Production of xylanase and protease by Penicillium janthinellum CRC 87M-115 from different agricultural wastes. *Bioresource Technology, 97*(6), 862–867.

Pandey, A., Nigam, P., Soccol, C. R., Soccol, V. T., Singh, D., & Mohan, R. (2000). Advances in microbial amylases. *Biotechnology and Applied Biochemistry, 31*(2), 135–152.

Pandey, A., Selvakumar, P., Soccol, C. R., & Nigam, P. (1999). Solid-state fermentation for the production of industrial enzymes. *Current Science, 77*, 149–162.

Pariza, M. W., & Johnson, E. A. (2001). Evaluating the safety of microbial enzyme preparations used in food processing: Update for a new century. *Regulatory Toxicology and Pharmacology, 33*(2), 173–186.

Patil, S. R., & Dayanand, A. (2006). Optimization of process for the production of fungal pectinases from deseeded sunflower head in submerged and solid-state conditions. *Bioresource Technology, 97*(18), 2340–2344.

Penella, J. S., Collar, C., & Haros, M. (2008). Effect of wheat bran and enzyme addition on dough functional performance and phytic acid levels in bread. *Journal of Cereal Science, 48*(3), 715–721.

Prakasham, R. S., Rao, C. S., & Sarma, P. N. (2006). Green gram husk—An inexpensive substrate for alkaline protease production by Bacillus sp. in solid-state fermentation. *Bioresource Technology, 97*(13), 1449–1454.

Rajagopalan, G., & Krishnan, C. (2008). Immobilization of malto-oligosaccharide forming α-amylase from Bacillus subtilis KCC103: Properties and application in starch hydrolysis. *Journal of Chemical Technology and Biotechnology, 83*(11), 1511–1517.

Ramachandran, S., Patel, A. K., Nampoothiri, K. M., Francis, F., Nagy, V., Szakacs, G., & Pandey, A. (2004). Coconut oil cake—A potential raw material for the production of α-amylase. *Bioresource Technology, 93*(2), 169–174.

Rebroš, M., Rosenberg, M., Mlichova, Z., & Krištofíková, L. (2007). Hydrolysis of sucrose by invertase entrapped in polyvinyl alcohol hydrogel capsules. *Food Chemistry, 102*(3), 784–787.

Rebroš, M., Rosenberg, M., Mlichová, Z., & Paluch, M. (2006). A simple entrapment of glucoamylase into LentiKats® as an efficient catalyst for maltodextrin hydrolysis. *Enzyme and Microbial Technology, 39*(4), 800–804.

Reshmi, R., Sanjay, G., & Sugunan, S. (2006). Enhanced activity and stability of α-amylase immobilized on alumina. *Catalysis Communications, 7*(7), 460–465.

Reshmi, R., Sanjay, G., & Sugunan, S. (2007). Immobilization of α-amylase on zirconia: A heterogeneous biocatalyst for starch hydrolysis. *Catalysis Communications, 8*(3), 393–399.

Rodrigues, B.S.S. (2011). *Production and purification of new microbial cellulases*.

Rodríguez-Nogales, J. M., & López, A. D. (2006). A novel approach to develop β-galactosidase entrapped in liposomes in order to prevent an immediate hydrolysis of lactose in milk. *International Dairy Journal, 16*(4), 354–360.

Rose, S. H., & Van Zyl, W. H. (2008). Exploitation of Aspergillus niger for the heterologous production of cellulases and hemicellulases. *The Open Biotechnol Journal, 2*, 167–175.

Roy, J. J., & Abraham, T. E. (2004). Strategies in making cross-linked enzyme crystals. *Chemical Reviews, 104*(9), 3705–3722.

Sandri, I. G., Fontana, R. C., Barfknecht, D. M., & da Silveira, M. M. (2011). Clarification of fruit juices by fungal pectinases. *LWT - Food Science and Technology, 44*(10), 2217–2222.

Santhi, R. (2014). Microbial production of protease by Bacillus cereus using cassava waste water. *European Journal of Experimental Biology, 4*(2), 19–24.

Seyis, I., & Aksoz, N. (2005). Xylanase production from Trichoderma harzianum 1073 D 3 with alternative carbon and nitrogen sources. *Food Technology and Biotechnology, 43*(1), 37–40.

Sheldon, R. A. (2007). Enzyme immobilization: The quest for optimum performance. *Advanced Synthesis & Catalysis, 349*(8–9), 1289–1307.

Singh, S. A., Ramakrishna, M., & Rao, A. G. A. (1999). Optimization of downstream processing parameters for the recovery of pectinase from the fermented broth of Aspergillus carbonarious. *Process Biochemistry, 35*, 411–417.

Siswoyo, T. A., Tanaka, N., & Morita, N. (1999). Effect of lipase combined with α-amylase on retrogradation of bread. *Food Science and Technology Research, 5*(4), 356–361.

Spök, A. (2006). Safety regulations of food enzymes. *Food Technology and Biotechnology, 44*(2), 197–209.

Spök, A., Aberer, W., Boyer, A., Getzinger, G., Krajnik, P., Kränke, B., Preiss, M., Scherbler, B.M., Steindl, J., Suschek-Berger, J. & Weinberger, F. (1998). *Enzyme in Wasch- und Reinigungsmitteln. Technikfolgenabschätzung und-bewertung unter besonderer Berücksichtigung der Anwendungen der Gentechnologie (Enzymes in detergents and cleansers. Technology assessment focussing on the application of genetic engineering methods)*. Series of the Federal Ministry for Environment, Youth and Family, 29.

Sukumaran, R. K., Singhania, R. R., & Pandey, A. (2005). Microbial cellulases-production, applications and challenges. *Journal of Scientific and Industrial Research, 64*(11), 832–844.

Sun, Y., & Cheng, J. (2002). Hydrolysis of lignocellulosic materials for ethanol production: A review. *Bioresource Technology, 83*(1), 1–11.

Titapoka, S., Keawsompong, S., Haltrich, D., & Nitisinprasert, S. (2008). Selection and characterization of mannanase-producing bacteria useful for the formation of prebiotic manno-oligosaccharides from copra meal. *World Journal of Microbiology and Biotechnology, 24*(8), 1425–1433.

Tumturk, H., Demirel, G., Altinok, H., Aksoy, S., & Hasirci, N. (2008). Immobilization of glucose isomerase in surface-modified alginate gel beads. *Journal of Food Biochemistry, 32*(2), 234–246.

Walde, P., & Ichikawa, S. (2001). Enzymes inside lipid vesicles: Preparation, reactivity and applications. *Biomolecular Engineering, 18*(4), 143–177.

Waseem, I., Gulsher, M., & Choudhry, S. (2014). Cellulase production from newly isolated bacterial strains from local habitat. *International Journal of Recent Scientific Research, 5*(8), 1454–1459.

Xu, Z. H., Bai, Y. L., Xu, X., Shi, J. S., & Tao, W. Y. (2005). Production of alkali-tolerant cellulase-free xylanase by Pseudomonas sp. WLUN024 with wheat bran as the main substrate. *World Journal of Microbiology and Biotechnology, 21*(4), 575–581.

Yang, W. Z., Beauchemin, K. A., & Rode, L. M. (2000). A comparison of methods of adding fibrolytic enzymes to lactating cow diets. *Journal of Dairy Science, 83*, 2512–2520.

Zhang, M., Chen, X. L., Zhang, Z. H., Sun, C. Y., Chen, L. L., He, H. L., … Zhang, Y. Z. (2009). Purification and functional characterization of endo-β-mannanase MAN5 and its application in oligosaccharide production from konjac flour. *Applied Microbiology and Biotechnology, 83*(5), 865–873.

Zurmiati, W., Abbas, M. H., & Mahata, M. E. (2017). Production of extracellular β-mannanase by Bacillus amyloliquefaciens on a coconut waste substrate. *Pakistan Journal of Nutrition, 16*(9), 700–707.

Advances in the Application of Food Proteins and Enzymes

Faiza Jhan, Nusrat Jan, Adil Gani, Nairah Noor, Mudasir Ahmad, Naseer Ahmad Bhat, and Bilal Ahmad Ashwar

Introduction

To maintain the nitrogen balance of body, sufficient protein intake is required by humans and permit for desirable deposition rates during growth and pregnancy. Ingestion of protein amounts greater than requirements leads to excess protein being metabolized and excreted. Conversely, in the case of inadequate dietary protein intake, the body utilizes its own proteins as a source of nitrogen. Therefore, a regular and adequate intake of proteins is necessary. A number of important functions are accomplished by proteins in the body such as building and repairing of tissues, cell signaling and the provision of energy (4 kcal/g protein). Proteins also perform enzymatic and structural functions.

Proteins consist of long chains of amino acid residues that fold into unique structures. These folds include one or more specific spatial conformations driven by a number of covalent and noncovalent interactions. Proteins are categorized as globular, membrane, or fibrous. The functional properties of most food proteins are due to their globular components, especially their solubility, which is attributed to the amphiphilicity of the molecules.

Food proteins are one of the most vital ingredients in the human diet due to their nutritional contribution and other specific functions. In recent years, the global demand for food with high protein content has dramatically increased (Mullen et al. 2017). Due to the rising demands and precarious supply of food proteins from conventional sources, various nonconventional sources have also been studied for their protein content, functionality, and potential applications. Various researchers investigated the process–structure–function aspects of proteins from oilseeds, grains,

F. Jhan · A. Gani (✉) · N. Noor · M. Ahmad · N. A. Bhat · B. A. Ashwar
University of Kashmir, Srinagar, Jammu and Kashmir, India

N. Jan
Skuast-K, Srinagar, Jammu and Kashmir, India

A. Gani, B. A. Ashwar (eds.), *Food biopolymers: Structural, functional and nutraceutical properties*, https://doi.org/10.1007/978-3-030-27061-2_15

legumes, fish, microorganisms, algae, and plant leaves (Joshi et al. 2012; Zeng et al. 2013). Proteins have also been added into various food products in order to improve the flavor, texture, and other sensory attributes of the processed foods.

Embarking upon a society that is increasingly aware about food and health, the science behind food and its constituents is an area that has inspired several scientists toward the bioactive nature of food proteins. Additionally, using as a source of energy and providing amino acids for the synthesis of body proteins, food proteins are essential biological entities that not only help sustain the overall growth, metabolism, and functioning of cellular metabolisms but also provide health attributes. It has been analyzed that increasing the dietary protein intake can modulate anabolic response (Tieland et al. 2012), improve muscle strength and physical function, cardiovascular health and weight management, fat and glucose metabolism, and modulate the immune system (Hartmann and Meisel 2007; Wolfe 2015). In this context, this chapter outlines proteins as hydrolysates, edible films and coatings and wall material for encapsulation and nano delivery systems for bioactive compounds. In addition, protein engineering and immobilization is also discussed.

Protein Hydrolysates

Protein hydrolysates are defined as the products obtained from cleavage of protein's peptide bonds, producing peptides with varying sizes and free amino acids. This type of protein structure modification has an effect on its physicochemical and functional properties (Severin and Xia 2006). The most notable modifications that affect functionality include a decrease in molecular weight of the peptide chain, an increase in polar groups ($-NH_4^+$, $-CO_2^-$), which increase hydrophilicity, and a change in molecular configuration. The cleavage of peptide bonds can be carried out via enzymatic or chemical methods. Hydrolysis using a chemical, alkaline or acidic process is more difficult to control and reduces the nutritional quality of products (Celus et al. 2007), destroying L-form of amino acids and producing toxic substances such as lysino-alanine (Lahl and Grindstaff 1989). Enzymatic hydrolysis works without destructing amino acids and therefore, the nutritional properties of the protein hydrolysates remain largely unaffected by avoiding the extreme temperatures and pH levels needed for chemical hydrolysis (Celus et al. 2007). Although more expensive than chemical hydrolysis, enzymatic hydrolysis is the preferred method.

Protein hydrolysates derived from food possess various physicochemical properties such as solubility, lipid binding, foaming, and emulsification properties depending on their composition, sequence, and length (Cho et al. 2014; Pokora et al. 2013). Hence, food derived protein hydrolysates are promising ingredients for developing functional foods (Chalamaiah et al. 2012). Several works in the past few decades have demonstrated that protein hydrolysates from various food sources exhibited various beneficial pharmacological properties namely, antioxidant (Udenigwe and Aluko 2012), hypotensive (He et al. 2013), anticancererous (Kannan et al. 2010), immunomodulatory (Nelson et al. 2007) and hypoglycemic activities (Nasri et al. 2015). These functions are related to biopeptides of 3–50 amino acid residues in length, present in the protein hydrolysates. Therefore, they may be used as alternative to artificial drugs. Indeed, most food proteins contain bioactive peptides that are inactive within the sequence of their parent proteins, and can be released by enzymatic hydrolysis, either during gastrointestinal digestion in the body by endogenous proteases or during food processing or by proteolytic processes using appropriate exogenous proteases (Clare and Swaisgood 2000).

Chemical Hydrolysis

Chemical hydrolysis is hydrolysis of peptide bonds with either acid or alkali solutions. However, chemical process is ecologically undesirable due to the involvement of strong acids and bases. Further, products obtained by chemical hydrolysis have less nutritional qualities and biological activities, since unwanted products could be produced during nonspecific chemical treatment. In addition, acid hydrolysis treatment oxidizes cysteine and methionine, destroys some serine and threonine, and may convert glutamine and asparagine to glutamate and aspartate, respectively (Bucci and Unlu 2000). On the other hand, chemical hydrolysis cannot be controlled to obtain reproducible bioactive protein hydrolysates, since cleavage of peptide bonds by chemical reagents is not specific. Hence, high variations during cleavage lead to high variations in bioactivity. These drawbacks significantly limit the high-value applications of these protein hydrolysates. However, acid hydrolysis is used in the production of flavor enhancers (Pasupuleti and Braun 2010).

Enzymatic Hydrolysis

Among the methods and according to the literature, in vitro hydrolysis of protein substrates is the most widely used process for the production of protein hydrolysates by using appropriate exogenous proteolytic enzymes. These methods produce peptides with desirable biological properties (Kristinsson and Rasco 2000). However, enzymatic hydrolysis is a valuable approach to produce protein hydrolysates compared to chemical hydrolysis due to milder process conditions required (pH 6.0–8.0; temperature 40–60 °C) and better control than enzymatic hydrolysis. On the other hand, bioactivities of protein hydrolysates obtained by enzymatic process can be reproducible as compared to chemical hydrolysis. Furthermore, in contrast to chemical process, the overall amino acid composition of enzymatic protein hydrolysates is nearly similar to that of the protein substrate, with slight modifications depending on the applied enzyme(s). Additionally, enzymatic digestion is suitable for the food and pharmaceutical industries due to non involvement of organic solvents or toxic chemicals (Kim and Wijesekara 2010). A schematic representation of protein hydrolysates production by in vitro enzymatic hydrolysis is illustrated in Fig. 1.

Protein hydrolysates can have a bitter taste and the reduction of this bitterness is necessary to make the hydrolysates acceptable to consumers. The most promising approach of protein hydrolysates for reducing bitter peptides is enzymatic hydrolysis by exopeptidases, including amino- and carboxy-peptidases without reducing yields (Saha and Hayashi 2001). Exopeptidases can selectively cleave peptide bonds

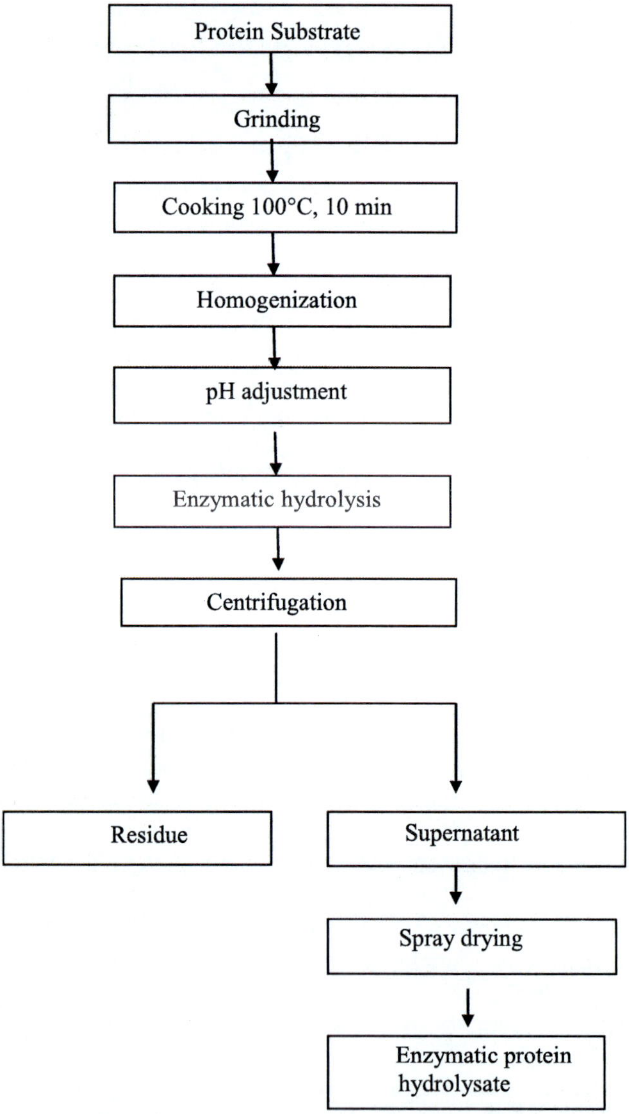

Fig. 1 Schematic representation of protein hydrolysates production by enzymatic hydrolysis

at the N– or C–termini of bitter peptides, releasing free hydrophobic amino acids and further reducing the bitter taste (Raksakulthai and Haard 2003). Sequential hydrolysis by endo- and exopeptidase of wheat gluten hydrolysates has been reported to reduce the bitterness (Liu et al. 2016). Cheung et al. (2015) claimed that whey protein hydrolysates produced by exopeptidase treatment decreased bitterness and increased the umami taste. Umami is a savoury taste, corresponding to the flavour of glutamates, especially monosodium glutamate. Recently, umami peptides suppressing bitterness via the human bitter taste receptor has been reported (Kim et al. 2015), which may serve as an alternative strategy to reduce bitter taste and raise consumer preference towards protein hydrolysates. Even though with the help of exopeptidase treatment, some low bitter protein hydrolysates have been prepared from dairy or plant proteins (FitzGerald and O'Cuinn 2006).

Bioactivity of Protein Hydrolysates

In vitro, animal or plant proteins release many peptides that are bioactive and have regulatory functions in humans beyond normal and adequate nutrition. A wide range of biological activities of protein hydrolysates has been reported as shown in Fig. 2.

Antioxidant Activity

Oxidation of fats and oil by reactive oxygen species (ROS) during food processing and storage is responsible for many degradation processes in foods because it results in the generation of off-flavors, odors, as well as potentially toxic products (Lin and Liang 2002). Furthermore, formation of free radicals and other ROS within the body can cause DNA mutations, protein malformations, and oxidation of phospholipids which are intimately involved in many degenerative diseases such as cancer, diabetes, coronary heart disease, atherosclerosis, hypertension, and Alzheimer's diseases (Diaz et al. 1997). To prevent lipid oxidation in foods and to provide

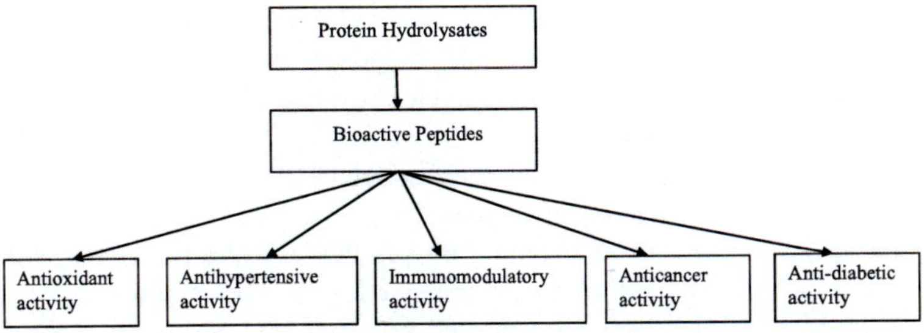

Fig. 2 Schematic representation of potential bioactivities of protein hydrolysates

protection against serious diseases, it is important to inhibit lipid oxidation and free radicals formation occurring in the foodstuff and living body.

Many food proteins, upon hydrolysis, have been shown to possess antioxidant activities. Antioxidant peptides from protein hydrolysates contain certain amino acid residues that help in preventing lipid-oxidation, as well as chelating oxidant metal ions. Yee et al. (1980) reported tha tsoy protein hydrolysates upon proteolysis exhibited antioxidant potential, as measured by the thiobarbituric (TBA) assay, which is a measure of lipid peroxidation. It was proposed that the liberation of bound antioxidant phenolics or copper chelating agents was responsible for the observed antioxidant activity. Similarly, soy protein hydrolysates manufactured with Flavourzyme or chymotrypsin had antioxidant potential greater than unhydrolysed soy protein isolate. However, it was observed that using enzymes such as papain, large-scale degradation unfavourably altered the antioxidant activity (Pena-Ramos and Xiong 2002). In addition, protein hydrolysates derived from wheat germ also possess radical scavenging abilities, with an antioxidant activity close to that of the well known, antioxidant α-tocopherol. Interestingly, these hydrolysates had low molecular weight <1500 Da (Zhu et al. 2006). Memarpoor-Yazdia et al. (2013) revealed that protein hydrolysates and purified peptides from fruits of *Zizyphus jujuba* could be included as antioxidant constituents in the food industry due to their role in food quality preservation and human health promotion. Furthermore, in rats fed a high-cholesterol diet, oral administration of boops boops protein hydrolysates has been reported to enhance antioxidant activity by strengthening antioxidant enzymes (Lassoued et al. 2014). Recently, Kumar et al. (2016a, b) reported that using enzymes such as Alcalase, α-chymotrypsin, and Papain, camel milk casein could be used to produce different protein hydrolysates with high antioxidant and antimicrobial activities.

Anti-Hypertensive Activity

High blood pressure is a precursor of the pathological condition called hypertension, which affects 30% of the adult population in the world and has been regarded as one of the major independent risk factors for cardiovascular diseases (CVDs) (Harris et al. 1985). Angiotensin-converting enzyme (ACE) inhibitors are an example of a drug treatment in regulating blood pressure. ACE reduces the conversion of angiotensin-1 (vasodilatory) to angiotensin-2 (vasoconstrictory) leading in a reduced blood pressure. Hence, there is great interest in novel compounds that can inhibit ACE. Among bioactive food protein hydrolysates, those that are ACE inhibitors are of particular interest for the prevention and treatment of hypertension. Indeed, the ACE-inhibitory peptides that modulate the renin–angiotensin system (RAS) are the most studied bioactive peptides.

Most of the recognized ACE-inhibitory peptides are short peptides with 2–20 amino acid residues and exhibiting different activities. This finding was supported by a study of Wu and Ding (2001) who reported that ACE-inhibitory activity increased with decreasing molecular weight of peptides in soy protein hydrolysate. In another study, wheat germ hydrolysate and its dominant peptide has been reported to be significantly reduced mean arterial pressure (MAP) in spontaneously hypertensive rats. It was found that the dominant bioactive peptide could be metabolised by an aminopeptidase to form an ACE inhibitor metabolite, indicating potential blood pressure lowering effects of the metabolite after absorption (Matsui et al. 2000). Similarly, a buckwheat protein hydrolysate was found to decrease systolic blood pressure in spontaneously hypertensive rats and also hamper ACE, especially when proteolysis was accomplished with pepsin followed by chymotrypsin and trypsin (Li et al. 2002). Casein and whey hydrolysates have also been observed to have strong antihypertensive activity, by way of ACE inhibition (Nongonierma and O'keeffe 2016). More recently, it was found that the enzymatic hydrolysate of flounder fish muscle includes novel ACE inhibitory peptides may be beneficial as a functional food for treating hypertension (Ju-Young et al. 2016).

Immunomodulatory Activity

It is well known that immune function plays an important role in preventing and controlling chronic diseases. However, there are few strategies available to effectively modulate the immune response. It is imperative, therefore, to find new and more efficiency immunomodulators that enhance immune function such as food protein-derived peptides.

In foodstuffs, immunomodulatory substances play key role as health-benefiting factors that protect the body from invading pathogens. Recently, immunomodulatory protein hydrolysates or peptides have been produced from a wide range of food protein sources. These include milk (Masotti et al. 2011; Pan et al. 2013), whey (Javier et al. 2014; Ma et al. 2014), egg (Lee et al. 2009; Lozano-Ojalvo et al. 2016), fish (Karnjanapratum et al. 2016; Chalamaiah et al. 2015), wheat (Horiguchi et al. 2005) and soybean (Vernaza et al. 2012; Yimit et al. 2012). In a study, Nelson et al. (2007) have demonstrated that peptidic fractions obtained from egg yolk induced a humoral immune response associated with an increase of IgA+ cells in lamina propria in BALB/c mice. In another study, Kong et al. (2008) reported that enzymatic hydrolysis of soy protein hydrolysates by Alcalase 2.4L released some low molecular weight and positively charged peptides which showed immune stimulatory activity. Furthermore, it has been shown that enzymatic hydrolysis of full-cream milk hydrolysates using papain and trypsin enhanced immunity in mice by increasing the weight of spleen and thymus and stimulating the phagocytosis of macrophages (Pan et al. 2013). Similarly, microencapsulated whey protein hydrolysate has been reported to be significantly enhanced splenocyte proliferation activity (Ma et al. 2014). More, recently, Tidarat et al. (2017) demonstrated that tilapia mince hydrolysate has a stimulatory effect on innate immunity through induction of IL-1β and cyclooxygenase-2 (COX-2) expressions in THP-1 macrophages.

Anti-Cancer Activity

Cancer (malignant tumor) is an abnormal growth and proliferation of cells in the body. The number of cancer cases is expected to rise from 14 million in 2012 to 22 million within the next two decades (WHO 2017). Therefore, to treat tumors, inhibition of deregulated cell multiplication is an important approach (Chi et al. 2015). Currently, the most promising anticancer approach is chemoprevention which is used to reduce the morbidity and mortality of cancer by decelerates the progression of carcinogenesis (Pan et al. 2016; Sheih et al. 2010). However, chemotherapy is costly and anticancer drugs have many unintended side effects which can cause damage to normal cells (Sah et al. 2015). Therefore, discovery of new anticancer agents from natural sources like food proteins can provide better alternatives for cancer prevention and management.

Several studies have been focused on anticancer activity of protein hydrolysates. Kim et al. (2000) isolated thermoase from soy protein hydrolysates, an anticancer hydrophobic peptide. In vitro, some protein hydrolysates obtained from several fish proteins exhibited a significant antiproliferative activity on human cancer cell lines (Picot et al. 2006). In another study, Kannan et al. (2010) purified and isolated anticancer pentapeptide, Glu-Gln-Arg-Pro-Arg, from rice bran. The peptide was reported to cause 84% inhibition to colon cancer cells (Caco-2, HCT-116) growth, 80% to breast cancer cells (MCF-7, MDA-MB-231) growth and 84% to liver cancer cells (HepG-2) growth at 600–700 μg/mL doses. Furthermore, administration of chickpea albumin hydrolysates to mice significantly increased the tumor inhibition rate and decreased tumor volume (Xue et al. 2012). Recently, Kim et al. (2012) isolated a new and potent anticancer peptide (Ala-Phe-Asn-Ile-His-Asn-Arg-Asn-Leu-Leu) from pepsin shellfish Mytilus coruscus protein hydrolysate, which was reported to induce cell death on prostate, breast, and lung cancer cells but not on normal liver cells.

Anti-Diabetic Activity

Diabetes mellitus (DM), characterized by a chronic glucose increase of blood level, is a widespread metabolic disease affecting millions of individuals in the world. DM is involved in several physiological processes, such as increasing utilization of the glucose by the peripheral tissues and lowering hepatic glucose output and adipocyte fat-accumulation (Zeng et al. 2016). Current synthetic antidiabetic drugs may result in risks of hypoglycemia, weight gain, high background risk of pancreatitis and gastrointestinal side effects while some patients may not tolerate them. For these reasons, the search of new, safe, and natural specific antihyperglycemic agents as alternative to synthetic ones is on the increase. In fact, various studies have been focused on the investigation of the hypoglycemic effects of protein hydrolysates or bioactive peptides from plant or animal sources (Yuan et al. 2008; Huang et al. 2014; Nasri et al. 2015).

In this context, Lee et al. (2007) reported that silk protein hydrolysates administration to rats significantly reduced blood glucose level by stimulating the insulin secretion and raising the serum leptin level. Similarly, protein hydrolysatesisolated from *M. charantia* L. Var. abbreviate Ser. (MCV) significantly lowered the BGL in alloxan-induced diabetic mice, using enzymes Alcalase and Pancreatin (Yuan et al. 2008) and Alcalase hydrolysate had stronger hypoglycemic effect than that of Pancreatin hydrolysate at the same dose, while the MCV undigested proteins had no effect on the BGL in diabetic mice. Furthermore, Yu et al. (2011) isolated eight potential antidiabetic peptides from egg white hydrolysate having α-glucosidase and α-amylase inhibitory activities. Among the eight peptides, Arg-Val-Pro-Ser-Leu-Met was the potent α-glucosidase inhibitor peptide with an IC50 value of 23.07 μmol/L. Similarly, Yu et al. (2012) described the purification and identification of novel albumin derived peptides. Among them, Lys-Leu-Pro-Gly-Phe showed α-glucosidase and α-amylase inhibitory activities. Authors concluded that this peptide has an antidiabetic potential. Recently, Roblet et al. (2016) have shown that peptides from salmon protein hydrolysates, separated by electrodialysis with filtration membrane, were reported to increase glucose uptake in L6 skeletal muscle cells.

Edible Films and Coatings

Edible films or coatings are defined as a thin layer of material which is used to coat different foods for extending shelf life and can be consumed together with that food. Recently, edible coatings and films have acquired good sized interest because of their advantages compared to artificial films. The most important is that they can be safely consumed with the packaged products (Bourtoom 2008). Edible coatings/films can effectively prevent moisture losses, gas aromas and solute migration out of the food, while selectively permitting for controlled exchange of important gases, such as oxygen, carbon dioxide, and ethylene, which are involved in food product respiration (Embuscado and Huber 2009). Additionally, the materials that are used for this purpose can completely coat the food or can be used as a continuous layer between food components (Guilbert 1995).

The motivation for the increase of interest in edible coatings is due to the rising consumer demands for safe, convenient, and stable foods, and also the awareness of the harmful environmental effects of non-biodegradable packaging. In most cases, Biofilms will degrade more readily than polymeric materials as they are produced completely from edible and generally renewable resources. Even if the films are not consumed, they generally will still contribute to the reduction of environmental pollution (Bourtoom 2008). Edible films and coatings provide physical protection to protect food products from mechanical damage, and from physical, chemical and microbiological activities (Min et al. 2005). They also can be edible, biocompatible, nontoxic, and perform as both a barrier and a carrier of food additives such as antioxidants and antimicrobials.

Although edible coatings and films have a similar definition, there is a difference. Generally, edible films are prepared separately and then applied to the surface of the food, whereas coatings are formed directly onto the food surfaces (Cordeiro de Azeredo 2012). Studies have shown that both methods when rightly formulated can improve the organoleptic characteristics of packaged food. Furthermore, they can function to retard oxidation and/or delay microbial spoilage by the incorporation of antibacterial and antioxidant agents. However, compared to artificial films, their permeability and mechanical properties are generally poorer (Kester and Fennema 1986) and this will limit their use to specific applications. With further research, some of these drawbacks will be overcome.

Edible coatings can keep fruits and vegetables fresh by improving the retention of color, acid, sugar, and flavor components to prolong shelf life and retain nutritional quality (Fakhouri et al. 2015). Most biopolymers which are used for the formation of Edible films and Edible coatings are polysaccharides, proteins, and lipids (Jahed et al. 2017). Among these, protein-based materials appear to be most attractive because they also provide nutritional value (Galus and Kadzinska 2016).

Protein Based Edible Films and Coatings

Proteins are good film formers exhibiting excellent oxygen, carbon dioxide, and lipid barrier properties, particularly at low relative humidities. Edible films made of proteins were reported to possess satisfactory mechanical properties (Peyron 1991), but their mechanical strength limit their application in food packaging compared toartificial polymers (Bourtoom 2009). Also, they are poor water barriers due to their predominantly hydrophilic character (Peyron 1991; McHugh 2000). Increasing cohesion among protein polypeptide chains by way of chemical (glutaraldehyde, formaldehyde, glyceraldehyde, glyoxal), enzymatic (transglutaminase), or physical (heating, irradiation) treatment was thought to be effective for modifying the barrier properties of films as well as the mechanical properties and resistance to proteolysis of films (Bourtoom 2009; Hernandez-Munoz et al. 2004a, b; Orliac et al. 2002; Ouattara et al. 2002a, b; Sabato et al. 2001). Edible films from different protein sources have been formulated. These include casein, gelatin, whey protein, corn zein, soy protein and wheat gluten (Bourtoom 2008; Saglam et al. 2013). Table 1 illustrates various advantages and disadvantages of some protein based edible films and coatings.

Zein

Zein, the major storage protein in corn accounts about 44–79% endosperm protein (Lawton 2002). Zein does not dissolve in water and dilute salt solutions due to the high composition of hydrophobic amino acids (such as leucine and alanine) and acidic amino acid (glutamic acid, 21–26%), but dissolve in 60–95% ethanol (Chen

Table 1 Advantages and disadvantages of some protein based films

Type	Advantages	Disadvantages	References
Zein	Good film forming Good barrier to water vapour and volatile components	Extremely brittle	Ozcalik and Tihminlioglu (2013)
Wheat gluten	Good film forming and visco-elastic properties Barrier to water vapour, oxygen and carbon dioxide at low relative humidities	Inherent water sensitivity and low mechanical strength	Gontard et al. (1992, 1996); Rocca-Smith et al. (2016)
Soy protein isolate	Film forming ability Low permeability to oxygen and carbon dioxide	Light beany flavor, film is rather brittle, and has relatively poor mechanical properties	Cao et al. (2007)
Gelatin	Excellent film-forming and good mechanical properties	More expensive than plant proteins	Cao et al. (2007)
Casein	Film formation ability Good barrier to the oxygen and other non polar molecules	Highly moisture sensitive, absorbing and releasing water molecules that greatly affect their mechanical and barrier properties	Picchio et al. (2018)
Whey Protein isolate	Film forming capacity Low barrier properties to oxygen, aromas and volatile compounds	Low mechanical properties and high water vapour permeability	Umaraw and Verma (2017)

et al. 2014). Biodegradable films and coatings are produced from zein due to its good film forming properties. Compared to other types of protein-based edible films, zein film exhibits good barrier against transmission of water vapor and volatile components (Ozcalik and Tihminlioglu 2013). However, brittleness of zein films is a major drawback. Therefore, to reduce brittleness, plasticizers have been incorporated in film forming formula (Lawton 2002; Paramawati et al. 2001). In addition, functional additives have been incorporated in zein to confer edible films with multiple activities. Phenolic compounds were incorporated into zein to modify the mechanical properties and to increase the antioxidant activities (Arcan and Yemenicioglu 2011). Cheng et al. (2015) demonstrated that zein/chitosan composite films exhibited excellent mechanical properties and water vapor barrier properties and with the addition of phenolic compounds, the composite films exert multiple functionality including antioxidant and antimicrobial activities. Moreover, mechanical properties can also be enhanced by incorporating polyvalent organic acids into zein forming mixture, which may be due to interaction between the terminal amino acids (serine or threonine) of zein molecules and the polyvalent organic acids (Selling and Sessa 2007).

Wheat Gluten

Wheat gluten, a hydrophobic protein of wheat flour is considered to be globular protein composed of a combination of polypeptide molecules. Wheat protein composed of two major components, i.e. gliadin and glutenin. Gliadin is soluble in 70% ethanol while glutenin is not soluble (Gennadios and Weller 1990). Even though wheat gluten can dissolve in aqueous solutions at low ionic strength of high or low pH but is not soluble in water (Krull and Inglett 1971). It has been shown that wheat gluten has interesting gel and film formation ability (Sun et al. 2009), acting as semi permeable barrier to water vapor, oxygen and carbon dioxide molecules (Gontard et al. 1996). Wheat gluten films could be thus applied as food coatings or edible films on naturally gluten containing foods (e.g. bakery products) in order to slow-down mass transfer phenomena such as water and oxygen, which are known to decrease the shelf life.

One of the main issues related to the utilization of wheat gluten films is their inherent water sensitivity. The mechanical and the barrier properties are highly modified in wet conditions due to water sorption and subsequent plasticization (Gontard et al. 1996; Lens et al. 2003). Incorporating a hydrophobic phase constituted by edible lipids into the protein or polysaccharide matrix is considered one of the most promising approaches for reducing water sorption and water transfer through edible films. Recently, Rocca-Smith et al. (2016) reported that by the addition of a lipid phase reduced the water sorption, water affinity (hydrophilicity) and water transfer of wheat gluten films. Moreover, by the lipid addition, mechanical properties are also affected with a decrease in rigidity and, at high aw, an increase in extensibility.

Legume

Legume proteins are called as "poor man's meat" due to their low price as compared to animal proteins. Legumes contain a protein level 2–3 times higher than cereals, ranging from 17 to 30% in chickpeas, lentils, dry peas, beans and 35–49.6% in soybeans (Boye et al. 2010). The 2 major storage proteins present in legumes are albumins and globulins (Gonzalez-Perez and Arellano 2009). Among them, globulins have the largest share (about 70%) of total legume-based storage proteins. In recent years, legume protein based edible films and coatings have acquired good attention due to their nutritional value as well as carrier of antimicrobial, antioxidant and other bioactives. Soy proteins are regarded as the most typical representative of legume proteins owing to high level of protein and decent functional characteristics. Soybean protein materials are suitable for edible coatings because of their low permeability to oxygen and carbon dioxide, and reasonable cost (Kang et al. 2013). It has been demonstrated that soybean protein materials combined with other materials have been used to prolong the shelf life of fresh-cut egg plants and walnut kernels (Ghidelli et al. 2014; Kang et al. 2013). In addition, soy protein-gelatin films could also act as carrier of different bioactives use for preservation of food (Denavi et al. 2009). Other legume proteins have also been investigated for preparation of edible films and coatings. Bamdad et al. (2006) prepared films from lentil protein concentrate (LPC) with the addition of glycerine showed good mechanical properties and water vapor permeability in concomitant with good solubility. Kowalczyk et al. (2014) prepared pea protein based films in combination with plasticizers (glycerol and sorbitol) and films plasticized with sorbitol exhibited higher tensile strength and elastic modulus. Kidney bean protein and chitosan based composite films containing antimicrobial (nisin) were fabricated. Antimicrobial activity results revealed that test bacteria were sensitive to composite films (Fan et al. 2014).

Gelatin

Gelatins are denatured protein, obtained from animal collagen protein through partial hydrolysis by acid or alkaline, or structural changes by enzymatic or thermal degradation. Gelatin has been one of the most studied biopolymers due to its film-forming ability and it is used as an outer film to protect food from drying and exposure to light and oxygen (Arvanitoyannis 2002). Films formed from gelatin sources as a primary biopolymer are more desirable to produce as they are of low cost and easily available (Hanani et al. 2013). Gelatin films or coatings exhibit good transparency, mechanical and barrier properties and can be manufactured by extrusion or casting processes (Hanani et al. 2013; Molinaro et al. 2015). It has been shown that gelatins form clear, flexible, strong and oxygen-impermeable films in presence of plasticizers when cast from aqueous solutions (Gennadios et al. 1994). Edible coatings based on gelatin reduce oxygen, moisture, and oil movement and serve as carriers of antioxidants or antimicrobial agents (Krochta and de Mulder-Johnston 1997). Gelatin films can act as good gas barriers but their hydrophilic nature also results in poor water barriers. According to Gomez-Guillen et al. (2011) the main factors affecting the physical and structural properties of gelatin are the molecular weight distribution and amino acid composition, and these characteristics could play a key role in the physicochemical properties of the resulting films.

Casein

Casein, a major class of protein found in milk, possesses a rheomorphic structure with a high degree of molecular flexibility and large number of polar groups, all of which provide for good film-forming properties. For these reasons, casein is considered as an excellent candidate for numerous applications such as paper coatings, adhesives, and food packaging materials (Ma et al. 2015). Caseins form films from aqueous solutions without further treatment due to their random-coil nature and ability to hydrogen bond extensively. It is accepted that electrostatic interactions also play a key role in the preparation of casein-based edible films (Gennadios et al. 1994). The casein films possess reasonably good strength and low permeability for oxygen and other nonpolar molecules. Due to presence of polar functional groups, such as carboxyl and amino groups, casein films combined with other packaging materials can also be used to protect products susceptible to oxidation (Bonnaillie et al. 2014). Apart from these, food additives, such as flavours, antimicrobial agents, antioxidants, vitamins, minerals and colors can also be added into these films to provide for controlled rate and site of release (Core 2005). In addition, casein also presents an excellent material for edible packaging due to the exceptional nutritional value (Mei and Zhao 2003). However, similar to other protein-based materials, casein possesses two major disadvantages: limited elasticity and water sensitivity, which may limit its applications (Bonnaillie et al. 2014; Ma et al. 2015).

Whey Protein

Whey proteins are valuable by-products from the cheese industry, which have been widely used in a variety of foods due to the superior gelling and emulsification properties. Due to the consumer demand for high-protein foods and supplements, whey proteins demand has been increased by the consumer awareness on its impact of food on human health, the documented nutritional and health properties of whey proteins (Krissansen 2007). Whey proteins contain a diverse amount of proteins with peculiar properties. According to the protein content, it is available as whey protein concentrate (WPC) or whey protein isolate (WPI), which ranges from 20 to 80% and>90%, respectively. They have good film forming capacity, and present as main advantages the low barrier properties to oxygen, volatile aromas and lipids (Ramos et al. 2012). WPC contains higher amounts of fat and lactose than WPI, and these compounds could influence the properties of edible coatings because they could act as discontinuities in the protein matrix of edible films and cause changes in barrier and mechanical properties (Oses et al. 2009).

Films based on whey protein isolates (WPI) are flexible, transparent and exerts excellent barrier function for gas, aroma compounds, and oil as compared to the films made with polysaccharides and lipids. However, due to the presence of hydrophilic amino acids in their structure, these films have low mechanical properties and high water vapor permeability (Umaraw and Verma 2017). Therefore, protein denaturation (physical, chemical, or enzymatic) is required for producing dense and strong films, thus, exhibits good mechanical resistance with excellent barrier to water vapours. Notably, when WPI is heated at low ionic strength and low pH, they can self assemble into semi flexible amyloid-like protein nanofibrils (WPNFs) (Loveday et al. 2012). Properties such as gelation, emulsification, foam stability, antioxidant activity, and digestibility become remarkably modified upon the formation of WPNFs, yielding a class of high-performance biomaterials (Loveday et al. 2012; Mohammadian and Madadlou 2016). In addition, hydrophilic plasticizers are generally added to the film-forming protein dispersion for reducing film brittleness, thus improving flexibility, extensibility, toughness, and tear resistance (Ramos et al. 2013). Glycerol (Gly) is the most widely used plasticizer for whey protein films, providing control of moisture transfer, respiration rate, and oxidation processes, and prolonging shelf life (Silva et al. 2015).

Encapsulation of Bioactive Compounds

Plants have antioxidant and antimicrobial properties due to the presence of many bioactive phenolic compounds, such as anthocyanins, flavonols, flavan-3-ols, curcumin, proanthocyanidins, and phenolic acid derivatives, thus, may help in preventing some diseases such as urinary tract infections (Blumberg et al. 2013), stomach ulcers and cancers (Pappas and Schaich 2009). However, at neutral pH and exposure to oxygen, these phenolic compounds are chemically unstable (Betz et al. 2012) and may deteriorate due to the oxidative degradation leading to the generation of free radicals (Chen et al. 2016). Furthermore, there is development of unpleasant tastes and off-odors in the fortified product and subsequently may result in a harmful impact on shelf stability, sensory characteristics and consumer acceptability of the product (Ariyarathna and Karunaratne 2015; Gomez-Mascaraque et al. 2017). Therefore, in order to preserve the quality of bioactive compounds, or to enhance their applicability to food, neutraceutical or biological formulations, encapsulation is considered as a feasible alternative. Encapsulation has been regarded as an attractive method to entrap bioactive compounds within a polymer material for the purpose of protecting and delivering bioactive compounds at the right time and to a targeted site (Ezhilarasi et al. 2013).

The demand for encapsulation systems continues to grow as the food industry needs to preserve the benefits of active compounds and deliver them at specific conditions. Encapsulation of bioactive compounds in food industry can be used to: (1) preserve functional properties, (2) increase the stability of compounds with low solubility in relevant (mostly aqueous) media, (3) mask undesirable flavours, (4) improve health benefits of food products (i.e. development of functional foods), (5) control the release of bioactive compounds at desired time and specific target, and (6) enhance the bioavailability of bioactive compounds (Davidov-Pardo et al. 2015; Gunasekaran et al. 2007; Huang et al. 2010; Kayitmazer et al. 2013; Livney 2010). In food applications, encapsulating materials must be recognized as "generally regarded as safe" (GRAS) materials (Robin and Sankhla 2013). Carbohydrate polymers, proteins, lipids, and other organic and inorganic materials are mostly used for encapsulation in the food sector. Due to the desirable drying properties of carbohydrates and ability to form matrices, various materials such as starches, maltodextrins and corn syrup are often used as microencapsulating agents (Gharsallaoui et al. 2007). However, carbohydrates usually have poor interfacial properties and must be chemically modified to improve their surface activity (Kanakdande et al. 2007). In recent years, an increasing interest in food protein based microencapsulation can be attributed to their excellent emulsifying, gel-and film-formation properties (Chen et al. 2006). Additionally, protein coatings are degradable by digestive enzymes, thus can be used in developing food applications for controlled-core release (Chen et al. 2006).

Phenolic Compounds in Plants

The main categories of bioactive compounds found in plants are anthocyanins, flavanols, flavan-3-ols (catechin monomers), curcumin, proanthocyanidins, and phenolic acid derivatives (Cote et al. 2010). Table 2 illustrates encapsulation of bioactive compounds in protein matrices.

Anthocyanins

Anthocyanins as water-soluble bioactive compounds are widely present in flowers, fruits, and vegetables (Flores et al. 2014a, b). Cyanidin, delphinidin, and peonidin are the most common anthocyanin molecules which bind to one or more sugar molecules. In anthocyanidin molecule, glycosylation almost always occurs at the C3 position (Cote et al. 2010). The high amount of anthocyanins contributes to the color of the plants (Blumberg et al. 2013). Their color depends on the pH: anthocyanins exist as red-colored flavylium cationat pH below 3 which is the stable form and they convert to quinoidal base (blue color)at pH ranging from 6 to 8 which is an unstable state and therefore, anthocyanins bioavailability is reduced (Oidtmann et al. 2012). In plants, anthocyanins are stable as the pH of the cells is acidic. However after extraction, anthocyanins tend to be unstable at high pH and in the presence of oxygen, therefore, susceptible to degradation (Betz et al. 2012; Flores et al. 2014a, b). Anthocyanins can also degraded by heat treatment, UV light exposure, and presence of specific enzymes or some metallic ions (Betz et al. 2012; Frank et al. 2011). Anthocyanins are absorbed by the stomach and the intestine after digestion (Zhang et al. 2014). However, the bioavailability of anthocyanins is very low as the environmental conditions trigger the degradation of anthocyanins in the gastrointestinal tract (Betz and Kulozik 2011). Hence, introducing a protective method to preserve them has acquired a lot of interest recently. Microencapsulation is the most promising approach for improving anthocyanins stability which aims at ensures the quality and physicochemical stability of heat- or photosensitive components (Vicente et al. 2017).

Oanceaa et al. (2018) investigated microencapsulation of the sour cherries anthocyanins within whey proteins isolate and gum acacia via freeze-drying method. In vitro digestibility tests revealed that the whey proteins isolate had the ability to protect the anthocyanins from the gastric digestion, their release being facilitated into the intestine. Similarly, Shaddel et al. (2018) encapsulated black raspberry anthocyanins followed by GE–GA (gelatine:gumacacia) complex coacervate matrix by double emulsion system, and results revealed that this treatment ismore efficient as it confers high loadingcapacity, good morphological features, high retention of anthocyanins, aswell as more color retention and greater stability during storage.

Table 2 Encapsulation of bioactive compounds in protein matrices

Encapsulating ingredient	Matrix	Observations	References
Black raspberry anthocyanins Blueberry anthocyanins	Gelatin/gumacacia Whey protein isolate	High loading capacity and high retention of anthocyanins Microcapsules showed sustained release of anthocyanins	Shaddel et al. (2018) Flores et al. (2014a, b)
Quercetin	Bovine serum albumin Zein nanoparticles	The antioxidant activity was protected within the nanoparticles The encapsulated quercetin had chemical stability against alkaline pH and UV light	Fang et al. (2011) Patel et al. (2012)
EGCG	β-lactoglobulin(β-Lg) Zein	Protect against oxidation, degradation and improved encapsulation efficiency (60–70%) The bitterness and astringency of EGCG was masked by the nanoparticles. The release of EGCG under simulated gastric digestion was limited The formed fibers had a diameter ranging from 150 to 600 nm	Shpigelman et al. (2010, 2012) Li et al. (2009)
Curcumin	Zein Gelatin	The encapsulation achieved an efficiency of 85–90% and curcumin was evenly distributed in the zein nanoparticles Through encapsulation, the water solubility of curcumin and the curcumin bioaccessibility in simulated gastrointestinal fluid was improved	Gomez-Estaca et al. (2012, 2015)
Sorghum condensed tannins (SCT) Pomegranate ellagitannins (PPE), mainly punicalagin	Kafirin Gelatin	The encapsulation efficiency of SCT was around 48% and only a limited amount of SCT was digested under the condition of simulated gastrointestinal fluids The nanoparticle had similar apoptotic effect as PPE on human promyelocytic leukemia cells HL-60	Links et al. (2015) Li et al. (2011)
Gallic acid Caffeic acid	Zein fibres Bovine serum albumin (BSA)	Retention of antioxidant activity Stability of the antioxidants improved with the presence of BSA	Neo et al. (2013) Almajano et al. (2007)

Flavanols

Flavanoids are polyphenolic secondary metabolites of plants Kawabata et al. 2015. Quercetin and kaempferol are the main compounds of this family, being often associated with different sugars, producing glucosides, galactosides, glucuronides, and diglycosides (Garrido and Borges 2013). The most abundant flavanol is quercetin, also known as 3,3,4,5,7-pentahydroxyflavone,occurring in a variety of foods such as onion, apple, broccoli, tea, and red wine (Lakhanpal and Rai 2007). Due to its several pharmacological effects like anti-obesity (Nabavi et al. 2015), antioxidant (Duenas et al. 2010), anti-inflammatory (Kleemann et al. 2011), and anti-cancer (Gibellini et al. 2011), current researches are directed towards formulating food products by adding quercetin. However, under heat treatment, exposure to oxygen and light, it is susceptible to degradation (Aceituno-Medina et al. 2015).

Quercetin has low aqueous solubility and bioavailability, chemical instability and short biological half-life, which are main drawbacks of its use in the food and pharmaceutical applications (Cai et al. 2013). Due to its lipophilic nature, it is moderately soluble in ethanol and highly soluble in dimethyl sulfoxide (Priprem et al. 2008; Ferry et al. 1996). It is therefore difficult to directly incorporate high levels of quercetin into water-based food matrix. Encapsulation of quercetin may enhance the physicochemical stability in liquid food products and also the final bioavailability within the human body.

Several studies proved that quercetin can bind with animal proteins, including human serum albumin (Sengupta and Sengupta 2002), bovine serum albumin (Fang et al. 2011), collagen (Yang et al. 2009), and β-casein (Mehranfar et al. 2013), which may enhance its solubility, stability, and bioavailability. Fang et al. (2011) discovered a nanoparticle using bovine serum albumin (BSA) as a matrix to encapsulate quercetin, and revealed that under both acidic and neutral conditions, BSA nanoparticles could improve the bioactive properties of quercetin for a long period. In another study, patel and co-workers discovered quercetin-loaded zein nanoparticles with the average particle size <200 nm and they reported that under the alkaline condition and exposition to UV-light irradiation, chemical stability of quercetin was improved entrapped in the colloidal particles (Patel et al. 2012).

Curcumin

Curcumin, also known as 1,7-bis (4-hydroxy-3-methoxyphenyl)-1,6-heptadiene-3,5-dione, is a natural phenolic compound found in the rhizome of the perennial herb, *Curcuma longa* (Sharma et al. 2005). *Curcuma longa* belonging to the group of curcuminoids, have four main types which are curcumin (77%), demethoxycurcumin (17%), bisdemethoxycurcumin (3%), and cyclocurcumin (Heger et al. 2014). Curcumin has been traditionally used as a natural food colorant owing to its yellow-orange color (Gomez-Estaca et al. 2015). In the last few decades, curcumin has acquired a good interest due to its potential health-benefitting properties, likeantioxidant, antimicrobial, anticancer, anti-inflammatory, anti-diabetic and wound healing properties (Gomez-Estaca et al. 2012; Rao and Khanum 2016). However, curcumin is difficult to incorporate in to some food products because it is a fat soluble compound and therefore has poor water solubility (Pinheiro et al. 2016). It is also susceptible to degradation under neutral and alkaline pH, high temperature, and exposition to light, metallic ions, enzymes, and oxygen (Kumar et al. 2016a, b). To overcome these drawbacks, encapsulation of curcumin might be the solution to this problem.

Encapsulation within or complexation with proteins has been regarded as most effective way for increasing the solubility of curcumin in aqueous solutions and enhancing its bioavailability. In this context, different food proteins including soy proteins (Chen et al. 2015), whey proteins (Liu et al. 2017), zein (Gomez-Estaca et al. 2012), and gelatin (Gomez-Estaca et al. 2015) have been used as carriers for improving the aqueous solubility and antioxidant activity of curcumin. Tapal and Tiku (2012) reported that the complexation with soy protein isolate, curcumin aqueous solubility drastically increased by 812-fold. Similarly, aqueous solubility of curcumin increased by about 1200-fold through the complexation with WPN, whereas its solubility is increased by approximately 180-fold through the complexation with WPI (Mohammadian et al. 2018). Isuru et al. (2016) investigated that curcumin encapsulated in a chickpea (*Cicer arietinum*) protein matrix improved the stability and bioavailability in the gut.

Flavan-3-Ols

Flavan-3-ols, also known as catechin monomers, occur naturally as aglycons of catechin and epicatechin without glycosylation in plant foods (Pappas and Schaich 2009). Flavan-3-ols are abundant in tea. In unfermented tea, the concentration of flavan-3-ols ranges from 35 to 115 mg/g of tea leaves, with (−)-epigallocatechin gallate (EGCG) being the most abundant (Socha et al. 2013). EGCG possesses several pharmacological properties such as antioxidant, anti-proliferative, anti-angiogenic and anti-carcinogenic effects (Singh et al. 2011).

The EGCG stability is affected by the pH, oxygen, and the temperature (Shpigelman et al. 2010). EGCG is easily oxidized at neutral and basic pH and forms a dimer, which is irreversible (Shpigelman et al. 2012) and it undergoes epimerization at high temperatures (Li et al. 2012). Due to the low pH in the stomach, EGCG is stable after ingestion while degradation of EGCG occurs in the intestine as the pH rises to a neutral pH (Haratifar et al. 2014). Therefore, EGCG bioavailability is very low. One effective way to prevent catechin degradation in an aqueous environment is microencapsulation

Among the catechins, epigallocatechin-3-gallate (EGCG) has been widely studied in microencapsulation. Catechins have been encapsulated in β-lactoglobulin (β-Lg) matrices by gelation and it was reported that the nanoparticles showed activities against oxidation and degradation. Moreover, these particles had good encapsulation efficiency (60–70%) for EGCG (Shpigelman et al. 2010). In another study EGCG encapsulating in native, heated, and desolvated β-Lg matrix respectively and the encapsulation efficiencies of all three nanoparticles were approximately 90% (Lestringant et al. 2014).

Proanthocyanidins (PACs)

Proanthocyanidins (PACs) are secondary metabolites of plants, which are oligomers of (−)-epicatechin and (+)-catechin (Zou et al. 2012). PACs are effective in preventing urinary tract infections (Zou et al. 2012). However, similar to other polyphenols, they are susceptible to degradation at high temperature, when exposed to oxygen, and also in the gastrointestinal tract due to the low pH and the presence of other compounds (Berendsen et al. 2015). Besides, PACs affects sensory attributes as it imparts bitterness and astringency (Berendsen et al. 2015). Therefore, encapsulation of PACs is of interest. Links et al. (2015) encapsulated sorghum condensed tannins (SCT) in kafirin matrix by coacervation. The encapsulation efficiency of SCT was around 48% and only a limited amount of SCT was digested undersimulated gastrointestinal fluid conditions. In another study, gelatin particles was used for the encapsulation of pomegranate ellagitannins (PPE) containing mainly punicalagin by gelation. The formed particles had a size of below 200 nm and exhibited a high loading efficiency (Li et al. 2011).

Phenolic Acid Derivatives

Phenolic acid derivatives include derivatives of hydroxylbenzoic acid (HBA) and hydroxycinnamic acid (HCA). They are rarely present in the free form as they are oftenbound to sugars by glycosidic linkage (Cote et al. 2010). Like HCA derivatives, HBA derivatives (ellagic, gallic and vanillic acid) are also bound with sugars. Ellagi-tannins are formed by the esterification of ellagic acid with flavan-3-ols (Cote et al. 2010). Phenolic acids are correlated to their characteristic flavour as well as their potent antioxidant and antimicrobial properties (Pappas and Schaich 2009). In general, HAs have low aqueous solubility and a bitter taste and therefore, in order to increase its solubility and bioavailability, protect from degradation, and mask unpleasant tastes, HAs may be encapsulated in polymeric nanoparticles for application in the food industry.

Because of the antioxidant activity of phenolic acids, several studies have been investigated to study their encapsulation. Gallic acid was encapsulated in zein fibers by electrospinning. The generated ultra-fine fibers had a diameter ranging from 327 to 387 nm. The antioxidant activity of gallic acid was retained in the fibers (Neo et al. 2013). Besides electrospinning, caffeic acid was encapsulated in an O/W emulsion at different pH and it was found thatthe antioxidant activity of caffeic acid was enhanced by the presence of BSA (Almajano et al. 2007). Some natural polyphenols, such as EGCG, tannic acid, and theaflavin, were respectively encapsulated in gelatin nanoparticles coated with an LbL shell of polyelectrolytes consisting mainly of a mixture of proteins and polysaccharides and it was revealed that the nanoparticles retained the biological activity of the polyphenols (Shutava et al. 2009).

Proteins as Nano Delivery Systems for Bioactive Compounds

During the last few years, food proteins has grabbed attention for development of encapsulation system due to their excellent functional properties including emulsification, gelation, foaming and water binding capacity, distinct biophysicochemical characteristics such as biocompatibility, structural versatility, offers ability to bind both hydrophobic and hydrophilic bioactive compounds for efficient delivery (Elzoghby et al. 2012; Chen et al. 2006). Moreover, by nanosizing, the protein particle increases bioavailability, absorption, bioadhesion, surface area-to-volume ratios, thereby increasing the beneficial effects (Cerqueira et al. 2014). The protein based nano-particles enhance the absorption and bioavailability of bioactive molecule mainly through the following pathways: (a) protection of the bioactive molecule from the harsh environment of the gastrointestinal tract (GIT), (b) prolongation of the residence time in the gut by mucoadhesion, and (c) endocytosis of the particle.

Protein Based Nanoparticles (PBNs)

Proteins are ideal because they are generally regarded as safe (GRAS) and contain a high nutritional value (Chen et al. 2006). Most proteins can be easily digested by the human gastrointestinal tract (Joye and McClements 2014). Furthermore, proteins are insoluble in acidic conditions as the isoelectric point (pI) of most proteins ranges between pH 3 and pH 5. Protein matrices are usually dissolved at alkaline pH to encapsulate the compounds of interest and the solution is thereafter acidified to form particles (Nesterenko et al. 2013). Both animal and plant proteins have been used as encapsulating matrices. The most common animal proteins used include gelatin, casein and whey proteins (mainly β-lactoglobulin). Plant-derived protein matrices include zein, soy protein, etc. (Joye and McClements 2014; Nesterenko et al. 2013). Plant proteins have an excellent hydrophobic character in comparison with animal proteins. They are non-toxic and reflect the present "green" symbol in food and pharmaceutical applications. They are also considered as less allergenic and less expensive than animal-derived proteins (Nesterenko et al. 2013). Therefore, plant-derived proteins have currently achieved an increasing interest over animal-based proteins.

Cereal Proteins

Zein, the major storage protein of corn, is one of the few hydrophobic, water-insoluble materials approved for food use by the FDA (Patel and Velikov 2014). In food applications, zein has become an extremely attractive delivery system for functional ingredients, not only because of its fully natural, biodegradable and food-grade origin, but also because of its ability to physically entrap a large number of hydrophobic compounds (Davidov-Pardo et al. 2015). Additionally, zein has been used to design functional organic-inorganic composite particles (Baars et al. 2015) and for the stabilization of foams and emulsions (Blanco et al. 2016). Moreover, zein particles are characterized by slower digestibility in the gastrointestinal tract in comparison with other proteins due to its resistance to digestive enzymes and therefore, enhanced capability of controlled release of the payload molecules (Patel and Velikov 2014). Therefore, zein is considered as a good candidate for the encapsulation of functional compounds and development of the delivery system in food and pharmaceutical industries (Yang et al. 2017). Phenolic phytochemical such as procyanidin with antioxidant activity was encapsulated into the zein NPs using an aqueous medium and it was found that the solubility of procyanidin enhanced when it was encapsulated (Zou et al. 2012). Tocopherol encapsulated in Zein-Gum Arabic biopolymer nanoparticles was investigated and found that the presence of GA not only prevented the precipitation of zein nanoparticles but also controlled the release of TOC from zein-GA nanoparticles during in vitro gastrointestinal digestion (Li et al. 2018a, b). More recently, it has been studied that zein/SSPS (soluble soybean polysaccharide) nanoparticles successfully used to encapsulate hydrophobic quercetin. The results disclosed that the encapsulation efficiency was greatly enhanced to 82.5% after SSPS coating, compared to 59.2% for the zein without coating, which suggests that zein/SSPS nanoparticles have great potential to be used as nano-delivery systems for bioactive compounds (Li et al. 2018a, b).

Wheat protein has successful application in encapsulation with alone or in combination with polysaccharides (Andreani et al. 2009; Liao et al. 2012). However, wheat gluten is allergenic and susceptible to celiac disease. Therefore, modification of wheat gluten could improve its properties. It has been reported that wheat gluten can be modified by acid heating deamidation which reduces its cytotoxicity and improves the overall functionalities (Berti et al. 2007). In subsequent, Liao et al. (2012) developed succinic acid deamidated wheat gluten microspheres for encapsulation of fish oil through double emulsion technique. The study revealed that the deamidated wheat gluten microspheres were suitable for the encapsulation, controlled release, and stability maintenance of fish oil.

Barley is a very adaptable crop. Barley grains and by-products are abundant and affordable protein sources which contain 8–13% and 20–30% (w/w) protein, respectively (Yalcin et al. 2008). The two major endosperm storage proteins of barley are hordein and glutelin which ranges from 35–55% and 35–40%, respectably, whereas albumin and globulin proteins are enriched in the bran and germ (Finnie and Svensson 2009). Barley proteins have excellent emulsification and film formation features as they are highly hydrophobic (Wang et al. 2010). Wang and coworkers first reported barley protein as a carrier material to encapsulate fish oil. The study disclosed high encapsulation efficiency and high loading capacity of fish oil. Barley protein microcapsules showed a strong protection to fish oil from oxidation and allowed the nano-encapsulated fish oil to be applicable in food systems (Wang et al. 2011b). It was also reported that barley protein nano-particles was successful to encapsulate β-carotene, and to the development of a targeted delivery system for lipophilic bioactive compounds (Wang et al. 2011a).

Legume Proteins

Legume proteins hold tremendous inherent physico-chemical properties that define the functional characteristics of food systems during processing, storage, and consumption. Solubility, foaming, gelling, and emulsification are the important functional properties (Liu et al. 2012), which dictate the encapsulation behavior of legume proteins. They have successfully been investigated for the encapsulation of bioactive compounds, e.g. lycopene using soy and pea protein-based matrix by oil-in-water emulsion (Ho et al. 2017), folate (vitamin B9) using chick pea by freeze drying (Ariyarathna and Karunaratne 2015), and omega-3 oil using lentil, pea and faba bean proteins by oil-in-water emulsion (Gumus et al. 2017). The application of soy protein isolate in encapsulation has widely been studied by a number of studies particularly for the encapsulation of algal oil, tomato oleoresin, curcumin, lycopene, and paprika oleoresin (Dai et al. 2017; Ho et al. 2017; Li et al. 2015; Rascón et al. 2011; Yuan et al. 2017). Soy protein isolate as an encapsulating material in combination with polysaccharides offers better protection; oxidative stability and drying properties (Rascón et al. 2011; Souza et al. 2017). Yuan and coworkers investigated complex coacervation of soy protein with chitosan for the microencapsulation of algal oil. The study revealed excellent encapsulation efficiency (up to 97.36%) and enhanced oxidative stability than single soy protein (Yuan et al. 2017). Soy protein conjugation with gum acacia revealed higher degree of emulsifying characteristics, better biocompatibility and encapsulation efficiency in the encapsulation of tomato oleoresin compared to soy protein alone (Li et al. 2015). The conjugates also revealed the ability in the protection of lycopene during storage from light, humidity and temperature, and the ability to control release profile of lycopene in gastric condition (Li et al. 2015). Recently, Cheng et al. (2017) developed conjugated nanoparticles by the conjugation of folic acid with modified soy protein isolate. The results showed that folic acid modified soy protein nanoparticles have superb stability and biocompatibility, and are the ideal carrier for tumor-target drug delivery. Moreover, Chang et al. (2016) reported that the combination of the lentil protein, maltodextrin and sodium alginate represented the best wall material to produce microcapsules with the highest entrapment efficiency (88%). They indicated that the resulting lentil protein-maltodextrin-alginate microcapsules showed better oxidative stabilityand had a stronger wall structure than the lentil protein-maltodextrin microcapsules.

Whey Proteins

Whey proteins are typically used as hydrogels, and in the development of nanoparticle systems and complex coacervates with several polysaccharides (Devi et al. 2017). Whey protein products are widely used food ingredients due to their broad range of functionality (Abd El-Salam et al. 2009), along with their exceptional nutritional and physiological properties. β-lactoglobulin and α-lactalbumin are the most common whey proteins which have been extensively studied for encapsulation purposes. β-lactoglobulin was used for encapsulation of sour cherries anthocyanins with a variation of its state (i.e. un-treated, pre-heat treated and cross-linked protein) (Oancea et al. 2017). Maximum encapsulation efficiency (64.69%) is provided by cross-linked protein matrix, while highest antioxidant activity showed by pre-heat treated protein. Compared to untreated protein, cross-linked and pre-treated β-lactoglobulin provided better protection of anthocyanins from the gastric digestion and thus, allowed possible release of anthocyanins into the gut (Oancea et al. 2017). β-lactoglobulin was also reported as an efficient nanocarrier for advanced delivery systems (Shafaei et al. 2017). Fucinos et al. (2017) developed α-lactalbumin nanotubes to encapsulate caffeine as a model molecule. The results disclosed an encapsulation efficiency of near about 100% and loading capacity of about 10%, which suggests a potential ability of α-lactalbumin nanotubes for the encapsulation of bioactive compounds. Ghasemi et al. (2018) investigated that d-limonene was nanoencapsulated with whey proteins and pectin and the encapsulation efficiency was determined to approximately 88%. Moreover, some other recent studies also revealed the potentiality of whey proteins in terms of encapsulation of bioactive compounds either as a single carrier or combination with other carrier compounds (Bilek et al. 2017; Devi et al. 2017; Esfanjani et al. 2017).

Caseins

In the food and pharmaceutical industries, casein protein is tremendously used as a natural vehicle for bioactive substances,especially for hydrophobic substances such as the anti-cancer drug flutamide, green tea flavonoids, and acetaminophen (Elzoghby et al. 2013). Caseins are the major milk proteins, which includes αs1-casein, αs2- casein, β-casein, and κ-casein. They have excellent emulsifying properties, low viscosity in solution, mild flavor and high nutritional value, and therefore they are used as shell matrix for encapsulation (Ho et al. 2017; Jain et al. 2016). Structural and physicochemical properties of β-casein facilitate its application in effective drug delivery (Bar-Zeev et al. 2016). Hydrophobicity and poor solubility of biomolecules are associated with poor absorption and bioavailability upon oral-administration (Sahu et al. 2008). Caseins have been utilised to improve the efficiency of such biomolecules. Camel β-casein was reported to have excellent characteristics as a carrier system for curcumin, a potent anticancer and antioxidant natural polyphenol poorly soluble in aqueous solutions (Esmaili et al. 2011). Curcumin molecules formed complexes with β-casein micelles via hydrophobic interactions which has consequently improved the bioavailability of curcumin by increasing its solubility nearly 2500-fold. Recently, β-casein micelles have been successfully developed to improve oral bioavailability of poorly water-soluble drugs such as celecoxib by encapsulation (Perlstein et al. 2014). Moreover, it was found that Cas-Ca was effectively used to encapsulate paprika red pigment (PRP) and the results revealed that the PRP embedded in the nanoparticles had better stability under the conditions of the high temperature, light, and addition of food additives as compared to PRP without embedded. At the same time, the copolymer nanoparticles can be used as an effective carrier of PRP and can effectively control their release behavior in simulated gastrointestinal fluid (Qiu et al. 2018).

Gelatins

Gelatins are considered as good materials for encapsulation due to good functional properties which includes biocompatibility, biodegradability, water retention ability, film formation ability and anti-carcinogenicity (Dang et al. 2017). However, in aqueous solutions gelatins are rapidly soluble that leads to the fast release of active compounds (Dang et al. 2017). Recently, the behaviour of fast dissolution of gelatins in the aqueous system has been introduced in the encapsulation of curcumin (Gomez-Estaca et al. 2017). Therefore, owing to the insolubility behavior of curcumin, its solubility in the aqueous system has been improved by the encapsulation in gelatin through elecrohydrodynamic atomization (Gomez-Estaca et al. 2017). The results indicated that the water solubility of gelatin encapsulated curcumin microparticles was improved by 38.6 times as compared to curcumin alone. The study also revealed that the gelatin encapsulated curcumin microparticles improved antioxidant and anti-microbiological properties (Gomez-Estaca et al. 2017). In contrast, the crosslinking process was introduced into gelatin in order to improve the controlled release behavior (Dang et al. 2017). Moreover, a number of recent studies also showed that gelatins have significant contribution in the encapsulation of bioactive compounds or development of fast release or controlled release system (Gomez-Mascaraque et al. 2017; Hani et al. 2017).

Techniques for Prepration of Protein-Based Nanodelivery Systems

To date, various techniques have been developed to produce protein-based nanodelivery systems as shown in Fig. 3.

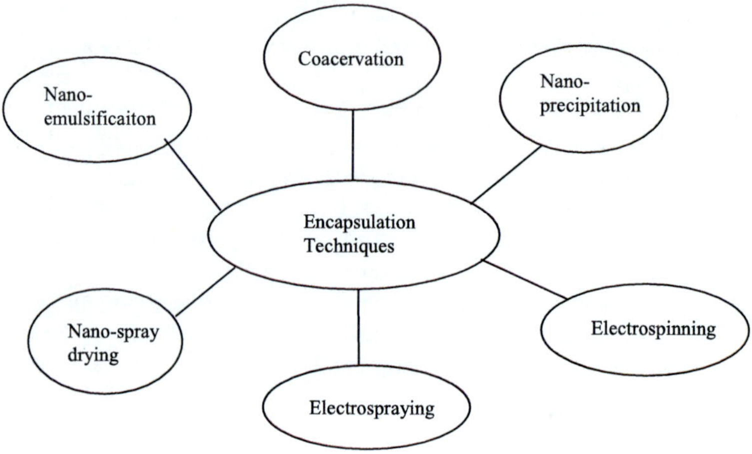

Fig. 3 Encapsulation techniques for bioactive compounds

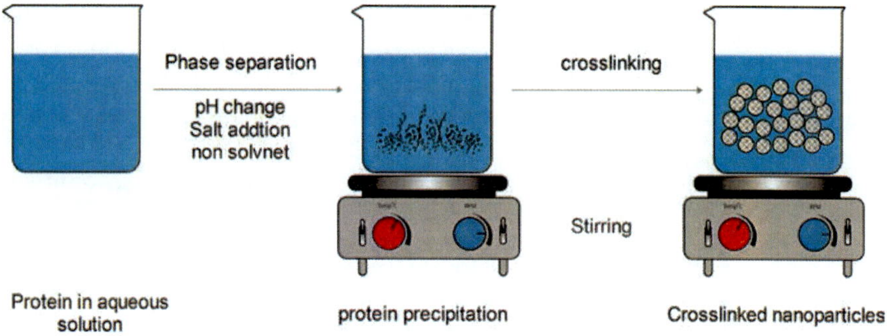

Fig. 4 Diagram of coacervation technique

Coacervation

It is chemical method used for separation of two liquid phases,one being the concentrated colloidal phase called as coacervate and another being dilute colloidal phase. By altering the temperature, dielectric, pH, constant and/or ionic strength of the protein mixture, results in the reduction of the solubility of the protein leading to phase separation (McClements 2015). In coacervation technique, when a salt is added or pH is changed of the protein solution, the liquid-liquid phase separation takes place, giving rise to a polymer rich dense phase at the bottom and a transparent solution at the top (Fig. 4), resulting in the formation of desired nanoparticles (Elzoghby et al. 2015). The process of coacerveation can be either simple or complex. In simple coacervation, only one polymer is involved while as, complex coacervation is brought about by the interactions of oppositely charged colloids at various conditions. Gelatin is most successfully used in nanoparticle production by coacervation but some proteins including zein and legumin nanoparticles were also produced by this method. Casein nanoparticles were also prepared by simple coacervation for the oral delivery of folic acid (Penalva et al. 2015). Gelatin and gum acacia nanoparticles were prepared by complex coacervation method to encapsulate tannins (Xing et al. 2005).

Nano-Precipitation

This method is also called as solvent displacement method. Nanoprecipitation occurs by a rapid desolvation of the protein when the nonsolvent is added to the protein solution, thereby causing the production of the active solute of interest (Joye and McClements 2013). This method require less shear forces to reduce droplet sizes to nanometers (Fig. 5), thus this process is considered advantageous compared to other conventional emulsification methods (Elzoghby et al. 2015). The various factors that affect the properties of nanoparticles are agitation speed, type and the amount of dispersing substance, viscosity of the phases and temperature,flow rate of the anti-solvent. Wheat gluten (gliadins) protein fractions were prepared by this method for the encapsulation of vitamin E by nano-precipitaion of ethanolic solution of vitamin E and gliadin in water (Duclairoir et al. 2002).

Nanoemulsification

Nanoemulsion involves the dispersion of one immiscible liquid into another for the production of nano-scale droplets, which have interesting applications in food industry. By reducing the microscale droplets into nanoscale emulsions have a number of potential advantages which include better stability to gravitational separation and particle aggregation due to an increase in elastic modulus (Tadros et al. 2004), weaker light scattering effects because the droplets are smaller than visible wavelengths so they are suitable for incorporation into products like soft drinks, water, soups (Mason et al. 2006), and increased bioavailability of encapsulated bioactive

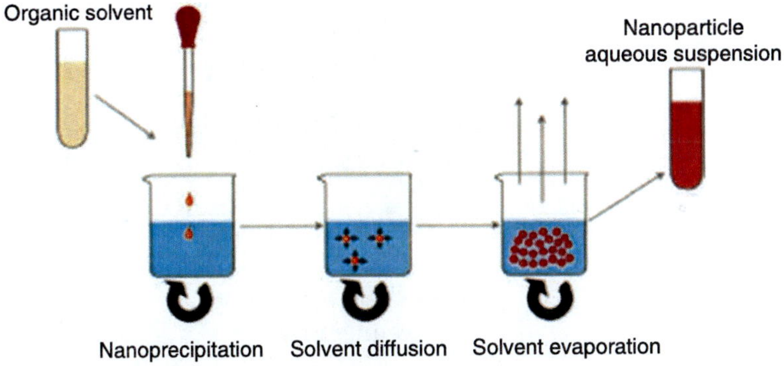

Fig. 5 Diagram of nano-precipitation technique

compounds(Acosta 2009),better product texture etc. Protein-based nanoemulsion can also be prepared using high-energy approaches including High Pressure Valve Homogenizer, Microfluidizers, Ultrasonic Homogenizers (McClements and Rao 2011).

Nanoemulsions Prepared by High Pressure Homogenisation

This method is the most commonly used for producing nano-emulsion in food industry now-a-days. An emulsion is directly fed into the high pressure homogeniser, as shown in Fig. 6 (McClements and Rao 2011). As the emulsion passes into the homogenizer, it experiences extremely high pressures which cause the disruption of larger droplets into smaller ones. The size of emulsion droplet further decreases to nano-level by increasing the number of passes through the homogeniser, homogeniser pressure and also by using nozzles of nanosize, this in turn increases the droplet disruption (Jafari et al. 2008). Mao et al. (2010) prepared whey protein-based nano-emulsion for encapsulation of bioactive compound (β-carotene) with a final particle diameter size 115–178 nm at the homogenisation pressure of >80 MPa with 3 passes.

Nanoemulsions Prepared by Microfluidisation

Conventionally Microfluidizers have been used in the pharmaceutical industry but now they are being used in the food and beverage industries to produce food emulsions. The basic working principle of Microfluidisation is somewhat similar to High Pressure Homogenisation, except the Homogenizer channel is different shown in Fig. 7 McClements and Rao 2011. Emulsion flows through fine channels into two streams and then meets at an interaction chamber creating acute disruptive forces,

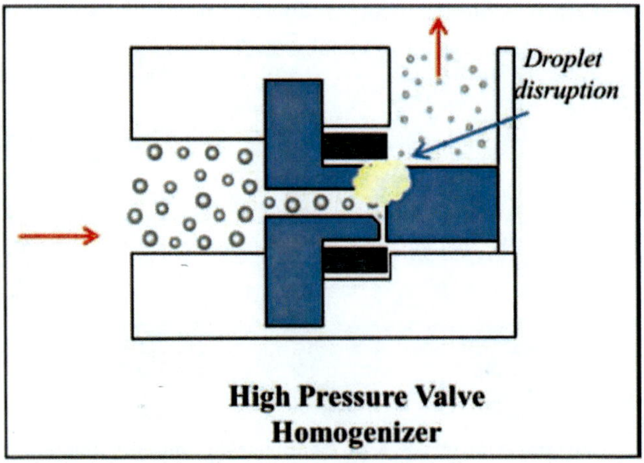

Fig. 6 Schematic diagram of high pressure homogenizer

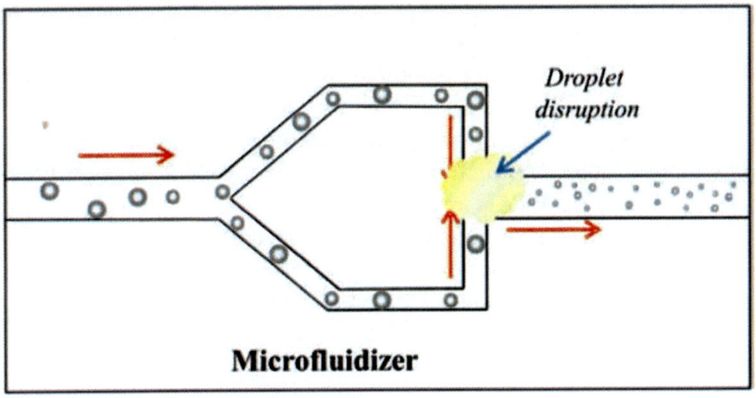

Fig. 7 Schematic diagram of microfluidizer

Fig. 8 Schematic diagram of ultrasonic probe homogenizer

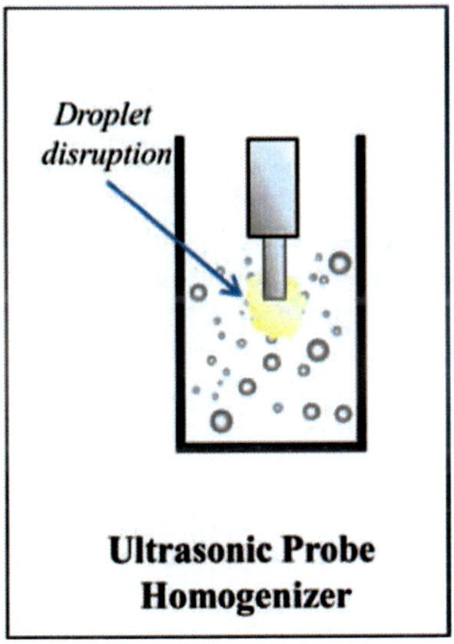

forming nanoemulsion. Qian and McClements (2011) successfully prepared globular protein (β-lactoglobulin) based nano-emulsion for encapsulation of bioactive compound (C*O*Q10) with a final particle diameter size 60–150 nm using microfluidiser.

Nanoemulsions Prepared by Ultrasonication

Ultrasonic homogenizers are being widely utilised for prepration of food based nano-emulsions. Ultrasonic homogenizers create the intense disruptive forces to breakup oil and water phases into very small droplets by utilizing high-intensity ultrasonic waves (Lin and Chen 2008; Leong et al. 2009). It consists of a sonicator probe as shown in Fig. 8 made of piezoelectric quartz crystals that expand and contract in response to an alternating electrical voltage; this provides energy input for the disruption of emulsion. Acute mechanical vibrations are produced within the emulsion by placing the tip of sonicator probe directly into the emulsion. This results in the generation of cavitation effects developing intense disruptive forces in that leads to droplet disruption. Greater the ultrasonic wave intensity smaller will be the emulsion droplet size (Maa and Hsu 1999). Whey protein concentrate-based nanoemulsion was prepared using this method for the encapsulation of d-limonene oil with droplet size below 250 nm (Jafari et al. 2006).

Nano-Spray Drying

This is most common technique used to dry liquids into fine powder efficiently. Spray drying is widely and commercially applied for the production of protein nanoparticles because of its ability to convert them into particles without undergoing chemical degradation due to its relatively short drying time. The process of spray drying can be divided into four steps: (1) atomization of the nanoparticles suspension into a spray, (2) spray-air contact, (3) drying of the spray and (4) separation of the dried end product from the drying gas (Haggag and Faheem 2015). Nowadays, the progress in nanotechnology of spray drying led to the development of a novel technology at the spray head, heating system and particle collector of the Nano Spray Dryer B-90 (Fig. 9) recently developed by BÜCHIL abortechnik AG, which resulted in high particle recovery rates upto milligram sample amounts of powder particles with particle sizes between 300 nm and 5 μm.

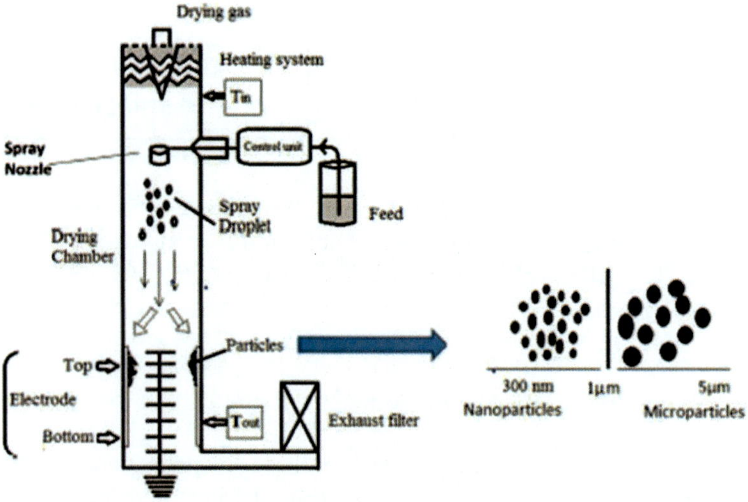

Fig. 9 Schematic diagram of the Nano Spray Dryer B-90

Electrospinning and Electrospraying

These are electro-hydrodynamic atomisation methods used for the production of nanoformulations of polymeric fibers or particles by the application of electrical forces (Jayaraman et al. 2015). When electric field is applied to the capillary nozzle, two major electrostatic forces act on the liquid jet at the needle tip namely the electrostatic repulsion of like charges and the coulombic force of the external electric field due to which the hemispherical droplet is distorted into a conical shape known as the Taylor cone. When electrostatic forces overcome the surface tension, a charged polymer jet is ejected from the tip of the Taylor cone (Reneker and Chun 1996). The main difference between these techniques lies in the solution concentration. In case of the electrospinnig technique, the concentration is high, the jet is stabilized due to whipping or bending instability and the output will be elongated fibers (Fig. 10a). Gomez-Estaca et al. (2012) reported the formation of curcumin-rich zein nanoparticles using electrospraying technique. On contrary, in case of electrospraying technique, the polymer concenration is low, the jet attached to cone destabilizes due to varicose instability, then the output is fine particles(Fig. 10b) (Bhushani and Anandharamakrishnan 2014). Fernandez et al. (2009) reported the nanoencapsulation of β-carotene in electrospun zein fibres which resulted in the improved oxidative stability of the bioactive compounds. In case of the electrospraying technique, the droplet size depends on the charge acquired due to voltage variation and solution flow rate while as in case of electrospinning method the main processing factors are voltage, solution flow rate, concentration, molecular weight of the polymer, and nozzle-to-ground distance (Dasgupta and Ranjan 2018).

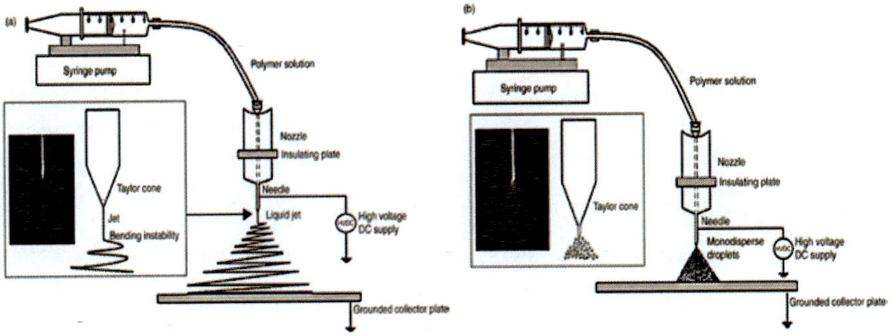

(A) (B)

Fig. 10 Schematic diagram of (**a**) Electrospinning and (**b**) Electrospraying

Other Applications

Protein Engineering and Immobilization

Biomolecules that are physically localized in a certain defined region of space with retention of their catalytic activities, and which can be used repeatedly and continuously are defined as immobilization (Brena and Batista-Viera 2006). In industrial field, enzyme immobilization is a progressing field due to its multiple functions including reuse of enzymes for the same reaction, longer half lives, low degradation and prevention of substrate contamination with enzymes or other compounds (Abdel-majeed et al. 2012). However, recent developments in protein engineering have revolutionized the development of commercially available enzymes into better industrial catalysts than provided by immobilization. Protein engineering and immobilization techniques are sequential and compatible strategies for the enhancement of enzyme properties. In many industrial processes, enzymes are immobilized to increase its stability and immobilized enzymes can be easily separated from reaction mixture and used again, thus, leading to economy of the process.

Immobilization have three most common methods viz; adsorption, entrapment, and crosslinking or covalently binding to a support. Proteins are immobilized either by physical adsorption to the surface of the nanoparticle or by covalent bonding to previously functionalized nanoparticles. Physical adsorption and covalent binding both decrease or avoid enzyme leaching, but binding to a planar surface can result in decreased stability or even protein denaturation (Brode et al. 1996). Crosslinking of enzymes typically enhances their stability at the expense of reduced activity. Microencapsulation into micelles or micellar polymers offers the highest potential to significantly improve enzyme lifetime and stop enzyme leaching, although problems of mass transfer may occur. Covalent binding of an enzyme to a carrier has the significance that the enzyme is tightly fixed. This is due to the fact that the multiple covalent bonds formation between the enzyme and the carrier decreases conformational flexibility and thermal vibrations, thus preventing protein unfolding and denaturation (Singh et al. 2010; Hanefeld et al. 2009). Multipoint and multi subunit covalent attachments of enzymes on appropriately functionalized supports via linkers provide rigidity to the immobilized enzyme structure, ultimately resulting in improved enzyme stability. However, there are few drawbacksrelated to enzyme immobilization including enzyme leakage owing to formation of weak bonds between enzyme and the carrier due to changes in temperatures, pH, ionic strength or even the mere presence of substrate (Dariush 2003). Another problem is restricted diffusion of high molecular weight substrates such as ribonuclease, trypsin and dextranase into entrapped enzymes Sankaran et al. 1989. Furthermore, selection of conditions for immobilization by covalent binding is more difficult than in other carrier binding methods.

Protein engineering however, can counteract these drawbacks by altering amino acid composition of enzymes that contain sequences of several hundred amino acids folded in a unique three-dimensional structure. Protein engineering is a faster, more controlled, targeted and more efficient technique then immobilization to optimize the properties of enzymes used in industrial processes and results in higher enzyme activity and stability. Protein engineered enzymes have other benefits such asdecreased consumption of raw materials and energy, use of alternative and renewable raw materials, reduced CO_2 and other greenhouse gas emissions, increased performance of industrial processes, enhanced quality of foods and animal feeds and cost-effective production and sustainable development (amfep.org 2009).

Protein engineering has been used to produce therapeutic pharmaceutical proteins with increased properties such as improved solubility and stability. The use of protein engineering for cancer treatment studies is a major area of interest. In clinical applications, protein engineering methods are used to modify antibodies to target cancer cells (Zafir-Lavie et al. 2007).

Conclusion

In conclusion, this study confirmed that the protein hydrolysates can be extracted through enzymatic digestion and/or chemical hydrolysis. However, a controlled enzymatic hydrolysis, targeting specific peptide bonds, is the preferred method. Bioactive properties of protein hydrolysates were presented in this chapter. It is suggesting that these food derived protein hydrolysates or peptides could be utilized as new materials to develop functional foods with antioxidant, antihypertensive, immunomodulatory, anticancer and antidiabetic activities. This chapter also presents edible films and coatings which have been received considerable interest over the last years due to their possibility to use as edible packaging materials over artificial ones and also contribute in reduction of environmental pollution. It has been shown that edible coatings improve the quality and increase the shelf-life of foods through inhibition of microbial growth and reduction of lipid oxidation. Different biopolymers such as polysaccharides, proteins, and lipids are applied to form edible films and coatings. Among these biopolymers, proteins appear to be most attractive in recent years not only due to their good barrier properties to oxygen, carbon dioxide, and lipids but also their nutritional value as well as satisfactory mechanical properties.

This chapter also reviewed about many phenolic compounds, such as anthocyanins, flavanols, curcumin, flavan-3-ols, proanthocyanidins, and phenolic acid derivatives present in plants. However, these compounds are prone to degradation. Encapsulation has been regarded as the protective method which not only prevents their degradation but also increases their bioavailability. Nano-encapsulation is a well-established way for the preservation of bioactive compounds. Nano-encapsulation exerts better functionality in terms of enhanced protection, increased stability and improved bioavailability of bioactive compounds. In recent years, an increasing interest in protein based nano delivery systems can be attributed in foodapplicationsdue to their nontoxicity and nutritional properties. Compared to animal proteins, plant proteins have attracted considerable interest due to their non-toxic and less-allergenicity. Proteins can also offer an advantageous formulation for the nano-encapsulation of bioactive compounds by using nano-spray drying, nano-precipitation, coacervation, electro-spinning and electro-spraying and emulsion techniques which can provide high encapsulation efficiency and excellent release profile.

References

Abd El-Salam, M. H., El-Shibiny, S., & Salem, A. (2009). Factors affecting the functional properties of whey protein products. *Food Reviews International, 25*, 251–270.

Abdel-majeed, N. A., Khelil, O. A., & Danial, E. N. (2012). Immobilization technology for enhancing bio-products industry. *African Journal of Biotechnology, 11*(71), 13528–13539.

Aceituno-Medina, M., Mendoza, S., Rodriguez, B. A., Lagaron, J. M., & Lopez-Rubio, A. (2015). Improved antioxidant capacity of quercetin and ferulic acid during in-vitro digestion through encapsulation within food-grade electrospun fibers. *Journal of Functional Foods, 12*, 332–341.

Acosta, E. (2009). Bioavailability of nanoparticles in nutrient and nutraceutical delivery. *Current Opinion in Colloid & Interface Science, 14*, 3–15.

Almajano, M. P., Carbo, R., Delgado, M. E., & Gordon, M. H. (2007). Effect of pH on the antimicrobial activity and oxidative stability of oil-in-water emulsions containing caffeic acid. *Journal of Food Science, 72*(5), C258–C263.

Amfep. (2009). Retrieved from www.amfep.org/sites/default/files/Amfep-List-of-Commercial-Enzymes

Andreani, L., Cercena, R., Ramos, B. G. Z., & Soldi, V. (2009). Development and characterization of wheat gluten microspheres for use in a controlled release system. *Materials Science and Engineering, 29*, 524–531.

Arcan, I., & Yemenicioglu, A. (2011). Incorporating phenolic compounds opens a new perspective to use zein films as flexible bioactive packaging materials. *Food Research International, 44*, 550–556.

Ariyarathna, I. R., & Karunaratne, D. N. (2015). Use of chickpea protein for encapsulation of folate to enhance nutritional potency and stability. *Food and Bioproducts Processing, 95*, 76–82.

Arvanitoyannis, I. S. (2002). Formation and properties of collagen and gelatine films and coating. In A. Gennadios (Ed.), *Protein-based films and coating* (pp. 275–304). Boca Raton, FL: CRC Press.

Baars, R. J., Van-Leeuwen, Y. M., Hendrix, Y., Velikov, K. P., Kegel, W. K., & Philipse, A. P. (2015). Morphology-controlled functional colloids by heterocoagulation of zein and nanoparticles. *Colloids and Surfaces A: Physicochemical and Engineering Aspects, 483*, 209–215.

Bamdad, F., Goli, A. H., & Kadivar, M. (2006). Preparation and characterization of proteinous film from lentil (*Lens culinaris*) Edible film from lentil (*Lens culinaris*). *Food Research International, 39*, 106–111.

Bar-Zeev, M., Assaraf, Y. G., & Livney, Y. D. (2016). β-Casein nanovehicles for oral delivery of chemotherapeutic drug combinations overcoming P-glycoprotein mediated multidrug resistance in human gastric cancer cells. *Oncotarget, 7*(17), 23322–23334.

Berendsen, R., Guell, C., & Ferrando, M. (2015). Aprocyanidin-rich extract encapsulated in water-in-oil-in-water emulsions produced by premix membrane emulsification. *Food Hydrocolloids, 43*, 636–648.

Berti, C., Roncoroni, L., Falini, M. L., Caramanico, R., Dolfini, E., & Bardella, M. T. (2007). Celiac-related properties of chemically and enzymatically modified gluten proteins. *Journal of Agricultural and Food Chemistry, 55*, 2482–2488.

Betz, M., & Kulozik, U. (2011). Whey protein gels for the entrapment of bioactive anthocyanins from bilberry extract. *International Dairy Journal, 21*(9), 703–710.

Betz, M., Steiner, B., Schantz, M., Oidtmann, J., Mader, K., Richling, E., & Kulozik, U. (2012). Antioxidant capacity of bilberry extract microencapsulated in whey protein hydrogels. *Food Research International, 47*(1), 51–57.

Bhushani, J. A., & Anandharamakrishnan, C. (2014). Electrospinning and electrosprayingtechniques: Potential food based applications. *Trends in Food Science and Technology, 38*(1), 21–33.

Bilek, S. E., Yılmaz, F. M., & Ozkan, G. (2017). The effects of industrial production on black carrot concentrate quality and encapsulation of anthocyanins in whey protein hydrogels. *Food and Bioproducts Processing, 102*, 72–80.

Blanco, E., Smoukov, S., Velev, O., & Velikov, K. P. (2016). Organic-inorganic patchy particles as a versatile platform for foam and emulsion stabilization. *Faraday Discussions, 191*, 73.

Blumberg, J. B., Camesano, T. A., Cassidy, A., Kris-Etherton, P., Howell, A., Manach, C., & Vita, J. A. (2013). Cranberries and their bioactive constituents in human health. *Advances in Nutrition, 4*(6), 618–632.

Bonnaillie, L. M., Zhang, H., Akkurt, S., Yam, K. L., & Tomasula, P. M. (2014). Casein films: The effects of formulation, environmental conditions and the addition of citric pectin on the structure and mechanical properties. *Polymers, 6*(7), 2018–2036.

Bourtoom, T. (2008). Edible films and coatings: Characteristics and properties. *International Food Research Journal, 15*(3), 237–248.

Bourtoom, T. (2009). Edible protein films: Properties enhancement. *International Food Research Journal, 16*, 1–9.

Boye, J., Zare, F., & Pletch, A. (2010). Pulse proteins: Processing, characterization, functional properties and applications in food and feed. *Food Research International, 43*(2), 414–431.

Brena, B. M., & Batista-Viera, F. (2006). Immobilization of enzymes. In J. M. Guisan (Ed.), *Immobilization of enzymes and cells* (2nd ed.). Totowa, NJ: Humana Press Inc, ISBN 1-59745-053-7.

Brode, P. F., Erwin, C. R., Rauch, D. S., Barnett, B. L., Armpriester, J. M., Wang, E. S. F., & Rubingh, D. N. (1996). Subtilisin BPN' variants: Increased hydrolytic activity on surface-bound substrates via decreased surface activity. *Biochemistry, 36*, 3162–3169.

Bucci, L. R., & Unlu, L. (2000). Protein and amino acid supplements in exercise and sport. In I. Wolinsky & J. A. Driskell (Eds.), *Energy-yielding macronutrients and energy metabolism in sports nutrition* (pp. 191–212). Boca Raton, FL: CRC Press.

Cai, X., Fang, Z., Dou, J., Yu, A., & Zhai, G. (2013). Bioavailability of quercetin: Problems and promises. *Current Medicinal Chemistry, 20*, 2572–2582.

Cao, N., Fu, Y., & He, J. (2007). Preparation and physical properties of soy protein isolate and gelatin composite films. *Food Hydrocolloids, 21*, 1153–1162.

Celus, I., Brijs, K., & Delcour, J. A. (2007). Enzymatic hydrolysis of brewers' spent grain proteins and techno-functional properties of the resulting hydrolysates. *Journal of Agricultural and Food Chemistry, 55*, 8703–8710.

Cerqueira, M. A., Pinheiro, A. C., Silva, H. D., Ramos, P. E., Azevedo, M. A., Flores-López, M. L., … Vicente, A. A. (2014). Design of bionanosystems for oral delivery of functional compounds. *Food Engineering Reviews, 6*, 1–19.

Chalamaiah, M., Dinesh-Kumar, B., Hemalatha, R., & Jyothirmayi, T. (2012). Fish protein hydrolysates: Proximate composition, amino acid composition, antioxidant activities and applications: A review. *Food Chemistry, 135*, 3020–3038.

Chalamaiah, M., Hemalatha, R., Jyothirmayi, T., Diwan, P. V., Bhaskarachary, K., & Vajreswari, A. (2015). Chemical composition and immunomodulatory effects of enzymatic protein hydrolysates from common carp (*Cyprinus carpio*) egg (roe). *Nutrition, 31*, 388–398.

Chang, C., Varankovich, N., & Nickerson, M. T. (2016). Microencapsulation of canola oil by lentil protein isolate-based wall materials. *Food Chemistry, 212*, 264–273.

Chen, F. P., Li, B. S., & Tang, C. H. (2015). Nanocomplexation of soy protein isolate with curcumin: Influence of ultrasonic treatment. *Food Research International, 75*, 157–165.

Chen, L. Y., Remondetto, G. E., & Subirade, M. (2006). Food protein-based materials as nutraceutical delivery systems. *Trends in Food Science Technology, 17*(5), 272–283.

Chen, W., Xu, Y., Zhang, L., Su, H., & Zheng, X. (2016). Blackberry subjected to in vitro gastrointestinal digestion affords protection against Ethyl Carbamate-induced cytotoxicity. *Food Chemistry, 212*, 620–627.

Chen, Y., Ye, R., & Liu, J. (2014). Effects of different concentrations of ethanol and isopropanol on physicochemical properties of zein-based films. *Industrial Crops and Products, 53*, 140–147.

Cheng, S. Y., Wang, B. J., & Weng, Y. M. (2015). Antioxidant and antimicrobial edible zein/chitosan composite films fabricated by incorporation of phenolic compounds and dicarboxylic acids. *LWT - Food Science and Technology, 63*, 115–121.

Cheng, Y. S., Lu, P. M., Huang, C. Y., & Wu, J. J. (2017). Encapsulation of lycopene with lecithin and α-tocopherol by supercritical antisolvent process for stability enhancement. *The Journal of Supercritical Fluids, 130*, 246–252.

Cheung, L. K. Y., Aluko, R. E., Cliff, M. A., & Li-Chan, E. C. Y. (2015). Effects of exopeptidase treatment on antihypertensive activity and taste attributes of enzymatic whey protein hydrolysates. *Journal of Functional Foods, 13*, 262–275.

Chi, C., Hu, F., Wang, B., Li, T., & Ding, G. (2015). Antioxidant and anticancer peptides from the protein hydrolysate of blood clam (*Tegillarca granosa*) muscle. *Journal of Functional Foods, 15*, 301–313.

Cho, D., Jo, K., Young-Cho, S., Man-Kim, J., Lim, K., Joo-Suh, H., & Oh, S. (2014). Antioxidant effect and functional properties of hydrolysates derived from egg-white protein. *Korean Journal for Food Science and Animal Resources, 34*, 362–371.

Clare, D. A., & Swaisgood, H. E. (2000). Bioactive milk peptides: A prospectus. *Journal of Dairy Science, 83*, 1187–1195.

Cordeiro de Azeredo, H. M. (2012). Edible coatings. In S. Rodrigues & F. A. N. Fernandes (Eds.), *Advances in fruit processing technologies* (pp. 345–361). Boca Raton, FL: CRC Press.

Core, J. (2005). Edible, water-resistant film from milk protein. *Agricultural Research, 53*(11), 21.

Cote, J., Caillet, S., Doyon, G., Sylvain, J. F., & Lacroix, M. (2010). Bioactive compounds in cranberries and their biological properties. *Critical Reviews in Food Science and Nutrition, 50*(7), 666–679.

Dai, L., Sun, C., Li, R., Mao, L., Liu, F., & Gao, Y. (2017). Structural characterization, formation mechanism and stability of curcumin in zein-lecithin composite nanoparticles fabricated by antisolvent co-precipitation. *Food Chemistry, 237*, 1163–1171.

Dang, X., Yang, M., Shan, Z., Mansouri, S., May, B. K., & Chen, X. (2017). On spray drying of oxidized corn starch cross-linked gelatin microcapsules for drug release. *Materials Science and Engineering, 74*, 493–500.

Dariush, N. R. (2003). Enzyme immobilization: The state of art in biotechnology. *Journal of Biotechnology, 1*(4), 197–206.

Dasgupta, N., & Ranjan, S. (2018). *An introduction to food grade nanoemulsions*. Singapore, Singapore: Springer.

Davidov-Pardo, G., Joye, I. J., & McClements, D. J. (2015). Chapter nine food-grade protein based nanoparticles and microparticles for bioactive delivery: Fabrication, characterization, and utilization. In D. Rossen (Ed.), *Advances in protein chemistry and structural biology* (pp. 293–325). Oxford, UK: Academic Press.

Denavi, G. A., Perez-Mateos, M., Anon, M. C., Montero, P., Mauri, A. N., & Gomez-Guillen, M. C. (2009). Structural and functional properties of soy protein isolate and cod gelatin blend films. *Food Hydrocolloids, 23*(8), 2094–2101.

Devi, N., Sarmah, M., Khatun, B., & Maji, T. K. (2017). Encapsulation of active ingredients in polysaccharide–protein complex coacervates. *Advances in Colloid and Interface Science, 239*, 136–145.

Diaz, M. N., Frei, B., Vita, J. A., & Keaney, J. F., Jr. (1997). Antioxidants and atherosclerotic heart disease. *New England Journal of Medicine, 337*, 408–416.

Duclairoir, C., Orecchioni, A., Depraetere, P., & Nakache, E. (2002). α-Tocopherol encapsulation and in vitro release from wheat gliadin nanoparticles. *Journal of Microencapsulation, 19*, 53–60.

Duenas, M., Gonzalez-Manzano, S., Gonzalez-Paramas, A., & Santos-Buelga, C. (2010). Antioxidant evaluation of o-methylated metabolites of catechin, epicatechin and quercetin. *Journal of Pharmaceutical and Biomedical Analysis, 51*(2), 443–449.

Elzoghby, A. O., Elgohary, M. M., & Kamel, N. M. (2015). Implications of protein- and peptide-based nanoparticles as potential vehicles for anticancer drugs. In *Protein and peptide nanoparticles for drug delivery* (1st ed.). Amsterdam, The Netherlands: Elsevier.

Elzoghby, A. O., Helmy, M. W., Samy, W. M., & Elgindy, N. A. (2013). Spray-dried casein-based micelles as a vehicle for solubilization and controlled delivery of flutamide: Formulation, characterization, and in vivo pharmacokinetics. *European Journal of Pharmaceutics and Biopharmaceutics, 84*, 487–496.

Elzoghby, A. O., Samy, W. M., & Elgindy, N. A. (2012). Albumin-based nanoparticles as potential controlled release drug delivery systems. *Journal of Controlled Release, 167*, 168–182.

Embuscado, M. E., & Huber, K. C. (2009). *Edible films and coatings for food applications*. Dordrecht, The Netherlands: Springer.

Esfanjani, A. F., Jafari, S. M., & Assadpour, E. (2017). Preparation of a multiple emulsion based on pectin-whey protein complex for encapsulation of saffron extract nanodroplets. *Food Chemistry, 221*, 1962–1969.

Esmaili, M., Ghaffari, S. M., Moosavi-Movahedi, Z., Atri, M. S., Sharifizadeh, A., & Farhadi, M. (2011). Beta casein-micelle as a nano vehicle for solubility enhancement of curcumin; food industry application. *LWT - Food Science and Technology, 44*(10), 2166–2172.

Ezhilarasi, P. N., Karthik, P., Chhanwal, N., & Anandharamakrishnan, C. (2013). Nanoencapsulation techniques for food bioactive components: A review. *Food and Bioprocess Technology, 6*(3), 628–647.

Fakhouri, F. M., Martelli, S. M., Caon, T., Velasco, J. I., & Mei, L. H. (2015). Edible films and coatings based on starch/gelatin: Film properties and effect of coatings on quality of refrigerated Red Crimson grapes. *Postharvest Biology and Technology, 109*, 57–64.

Fan, J. M., Ma, W., Liu, G. Q., Yin, S. W., Tang, C. H., & Yang, X. Q. (2014). Preparation and characterization of kidney bean protein isolate (KPI)-chitosan (CH) composite films prepared by ultrasonic pretreatment. *Food Hydrocolloids, 36*, 60–69.

Fang, R., Hao, R., Wu, X., Li, Q., Leng, X., & Jing, H. (2011). Bovine serum albumin nanoparticle promotes the stability of quercetin in simulated intestinal fluid. *Journal of Agricultural and Food Chemistry, 59*, 6292–6298.

Fernandez, A., Torres-Giner, S., & Lagaron, J. M. (2009). Novel route to stabilization of bioactive antioxidants by encapsulation in electrospun fibers of zein prolamine. *Food Hydrocolloids, 23*, 1427–1432.

Ferry, D. R., Smith, A., Malkhandi, J., Fyfe, D. W., Anderson, D., & Baker, J. (1996). Phase I clinical trial of the flavonoid quercetin: Pharmacokinetics and evidence for in vivo tyrosine kinase inhibition. *Clinical Cancer Research, 2*, 659–668.

Finnie, C., & Svensson, B. (2009). Barley seed proteomics from spots to structures. *Journal of Proteomics, 72*, 315–325.

FitzGerald, R. J., & O'Cuinn, G. (2006). Enzymatic debittering of food protein hydrolysates. *Biotechnology Advances, 24*(2), 234–237.

Flores, F. P., Singh, R. K., Kerr, W. L., Pegg, R. B., & Kong, F. B. (2014a). Total phenolics content and antioxidant capacities of microencapsulated blueberry anthocyanins during in vitro digestion. *Food Chemistry, 153*, 272–278.

Flores, F. P., Singh, R. K., & Kong, F. B. (2014b). Physical and storage properties of spray-dried blueberry pomace extract with wheyprotein isolate as wall material. *Journal of Food Engineering, 137*, 1–6.

Frank, K., Kohler, K., & Schuchmann, H. P. (2011). Formulation of labile hydrophilic ingredients in multiple emulsions: Influence of the formulation's composition on the emulsion's stability and on the stability of entrapped bioactives. *Journal of Dispersion Science and Technology, 32*(12), 1753–1758.

Fucinos, C., Míguez, M., Fuciños, P., Pastrana, L. M., Rúa, M. L., & Vicente, A. A. (2017). Creating functional nanostructures: Encapsulation of caffeine into α-lactalbumin nanotubes. *Innovative Food Science and Emerging Technologies, 40*, 10–17.

Galus, S., & Kadzinska, J. (2016). Whey protein edible films modified with almond and walnut oils. *Food Hydrocolloids, 52*, 78–86.

Garrido, J., & Borges, F. (2013). Wine and grape polyphenols - A chemical perspective. *Food Research International, 54*, 1844–1858.

Gennadios, A., McHugh, T., Weller, C., & Krochta, J. M. (1994). Edible coatings and films based on proteins. In J. M. Krochta, E. A. Baldwin, & M. Nisperos-Carriedo (Eds.), *Edible coatings and films to improve food quality* (pp. 231–247). Chicago, IL: Technomic.

Gennadios, A., & Weller, C. L. (1990). Edible films and coatings from wheat and corn proteins. *Food Technology, 44*(10), 63–69.

Gharsallaoui, A., Roudaut, G., Chambin, O., Voilley, A., & Saurel, R. (2007). Application of spray-drying in microencapsulation of food ingredients: An overview. *Food Research International, 40*, 1107–1121.

Ghasemi, S., Jafari, S. M., Assadpour, E., & Khomeiri, M. (2018). Nanoencapsulation of D-limonene within nanocarriers produced by pectin-whey protein complexes. *Food Hydrocolloids, 77*, 152–162.

Ghidelli, C., Mateos, M., Rojas-Argudo, C., & Pérez-Gago, M. B. (2014). Extending the shelf life of fresh-cut eggplant with a soy protein–cysteine based edible coating and modified atmosphere packaging. *Postharvest Biology and Technology, 95*, 81–87.

Gibellini, L., Pinti, M., Nasi, M., Montagna, J. P., De Biasi, S., & Roat, E. (2011). Quercetin and cancer chemoprevention. *Evidence-based Complementary and Alternative Medicine, 2011*, 1–15.

Gomez-Estaca, J., Balaguer, M. P., Gavara, R., & Hernandez-Munoz, P. (2012). Formation of zein nanoparticles by electrohydrodynamic atomization: Effect of the main processing variables and suitability for encapsulating the food coloring and active ingredient curcumin. *Food Hydrocolloids, 28*(1), 82–91.

Gomez-Estaca, J., Balaguer, M. P., López-Carballo, G., Gavara, R., & Hernández-Muñoz, P. (2017). Improving antioxidant and antimicrobial properties of curcumin by means of encapsulation in gelatin through electrohydrodynamic atomization. *Food Hydrocolloids, 70*, 313–320.

Gomez-Estaca, J., Gavara, R., & Hernandez-Munoz, P. (2015). Encapsulation of curcumin in electrosprayed gelatine microspheres enhances its bioaccessibility and widens its uses in food applications. *Innovative Food Science Emerging Technologies, 29*, 302–307.

Gomez-Guillen, M. C., Gimenez, B., Lopez-Caballero, M. E., & Montero, M. P. (2011). Functional and bioactive properties of collagen and gelatin from alternative sources: A review. *Food Hydrocolloids, 25*, 1813–1827.

Gomez-Mascaraque, L. G., Hernandez-Rojas, M., Tarancon, P., Tenon, M., Feuillere, N., & Velez Ruiz, J. F. (2017). Impact of microencapsulation within electrosprayed proteins on the formulation of green tea extract-enriched biscuits. *LWT - Food Science and Technology, 81*, 77–86.

Gontard, N., Thibault, R., Cuq, B., & Guilbert, S. (1996). Influence of relative humidity and film composition on oxygen and carbon dioxide permeabilities of edible films. *Journal of Agricultural and Food Chemistry, 44*, 1064–1069.

Gonzalez-Perez, S., & Arellano, J. B. (2009). Vegetable protein isolates. In G. O. Phillips & P. A. Williams (Eds.), *Handbook of hydrocolloids*. Oxford, UK: Woodhead Publishing.

Gontard, N., Guilbert, S., & Cuq, J. L. (1992). Edible wheat gluten films: Influence of the main process variables on film properties using response surface methodology. *Journal of Food Science, 57*(1).

Guilbert, S. (1995). Technology and application of edible protective films. In M. Mathlouthi (Ed.), *Food packaging and preservation* (pp. 371–394). London, UK: Elsevier Applied Science.

Gumus, C. E., Decker, E. A., & McClements, D. J. (2017). Gastrointestinal fate of emulsion-based ω-3 oil delivery systems stabilized by plant proteins: Lentil, pea, and faba bean proteins. *Journal of Food Engineering, 207*, 90–98.

Gunasekaran, S., Ko, S., & Xiao, L. (2007). Use of whey proteins for encapsulation and controlled delivery applications. *Journal of Food Engineering, 83*(1), 31–40.

Haggag, Y. A., & Faheem, A. M. (2015). Evaluation of nano spray drying as a method for drying and formulation of therapeutic peptides and proteins. *Frontiers in Pharmacology, 6*, 140.

Hanani, N. Z. A., McNamara, J., Roos, Y. H., & Kerry, J. P. (2013). *Food Hydrocolloids, 31*(2), 264–269.

Hanefeld, U., Gardossi, L., & Magner, E. (2009). Understanding enzyme immobilisation. *Chemical Society Reviews, 38*(2), 453–468.

Hani, N. M., Torkamani, A. E., Azarian, M. H., Mahmood, K. W. A., & Ngalim, S. H. (2017). Characterisation of electrospun gelatine nanofibres encapsulated with Moringa oleifera bioactive extract. *Journal of the Science of Food and Agriculture, 97*, 3348–3358.

Haratifar, S., Meckling, K. A., & Corredig, M. (2014). Bioefficacy of tea catechins encapsulated in casein micelles tested on a normal mouse cell line (4D/WT) and its cancerous counterpart (D/v-src) before and after in vitro digestion. *Food & Function, 5*(6), 1160–1166.

Harris, T., Cook, E. F., Kannel, W., Schatzkin, A., & Goldman, L. (1985). Blood pressure experience and risk of cardiovascular disease in the elderly. *Hypertension, 7*, 118–124.

Hartmann, R., & Meisel, H. (2007). Food-derived peptides with biological activity: From research to food applications. *Current Opinion in Biotechnology, 18*(2), 163–169.

He, H. L., Liu, D., & Ma, C. B. (2013). Review on the angiotensin-I-converting enzyme (ACE) inhibitor peptides from marine proteins. *Applied Biochemistry and Biotechnology, 169*, 738–749.

Heger, M., Golen, R. F. V., Broekgaarden, M., & Michel, M. C. (2014). The molecular basis for the pharmacokinetics and pharmacodynamics of curcumin and its metabolites in relation to cancer. *Pharmacological Reviews, 66*, 222–307.

Hernandez-Munoz, P., Lopez-Rubio, A., Lagaron, J. M., & Gavara, R. (2004a). Formaldehyde cross-linking of gliadin films: Effects on mechanical and water barrier properties. *Biomacromolecules, 5*(2), 415–421.

Hernandez-Munoz, P., Villalobos, R., & Chiralt, A. (2004b). Effect of cross-linking using aldehydes on properties of glutenin-rich films. *Food Hydrocolloids, 18*, 403–411.

Ho, K. K. H. Y., Schroën, K., San Martín-González, M. F., & Berton-Carabin, C. C. (2017). Physicochemical stability of lycopene-loaded emulsions stabilized by plant or dairy proteins. *Food Structure, 12*, 34–42.

Horiguchi, N., Horiguchi, H., & Suzuki, Y. (2005). Effect of wheat gluten hydrolysate on the immune system in healthy human subjects. *Bioscience, Biotechnology, and Biochemistry, 69*, 2445–2449.

Huang, Q., Yu, H., & Ru, Q. (2010). Bioavailability and delivery of nutraceuticals using nanotechnology. *Journal of Food Science, 75*(1), R50–R57.

Huang, S. L., Hung, C. C., Jao, C. L., Tung, Y. S., & Hsu, K. C. (2014). Porcine skin gelatine hydrolysate as a dipeptidyl peptidase IV inhibitor improves glycemic control in streptozotocin-induced diabetic rats. *Journal of Functional Foods, 11*, 235–242.

Isuru, R., Ariyarathna, D., & Karunaratne, N. (2016). Microencapsulation stabilizes curcumin for efficient delivery in food applications. *Food Packaging and Shelf Life, 10*, 79–86.

Jafari, S. M., Assadpoor, E., He, Y. H., & Bhandari, B. (2008). Re-coalescenceof emulsion droplets during high-energy emulsification. *Food Hydrocolloids, 22*(7), 1191–1202.

Jafari, S. M., He, Y., & Bhandari, B. (2006). Nano-emulsion production by sonication and microfluidization - A comparison. *International Journal of Food Properties, 9*, 475–485.

Jahed, E., Khaledabad, M. A., Almasi, H., & Hasanzadeh, R. (2017). Physicochemical properties of Carum copticum essential oil loaded chitosan films containing organic nanoreinforcements. *Carbohydrate Polymers, 164*, 325–338.

Jain, A., Thakur, D., Ghoshal, G., Katare, O. P., & Shivhare, U. S. (2016). Characterization of microcapsulated β-carotene formed by complex coacervation using casein and gum tragacanth. *International Journal of Biological Macromolecules, 87*, 101–113.
Javier, R., Ayoa, F., Francisco, A. R., & Ana, S. (2014). Immunomodulatory activities of whey β-lactoglobulin tryptic-digested fractions. *International Dairy Journal, 34*, 65–73.
Jayaraman, P., Gandhimathi, C., Venugopal, J. R., Becker, D. L., Ramakrishna, S., & Srinivasan, D. K. (2015). Controlled release of drugs in electrosprayed nanoparticles for bone tissue engineering. *Advanced Drug Delivery Reviews, 94*, 77–95.
Joshi, M., Adhikari, B., Aldred, P., Panozzo, J. F., Kasapis, S., & Barrow, C. J. (2012). Interfacial and emulsifying properties of lentil protein isolate. *Food Chemistry, 134*, 1343–1353.
Joye, I. J., & McClements, D. J. (2013). Production of nanoparticles by anti-solvent precipitation for use in food systems. *Trends in Food Science and Technology, 34*, 109–123.
Joye, I. J., & McClements, D. J. (2014). Biopolymer-based nanoparticles and micro-particles: Fabrication, characterization, and application. *Current Opinion in Colloid & Interface Science, 19*(5), 417–427.
Ju-Young, K., Kanga, N., Ji-Hyeok, L., Jin-Soo, K., Won-Suck, K., Sun-Joo, P., … You-Jin, J. (2016). Angiotensin I-converting enzyme inhibitory peptides from anenzymatic hydrolysate of flounder fish (*Paralichthys olivaceus*) muscleas a potent anti-hypertensive agent. *Process Biochemistry, 51*, 535–541.
Kanakdande, D., Bhosale, R., & Singhal, R. S. (2007). Stability of cumin oleoresin microencapsulated in different combination of gum Arabic, maltodextrin and modified starch. *Carbohydrate Polymers, 67*, 536–541.
Kang, H. J., Kim, S. J., You, Y. S., Lacroix, M., & Han, J. (2013). Inhibitory effect of soy protein coating formulations on walnut (*Juglans regia* L.) kernels against lipid oxidation. *LWT - Food Science and Technology, 51*, 393–396.
Kannan, A., Hettiarachchya, N., Lay, J. O., & Liyanage, R. (2010). Human cancer cell proliferation inhibition by a pentapeptide isolated and characterized from rice bran. *Peptides, 31*, 1629–1634.
Karnjanapratum, S., O'Callaghan, Y. C., Benjakul, S., & O'Brien, N. (2016). Antioxidant, immunomodulatory and antiproliferative effects of gelatin hydrolysate from Unicorn leather jacket skin. *Journal of the Science of Food and Agriculture, 96*, 3220–3226.
Kawabata, K., Mukai, R., & Ishisaka, A. (2015). Quercetin and related polyphenols: New insights and implications for their bioactivity and bioavailability. *Food & Function, 6*(5), 1399–1417.
Kayitmazer, A. B., Seeman, D., Minsky, B. B., Dubin, P. L., & Xu, Y. (2013). Protein-polyelectrolyte interactions. *Soft Matter, 9*(9), 2553–2583.
Kester, J. J., & Fennema, O. R. (1986). Edible films and coatings: A review. *Food Technology, 40*(12), 47–59.
Kim, E. K., Joung, H. J., Kim, Y. S., Hwang, J. W., Ahn, C. B., & Jeon, Y. J. (2012). Purification of a novel anticancer peptide from enzymatic hydrolysate of Mytilus coruscus. *Journal of Microbiology and Biotechnology, 22*, 1381–1387.
Kim, M. J., Son, H. J., Kim, Y., Misaka, T., & Rhyu, M. R. (2015). Umami-bitter interactions: The suppression of bitterness by umami peptides via human bitter taste receptor. *Biochemical and Biophysical Research Communications, 456*(2), 586–590.
Kim, S. E., Kim, H. H., Kim, J. Y., Kang, Y. I., Woo, H. J., & Leu, H. J. (2000). Anticancer activity of hydrophobic peptides from soy proteins. *Biofactors, 12*, 151–155.
Kim, S. K., & Wijesekara, I. (2010). Development and biological activities of marine-derived bioactive peptides: A review. *Journal of Functional Foods, 2*, 1–9.
Kleemann, R., Verschuren, L., Morrison, M., Zadelaar, S., & van-Erk, M. J. (2011). Anti-inflammatory, anti-proliferative and anti-atherosclerotic effects of quercetin in human in vitro and in vivo models. *Atherosclerosis, 218*(1), 44–52.
Kong, X., Guo, M., Hua, Y., Cao, D., & Zhang, C. (2008). Enzymatic preparation of immunomodulating hydrolysates from soy proteins. *Bioresource Technology, 99*, 8873–8879.

Kowalczyk, D., Gustaw, W., Swieca, M., & Baraniak, B. (2014). A study on the mechanical properties of pea protein isolate films. *Journal of Food Processing and Preservation, 38*(4), 1726–1736.

Krissansen, G. W. (2007). Emerging health properties of whey proteins and their clinical implications. *Journal of the American College of Nutrition, 26*, 713S–723S.

Kristinsson, H. G., & Rasco, B. A. (2000). Fish protein hydrolysates: Production, biochemical, and functional properties. *Critical Reviews in Food Science and Nutrition, 40*, 43–81.

Krochta, J. M., & de Mulder-Johnston, C. (1997). Edible and biodegradable polymer films: Challenges and opportunities. *Food Technology, 51*, 60–74.

Krull, L. H., & Inglett, G. E. (1971). Industrial uses of gluten. *Cereal Science Today, 16*(8), 232–261.

Kumar, D., Chatli, M. K., Singh, R., & Mehta, N. (2016a). Antioxidant and antimicrobial activity of camel milk casein hydrolysates and its fractions. *Small Ruminant Research, 139*, 20–25.

Kumar, D. D., Mann, B., Pothuraju, R., Sharma, R., Bajaj, R., & Minaxi, M. (2016b). Formulation and characterization of nanoencapsulated curcumin using sodium caseinate and its incorporation in icecream. *Food & Function, 7*(1), 417–424.

Lahl, W. J., & Grindstaff, D. A. (1989). Spices and seasonings: Hydrolyzed proteins. In *Proceedings of the 6th SIFST Symposium on Food Ingredients—Applications, Status and Safety*, 27–29 April 1989 (pp. 51–65). Singapore, Singapore: Singapore Institute of Food Science and Technology.

Lakhanpal, P., & Rai, D. K. (2007). Quercetin: A versatile flavonoid. *Internet Journal of Medical Update, 2*(2), 1–16.

Lassoued, I., Trigui, M., Ghlissi, Z., Nasri, R., Jamoussi, K., & Kessis, M. (2014). Evaluation of hypocholesterolemic effect and antioxidant activity of Boops boops proteins in cholesterol-fed rats. *Food & Function, 5*, 1224–1231.

Lawton, J. W. (2002). Zein: A history of processing and use. *Cereal Chemistry, 79*, 1–18.

Lee, M., Kovacs-Nolan, J., Archbold, T., Fan, M. Z., Juneja, L. R., Okubo, T., & Mine, Y. (2009). Therapeutic potential of hen egg white peptides for the treatment of intestinal inflammation. *Journal of Functional Foods, 1*, 161–169.

Lee, Y. S., Park, M. J., Choi, J. E., Kim, J. Y., Nam, M. S., & Jeong, Y. H. (2007). Effects of silk protein hydrolysates on blood glucose level, serum insulin and leptin secretion in OLETF rats. *Journal of the Korean Society of Food Science and Nutrition, 36*, 703–707.

Lens, J. P., de Graaf, L. A., Stevels, W. M., Dietz, C. H. J. T., Verhelst, K. C. S., Vereijken, J. M., & Kolster, P. (2003). Influence of processing and storage conditions on the mechanical and barrier properties of films cast from aqueous wheat gluten dispersions. *Industrial Crops and Products, 17*, 119–130.

Leong, T. S. H., Wooster, T. J., Kentish, S. E., & Ashokkumar, M. (2009). Minimising oildroplet size using ultrasonic emulsification. *Ultrasonics Sonochemistry, 16*(6), 721–727.

Lestringant, P., Guri, A., Gulseren, I., Relkin, P., & Corredig, M. (2014). Effect of processing on physicochemical characteristics and bioefficacy of beta-lactoglobulin-epigallocatechin-3-gallate complexes. *Journal of Agricultural and Food Chemistry, 62*(33), 8357–8364.

Li, B., Du, W., Jin, J., & Du, Q. (2012). Preservation of (−)-epigallocatechin-3-gallate antioxidant properties loaded in heat treated beta-lactoglobulin nanoparticles. *Journal of Agricultural and Food Chemistry, 60*(13), 3477–3484.

Li, C., Wang, J., Shi, J., Huang, X., Peng, Q., & Xue, F. (2015). Encapsulation of tomato oleoresin using soy protein isolate-gum aracia conjugates as emulsifier and coating materials. *Food Hydrocolloids, 45*, 301–308.

Li, H., Wang, D., Liu, C., Zhu, J., Fan, M., Sun, X., … Cao, Y. (2018a). Fabrication of stable zein nanoparticles coated with soluble soybean polysaccharide for encapsulation of quercetin. *Food Hydrocolloids, 87*, 342–351. https://doi.org/10.1016/j.foodhyd.2018.08.002

Li, J., Xu, X., Chen, Z., Wang, T., Wang, L., & Zhong, Q. (2018b). Biological macromolecule delivery system fabricated using zein and gum arabic to control the release rate of encapsulated tocopherol during in vitro digestion. *Food Research International, 114*, 251. https://doi.org/10.1016/j.foodres.2018.08.073

Li, Y., Lim, L. T., & Kakuda, Y. (2009). Electrospun zein fibers as carriers to stabilize (−)- epigallocatechingallate. *Journal of Food Science, 74*(3), C233–C240.
Li, Z., Percival, S. S., Bonard, S., & Gu, L. W. (2011). Fabrication of nanoparticles using partially purified pomegranate ellagitannins and gelatine and their apoptotic effects. *Molecular Nutrition & Food Research, 55*(7), 1096–1103.
Li, C. H., Matsui, T., Matsumoto, K., Yamasaki, R., & Kawasaki, T. (2002). Latent production of angiotensin I-converting enzyme inhibitors from buckwheat protein. *Peptide Science, 8*(6), 267–274.
Liao, L., Luo, Y., Zhao, M., & Wang, Q. (2012). Preparation and characterization of succinic acid deamidated wheat gluten microspheres for encapsulation of fish oil. *Colloids and Surfaces B: Biointerfaces, 92*, 305–314.
Lin, C. C., & Liang, J. H. (2002). Effect of antioxidants on the oxidative stability of chicken breast meat in a dispersion system. *Journal of Food Science, 67*, 530–533.
Lin, C. Y., & Chen, L. W. (2008). Comparison of fuel properties and emission characteristicsof two-and three-phase emulsions prepared by ultrasonically vibrating andmechanically homogenizing emulsification methods. *Fuel, 87*, 2154–2161.
Links, M. R., Taylor, J., Kruger, M. C., & Taylor, J. R. N. (2015). Sorghum condensed tannins encapsulated in kafirin microparticles as a nutraceutical for inhibition of amylases during digestion to attenuate hyperglycaemia. *Journal of Functional Foods, 12*, 55–63.
Liu, B. Y., Zhu, K. X., Peng, W., Guo, X. N., & Zhou, H. M. (2016). Effect of sequential hydrolysis with endo- and exo-peptidase on bitterness properties of wheat gluten hydrolysates. *RSC Advances, 6*(33), 27659–27668.
Liu, J., Jiang, L., Zhang, Y., Du, Z., Qiu, X., Kong, L., & Zhang, H. (2017). Binding behaviors and structural characteristics of ternary complexes of β-lactoglobulin, curcumin, and fatty acids. *RSC Advances, 7*(73), 45960–45967.
Liu, J., Ru, Q., & Ding, Y. (2012). Glycation a promising method for food protein modification: Physicochemical properties and structure - A review. *Food Research International, 49*(1), 170–183.
Livney, Y. D. (2010). Milk proteins as vehicles for bioactives. *Current Opinion in Colloid & Interface Science, 15*(1-2), 73–83.
Loveday, S. M., Su, J., Rao, M. A., Anema, S. G., & Singh, H. (2012). Whey protein nanofibrils: Kinetic, rheological and morphological effects of group IA and IIA cations. *International Dairy Journal, 26*, 133–140.
Lozano-Ojalvo, D., Molina, E., & Lopez-Fandino, R. (2016). Hydrolysates of egg white proteins modulate T- and B-cell responses in mitogen-stimulated murine cells. *Food & Function, 17*, 1048–1056.
Ma, J., Gan, C., Xu, Q., Zhou, J., & Zhang, J. (2015). Amphiphilic copolymer stabilized coreeshell structural casein-based emulsion. *Colloids and Surfaces A: Physicochemical and Engineering Aspects, 471*, 65–72.
Ma, J., Mao, X., Wang, Q., Yang, S., Zhang, D., Chen, S., & Li, Y. (2014). Effect of spray drying and freeze drying on the immunomodulatory activity, bitter taste and hygroscopicity of hydrolysate derived from whey protein concentrate. *LWT - Food Science and Technology, 56*, 296–302.
Maa, Y. F., & Hsu, C. C. (1999). Performance of sonication and microfluidization for liquid-liquid emulsification. *Pharmaceutical Development and Technology, 4*, 233–240.
Mao, L., Yang, J., Xu, D., Yuan, F., & Gao, Y. (2010). Effects of homogenization models and emulsifiers on the physicochemical properties of b-carotene nanoemulsions. *Journal of Dispersion Science and Technology, 31*, 986–993.
Mason, T. G., Wilking, J. N., Meleson, K., Chang, C. B., & Graves, S. M. (2006). Nanoemulsions: Formation, structure, and physical properties. *Journal of Physics. Condensed Matter, 18*(41), R635.

Masotti, A. I., Buckley, N., Champagne, C. P., & Green-Johnson, J. (2011). Immunomodulatory bioactivity of soy and milk ferments on monocyte and macrophage models. *Food Research International, 44*, 2475–2481.

Matsui, T., ChunHui, L., Tanaka, T., Maki, T., Osajima, Y., & Matsumoto, K. (2000). Depressor effect of wheat germ hydrolysate and its novel angiotensin I-converting enzyme inhibitory peptide, Ile-Val-Tyr, and the metabolism in rat and human plasma. *Biological and Pharmaceutical Bulletin, 23*, 427–431.

McClements, D. J. (2015). Nanoscale nutrient delivery systems for food applications: Improving bioactive dispersibility, stability, and bioavailability. *Journal of Food Science, 80*(7), N1602–N1611.

McClements, D. J., & Rao, J. (2011). Food-grade nanoemulsions: Formulation, fabrication, properties, performance, biological fate, and potential toxicity. *Critical Reviews in Food Science and Nutrition, 51*(4), 285–330.

McHugh, T. H. (2000). Protein-lipid interactions in edible films and coatings. *Nahrung, 44*, 148–151.

Mehranfar, F., Bordbar, A. K., & Parastar, H. (2013). A combined spectroscopic, molecular docking and molecular dynamic simulation study on the interaction of quercetin with beta-casein nanoparticles. *Journal of Photochemistry and Photobiology B: Biology, 127*, 100–107.

Mei, Y., & Zhao, Y. (2003). Barrier and mechanical properties of milk protein-based edible films containing nutraceuticals. *Journal of Agricultural and Food Chemistry, 51*(7), 1914–1918.

Memarpoor-Yazdia, M., Mahakia, H., & Zare-Zardinib, H. (2013). Antioxidant activity of protein hydrolysates and purified peptides from *Zizyphus jujuba* fruits. *Journal of Functional Foods, 5*, 62–70.

Min, S., Harris, L. J., & Krochta, J. M. (2005). Listeria monocytogenes inhibition by whey protein films and coatings incorporating the lactoperoxidase system. *Journal of Food Science, 70*, 317–324.

Mohammadian, M., & Madadlou, A. (2016). Characterization of fibrillated antioxidant whey protein hydrolysate and comparison with fibrillated protein solution. *Food Hydrocolloids, 52*, 221–230.

Mohammadian, M., Salami, M., Momen, S., Alavi, F., Emam-Djomeh, Z., & Moosavi-Movahedi, A. A. (2018). Enhancing the aqueous solubility of curcumin at acidic condition through the complexation with whey protein nanofibrils. *Food Hydrocolloids, 87*, 902. https://doi.org/10.1016/j.foodhyd.2018.09.001

Molinaro, S., Cruz-Romero, M., Sensidoni, A., Morris, M., Lagazio, C., & Kerry, J. P. (2015). *Innovative Food Science and Emerging Technologies, 30*(12), 15–23.

Mullen, A. M., Álvarez, C., Zeugolis, D. I., Henchion, M., O'Neill, E., & Drummond, L. (2017). Alternative uses for co-products: Harnessing the potential of valuable compounds from meat processing chains. *Meat Science, 132*(Suppl. C), 90–98.

Nabavi, S. F., Russo, G. L., Daglia, M., & Nabavi, S. M. (2015). Role of quercetin as an alternative for obesity treatment: You are what you eat! *Food Chemistry, 179*, 305–310.

Nasri, R., Abdelhedi, O., Jemil, I., Daoued, I., Hamden, K., & Kallel, C. (2015). Ameliorating effects of goby fish protein hydrolysates on high-fat-high-fructose diet-induced hyperglycemia, oxidative stress and deterioration of kidney function in rats. *Chemico-Biological Interactions, 242*, 71–80.

Nelson, R., Katayama, S., Mine, Y., Duarte, J., & Matar, C. (2007). Immunomodulating effects of egg yolk low lipid peptic digests in a murine model. *Food and Agricultural Immunology, 18*, 1–15.

Neo, Y. P., Ray, S., Jin, J., Gizdavic-Nikolaidis, M., Nieuwoudt, M. K., Liu, D. Y., & Quek, S. Y. (2013). Encapsulation of food grade antioxidant in natural biopolymer by electrospinning technique: A physicochemical study based on zein-gallicacid system. *Food Chemistry, 136*(2), 1013–1021.

Nesterenko, A., Alric, I., Silvestre, F., & Durrieu, V. (2013). Vegetable proteins in microencapsulation: Are view of recent interventions and their effectiveness. *Industrial Crops and Products, 42*, 469–479.

Nongonierma, A., & O'keeffe, M. (2016). Milk protein hydrolysates and bioactive peptides. In *Advanced Dairy Chemistry* (pp. 417–482). New York, NY: Springer.

Oancea, A. M., Aprodu, I., Ghinea, I. O., Barbu, V., Ioniţă, E., & Bahrim, G. (2017). A bottom-up approach for encapsulation of sour cherries anthocyanins by using β- lactoglobulin as matrices. *Journal of Food Engineering, 210*, 83–90.

Oanceaa, A. M., Hasan, M., Vasile, A. M., Barbu, V., Enachi, E., Bahrim, G., ... Stănciuc, N. (2018). Functional evaluation of microencapsulated anthocyanins from sour cherries skins extract in whey proteins isolate. *LWT - Food Science and Technology, 95*, 129–134.

Oidtmann, J., Schantz, M., Mader, K., Baum, M., Berg, S., Betz, M., & Richling, E. (2012). Preparation and comparative release characteristics of three anthocyanin encapsulation systems. *Journal of Agricultural and Food Chemistry, 60*(3), 844–851.

Orliac, O., Rouilly, A., Silvestre, F., & Rigal, L. (2002). Effect of additives on the mechanical properties, hydrophobicity and water uptake of thermo-moulded films produced from sunflower protein isolate. *Polymer, 43*, 5417–5425.

Oses, J., Fabregat-Vazquez, M., Pedroza-Islas, R., Tomas, S. A., Cruz-Orea, A., & Mate, J. I. (2009). Development and characterization of composite edible films based on whey protein isolate and mesquite gum. *Journal of Food Engineering, 92*, 56–62.

Ouattara, B., Giroux, M., Smoragiewicz, W., Saucier, L., & Lacroix, M. (2002b). Combined effect of gamma irradiation, ascorbic acid and edible film on the improvement of microbial and biochemical characteristics of ground beef. *Journal of Food Protection, 65*(6), 981–987.

Ouattara, B., Le, T. C., Vachon, C., Mateescu, M. A., & Lacroix, M. (2002a). Use of gamma irradiation cross-linking to improve the water vapor permeability and the chemical stability of milk protein films. *Radiation Physics and Chemistry, 63*, 821–825.

Ozcalik, O., & Tihminlioglu, F. (2013). Barrier properties of corn zein nanocomposite coated polypropylene films for food packaging applications. *Journal of Food Engineering, 114*, 505–513.

Pan, D. D., Wu, Z., Liu, J., Cao, X. Y., & Zeng, X. Q. (2013). Immunomodulatory and hypoallergenic properties of milk protein hydrolysates in ICR mice. *Journal of Dairy Science, 96*, 4958–4964.

Pan, X., Zhao, Y., Hu, F., Chi, C., & Wang, B. (2016). Anticancer activity of a hexapeptide from skate (Raja porosa) cartilage protein hydrolysate in HeLa cells. *Marine Drugs, 14*, 153–164.

Pappas, E., & Schaich, K. M. (2009). Phytochemicals of cranberries and cranberry products: Characterization, potential health effects, and processing stability. *Critical Reviews in Food Science and Nutrition, 49*(9), 741–781.

Paramawati, R., Yoshino, T., & Isobe, S. (2001). Properties of plasticized-zein film as affected by plasticizer treatments. *Food Science and Technology Research, 7*, 191–194.

Pasupuleti, V. K., & Braun, S. (2010). State of the art manufacturing of protein hydrolysates. In *Protein hydrolysates in biotechnology* (pp. 11–32). Dordrecht, The Netherlands: Springer.

Patel, A. R., Heussen, P. C., Hazekamp, J., Drost, E., & Velikov, K. P. (2012). Quercetin loaded biopolymeric colloidal particles prepared by simultaneous precipitation of quercetin with hydrophobic protein in aqueous medium. *Food Chemistry, 133*, 423–429.

Patel, A. R., & Velikov, K. P. (2014). Zein as a source of functional colloidal nano- and microstructures. *Current Opinion in Colloid & Interface Science, 19*(5), 450–458.

Penalva, R., Esparza, I., Agüeros, M., Gonzalez-Navarro, C. J., Gonzalez-Ferrero, C., & Irache, J. M. (2015). Casein nanoparticles as carriers for the oral delivery of folic acid. *Food Hydrocolloids, 44*, 399–406.

Pena-Ramos, E. A., & Xiong, Y. (2002). Antioxidant activity of soy protein hydrolysates in a liposomal system. *Journal of Food Science, 67*, 2952–2956.

Perlstein, H., Bavli, Y., Turovsky, T., Rubinstein, A., Danino, D., & Stepensky, D. (2014). Beta-casein nanocarriers of celecoxib for improved oral bioavailability. *European Journal of Nanomedicine, 6*(4), 217–226.

Peyron, A. (1991). Lenrobage et les produits filmogenes: un nouveau mode demballage. *Viandes et Produits Carnés, 12*(2), 41–46.

Picchio, M. L., Linck, Y. G., Monti, G. A., Gugliotta, L. M., Minari, R. J., & Alvarez-Igarzabal, C. I. (2018). Casein films crosslinked by tannic acid for food packaging applications. *Food Hydrocolloids, 84*, 424–434.

Picot, L., Bordenave, S., Didelot, S., Fruitier-Arnaudin, I., Sannier, F., & Thorkelsson, G. (2006). Antiproliferative activity of fish protein hydrolysates on human breast cancer cell lines. *Process Biochemistry, 41*, 1217–1222.

Pinheiro, A. C., Coimbra, M. A., & Vicente, A. A. (2016). In vitro behaviour of curcumin nanoemulsions stabilized by biopolymer emulsifiers – effect of interfacial composition. *Food Hydrocolloids, 52*, 460–467.

Pokora, M., Eckert, E., Zambrowicz, A., Bobak, L., Szołtysik, M., Dąbrowska, A., & Trziszka, T. (2013). Biological and functional properties of proteolytic enzyme modified egg protein byproducts. *Food Science and Nutrition, 1*, 184–195.

Priprem, A., Watanatorn, J., Sutthiparinyanont, S., Phachonpai, W., & Muchimapura, S. (2008). Anxiety and cognitive effects of quercetin liposomes in rats. *Nanomedicine: Nanotechnology, Biology and Medicine, 4*(1), 70–78.

Qian, C., & McClements, D. (2011). Formation of nanoemulsions stabilized by model food-grade emulsifiers using high pressure homogenization: Factors affecting particle size. *Food Hydrocolloids, 25*, 1000–1008.

Qiu, J., Zheng, Q., Fang, L., Wang, Y., Min, M., Shen, C., … Xiong, C. (2018). Preparation and characterization of casein-carrageenan conjugates and selfassembled microcapsules for encapsulation of red pigment from paprika. *Carbohydrate Polymers, 196*, 322–331.

Raksakulthai, R., & Haard, N. F. (2003). Exopeptidases and their application to reduce bitterness in food: A review. *Critical Reviews in Food Science and Nutrition, 43*(4), 401–445.

Ramos, Ó. L., Pereira, J. O., Silva, S. I., Fernandes, J. C., Franco, M. I., Lopes-da-Silva, J. A., & Malcata, F. X. (2012). Evaluation of antimicrobial edible coatings from a whey protein isolate base to improve the shelf life of cheese. *Journal of Dairy Science, 95*(11), 6282–6292.

Ramos, O. L., Reinas, I., Silva, S. I., Fernandes, J. C., Cerqueira, M. A., & Pereira, R. N. (2013). Effect of whey protein purity and glycerol content upon physical properties of edible films manufactured therefrom. *Food Hydrocolloids, 30*, 110–122.

Rao, P. J., & Khanum, H. (2016). A green chemistry approach for nanoencapsulation of bioactive compound – Curcumin. *LWT - Food Science and Technology, 65*, 695–702.

Rascón, M. P., Beristain, C. I., Garcia, H. S., & Salgado, M. A. (2011). Carotenoid retention and storage stability of spray-dried encapsulated paprika oleoresin using gum Arabic and Soy protein isolate as wall materials. *LWT- Food Science and Technology, 44*, 549–557.

Reneker, D. H., & Chun, I. (1996). Nanometre diameter fibres of polymer, produced by electrospinning. *Nanotechnology, 7*(3), 216–223.

Robin, A. L., & Sankhla, D. (2013). European legislative framework controlling the use of food additives. In M. Saltmarsh (Ed.), *Essential guide to food additives* (pp. 44–53). Cambridge, UK: RSC Publishing.

Roblet, C., Akhtar, M. J., Mikhaylin, S., Pilon, G., Gill, T., Marette, A., & Bazinet, L. (2016). Enhancement of glucose uptake in muscular cell by peptide fractions separated by electrodialysis with filtration membrane from salmon frame protein hydrolysate. *Journal of Functional Foods, 22*, 337–346.

Rocca-Smith, J. R., Marcuzzo, E., Karbowiak, T., Centa, J., Giacometti, M., Scapin, F., … Debeaufort, F. (2016). Effect of lipid incorporation on functional properties of wheat gluten based edible films. *Journal of Cereal Science, 69*, 275–282.

Sabato, S. F., Ouattara, B., Yu, H., D'Aprano, G., & Lacroix, M. (2001). Mechanical and barrier properties of cross-linked soy and whey protein based films. *Journal of Agricultural and Food Chemistry, 49*, 1397–1403.

Saglam, D., Venema, P., De Vries, R., Shi, J., & Van der Linden, E. (2013). Concentrated whey protein particle dispersions: Heat stability and rheological properties. *Food Hydrocolloids, 30*, 100–109.

Sah, B. N. P., Vasiljevic, T., McKechnie, S., & Donkor, O. N. (2015). Identification of anticancer peptides from bovine milk proteins and their potential roles in management of cancer: A critical review. *Comprehensive Reviews in Food Science and Food Safety, 14*, 123–138.

Saha, B. C., & Hayashi, K. (2001). Debittering of protein hydrolyzates. *Biotechnology Advances, 19*(5), 355–370.

Sahu, A., Kasoju, N., & Bora, U. (2008). Fluorescence study of the curcumin-casein micelle complexation and its application as a drug nanocarrier to cancer cells. *Biomacromolecules, 9*(10), 2905–2912.

Sankaran, K., Godbole, S. S., & D'Souza, S. F. (1989). Preparation of spray-dried, sugar free egg powder using glucose oxidase and catalase co-immobilized on cotton cloth. *Enzyme and Microbial Technology, 11*(9), 617–619.

Selling, G., & Sessa, D. J. (2007). Multivalent carboxylic acids to modify the properties of zein. *Industrial Crops and Products, 25*, 63–69.

Sengupta, B., & Sengupta, P. K. (2002). The interaction of quercetin with human serum albumin: A fluorescence spectroscopic study. *Biochemical and Biophysical Research Communications, 299*, 400–403.

Severin, S., & Xia, W. S. (2006). Enzymatic hydrolysis of whey proteins by two different proteases and their effect on the functional properties of resulting protein hydrolysates. *Journal of Food Biochemistry, 30*, 77–97.

Shaddel, R., Hesari, J., Azadmard-Damirchi, S., Hamishehkar, H., Fathi-Achachlouei, B., & Huang, Q. (2018). Use of gelatin and gum Arabic for encapsulation of black raspberry anthocyanins by complex coacervation. *International Journal of Biological Macromolecules, 107*, 1800–1810.

Shafaei, M., Ghalandari, B., Vaseghi, A., Divsalar, A., Haertlé, T., & Saboury, A. A. (2017). β-Lactoglobulin: An efficient nanocarrier for advanced delivery systems. *Nanomedicine: Nanotechnology, Biology and Medicine, 13*, 1685–1692.

Sharma, R. A., Gescher, A. J., & Steward, W. P. (2005). Curcumin: The story so far. *European Journal of Cancer, 41*, 1955–1968.

Sheih, I. C., Fang, T. J., Wu, T., & Lin, P. (2010). Anticancer and antioxidant activities of the peptide fraction from algae protein waste. *Journal of Agricultural and Food Chemistry, 58*, 1202–1207.

Shpigelman, A., Cohen, Y., & Livney, Y. D. (2012). Thermally-induced β-lactoglobulin–EGCG nano vehicles: Loading, stability, sensory and digestive-release study. *Food Hydrocolloids, 29*(1), 57–67.

Shpigelman, A., Israeli, G., & Livney, Y. D. (2010). Thermally-induced protein–polyphenol co-assemblies: Beta lactoglobulin-based nano complexes as protective nano vehicles for EGCG. *Food Hydrocolloids, 24*(8), 735–743.

Shutava, T. G., Balkundi, S. S., Vangala, P., Steffan, J. J., Bigelow, R. L., Cardelli, J. A., & Lvov, Y. M. (2009). Layer-by-layer-coated gelatine nanoparticles as a vehicle fordelivery of natural polyphenols. *ACS Nano, 3*(7), 1877–1885.

Silva, K. S., Garcia, C. C., Amado, L. R., & Mauro, M. A. (2015). Effects of edible coatings on convective drying and characteristics of the dried pineapple. *Food and Bioprocess Technology, 8*, 1465–1475.

Singh, B. N., Shankar, S., & Srivastava, R. K. (2011). Green tea catechin, epigallocatechin-3-gallate (EGCG): Mechanisms, perspectives and clinical applications. *Biochemical Pharmacology, 82*(12), 1807–1821.

Singh, R. K., Zhang, Y. W., Nguyen, N. P., Jeya, M., & Lee, J. K. (2010). Covalent immobilization of β-1, 4-glucosidase from *Agaricus arvensis* onto functionalized silicon oxide nanoparticles. *Applied Microbiology and Biotechnology, 89*, 337–344.

Socha, R., Baczkowicz, M., Fortuna, T., Kura, A., Labanowska, M., & Kurdziel, M. (2013). Determination of free radicals and flavan-3-ols content in fermented and unfermented teas and properties of their infusions. *European Food Research and Technology, 237*(2), 167–177.

Souza, A. C. P., Gurak, P. D., & Marczak, L. D. M. (2017). Maltodextrin, pectin and soy protein isolate as Carrier agents in the encapsulation of anthocyanins-rich extract from jaboticaba pomace. *Food and Bioproducts Processing, 102*, 186–194.

Sun, S., Song, Y., & Zheng, Q. (2009). Rheological behavior of heat-induced wheat gliadin gel. *Food Hydrocolloids, 23*, 1054–1056.

Tadros, T., Izquierdo, P., Esquena, J., & Solans, C. (2004). Formation and stability of nano-emulsions. *Advances in Colloid and Interface Science, 108-109*, 303–318.

Tapal, A., & Tiku, P. K. (2012). Complexation of curcumin with soy protein isolate and its implications on solubility and stability of curcumin. *Food Chemistry, 130*(4), 960–965.

Tidarat, T., Jurriaan, J. M., Harry, J. W., & Jirawat, Y. (2017). Immunomodulatory activity of protein hydrolysates derived from Virgibacillus halodenitrificans SK1-3-7 proteinase. *Food Chemistry, 224*, 320–328.

Tieland, M., Borgonjen-Van Den Berg, K. J., Van-Loon, L. J. C., & De-Groot, L. C. P. G. M. (2012). Dietary protein intake in community-dwelling, frail, and institutionalized elderly people: Scope for improvement. *European Journal of Nutrition, 51*(2), 173–179.

Udenigwe, C. C., & Aluko, R. E. (2012). Food protein-derived bioactive peptides: Production, processing and potential health benefits. *Journal of Food Science, 71*, R11–R24.

Umaraw, P., & Verma, A. K. (2017). Comprehensive review on application of edible film on meat and meat products: An eco-friendly approach. *Critical Reviews in Food Science and Nutrition, 57*, 1270–1279.

Vernaza, M. G., Dia, V. P., Mejia, E. G., & Chang, Y. K. (2012). Antioxidant and anti inflammatory properties of germinated and hydrolysed Brazilian soybean flours. *Food Chemistry, 134*, 2217–2225.

Vicente, J., Cezarino, T. D. S., Pereira, L. J. B., da Rocha, E. P., Raymundo Sa, G., Gamallo, O. D., … Garcia-Rojas, E. E. (2017). Microencapsulation of sachainchi oil using emulsion-based delivery systems. *Food Research International, 99*, 612–622.

Wang, C., Tian, Z., Chen, L., Temelli, F., Liu, H., & Wang, Y. (2010). Functionality of barley proteins extracted and fractionated by alkaline and alcohol methods. *Cereal Chemistry, 87*, 597–606.

Wang, R., Tian, Z., & Chen, L. (2011a). Nano-encapsulations liberated from barley protein microparticles for oral delivery of bioactive compounds. *International Journal of Pharmaceutics, 406*, 153–162.

Wang, R., Tian, Z., & Chen, L. (2011b). A novel process for microencapsulation of fish oil with barley protein. *Food Research International, 44*, 2735–2741.

WHO. (2017). Retrieved from http://www.who.int/mediacentre/factsheets/fs297/en/

Wolfe, R. R. (2015). Update on protein intake: Importance of milk proteins for health status of the elderly. *Nutrition Reviews, 73*(Suppl. 1), 41–47.

Wu, J., & Ding, X. (2001). Hypotensive and physiological effect of angiotensin converting enzyme inhibitory peptides derived from soy protein on spontaneously hypertensive rats. *Journal of Agricultural and Food Chemistry, 49*, 501–506.

Xing, F., Cheng, G., Yi, K., & Ma, L. (2005). Nanoencapsulation of capsaicin by complex coacervation of gelatin, acacia, and tannins. *Journal of Applied Polymer Science, 96*(6), 2225–2229.

Xue, Z., Gao, J., Zhang, Z., Yu, W., Wang, H., & Kou, X. (2012). Antihyperlipidemic and antitumor effects of chickpea albumin hydrolysate. *Plant Foods for Human Nutrition, 67*, 393–400.

Yalcin, E., Celik, S., & Ibanoglu, E. (2008). Foaming properties of barley protein isolates and hydrolysates. *European Food Research and Technology, 226*, 967–974.

Yang, H., Feng, K., Wen, P., Zong, M. H., Lou, W. Y., & Wu, H. (2017). Enhancing oxidative stability of encapsulated fish oil by incorporation of ferulic acid into electrospun zein mat. *LWT - Food Science and Technology, 84*, 82–90.

Yang, X., Wu, D., Du, Z., Li, R., Chen, X., & Li, X. (2009). Spectroscopy study on the interaction of quercetin with collagen. *Journal of Agricultural and Food Chemistry, 57*, 3431–3435.
Yee, J., Shipe, W., & Kinsella, J. (1980). Antioxidant effects of soy protein hydrolysates on copper catalyzed methyl linoleate oxidation. *Journal of Food Science, 45*, 1082–1083.
Yimit, D., Hoxur, P., Amat, N., Uchikawa, K., & Yamaguchi, N. (2012). Effects of soybean peptide on immune function, brain function, and neurochemistry in healthy volunteers. *Nutrition, 28*, 154–159.
Yu, Z., Yin, Y., Zhao, W., Liu, J., & Chen, F. (2012). Anti-diabetic activity peptides from albumin against α-glucosidase and α-amylase. *Food Chemistry, 135*, 2078–2085.
Yu, Z., Yin, Y., Zhao, W., Yu, Y., Liu, B., & Liu, J. (2011). Novel peptides derived from egg white protein inhibiting alpha-glucosidase. *Food Chemistry, 129*, 1376–1382.
Yuan, X. Q., Gu, X. H., Tang, J., & Wasswa, J. (2008). Hypoglycemic effect of semipurified peptides from *Momordica charantia* L. Var. abbreviata Ser. in alloxan-induced diabetic mice. *Journal of Food Biochemistry, 32*, 107–121.
Yuan, Y., Kong, Z. Y., Sun, Y. E., Zeng, Q. Z., & Yang, X. Q. (2017). Complex coacervation of soy protein with chitosan: Constructing antioxidant microcapsule for algal oil delivery. *LWT- Food Science and Technology, 75*, 171–179.
Zafir-Lavie, I., Michaeli, Y., & Reiter, Y. (2007). Novel antibodies as anticancer agents. *Oncogene, 26*, 3714–3733.
Zeng, M., Adhikari, B., He, Z., Qin, F., Huang, X., & Chen, J. (2013). Improving the foaming properties of soy protein isolate through partial enzymatic hydrolysis. *Drying Technology, 31*(13-14), 1545–1552.
Zeng, Z., Shuai, T., Yi, L. J., Wang, Y., & Song, G. M. (2016). Effect of case management on patients with type 2 diabetes mellitus: A meta-analysis. *Chinese Nursing Research, 3*, 71–76.
Zhang, T., Lv, C. Y., Chen, L. L., Bai, G. L., Zhao, G. H., & Xu, C. S. (2014). Encapsulation of anthocyanin molecules within a ferritin nanocage increases their stability and cell uptake efficiency. *Food Research International, 62*, 183–192.
Zhu, K., Zhou, H. M., & Qian, H. F. (2006). Antioxidant and free radical-scavenging activities of wheat germ protein hydrolysates (WGPH) prepared with Alcalase. *Process Biochemistry, 41*(6), 1296–1302.
Zou, T., Li, Z., Percival, S. S., Bonard, S., & Gu, L. (2012). Fabrication, characterization, and cytotoxicity evaluation of cranberry procyanidins-zein nanoparticles. *Food Hydrocolloids, 27*(2), 293–300.

Part IV
Lipids and Oils: Nutraceutical Properties

Lipids and Oils: An Overview

Sangamithra Asokapandian, S. Sreelakshmi, and Gopirajah Rajamanickam

Introduction

The word lipid refers to diverse group of heterogeneous compounds occurring in nature. Lipids are defined mainly on the basis of their solubility, not primarily their chemical structure (Gurr et al. 2002). Lipids are relatively insoluble in aqueous solution and highly soluble in non-polar solvents such as alcohol, acetone, ether, chloroform. Chemically lipids are defined as esters of glycerol and fatty acids or as the triglycerides of fatty acids.

As there is no standard definition for lipids, different authors have defined lipids. Kates (1986) defined lipids as "*those substances which are (a) insoluble in water; (b) soluble in organic solvents such as chloroform, ether or benzene; (c) contain long-chain hydrocarbon groups in their molecules; and (d) are present in or derived from living organisms.*"

As per Gurr et al. (2002), lipids are defined as "*a chemically heterogeneous group of substances, having in common the property of insolubility in water, but solubility in nonpolar solvents such as chloroform, hydrocarbons or alcohols.*"

Christie and Han (2010) termed lipids as "*a wide variety of natural products including fatty acids and their derivatives, steroids, terpenes, carotenoids, and bile acids, which have in common a ready solubility in organic solvents such as diethyl ether, hexane, benzene, chloroform, or methanol.*"

S. Asokapandian (✉)
Department of Food Technology, Kongu Engineering College, Erode, Tamil Nadu, India

S. Sreelakshmi
Department of Food Processing Technology, PSG College of Arts and Science, Coimbatore, Tamil Nadu, India

G. Rajamanickam
Department of Food Science, Cornell University, Ithaca, NY, USA

A. Gani, B. A. Ashwar (eds.), *Food biopolymers: Structural, functional and nutraceutical properties*, https://doi.org/10.1007/978-3-030-27061-2_16

Lipids are one of the biologically essential compounds apart from carbohydrates, proteins and nucleic acids. They are small bio-molecules with molecular weight ranges between 100 and 5000 and also vary in polarity. Lipids are stored in adipose tissue and function as the best energy reserve. Lipids provide the major part of calories, on average, 9 kcal/g whereas carbohydrates and proteins give 4 kcal/g on an average. Lipids serve as the structural component of living organisms. It is utilized for the construction of biological membranes and contributes as a barrier in separating intracellular and extracellular contents. It aids in maintaining the integrity of the cellular compounds. Lipids also serve as essential nutrients (Vitamin A, D, E, K) and most of the hormones are lipids. It helps to facilitate the digestive process, depressing gastric secretion, slowing gastric emptying and stimulating biliary and pancreatic flow. In foods, they affect the texture, flavor and palatability of foods.

Lipids in Foods

Lipids that are consumed by humans are either from plant or animal sources. They are extracted in the industries by pressing or solvent extraction and heating in case of plant and animals lipids respectively. Except few plant sources, all other plant based raw materials contain notable quantity of lipids in it.

Food products that are devoid of lipids are honey, sugar, juices etc. Most of the fruits and vegetables belong to category of foods with low amount of fat. A well-known exception is the avocado also called as butter fruit contains about 15% lipids. Table 1 represents the total lipid (g/100 g) in different plant and animal based food

Table 1 Total lipid content in plant and animal based food products

Fruits and vegetables		*Cereal products*		*Nuts and seeds*	
Avocado	15.41	Rice bran	20.85	Macadamia nuts	75.77
Potato	14.95	Wheat germ	9.72	Pecan nut	71.97
Plantain	11.81	Oats	6.90	Walnut	65.21
Sweet potato	11.10	Quinoa	6.07	Sunflower seeds	51.46
Soybean	6.80	Millet	4.22	Almond	49.93
Sweet corn	1.50	Sorghum	3.46	Sesame seeds	49.67
Pomegranate	1.17	Cornstarch	0.05	Pistachio nut	45.32
Fig	0.93	*Meat products*		Cashew nut	43.85
Beans	0.70	Pork	94.16	Coconut	33.49
Mushroom	0.57	Beef	70.89	Chia seeds	30.74
Spinach	0.39	Turkey	38.93	*Dairy products*	
Papaya	0.26	Egg yolk	26.54	Butter oil	99.48
Eggplant	0.18	Lamb	23.41	Butter	81.11
Orange juice	0.12	Smoked fish	20.14	Condensed milk	8.70
Cabbage	0.10	Quail	12.05	Buffalo milk	6.89
Fruit cocktail	0.01	Raw Salmon	10.43	Dried whey	1.07

products. The different lipid content in food materials are due to the changes in species, genotype, variety of the plant or animal, and also part of the plant or animal (Sikorski and Kolakowska 2010).

Classification of Lipids

The most common classification of lipids is based on its structure. The other possible classifications are based on its physical properties at ambient temperature, polarity and essentiality for humans. Figure 1 represents the broad classifications of lipids. Based on structure, lipids are classified into three types. They are

- Simple lipids
- Complex lipids
- Derived lipids

Simple Lipids

Simple lipids or homo lipids are the ester of fatty acid with various alcohols. These fatty acids contain even number of carbon atoms hence the name homo lipids. Simple lipids are further classified into

- Fats and oils
- Waxes

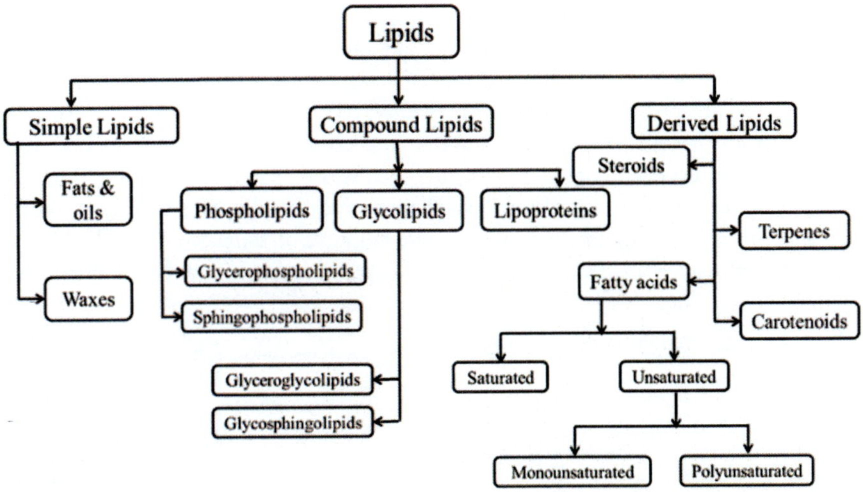

Fig. 1 Classification of lipids

Fats and Oils

Fats and oils are made of three elements such as carbon (C), hydrogen (H), and oxygen (O). These three elements combine together to form fatty acid. These three fatty acid units linked with glycerol, a trihydroxy alcohol to form triglycerides (Fig. 2). Glycerol is a trihydric alcohol (containing three –OH hydroxyl groups) that can combine with up to three fatty acids to form monoglycerides, diglycerides, and triglycerides.

The main and most important simple lipids are fats and oils (triglycerides or triacylglycerols). Lipids which are solid at room temperature are termed as fats while oils are liquid at room temperature. Table 2 represents the difference between fats and oils.

Generally, lipids obtained from animal sources are usually solids, while those from plant origin are generally liquids. Hence it is commonly referred as animal fats and vegetable oils. Triglycerides are more abundant which constitutes about 98% of all dietary lipids. Both oils and fats differ in their texture, appearance and functionality due to the difference in type of fatty acid chains which are joined to form triglycerides.

The plant fats, on the other hand, contain a large proportion of unsaturated fatty acids especially polyunsaturated fatty acids. These unsaturated fatty acids have low melting point and contribute liquid state to the plants fats.

Most animal fats are relatively rich in saturated fatty acids but contain a very low amount of polyunsaturated fatty acids. The large proportion of saturated fatty acids especially palmitic and stearic with high melting point confers solid state to the animal fats.

Waxes

Waxes are esters like fats, but it uses one long chain alcohol (16–36 carbon atoms) instead of glycerol. They are the esters of long chain fatty acids (14–36 carbon atoms) with high molecular weight mono hydroxy alcohols. The term 'wax' was derived from the old English word 'weax' representing honey comb of bee hive. Thus bee wax has been considered as reference wax.

Waxes are present naturally on leaves and stems of plants as a coating on the surface, thus preventing the loss of excessive amounts of water from the plants. Few examples of waxes are shown in Table 3. Waxes are also secreted in animals and

Fig. 2 Structure of triglyceride

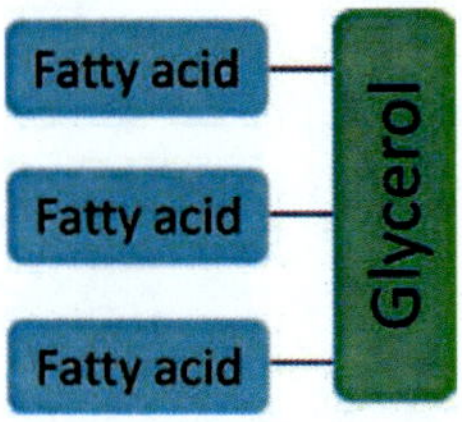

Table 2 Difference between fats and oils

Fats	Oils
Remain solid at room temperature	Remain liquid at room temperature
Mostly from animal source—Stored in liver and bone marrow	Mostly from plant source—Stored in seeds and fruits
Rich in saturated fatty acids—No C=C bonds	Rich in unsaturated fatty acids—One or more C=C bonds
Higher melting point	Relatively lower melting point
Have low iodine number	Have high iodine number
Have high Reichert-Meissl number	Have low Reichert-Meissl number
More stable	Less stable
Example: Butter, tallow, lard	Example: Palm oil, coconut oil, corn oil

Table 3 Types of plant and animal waxes

Wax type	Sources
Plant wax	
Carnauba wax	Leaves of Brazilian palm tree
Castor wax	Produced by hydrogenation of castor oil
Rice bran wax	Obtained from rice bran
Soy wax	Obtained from soybean oil
Candelilla wax	Leaves of the small Candelilla shrub
Japan wax	From the berries and stems of the Rhus succedanea tree
Animal wax	
Bee wax	Secreted by bees and present in honey comb
Spermaceti wax	Head cavities and blubber of sperm whale
Lanolin wax	Obtained from wool of lamb
Shellac wax	Secreted by female lac bug on the bark of trees
Chinese wax	Secreted by *Ceroplastes ceriferus* insect

insects. In case of vertebrates, waxes help to maintain the skin soft, flexible, lubricated, and water proof. The hair, fur and wool of the animals are also coated with wax. Similarly, the wax secreted in the body of birds, keep their feathers water repellent.

Compound Lipids

Compound lipids are ester of fatty acids and alcohol along with other additional groups such as phosphorus, amino acid or amine derivative, carbohydrates, protein. Hence it is also referred as complex or conjugated lipids. Based on the group, compound lipids are classified into

- Phospholipids
- Glycolipids
- Lipoproteins

Phospholipids

Phospholipids are the major components of cell membranes of plants, animals, and microorganisms. Phospholipids are modified triglycerides with one of the fatty acid chains replaced with a phosphate group. In simpler terms, phospholipids composed of fatty acids, nitrogenous base, phosphoric acid, and alcohol. The basic structure of the phospholipid molecule composed of a hydrophilic "head" consisting of a phosphate group and two hydrophobic fatty acid "tails". These two components are joined together by a glycerol molecule.

Based on the type of alcohol attached, it is divided into two classes. They are

- Glycerophospholipids
- Sphingophospholipids

Glycerophospholipids

Glycerophospholipids or phosphoglycerides are the most common phospholipids found in membranes. It is composed of two molecules of fatty acids, glycerol, phosphate group and alcohol amine. The common alcohol moieties of phosphoglycerides (Fig. 3) are the amino acid serine, ethanolamine, choline, glycerol, and the inositol. The presence of different alcohol head groups gives different phosphoglycerides (Pollard et al. 2016).

- No head group—Phosphatidic acid
- Glycerol head group—Phosphatidylglycerol
- Ethanolamine head group—Phosphatidylethanolamine
- Choline head group—Phosphatidylcholine
- Serine head group—Phosphatidylserine
- Inositol head group—Phosphatidylinositol

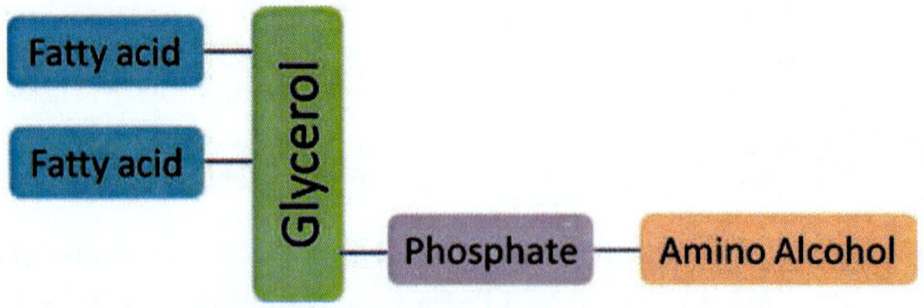

Fig. 3 Structure of glycerophospholipid

Phosphatidylcholine are lecithin. Lecithin is the most common phosphoglycerides.

Sphingophospholipids

In sphingophospholipids, the phosphate and fatty acids are attached to the alcohol sphingosine instead of glycerol. Sphingophospholipids (Fig. 4) are found in the membrane of nerve cells.

Glycolipids

Glycolipids differ from phospholipids in that instead of phosphate containing head, glycolipid consists of sugars such as glucose or galactose. The sugar can be a monosaccharide, oligosaccharide, or polysaccharide. Similar to phospholipids, glycolipids are composed of a hydrophobic region, containing two long hydrocarbon tails, and a polar region, which now contains one or more sugar residues but no phosphate group.

Glyceroglycolipids—sugar and fatty acid residues are attached to a glycerol backbone.

Glycosphingolipids—sugar and fatty acid residues are attached to a sphingosine backbone.

Glycolipids are structural lipid generally found in the extracellular face of eukaryotic cellular membrane. It can also act as receptors for viruses and other pathogens to enter cells. Glycosphingolipids typically account for about 5–10% of lipid in plasma membranes, and are important in intercellular communication (Engelking 2010). The common glycolipids are cerebrosides, gangliosides, and ceramide oligosaccharides.

Lipoprotein

Lipoprotein is the complexes of lipids and protein. They are the simple soluble proteins and non-soluble fats bonded together to facilitate transport of the fats through blood and lymphatic system to various cells throughout the body. Each molecule of lipoprotein consists of a hydrophobic non polar lipid core surrounded by a single

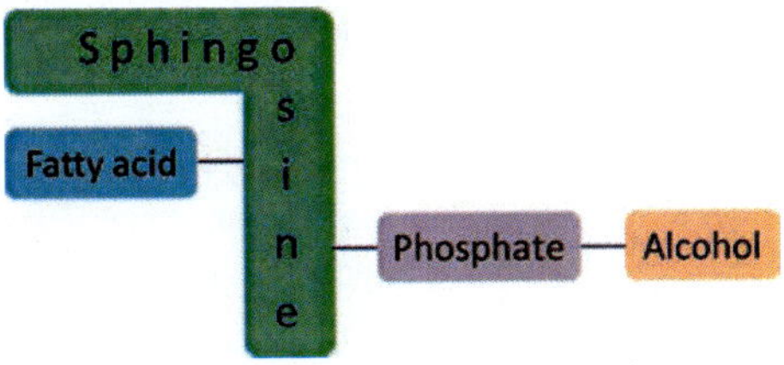

Fig. 4 Structure of Sphingophospholipid

surface layer of amphipathic lipid and polar proteins. The core of the lipoprotein particle consists of triacylglycerol and/or cholesteryl esters, fatty acids and fat-soluble vitamins like Vitamin E. The surface single layer is made up of phospholipids, unesterified cholesterol and specific type of water soluble apolipoproteins. The protein moiety of a lipoprotein is called apolipoprotein. Lipoproteins are further classified into five categories based on its density. Table 4 shows the classification of lipoprotein along with its density and mean diameter.

Derived Lipids

Derived lipids are hydrolysis products of simple and compound lipids. Derived lipids include fatty acids, alcohols, monoglycerides and diglycerides, steroids, terpenes, carotenoids, vitamin A, D, E & K, essential oils. They are non-polar, non-saponifiable, fat soluble substances with complex structure. The most common derived lipids are steroids, terpenes and carotenoids.

Fatty Acids

Fatty acids are organic acids, comprised of straight, long chain hydrocarbon with atleast one carboxyl group. Therefore R–COOH is the general formula of fatty acid where R is a hydrocarbon chain. Fatty acids and their associated derivatives are the primary components of lipids. Fatty acids are generally derived from triglycerides and phospholipids. Dietary fats are generally fatty acids. Fatty acids in any living organisms are comprised of unbranched hydrocarbon chain with number of carbon atoms generally from 12 to 24 carbon atoms. Fatty acids from plant sources have even numbered carbon chains whereas some animal fats and marine oils contain odd numbered chain fatty acids in addition to the even numbered chains (O'brien 2008).

Fatty acids can be further classified based on the degree of saturation. The degree of saturation refers to the presence or absence of C=C double bonds in the hydrocarbon region. Based on the degree of saturation, they are classified into two groups.

- Saturated fatty acids
- Unsaturated fatty acids

Table 4 Types of lipoprotein

S. No	Types of lipoprotein	Density (g/mL)	Mean diameter (nm)
1.	Chylomicrons	<0.95	100–500
2.	Very-low-density lipoprotein (VLDL)	0.95–1.006	30–80
3.	Intermediate density lipoprotein (IDL)	1.006–1.019	25–50
4.	Low-density lipoprotein (LDL)	1.019–1.063	18–28
5.	High-density lipoprotein (HDL	1.063–1.121	5–15

Saturated Fatty Acids

Saturated fatty acids are fatty acids that predominantly contain only single bonds between carbon atoms. They are completely saturated with hydrogen molecules; hence they contain no carbon-carbon double bonds. Majority of the naturally occurring saturated fatty acids are unbranched with even number of carbon atoms (Chow 2007). Saturated fats are having higher melting point and less reactive than unsaturated fats. Coconut oil, palm oil, butter, tallow are few of the sources of saturated fatty acids. Table 5 represents few examples of saturated fatty acids.

Unsaturated Fatty Acids

Fatty acids that comprise of carbon-carbon double bonds are termed as unsaturated fatty acids and are mostly liquid at room temperature. The degree of unsaturation is determined by iodine number. Hence, fatty acid with greater number of double bond has higher iodine number. Compared to fats, oil has higher iodine number. This is due to the fact that oils have higher percentages of carbon- carbon double bonds. Unsaturated fatty acids are less stable than saturated fatty acids, therefore more vulnerable to rancidity (Marcus 2013). Based on the number of double bonds, unsaturated fatty acids are further classified into monounsaturated (one double bond) and polyunsaturated fatty acids (more than one double bond). Examples of oils rich in monounsaturated fatty acids are olive oil, canola oil peanut oil, walnut oil, almond oil. Examples of sources rich in polyunsaturated fatty acids are safflower oil, sesame oil, corn oil, sunflower oil, fish oil. Table 6 represents the examples of mono and polyunsaturated fatty acids.

Table 5 Examples of saturated fatty acids

Saturated fatty acids	Structure
Caprylic acid	$CH_3(CH_2)_6COOH$
Capric acid	$CH_3(CH_2)_8COOH$
Lauric acid	$CH_3(CH_2)_{10}COOH$
Myristic acid	$CH_3(CH_2)_{12}COOH$
Palmitic acid	$CH_3(CH_2)_{14}COOH$
Stearic acid	$CH_3(CH_2)_{16}COOH$
Arachidic acid	$CH_3(CH2)_{18}COOH$
Behenic acid	$CH_3(CH_2)_{20}COOH$
Lignoceric acid	$CH_3(CH_2)_{22}COOH$
Cerotic acid	$CH_3(CH_2)_{24}COOH$

Table 6 Examples of unsaturated fatty acids

Unsaturated fatty acids	Structure
Monounsaturated fatty acids	
Palmitoleic acid	$CH_3(CH_2)_5CH{=}CH(CH_2)_7COOH$
Oleic acid	$CH_3(CH_2)_7CH{=}CH(CH_2)_7COOH$
Erucic acid	$CH_3(CH_2)_7CH{=}CH(CH_2)_{11}COOH$
Polyunsaturated fatty acids	
Linoleic acid	$CH_3(CH_2)_4CH{=}CHCH_2CH(CH_2)_7COOH$
Linolenic acid	$CH_3CH_2(CH{=}CHCH_2)_3(CH_2)_6COOH$
Arachidonic	$CH_3(CH_2)_4(CH{=}CHCH_2)_4(CH_2)_2COOH$

Steroids

Steroids are one of the most important biomolecules. They are insoluble in water but highly soluble in non-polar organic solvents. The structure of steroids is different from other lipids. They do not have any fatty acid but categorized under lipids due to their similar characteristics as lipids. They are non-saponifiable. Steroids can be separated from fat after saponification. Steroids are made of fused and fully saturated ring called cycloperhydrophenanthrene or sterane. It consists of 17 carbon atoms bonded in four fused rings: three cyclohexane ring (six-carbon) and one cyclopentane ring (five-carbon). The hexagonal ring structure of steroids is different from benzene ring as the carbon atom is fully saturated with hydrogen atom and the double bonds. Steroids have a wide range of biological activity and occur in higher animals. Cholesterol is one of the common steroids in animals and found in cell membrane of all human body and transported in the blood plasma of all animals (Mu and Plummer 1988).

Terpenes

Terpenes are also non-saponifiable lipids that include certain fat soluble vitamins, carotenoids and certain co-enzymes. Terpenes are the diverse group of plant secondary compounds and a major component of essential oils. The basic structure of terpenes is the isoprene units. The larger structures are assembled from several isoprene units, usually by "head-to-tail" linked isoprene units.

Carotenoids

Carotenoids are fat soluble pigments widely present in plants, animals, algae and bacteria. They are long aliphatic polyene chain composed of eight isoprene units. It is composed of 40 carbon atoms (tetraterpenes), with a structure characterized by an extensive conjugated double-bond system which is responsible for its color. As the number of conjugated double-bond increases, color changes from pale yellow, to orange, to red. There are about 600 types of carotenoids are available. The most common ones carotenoids are alpha-carotene, beta-carotene, beta-cryptoxanthin, lutein, zeaxanthin and lycopene.

Physicochemical and Structural Characterization of Lipids

Lipids are one among the major food constituents with their characteristic physicochemical properties, such as mouth-feel, flavor, texture and appearance. Edible oils and fats are important nutrients in the human diet, providing essential fatty acids and energy. Several analytical techniques are available to characterize the lipids based on their crystallization, melting point, smoke point, rheology, density and color.

Lipids are an extremely diverse group of compounds consisting of tri-, di- and mono acylglycerols, free fatty acids, phospholipids, sterols, carotenoids and vitamins A and D. In addition, most of these sub-groups are themselves chemically complex. All triacylglycerols are esters of glycerol and three fatty acid molecules, nevertheless, the fatty acids can have different chain lengths, branching, unsaturation, and positions on the glycerol molecule. Thus even a lipid which consists of only triacylglycerols may contain a huge number of different chemical species. It is often important for food scientists to either know or to be able to specify the concentration of the different types of lipid molecules present, as well as the total lipid concentration.

Physicochemical Characterization of Lipids

Crystallization

The crystallization process includes the nucleation (formation of molecular aggregates) and crystal growth (incorporation of molecules from the adjacent liquid layer). The final structure of fat is influenced by the crystallization kinetics and linked to its rheological and plasticity properties (Ribeiro et al. 2015). The crystallization kinetics can be determined by monitoring the formation of crystalline solid material as a function of time. The solid fat content (SFC) is the percentage of the total lipid that is solid at a particular temperature. The solid fat content has its impact on the sensory and physical properties, such as spreadability, firmness, mouth feel, processing and stability. Several methods have been developed to measure the solid fat content as affected by temperature (Vasic 1965; Torbica et al. 2016). The density of solid fat being higher than the density of liquid oil, there is an increase in density when a fat crystallizes and a decrease when it melts. Thus, the solid fat content-temperature profile can be determined by measuring the density over a range of temperatures as given in Eq. (1).

$$SFC = \frac{(\rho - \rho_L)}{(\rho_S - \rho_L)} \times 100 \tag{1}$$

where ρ is the density of the lipid at a particular temperature, and ρ_L and ρ_S are the densities of the lipid when liquid or solid at the same temperature.

Recent techniques such as differential scanning calorimetry (DSC), and nuclear magnetic resonance (NMR) are used to replace density measurements for quick and easy measurement of changes in SFC. DSC measure the heat evolved or absorbed by a lipid when it crystallizes or melts. NMR basically involves the application of radio frequency pulse, inducing a NMR signal on the sample with a decay rate depending on whether the lipid is solid or liquid. The signal from the solid fat decays much more rapidly than the signal from the liquid oil and therefore it is possible to distinguish between these two components. Also, the melting point of fat can be determined using this SFC-temperature profile measurement (Braipson-Danthine and Deroanne 2006).

Melting Point

The melting profile of a fat plays a significant role in determining the mechanical characteristics of fat rich product such butter, and ice cream. In most situations, measuring the melting point (at which it starts and ends) gives needed information instead of determining SFC over the whole range of temperature. Melting point (Table 7) is unique and varies with different triacylglycerol at a specific temperature (Neidleman 1987).

Based on the amount of remaining SFC, melting point of a lipid can be defined in terms of three aspects such as clear point, slip point, and Wiley melting point. The transparent state of a completely melted fat at a particular temperature is called the "clear point". It is measured by heating a capillary tube containing a small amount of fat. Slip point is the temperature at which the fat just starts to move downwards because of its weight. The heat is applied to fat at a controlled rate using a capillary tube. The point at which a disc of fat suspended in an alcohol-water mixture of similar density changes shape to a sphere, while heating at a controlled rate is called Wiley melting point.

Table 7 Melting point of some dietary fats (Hein et al. 2015)

Name	Source	Melting point (°C)
Lauric acid	Butter	43.8
Myristic acid	Coconut	57.7
Palmitic	Palm	62.7
Stearic	Beef	70
Oleic	Nuts	16
Linoleic	Flax	−5

Smoke, Flash and Fire Point

The smoke point also known as burning point is the temperature at which the oil or fat begins to produce a bluish smoke under specified conditions. They are measured by pouring lipid on a metal container and applying heat at a controlled rate in an oven. The test reflects the amount of volatile organic material in oils and fats such as free fatty acids. Also, smoke test helps to select lipids that will be used at high temperatures, e.g. during baking or frying. Fat is considered no longer good for consumption once it has exceeded its smoke point, due to the large quantity of free radicals. Similar to smoke point, the flash point and fire point is determined based on the volatile organic material in a lipid. The appearance of flash on the surface of the sample due to the ignition of volatile gaseous products, at a particular temperature denotes the flash point. The fire point denotes the temperature at which continuous combustion occurs (a fire) due to the thermal decomposition of the volatile lipids. Oils like pongamia (*Pongamia glabra)*oil and nahor (*Mesua ferrea L., Guttiferae*) oilused to produce bio-diesel has a flash point of 134 and 142 °C respectively (De and Bhattacharyya 1999).

Rheology

The rheology of lipids is important in many food applications. Rheology is the science that deals with the deformation and flow of matter. The rheology is measured by applying a force to a material and measuring its flow or change in deformation of shape. Rheological models for oil fields are classified as Newtonian or non-Newtonian. Newtonian fluids shows a linear relationship when shear rate plotted against shear rate against shear stress, whereas a non-linear relationship obtained for non-Newtonian fluids that are sensitive to shear rate. Plasticity is one of the significant rheological characteristics to determine the spreadability of lipids. Plasticity of any lipid rises when the fat crystal forms a three-dimensional network giving the product solid-like characteristics. The flow properties of liquid oils are determined based on their critical stress limits. The product behaves like a shear-thinning liquid above the critical stress (known as the "elastic modulus") due to the disruption of crystal network. At the same time, below the critical stress limit (known as the "yield stress"), where the crystal network is not disrupted, and the product remain a solid. Rheological techniques are hence needed to measure the change in deformation of a lipid under applied stress. The stiffness of an emulsion filled gel is determined by the hardness of fat and its structural organization (Oliver et al. 2015). Increasing the hardness of fat stiffens the emulsion filled gels with high rigidity. Thus, the gel stiffness increases as a function of the solid fat content, due to inhomogeneous distribution of the fat droplets, droplet aggregation and/or concentration of the droplets.

Structural Characterization of Lipids

Oils and fats are of great economic importance as agricultural products, and they are major items of international commerce. They are ingredient in most of foods. Fat amounts to almost 40% of dietary calories in developed countries, and its composition is obviously vital to good nutrition. In addition, fats contribute greatly to the palatability, taste and structure of foods (Christie and Han 2010). The term lipids includes organic compounds such as fats, vegetable oils, waxes, cholesterol, phospholipids, steroids, and fat-soluble vitamins (A, D, E, and K). Though lipids are group of amphiphatic molecules (containing both components of hydrophilic and hydrophobic regions within the molecule), they are generally hydrophobic due largely in part to their large proportion of hydrocarbons to polar regions (due to oxygen containing functional groups). Therefore, lipids are insoluble in water but soluble in non-polar solvents such as benzene and chloroform. The hydrophobic nature of the molecule strongly determines the structures these molecules assume in nature. Being structurally diverse, lipids have a variety of biochemical functions, including structure, energy, energy storage, and signaling (steroids and eicosanoids).

Lipids are an extremely varied group of compounds consisting of tri-, di- and mono acylglycerols, free fatty acids, phospholipids, sterols, carotenoids and vitamins A and D. Fatty acids differ in their structure due to chain lengths, branching, unsaturation, and positions on the glycerol molecule. Even a lipid which consists of only triacylglycerols may contain enormous number of different chemical species. It is essential for any food scientists to know and to be able to specify the concentration of different types of lipid molecules present, as well as the total lipid concentration.

The study of lipids has assumed considerable importance with the recognition that they are involved in many vital biological processes in animals, plants and microorganisms. Organic solvents are employed to separate lipids from proteins, carbohydrates and water soluble metabolites. The lipid composition of cells and membranes can differ significantly with respect to lipid group, phospholipids class and acyl group identity. Lipids are made of diverse combination of chemicals, they also vary in terms of the lipid type and distribution and these play a important role in determining their nutritional value, purity and its various other characteristic. Therefore, it has become a necessity for scientists and legal bodies to acquire a thorough knowledge about the various classes of lipids (McClements 2003).

Sample preparation is the vital part in most method to analyze lipids. It is imperative that during the extraction of lipids due care needs to be taken to prevent lipid conversion to other forms. Production of free fatty acids due to hydrolysis needs to be prevented as quickly as possible. Extraction should cause as less oxidation as possible (Greenfield and Southgate 2003). Extraction in chloroform, chloroform-methanol and hexane-isopropanol is preferred. Storage of sample in cold at −20 °C is preferred (AAFCO 2014). The choice of the method depends on the type of food and lipid to be examined, however every available method has its advantage and disadvantage (Greenfield and Southgate 2003).

Lipid classes may be identified by their reaction with specific chemical reagents, by spectroscopic techniques, or by their chromatographic behaviour relative to authentic standards, and the amounts of each determined by appropriate methods. The fatty acid composition of each lipid class can then be determined by gas chromatography (GC). Simpler molecular species of lipids can be isolated by some form of partition or complexation chromatography. It is a very powerful tool for lipid analysis. Complete lipid profiling of a given sample is possible by this method. Different components of lipid are separated in fractions by chromatography. The various fractions are then subject to spectrometric analysis to assess the molecular identity and their relative concentration. IR, NMR and Mass Spectrometry are most commonly used (McClements 2003). In spite of their reliability, chromatographic analysis is done only when complete molecular identification of the lipid is required. Factors like high cost of the instrument and reagent also limit the usage of this method (Greenfield and Southgate 2003).

Methods for the analysis of lipids are therefore essential for many research, clinical and quality control applications. Lipid analysts have been at the forefront in the development of gas chromatography (GC) (or gas-liquid chromatography (GLC)), thin-layer chromatography (TLC), and high-performance liquid chromatography (HPLC), Mass spectroscopy. Chromatography is the most frequently used method of separating lipids into groups.

Chromatographic Analysis of Lipids

Chromatography technique is the powerful analytical procedures for separating and analyzing the properties of lipids, especially when it is combined with techniques like mass spectrometry or NMR which can analyse the chemical structure of the peaks. In chromatographic analysis mixture of the molecules to be separated is passed through a column that contains a matrix capable of selectively retarding the flow of the molecules. Based on their differing affinities for the matrix in the column the molecules in the mixture are separated because of. The stronger the affinity between a specific molecule and the matrix, its movement is more retarded, and passes through the column at a reduced rate. Based on the strength of their interaction with the matrix different molecules can be separated. After being separated by the column, suitable detectors like UV-visible, fluorescence, or flame ionization can be utilized to analyse the concentration of each molecules. Chromatography helps in complete lipid profiling, by which lipid characteristics such as the amount of saturation, cholesterol and other factors such as adulteration, presence of antioxidants could also be determined. The various types of chromatography that are available for lipid analysis of foods are thin layer chromatography (TLC), gas chromatography (GC), and high pressure liquid chromatography (HPLC).

Lipid samples obtained from natural sources are complex mixtures of individual lipid classes. In most of the cases a combination of techniques are required to separate and analyse all of these rather than a single technique. Chromatography is one of the most preferred method by lipid analysts. There are six main modes of separation in chromatography they are adsorption, normal-phase liquid partition, liquid-liquid or gas-liquid partition (in which reversed-phase chromatography is a special case), ion exchange, gel-permeation and chiral-phase chromatography. HPLC or TLC is generally used to separate each of the various simple lipid classes from the complex lipids. The latter may be further fractionated by adsorption or ion-exchange chromatography or by combinations of both, until the necessary separations are achieved. The following are the three types of chromatography that is preferred in lipid analysis.

Thin Layer Chromatography is used to find concentration of different lipid groups. The TLC plate is prepared with suitable adsorbent and kept in proper solvent. A drop of sample is placed on one end and let to flow. The plate after separation to different fractions is compared to standard plates to identify the lipids. GC, MS, NMR are used for further analysis of the spots (McClements 2003). The use of TLC has decreased considerably after the advent of TLC. However, the use of TLC for instant result is still carried out with the help of pre-coated plates. It is cheaper than HPLC. However, care during experimentation is required (Christie 2011). Different types of phospholipids cannot be separated by this method (Zaima et al. 2011). TLC is used mainly to separate and determine the concentration of different types of lipid groups in foods, such as tri, di, mono—acylglycerols, phospholipids and cholesterol. TLC plate is coated with a suitable absorbing material and placed into an appropriate solvent. This method involves spotting a small amount of the lipid sample on TLC plate. Over a period of time, due to capillary force the solvent moves up the plate and separates into various lipid fractions depending on their affinity for the absorbing material. At the end of the separation process dye is sprayed on the plate so as to make the spots visible. Identification of the lipids is done by comparing the distance that the spots move with that of known composition of standards. Further analysis of the scraped off spots can be done by using techniques such as GC, NMR or mass spectrometry. This is a cheap and rapid method for analyzing various lipids in fatty foods.

High Performance Liquid Chromatography is now a preferred method for lipid analysis because of its versatility over TLC at operating at room temperature and hence can be used for analysis of labile groups which cannot be determined by using GC (Christie 2011).

Gas Chromatography is the preferred method for analysis of trans fatty acid. It can also be used for triglycerides and fatty acids; however, methylation is necessary (Greenfield and Southgate 2003). Before carry out gas chromatography for fatty acids that are non-volatile, it is imperative that the lipids are made into volatile fatty acid methyl esters (FAME) by saponification and methylation (McClements 2003). Gas - liquid chromatography (GLC) can only be used for compounds which can be volatilized without decomposition. In practice this normally limits its application to molecules with molecular mass of less than 800 Da. Lipids such as phosphoglycerides which do not remain intact are not amenable to this method. Perhaps the most important use for the lipid biochemist is in the analysis of the fatty acid composition of lipids or tissues. In addition, gas chromatography may be used for steroid, glycerol, sphingosine, inositol and carbohydrate analysis. Molecular species of partial glycerides or, on highly thermostable silicone liquid phases, even triacylglycerols may be analysed also.

Other Methods

Apart from chromatography other techniques for fat analysis include NMR, IR, MS however they are usually coupled with chromatography. Measurement of density and refractive index can be used to measure change in chain length and unsaturation (McClements 2003). Modern mass spectrometry methods involving ionization by electrospray (ESI), fast atom bombardment (FAB), atmospheric pressure chemical-ionization (APCI), atmospheric pressure photo-ionization (APPI), and matrix-assisted laser desorption (MALDI) techniques are highly sensitive and can produce excellent quantitative data. The newest generation of mass spectrometers include ion trap, mass quadrupole:linear ion trap (Q:LIT), quadrupole:time of flight (QToF), linear ion trap:Orbitrap (LIT:Orbitrap), and Fourier transform ion cyclotron resonance (FT-ICR) and enable highly detailed analyses, though often at substantially greater capital cost. All of these methods give a wealth of structural information about analytes.

Of the various ionization methods, electrospray ionization appears to be the most popular, as it offers both high sensitivity and good quantification with less need for substantial structure-dependent response factors than the other ionization methods. Excellent results are obtained for phospholipids, sphingolipids and simple lipids such as triacylglycerols.

Modification of Lipids

With the ever growing demand for oils and fats in countries like China and India, modifying lipids has become a necessity. The consumption of oils are limited to lesser than 20 oils, however these are not always suited for food use and hence the need for modifying it for the purpose of consumption. It is estimated that approximately 80% of total oil and fat produced is used for food and the balance for animal feed and by the oleochemical industry. The major food uses include frying oils, baking fats, cooking fats, shortenings, spreads, salad oils, mayonnaise, confectionery fats and ice cream. The changing nutritional requirements of the growing population have also led to new vistas in the modification process of lipids. Lipid Modification processes is focused on specific areas such as developing target specific lipids that are effective in preventing and curing diseases in children and adults likewise. The primary area of focus has been in developing structured lipids with the help of medium-chain acids and long-chain polyunsaturated fatty acids (PUFAs). Specific long chain PUFAs like docosahexaenoic acid and arachidonic acid, are important both in the growth and development of infants. Based on the requirement of the segment of the population the ideal lipids could be formulated using various lipid modification processes. Target - specific lipid modification process includes using novel fat sources and altering the concentrations of medium-chain, long-chain saturated, and long-chain polyunsaturated fatty acids as well as cholesterol.

Lipid modification also includes methods such as genetic modification of oil seeds, chemical and lipase catalyzed interesterification reactions and fractionation of fats and oils. However the choice of selecting a suitable modification process depends largely on the end use and target population and to achieve this, a combination of one or more of these techniques may be needed. The choice of refined commodity oils for the use in food industry is very limited and this has fuelled the need for modifying oils. It is essential to know why the existing commodity oils are less suitable than modified oils.

Methods of Modification of Lipids

Methods of lipid modification may be either technological or biological. Technological procedures include techniques like blending, fractionation, hydrogenation, and interesterification with chemical or enzymatic catalysts. Biological procedures includes development of new crops and hybrids to make them more suitable for commercial growing and harvesting, seed breeding by conventional or newer procedures so as to produce oils with a favorable and desired fatty acid composition, and production of single cell oils rich in valuable polyunsaturated fatty acids not otherwise easily available (Gunstone 2006).

The usage of the four technological procedures listed above consideration has to be given to their effectiveness in achieving the desired physical, chemical, and nutritional properties and to their relative costs in production. Among the four methods blending is the cheapest and hydrogenation is the most expensive in terms of both equipment and consumable costs for catalyst and hydrogen. Dry fractionation does not require additional materials and there is no yield loss but there has to be an economic use for all the fractions. Interesterification can be carried out with lower capital cost than fractionation but operating costs are higher because of yield loss. Enzymatic interesterification is more expensive because the cost of enzyme is still high. It could be seen that costs rise through the blending, dry fractionation, chemical interesterification, and hydrogenation. It is practically very difficult for smaller producers to carry out these processes and hence they prefer to buy refined and modified oils and fats from larger food producers and specialist suppliers. However, to fully understand the properties of these oils it is essential to know the advantages and disadvantages of these processes (Gunstone 2009).

In technological procedures, the changes are made only in fatty acid and/or triacylglycerol composition by modifying nutritional, chemical, and physical properties to make them more appropriate for their end-use and by developing improved techniques and sophisticated equipment. However, in the biological procedures we interfere at an earlier stage by using conventional techniques of seed breeding or by exploiting newer methods based on increasing genetic understanding to seek new sources or to modify the composition of the oils in plants which already produce large quantities of oil efficiently.

Blending

Blending is the oldest process of mixing of oils and fats to produce blends with improved nutritional or physical properties. In the preparation of most spreads, blends of two or more oils are used to achieve desirable nutritional and essential physical properties. Interesterification is usually carried out on oil blends. Inorder to obtain oils with desired mix, the oils are blended at minimum cost and by developing computer programs that aid to give the best solution. For example, oil with high oxidative stability (such as sesame oil or rice bran oil) may be added to a less stable commodity oil to enhance its stability. Frequently lauric oils are usually blended with non-lauric oils.

Fractionation

Fractionation is a process by which oil is divided into two or more fractions which differ in fatty acid and triacylglycerol composition. This is achieved without loss of material and without need for further refining. The two fractions extend the range of usage of the original oil. This is achieved by a process called dry fractionation. Dry fractionation requires slow cooling of liquid oil without solvent followed by efficient separation of the solid (stearin) and liquid phases (olein) by suction or by pressure. Poor separation leaves olein in the stearin and results in less olein and in more stearin with higher iodine value than is desired for a stearin. This technique has wide application and usage in palm oil apart from other applications. However one of the major drawbacks is the formation of unstable compounds that happens to minor components and impurities due to changes during fractionation. A simple form of fractionation is referred to as winterization, which originally consisted of storing oil such as cottonseed in tanks during the winter. The more saturated triacylglycerols crystallized leaving a more unsaturated liquid fraction less likely to crystallize when stored in the domestic refrigerator. Some oils like sunflower may contain wax which makes the oil cloudy during storage, especially at sub-ambient temperatures. Such oils may be dewaxed by storage at appropriate temperatures and subsequent filtration before release for sale in order to overcome this.

Hydrogenation

Hydrogenation can be conducted at three different levels depending on the end use and desired trait. Very light or partial hydrogenation also known as brush hydrogenation is a technique applied to oils containing linolenic acid such as soybean oil and rapeseed oil inorder to reduce the level of triene acid to about one half its normal value thereby extending the shelf life of foods containing these oils. At the other extreme, certain oils are virtually completely hydrogenated to iodine values less than 2. Of greater importance is partial hydrogenation applied to soybean and other linoleic rich oils to raise the content of solid triacylglycerols. This is achieved through formation of saturated acids and of transunsaturated acids. Partial hydrogenation of unsaturated oil gives a product of higher melting point (more suitable for spreads and cooking fats) and of enhanced oxidative stability.

Interesterification Using a Chemical Catalyst

The process of rearranging the fatty acids in oils or in combination of oils thereby changing the triacylglycerol composition is called interesterification. However this does not affect the fatty acid composition of the single oil or the blend and it remains the same. Usually the blend will contain two or more oils differing in chain length and/or in patterns of unsaturation. The interesterification process could be achieved with the help of alkaline catalyst such as NaOH wherein the distribution of the fatty acids is randomly distributed in the product. This is in contrast to the natural vegetable oils that are produced by enzymatically catalysed biological processes and in the fatty acids are not randomly distributed. It is these changes that are made in triacylglycerol composition that affect thermal behaviour and may also affect bioavailability. This process also results in loss of some oil through conversion to acid (with NaOH) or ester (with sodium methoxide). Inspite of the loss, this procedure is also being used to produce fats suitable for use as spreads without hydrogenation.

Interesterification Using an Enzymatic Catalyst

Interesterification can also be carried out with lipases. The enzymatic processes offers advantages over the chemical process in terms of occurring under milder conditions, requirement of less costly equipment, and produce less by-product, thereby reducing wastage and reducing purification process. The major benefit of using a lipase is the added control over the nature of the product as a consequence of the specificity shown by many lipases. However, chemical or enzymatic catalysts interesterified products are generally less stable than the original oils. There is a perception in some quarters that the enzymatic reaction is superior because it is 'more natural' and avoids the use of 'chemicals'.

Oilseeds Modified by Seed Breeding

Growing wild plants need greater effort and in most cases have a lot number of difficulties. To overcome these modern agricultural practices are being employed to grow and commercialize them. This method requires reducing the use of chemicals, apart from which altering the fatty acid composition and levels of other important components. This must be achieved without sacrificing other parameters such as yield and stability of the crop from season to season. The modified seed must not be mixed with the conventional seed. In the production of modified seed it is important to know the methods employed in producing such seeds, since there is a growing concern over the use of these modified seeds in certain population. Inspite of these advances in production of modified seeds, the commercial development of oils from such seeds will definitely take a long time. The high demand for oils and fats for food and non-food uses has coupled the need to increase the production of such seeds. This could also be attained by seed breeding or by developing seeds to give higher yields, drought-resistant and provide better yields under adverse and harsh conditions.

Animal Fats Modified Through Nutritional Changes

Over the years, animal fats from nutritional stand point have always been seen with lot of disadvantages. This is mainly because these animal fats being loaded with saturated fats coupled with transfat that could lead to high cholesterol levels. The other factors include low levels of essential fatty acids and antioxidants. This apart, the use of animal fat is restricted to certain groups in the population as few ethnic groups prefer fat from plant sources. The advantage that animal fat has over fat from plant sources is that it is valuable source of long-chain PUFA in the diet especially for people who do not consume fish. Apart from these, altering the fatty acid composition of chicken eggs and meat by making changes in the diet of chicken is also fast gaining momentum. However great care is required to be taken so that these changes in fatty acid composition do not reduce the shelf life of the products and during cooking it does not lead to unexpected flavours and odours.

Conclusion

The chapter on lipids highlights on the general classification, properties, structural classification and methods of modification of lipids. The term lipids has a plethora of definitions; however lipids are most commonly referred to as substances that are insoluble in water and having composed of fatty acids and its derivatives that are soluble in organic solvents.

In terms of classification of lipids, the lipids could be classified based on their structure as simple lipids, complex lipids and derived lipids. These are further classified into numerous types depending on their composition. Apart from the structural characterization, the physicochemical properties of the lipids also greatly determine the end-use. Properties such as crystallization greatly influence the characteristics of solid fat and this in turn has an impact on physical properties such as spread ability, stability and processing characteristics. Properties like melting point determine the mechanical characteristics of butter and ice-cream.

The lipid could be characterized structurally by using techniques such as chromatography and spectroscopy. Chromatography techniques such as HPLC and TLC combined with NMR and mass spectroscopy is among the most commonly used procedure for separating and analyzing the properties of lipids.

With the ever-growing demand for healthier lipids, there is a constant need for modifying lipids. Methods of modification are limited to technological or biological, depending on the end use the lipids could be modified to suit the needs of the consumer but, it is imperative that the modification process does not lead to unwanted changes during cooking or during storage.

References

AAFCO. (2014). *Crude fat methods—Considerations*. Atlanta, GA: Association of American Feed Control Officials Incorporated.

Braipson-Danthine, S., & Deroanne, C. (2006). Determination of solid fat content (SFC) of binary fat blends and use of these data to predict SFC of selected ternary fat blends containing low-erucic rapeseed oil. *Journal of the American Oil Chemists' Society, 83*(7), 571–581.

Chow, C. K. (2007). *Fatty acids in foods and their health implications*. Boca Raton, FL: CRC Press.

Christie, W. W. (2011). *Thin-layer chromatography of lipids*. Urbana, IL: AOCS Lipid Library.

Christie, W., & Han, X. (2010). *Lipid analysis-isolation, separation, identification and lipidomic analysis* (p. 446). Bridgwater, UK: Oily Press.

De, B., & Bhattacharyya, D. (1999). Biodiesel from minor vegetable oils like karanja oil and nahor oil. *Lipid/Fett, 101*(10), 404–406.

Engelking, L. R. (2010). *Textbook of veterinary physiological chemistry, updated 2/e*. Amsterdam, Netherlands: Academic Press.

Greenfield, H., & Southgate, D. (2003). Review of methods of analysis. In *Food composition data-production, management, and use* (2nd ed.). Rome, Italy: Food and Agriculture Organization of the United Nations.

Gunstone, F. D. (2006). *Modifying lipids for use in food*. Cambridge, UK: Woodhead Publishing.

Gunstone, F. (2009). *Oils and fats in the food industry*. New York, NY: Wiley.

Gurr, M. I., Harwood, J. L., & Frayn, K. N. (2002). *Lipid biochemistry* (3rd ed.). Dordrecht, Netherlands: Springer.

Hein, M., Pattison, S., Arena, S., & Best, L. R. (2015). *Introduction to general, organic, and biochemistry*. Hoboken, NJ: Wiley.

Kates, M. (1986). *Techniques of lipidology: Isolation, analysis and identification of lipids*. Amsterdam, Netherlands: Elsevier.

Marcus, J. B. (2013). Lipids basics: Fats and oils in foods and health. In *Culinary nutrition: The science and practice of healthy cooking*. Waltham, MA: Academic Press.

McClements, D. (2003). *Analysis of lipids*. Umass Amherst. Retrieved June 20, 2018, from http://people.umass.edu/~mcclemen/581Lipids.html

Mu, P., & Plummer, D. T. (1988). *Introduction to practical biochemistry*. New Delhi, India: Tata McGraw-Hill.

Neidleman, S. L. (1987). Effects of temperature on lipid unsaturation. *Biotechnology and Genetic Engineering Reviews, 5*(1), 245–268.

O'brien, R. D. (2008). *Fats and oils: Formulating and processing for applications*. Boca Raton, FL: CRC Press.

Oliver, L., Scholten, E., & van Aken, G. A. (2015). Effect of fat hardness on large deformation rheology of emulsion-filled gels. *Food Hydrocolloids, 43*, 299–310.

Pollard, T. D., Earnshaw, W. C., Lippincott-Schwartz, J., & Johnson, G. (2016). *Cell biology E-book*. Amsterdam, Netherlands: Elsevier.

Ribeiro, A. P. B., Masuchi, M. H., Miyasaki, E. K., Domingues, M. A. F., Stroppa, V. L. Z., de Oliveira, G. M., & Kieckbusch, T. G. (2015). Crystallization modifiers in lipid systems. *Journal of Food Science and Technology, 52*(7), 3925–3946.

Sikorski, Z. E., & Kolakowska, A. (2010). *Chemical and functional properties of food lipids*. Boca Raton, FL: CRC Press.

Torbica, A., Jambrec, D., Tomić, J., Pajin, B., Petrović, J., Kravić, S., & Lončarević, I. (2016). Solid fat content, pre-crystallization conditions, and sensory quality of chocolate with addition of cocoa butter analogues. *International Journal of Food Properties, 19*(5), 1029–1043.

Vasic, J. (1965). Effect of temperature history on the solid fat content of milk fat. *Journal of Dairy Science, 48*(10), 1277–1281.

Zaima, N., Goto-Inoue, N., Adachi, K., & Setou, M. (2011). Selective analysis of lipids by thin-layer chromatography blot matrix-assisted laser desorption/ionization imaging mass spectrometry. *Journal of Oleo Science, 60*(2), 93–98.

Nutraceutical Properties of Lipids

Gabriela John Swamy, Ezhilarasi Perumal Natarajan, and Gopirajah Rajamanickam

Introduction

Nutraceutical have emerged as beneficial health products obtained from food, herbal, and pharmaceutical manufacturing industries. Nutraceuticals have gained worldwide popularity and defined as purified or concentrated food and food products with health-promoting and/or disease-preventing properties beyond their basic nutritional function (Gunther et al. 2004). The significance of nutraceuticals has been associated with the treatment of many disorders such as obesity, cancer, metabolic problems, cold and cough, depression, coronary heart disease, osteoporosis, diabetes, cardiovascular problems, delayed gastrointestinal emptying, and other degenerative diseases (Kaur et al. 2015). Nutraceutical has led to the new era of medicine and health, in which the food industry has become a research-oriented sector. Although the use of nutraceuticals by people has a long history, recently it has been scientifically supported by nutritional and medical evidence that has allowed nutraceuticals to emerge as being potentially effective (Dillard and German 2000).

Foods can be regarded as functional if they help in reducing the risk of disease and promote good health. Fats and lipids are common components of food and perform essential roles in human health. Generally, the high calorific value of fats and oils has earned them a bad reputation among the public and an average consumer might wrongly believe that lipids are intrinsically harmful. Fortunately, the progress in our understanding of the various dietary lipids and its functions in our human

G. J. Swamy
Department of Agricultural Engineering, South Dakota State University, Brookings, SD, USA

E. P. Natarajan
James L. Winkle College of Pharmacy, University of Cincinnati, Cincinnati, OH, USA

G. Rajamanickam (✉)
Department of Food Science, Cornell University, Ithaca, NY, USA

A. Gani, B. A. Ashwar (eds.), *Food biopolymers: Structural, functional and nutraceutical properties*, https://doi.org/10.1007/978-3-030-27061-2_17

body helps to deny the misconceptions about dietary fats and creates new opportunities for the development of lipid-enhanced functional foods, nutraceuticals, and dietary supplements. The lipids such as omega-3 fatty acids (alpha-linolenic acid (ALA), eicosapentaenoic acid (EPA), docosahexaenoic acid (DHA)), omega-6 fatty acids (gamma linoleic acid (GLA), and linoleic acid (LA)), conjugated linoleic acid (CLA), medium chain triglyceride oils, and phytosterols (mainly b-sitosterol, campesterol and stigmasterol) are considered as functional lipids due to their beneficial effects on human health. Additionally, the fat-soluble vitamins (Vitamin A, D, E, K) essential oils, lipophilic nutraceuticals like polyphenols, flavonoids are lipid-based components that impart essential health benefits on humans can also be considered. Functional lipids have many beneficial effects on human health such as treating obesity, bone health and managing depression, blood pressure, cardiovascular health, cancer etc. (Alabdulkarim et al. 2012). Many types of research have linked functional lipids to the prevention and treatment of many diseases.

The ALA and LA are considered "essential fatty acids" (EFAs) need to be included in the diet. ALA acts as precursors for the synthesis of more highly unsaturated and longer-chain omega-3 fatty acids such as DHA and EPA. EPA and DHA are found mainly in fatty fishes such as mackerel, salmon, herring, trout, blue fin tuna and in fish-oils. Principal sources of ALA are mainly flaxseed, soybeans, canola, some nuts (e.g. walnuts) and red/black currant seeds (Institute of Medicine 2002). LA occurs mainly in vegetable oils e.g. corn, safflower, soya bean, and sunflower. ARA is found in animal products such as meat, poultry, and eggs. Phytosterols (plant sterols) are common plant and vegetable constituents, part of the human diet and are biologically active (Pirronen et al. 2000). They are structurally related to cholesterol but differ from cholesterol in the structure of the side chain. Polyunsaturated fatty acids (PUFAs) (which include the omega-3 and omega-6 fatty acids) and phytochemicals also play an important role as healthy dietary bioactive compounds. The objective of this chapter is to review and summarize the research evidence on the nutraceutical properties of functional lipids. The major nutraceutical benefits of lipids highlighted in the chapter are prevention of cancer, hypoglycemic effects, and reduction of bile stone formation, mineral absorption, hypocholesterolemic effects and other properties.

Prevention of Cancer

Generally, Nutraceuticals, have chemopreventive activities, as reported by many epidemiological and animal model studies. Their ability to reduce cancer incidence in these studies is likely related to apoptosis. The potential of using nutraceuticals as chemopreventive reagents has been proved through *in vitro* and *in vivo* studies. Essential fatty acids (EFA) may play an important role in inhibiting the prostate cancer cell proliferation. Pandalai et al. (1996) demonstrated the effects of omega-6 and omega-3 fatty acids in prostate cancer cell lines. They investigated the effects of dietary fats on nontumorigenic prostate cell growth through in vitro studies with human metastatic PC-3, LNCaP and TSU prostate cell lines, the rat metastatic Mat-Ly-Lu cell line and rat non-metastatic epithelial cell lines EPYP1, EPYP2 and EPYP3. Cell lines were treated with linoleic acid (LA), an omega-6 FA (n-6), as well as linolenic (LLA) and eicosapentaenoic (EPA) acids, which are both omega-3 FAs (n-3). All cell lines were treated with 10% and 0.5% serum supplemented media plus fatty acid for comparison. The results demonstrate that linoleic acid (n-6) has promotional effects at doses of 1–100 ng/ml in all cell lines with the exception of EPYPl. Experiments with linolenic acid (n-3) demonstrated consistent growth promotion in all cell lines except EPYP2. EPA demonstrated promotion at low concentrations and inhibited prostate cell growth at higher concentrations. Recently, Palakurthi et al. (2000) demonstrated the anticancer activity of eicosapentaenoic acid (EPA), through in vitro and in vivo models of experimental cancer. The molecular basis for the anticancer effects is EPA inhibits cell division by inhibiting translation initiation. This is a consequence of the ability of EPA to release Ca^{2+} from intracellular stores while inhibiting their refilling via capacitative Ca^{2+} influx that results in partial emptying of intracellular Ca^{2+} stores and thereby activation of protein kinase R. In a KLN-205 squamous cell carcinoma mouse model, daily oral administration of EPA resulted in a significant reduction of tumor size and malignant hypertetraploid cells. Recently, Simopoulos (2008) stated that excessive amounts of omega-6 polyunsaturated fatty acids (PUFA) and a very high omega-6/omega-3 ratio, promote the pathogenesis of many diseases, including cardiovascular disease, cancer, and inflammatory and autoimmune diseases, whereas increased levels of omega-3 PUFA (a lower omega-6/omega-3 ratio), exert suppressive effects. A ratio of omega-6/omega-3 at 2.5/1 reduced rectal cell proliferation in patients with colorectal cancer, whereas a ratio of 4/1 with the same amount of omega-3 PUFA had no effect. A lower ratio of omega-6/omega-3 fatty acids is more desirable in reducing the risk of cancer.

The fat-soluble vitamins (Vitamin A, D, E, and K) present in the plant and animal sources and cooking oils, has been found to inhibit proliferation of several types of cancer cells, including those of breast, lung, and endometrium. Amir et al. (1998) evaluated the mechanistic aspects of vitamin A (lycopene) treatment, which resulted in a concentration-dependent reduction in HL-60 cell growth as measured by [$_3$H] thymidine incorporation and cell counting. This effect was accompanied by inhibition of cell cycle progression in the G_0/G_1 phase as measured by flow cytometry. Furthermore, the combination of low concentrations of lycopene with 1,25-dihydroxy vitamin D_3 exhibited a synergistic effect on cell proliferation and differentiation and an additive effect on cell cycle progression. Likewise, Weinstein et al. (2007) reported that Alpha-tocopherol supplementation (50 mg daily for 5–8 years) reduced prostate cancer incidence by 32% in the alpha-Tocopherol, Beta-Carotene Cancer Prevention Study. They also investigated the effect of serum alpha-tocopherol or intake of vitamin E (eight tocopherols and tocotrienols) on prostate cancer risk with up to 19 years of follow-up study. Higher serum alpha-tocopherol was associated with reduced risk of prostate cancer.

Likewise, Noto et al. (1989) reported the anticancer activity of vitamin K3 in combination with Vitamin C. VitaminK3 (Menadione) in combination with vitamin

C (ratio of 100:1, vitamin C: vitamin K3) achieved cytotoxic dosages 10–50 times lower than when singly administered. Vitamin K3 at 13.8 μg/mL produced a 50% inhibition of breast cancer (MCF-7) cell growth. When combined with vitamin C (99.01 μg/mL), K3 (1.38 μg/mL) increased inhibition by 74%. Increasing the concentrations of vitamins C and K (10^4 μmol/L and 10^5 nmol/L, respectively) produced a 93% inhibition. Using KB cells, a 50% inhibition of growth was seen with K3 (13.8 μg/mL). When KB cells were subjected to vitamin C at 10^3 mmol/L or K3 at 10^4 nmol/L separately, no in vitro cytotoxic effect resulted. When combined at the same concentrations, a synergistic inhibition of 100% was achieved. Similarly, Vitamin K2 also exhibited anticancer effects with an ID50 value from 0.8–2 mM, which, is higher than inhibitory levels of K3 (18–45 μM) (Wu et al. 1993). Similarly, Nishikawa et al. (1999) have reported that lower concentrations of vitamin K2 are effective anticancer agents for a number of cancer cell types through in vitro studies. The HOS TE85 human osteosarcoma cell line and the MC3T3-E1 mouse osteoblastic cell line were cultured for three days in a medium containing various concentrations of MK-4. The proliferation of HOS cells was suppressed by vitamin K2 in a dose-dependent manner up to 56% of control by 10^{-7} M of K2, and that of MC3T3-E1 cells was suppressed to 84% of control by 10^{-6} M of vitamin K2. MK-3 had an ID50 of 112 μM for the human hepatoma cell line Hep3B.

The phytosterols are also effective against various cancers like colon, prostate, and breast cancer. Awad et al. (2002) evaluated the dietary effect of phytosterols versus cholesterol on the growth and metastasis of the PC-3 human prostate cancer cells in mice. Also, their direct effect on the growth and migration of these cells in vitro was analyzed. Animals fed with the phytosterol diet had tumors 40–43% smaller than those fed with the cholesterol diet. Furthermore, the number of mice with lymph node and lung metastasis was almost one-half that of the cholesterol-fed group. In the *in vitro* studies, both β-sitosterol and campesterol inhibited the growth of PC-3 cells by 70% and 14%, respectively, while cholesterol supplementation increased the growth by 18% when compared with controls. It was concluded that phytosterols indirectly (in vivo as a dietary supplement) and directly (in tissue culture media) inhibited the growth and metastasis of PC-3 cells. β-Sitosterol was more effective than campesterol in offering this protection in most of the parameters studied.

There are few lipophilic flavonoids and polyphenols, that are more effective against various cancer. Curcumin is a polyphenol extracted from turmeric, with anti-oxidation, anti-inflammation, and anti-cancer properties. However, the effects of curcumin on human cancer cells remain largely unknown. Wang et al. (2017) investigated the anticancer effects and the mechanisms by which curcumin affects breast cancer cells. The anticancer effect of curcumin on cell viability and cytotoxicity on human breast cancer MCF-7 cells was analyzed. The results demonstrated that curcumin inhibited cell viability and induced cytotoxicity of MCF-7 cells in a concentration- and time-dependent manner, by inducing apoptosis and increasing caspase-3/9 activities. Yu and Huang (2010) encapsulated curcumin in hydrophobically modified starch (HMS), a food-grade biopolymer, and formed micelles. Upon encapsulation, curcumin showed increased solubility by about 1670-folds, which improves bioaccessibility and bioavailability. Encapsulated curcumin had enhanced in vitro anticancer activity compared to free curcumin.

The green tea and its polyphenols were also reported to Inhibit various carcinogenesis in many in vitro and animal models. The mechanisms of action have been extensively investigated mostly in cell culture systems with (-)-epigallocatechin-3-gallate (EGCG), the most active and major polyphenolic compound from green tea. Experimental evidence from animal models also suggests that tea plays a significant role in inhibition of carcinogenesis in organ sites in skin, lung, esophagus, stomach, liver, small intestine, pancreas, colon, bladder, and mammary gland. (role of nutraceuticals inhuman health). The chemopreventive effects of green tea depend are mainly due to its antioxidant action; the specific induction of detoxifying enzymes; its molecular regulatory functions on cellular growth, development and apoptosis; and a selective improvement in the function of the intestinal bacteria flora. Zhang et al. (2002) reported that ovarian cancer risk declined with increasing frequency and duration of green tea consumption. In the same line of research, Yu et al. (2004) reported that EGCG inhibited the growth of prostate cancer adenoma cells and induced apoptosis.

Like polyphenols, flavonoids are more effective against various cancer and their mechanisms of action have been identified, including carcinogen inactivation, antiproliferation, cell cycle arrest, induction of apoptosis, inhibition of angiogenesis, antioxidation, and reversal of multidrug resistance or a combination of these mechanisms. Benavente-García and Castillo (2008) stated that the pharmacological properties of *Citrus* flavonoids can be linked to the abilities of these compounds to inhibit enzymes involved in cell activation. The mechanism for cancer control includes the use of suppressing, blocking, and transforming agents. Suppressing agents prevent the formation of new cancers from procarcinogens, and blocking agents prevent carcinogenic compounds from reaching critical initiation sites, while transformation agents facilitate the metabolism of carcinogenic components into less toxic materials or prevent their biological actions.

Various studies demonstrated that functional lipids are effectively inhibiting the cancerous cell growth. EPA inhibited prostate cell growth at higher concentrations and significantly reduced the tumor size and carcinoma cells in a mouse model. Moreover, omega-6/omega-3 fatty acids (2.5/1) reduced the cancer risk at a lower ratio. Likewise, the vitamin A, E, K2, and K3 and phytosterols exhibited chemoprevention activity by suppressing the proliferation of cancerous cells. This shows that functional lipids play a major role in reducing the incidence of cancer and could be an avenue for therapeutic intervention. However, an understanding of the chemoprevention mechanism, biomarkers, clinical trials, and long-term effects of these lipids in humans awaits further study.

Hypoglycemic Effect

Shimura et al. (1997) investigated the hypoglycemic and hypolipidemic effect of docosahexaenoic acid (DHA) ethyl ester in KK-Ay mice and neonatal streptozotocin-induced diabetic mice, and in the normal mice. A single administration of DHA (500 mg/kg body weight) to KK-Ay mice significantly reduced the blood glucose levels and plasma free fatty acid levels at 10 h after oral administration when compared with the control group. DHA treated neonatal diabetic mice and normal mice, however, showed no change in these parameters. In addition, repeated administration of DHA (100 mg/kg) to KK-Ay mice significantly suppressed the increment of blood glucose and plasma triglyceride levels, and significantly decreased free fatty acid at 30 days compared with control group. DHA also significantly decreased the blood glucose at 60 and 120 min on the insulin tolerance test. It clearly shows that DHA exhibits its hypoglycemic effects by increasing insulin sensitivity.

Similarly, Morishita et al. (2000) demonstrated the dose-related pharmacological effects of insulin emulsion incorporating highly purified eicosapentaenoic acid (EPA) and docosahexaenoic acid (DHA). Water-in-oil-in-water multiple emulsions (insulin dose, 0, 10, 25 and 50 IU/kg) incorporating 2% DHA or EPA were administered directly into the colonic and rectal loops in situ. Serum insulin levels rose and serum glucose levels decreased in an insulin dose-related fashion. The relationship of insulin dose and Cmax or AUC insulin was linear at the rectum, but a non-linear relationship was observed at the colon. The trend was more predominant in DHA. In the in vivo rectal absorption experiment using emulsions incorporating 2% DHA, 5 IU/kg of insulin emulsion produced a rapid, transitory increase in serum insulin levels and strong reduction of serum glucose levels. Mucosal irritation caused by administration of emulsions incorporating 2% EPA or DHA was evaluated by a lactate dehydrogenase (LDH) release study and compared with those of the emulsion incorporating 2% oleic or linolenic acid. The results indicate that emulsion incorporating highly purified long-chain polyunsaturated fatty acid, especially DHA, has the potential of becoming the formulation for enteral delivery of insulin.

Recently, Li et al. (2014) investigated the role of DHA in the prevention and treatment of type 1 diabetes in a murine model. Streptozotocin-induced diabetic mice were divided into control with diabetes, diabetes prevention (500 μg/kg DHA orally for 5 days) or diabetes treatment groups (DHA solvent in DMSO into the colon for 5 days). The groups were observed for 25 days after the administration of DHA. Treatment with DHA greatly improved polyphagia and polydipsia, indicating improvement in diabetic conditions. Besides that DHA also showed significant improvements in body weight. The colonic administration of DHA caused a significant increase in insulin level in the prevention and therapy groups. Histological assessment of the pancreatic sections revealed that pancreatic cells from the DHA-treated mice were less damaged (no vacuolization and less degranulation) compared with those in the diabetic control, suggesting that there were more functional pancreatic cells in the DHA-treated group. This may explain the higher level of insulin that was found. Taghizadeh et al. (2016) determined the effects of omega-3 fatty acids and vitamin E co-supplementation on glucose homeostasis parameters and lipid concentrations among women with gestational diabetes. Patients were randomly allocated to take either 1000 mg of omega-3 fatty acids from flaxseed oil plus 400 IU of vitamin E supplements or placebo for 6 weeks. After 6 weeks of intervention, changes in fasting plasma glucose serum insulin concentrations, estimated insulin resistance, estimated beta cell function, and quantitative insulin sensitivity check index in the omega-3 fatty acids plus vitamin E group were significantly different from the placebo group. Overall, it was demonstrated that omega-3 fatty acids and vitamin E co-supplementation in gestational diabetic women had beneficial effects on glucose homeostasis parameters, serum triglycerides, VLDL-cholesterol, and HDL-cholesterol concentrations, but it did not influence total-cholesterol and LDL-cholesterol levels.

Like the Polyunsaturated fatty acids, the fat-soluble vitamins, essential oils, flavonoids also exhibited the hypoglycemic effect and anti diabetic activity. Singh and Verma (2016) studied the effect of Vitamin A on blood glucose level and its possible interactions with oral antidiabetic agents such as glimepiride, gliclazide, and metformin in hyperglycemia-induced albino rats. Vitamin A significantly lowered the blood glucose level but antagonized the effect of oral antidiabetic agents in co-administration therapy. High doses of Vitamin A produced antioxidant effects in diabetic patients. Similarly, Suksomboon et al. (2011) evaluated the effect of vitamin E supplementation on glycemic control in type 2 diabetes. The treatment effect was estimated with the mean difference in the changes of HbA1C (glycated hemoglobin) from baseline to final assessment between the vitamin E and the control groups. It was effective only in a subgroup of patients with inadequate glycemic control at baseline (HbA1C) $\geq$ 8%) and in those whose baseline serum vitamin E levels were below normal ranges. The pooled mean difference in the changes of HbA1C was −0.58% which was the same for both subgroups of patients. The evidence suggested no beneficial effect of vitamin E supplementation in improving glycemic. However, HbA1C may decrease with vitamin E supplementation in patients with inadequate glycemic control or low serum levels of vitamin E. Udupa et al. (2013) evaluated the effects of antioxidants; alpha lipoic acid (ALA), omega 3 fatty acid and vitamin E on parameters of insulin sensitivity (blood glucose and HbA1c) in patients of type 2 diabetes mellitus with documented insulin resistance. Fasting blood glucose and HbA1c were measured at the first visit and after 90 days of intervention. There was a statistically significant decrease in HbA1C levels in Group I (alpha lipoic Acid), Group II (omega 3 fatty acid) and Group III (vitamin E) after 90 days compared to visit 1.

Kangralkar et al. (2010) elucidated the hypoglycemic effect of the essential oil of Citrus reticulata and its interaction with glibenclamide in euglycemic and hyperglycemic rats. Euglycemic and alloxan induced hyperglycemic rats were orally administered with the essential oil and glibenclamide (0.9 mg/kg), and at subhypoglycemic dose (0.45 mg/kg). The blood glucose was estimated using glucometer at 0, 2, 4, 6, 8 and 12 hrs. The essential oil of Citrus reticulata exhibited significant hypoglycemic activity, while it did not potentiate the hypoglycemic activity of glibenclamide.

Likewise, Boukhris et al. (2012) demonstrated the hypoglycemic effect of essential oil extracted from Pelargonium graveolens. The essential oil of P. graveolens was administered daily and orally to the rats at two doses of 75 mg/kg and 150 mg/kg body weight (b.w.) for 30 days. The hypoglycemic effect of P. graveolens oil was significantly more effective than that of diabetic drug glibenclamide which was confirmed through the histological findings in hepatic and renal tissues of diabetic rats as well. Recently, Tahir et al. (2016) evaluated the antidiabetic potential of Syzygium aromaticum and Cuminum cyminum essential oils and their emulsions by alpha-amylase inhibition assay in dose-dependent mode (1 to 100 μg/mL). Five emulsions (essential oil + surfactant [tween 80] + co-surfactant [ethanol] + water) of different concentrations for S. aromaticum (A1–A5) and C. cyminum (B1–B5) essential oils were formulated. Among different emulsions, A5 of S. aromaticum (100 μg/mL) and B5 of C. cyminum essential oil (100 μg/mL) exhibited a maximum antidiabetic activity with 95.30 and 83.09% inhibition of α-amylase, respectively. Moreover, the analysis of essential oils showed that eugenol (18.7%) and α-pinene (18.8%) were the major components of S. aromaticum and C. cyminum essential oils, respectively.

De Sousa et al. (2004) investigated the effects of kaempferol-3,7-O-(α)-dirhamnoside (kaempferitrin), a major flavonoid compound of the n-butanol fraction from Bauhinia forficata leaves, on serum glucose levels, as well as its antioxidant potential through In vivo and in vitro treatments. Oral administration of kaempferitrin led to a significant hypoglycemic effect in normal and in alloxan-induced diabetic rats. In normal rats, blood glucose was lowered only with the higher dose of kaempferitrin (200 mg/kg) at 1 h after treatment. However, the hypoglycemic effect of kaempferitrin in diabetic rats was evident at all doses tested (50, 100, and 200 mg/kg), and this profile was maintained throughout the period studied for both higher doses. Similarly, Tanaka et al. (2006) evaluated the anti-hyperglycemic effect of Aloe vera gel and its isolated compounds such as lophenol, 24-methyl-lophenol, 24-ethyl-lophenol, cycloartenol, and 24-methylene-cycloartenol from the gel. In vehicle-treated mice, statistically significant decreases of 15–18% in HbA1c levels were observed in treatment with 1 μg of the five phytosterols. Considering the ability to reduce blood glucose in vivo, there were no differences among the five phytosterols. After administration of the five phytosterols for 28 days, fasting blood glucose levels decreased to approximately 28–64% than control levels. These findings suggested that Aloe vera gel and its derived phytosterols have a long-term blood glucose level control effect. Likewise, Sharma et al. (2008) evaluated the hypoglycemic and hypolipidemic effects of flavonoid-rich extract obtained from the seeds of *Eugenia jambolana* in diabetic rats. Hypoglycemic activity was assessed by reduction in fasting blood glucose and peak blood glucose level within 60 min of glucose tolerance test in mild and severely diabetic rats. Different biochemical parameters like glycogen biosynthesis, glucose homeostatic enzyme (glucose-6-phosphatase, hexokinase) activities demonstrated significant improvement as compared to diabetic counterparts. Further, the flavonoids also stimulated a 16% increase in insulin release in vitro from pancreatic islets.

DHA exhibits its hypoglycemic effects by increasing insulin sensitivity and suppressing the blood glucose level. Moreover, the DHA in the insulin emulsion formulation effectively reduced the blood glucose level and DHA has the potential to become a delivery vehicle. Moreover, omega-3 fatty acids and vitamin E co-supplementation had beneficial effects on blood glucose level. Likewise, vitamin A and E supplementation effectively lowered the blood glucose level. The essential oils also showed higher antidiabetic activity than the diabetic drug. However, more studies are required to investigate and fully elucidate the mechanisms of various functional lipids.

Reduction of Bile Stone Formation

In the current state of affairs, people with gallstone disease are advised to follow a low fat, low cholesterol diet, or maybe both (Madden et al. 2017). Though the grounds emerge to be vague, a dietary intervention is essential for the treatment of this disease. Literature shows that there is no published data on the benefits of a low-fat diet as compared to a standard diet. However, with the greater than ever cases of obesity, there is adequate evidence that people with obesity, who are suggested to go after weight-reducing diets that include a very low-fat diet, are likely to develop gallstones, and that diets higher in fat may reduce gallstone risk in adults losing weight.

The basis for limiting or altering dietary fat in the treatment of gallstone disease has two putative mechanisms. Firstly, as dietary fat is a strong stimulator of gallbladder contraction, dietary fat may exacerbate post-prandial pain. Consequently, limiting dietary fat might reduce pain. Nevertheless, the gallbladder also contracts spontaneously, in reaction to an intake of mixed meals, protein, or cephalic stimulation. Additionally, if limiting dietary fat does direct to a decrease in gallbladder contractions and emptying, it may also enhance the possibility of gallstone deposition. The lithogenic bile would be held longer in the gallbladder, thus potentially aggravating the problems.

Secondly, giving up total dietary fat, and predominantly saturated fat, leads to a reduction in plasma cholesterol. Lower plasma cholesterol levels go along with a corresponding decrease in biliary cholesterol concentration. This would reduce the precipitation of cholesterol in the bile, and decrease the risk of cholesterol-rich gallstones emergence. The system is complex considering the fact that circulating cholesterol levels are subjective to endogenous cholesterol synthesis rather than the intake of dietary cholesterol. If this affords a rationale for the possible treatment of cholesterol-rich stones, it is improbable to be pertinent to the management of stones composed mainly of pigment.

Proof for the function of dietary intervention in the prime prevention of gallbladder stones in adults is presently under review (Stokes et al. 2014). Dietary recommendation to control or adjust fat intake, which is now promoted as a treatment for people with gallstones, does not materialize to be based on rationalized substantiation. While there are general health benefits linked with avoiding too much dietary fat, i.e. less risk of obesity and cardiovascular disease, exact benefits for the treatment of gallstone disease require explanation. At first, it is essential to find out if there are advantages from modifying fat intake, or harmful effects from reduced gallbladder emptying. Then, it would be helpful to measure the amount of fat cut-back required, so that customized advice could be given, particularly to the minority of patients with gallstone disease who are underweight and potentially at the threat of malnutrition.

Mineral Absorption

With all the low-fat and fat-free products on the market, it is assumed that reducing fat intake would be the ideal way to greater health. However, eating fat-free can lead to health problems. Specifically, it can lead to vitamin deficiencies. The first fat-soluble vitamin was discovered in 1909; it was vitamin A. Ten years later, vitamin D3 was identified; and in 1922, Evans and Bishop discovered vitamin E. The last of the fat-soluble vitamins, vitamin K1, was identified in the 1930s by Dam and Doisy, who achieved worldwide recognition of their studies on vitamins and received the Nobel Prize for these efforts. Vitamins do not constitute any source of energy for the human body, nor are they considered to be a structural material. But in cells and tissues, they have various regulatory functions that determine the development, physical efficiency, and health status of humans.

Vitamins require fat to be absorbed into the body and to perform their role, which includes providing energy, keeping cells functioning and supporting the immune system. Vitamins are essential, which means fats are essential too. Vitamin A is essential for good vision, vitamin D for bone health, K for blood clotting, and E for limiting the formation of harmful free radicals. Dietary fat, which comes from the food you eat, is crucial to the absorption of the fat-soluble vitamins, which includes vitamins A, D, K and E. The process calls for fat-digesting bile acids coming from the liver to stay in the small intestine. While the bile acids collapse the fat, the vitamins are dissolved in and travel with the fat through the intestinal wall, into the body. Ultimately they end up in the liver and in body fat, where they are stored.

Vitamin A

Vitamin A in foods is present in its ester form and, after enzyme-catalyzed hydrolysis, it is absorbed into the gastrointestinal tract. After forming an emulsion in the presence of bile salts and pancreatic juices, it is incorporated into molecules of chylomicrons. Other forms of carotenoids are oxidatively converted to retinal in the intestinal mucosa. Then, via lymphatic vessels, it is transported from the bloodstream to the liver. Vitamin A is primarily stored in the liver and is also found in ester form in lipocytes. Transport from the liver depends on the level of retinol binding proteins (RBP). Vitamin A is well absorbed from the alimentary tract—its absorption efficiency is assessed as 80%. Retinol levels should exceed 300 μg/L of blood plasma.

Vitamin A deficiency may manifest itself by decreased immunity to infections, growth inhibition, and reproduction disorders. Typical pathological signs of vitamin A avitaminosis include lesions of the mucous membranes of the gastrointestinal tract, air passages, and urinary system. Skin keratosis can be seen, especially around outlets of the hair sheaths (phrynoderma). Vitamin A can be toxic when ingested in overly large doses. The most common sign of excessive vitamin A intake is the so-called dermal paralysis manifested by changes in skin pigmentation and dryness, accompanied by nail brittleness and hair loss. Vitamin A excess may be the reason for joint and bone pain as well as weight loss. Enlargement of the liver and spleen usually occurs (hepatosplenomegaly). Teratogenic activity has also been suggested, notably birth defects. Symptoms of acute toxicity include a headache, nausea and vomiting, convulsions, and light intolerance.

Vitamin D

Vitamin D is considered a hormone-like compound. Ergosterol is a vitamin D_2 precursor, a phytosterol that occurs mainly in plants. Ergocalciferol results from UV irradiation of ergosterol. 7-Dehydrocholesterol can be converted into the most active form—vitamin D_3 (cholecalciferol). Ergosterol and 7-dehydrocholesterol are clearly identified as provitamin D because, after irradiation of the skin with UV light or exposure to sunlight, they can be converted into the active vitamin D_2 and D_3 forms.

Vitamins synthesized in the skin (D_3) and those supplied with food undergo the same transformations, finally leading to the generation of their active metabolites. They are deposited in the liver, lungs, and kidneys as well as in fatty tissue. The amount of vitamin D in the organism depends on exposure to sunshine and diet, and it ranges from 65 to 165 IU/100 mL blood. All forms of this vitamin are being absorbed in the small intestine with a bile salt, fatty acids, and acylglycerols present, and then delivered to the liver bound to a specific vitamin D-binding protein. From the liver, the vitamin is transported via chylomicrons to the bloodstream. Molecules of vitamin D_2 and D_3 are activated in the liver by hydroxylation; the OH group appears at carbon-25 to give 25-hydroxy-D_3[25-(OH)]D. Further reaction occurs in the kidneys, bone, cartilage, and/or in the placenta under the influence of the specific D_3–1-hydroxylase. The next two stages produce the most potent and physiologically active molecule, 1,25-dihydroxy vitamin D_3 [1,25-$(OH)_2D_3$]. According to the most recent nomenclature, it is called calcitriol. 25-(OH)-D is activated under the influence of the specific D3–24-hydroxylase at carbon-24 [24,25-(OH)2D3]. Both circulate in the bloodstream, and their main function as active metabolites is maintaining the homeostasis of calcium and phosphorus, which is vital for many body functions.

In human and animal organisms, vitamin D deficiency can result from its decreased synthesis in the skin or disturbances in its metabolism related to malabsorption in the alimentary tract. Certain liver and kidney disorders may result in excessive excretion of vitamin D in the urine and feces, thus leading to vitamin D deficiency. If calcium and phosphorus homeostasis is disturbed, bone changes quickly follow. Clinical studies reveal that vitamin D deficiency plays a role in tuberculosis, stroke, high blood pressure, and inflammatory bowel disease. Low vitamin D levels could be a risk factor for arteriosclerosis, rheumatoid arthritis, or osteoarthritis. Ingestion of excessive quantities of vitamin D may be toxic, and can result in bone and/or pain, diffuse demineralization of bones, muscle fatigue, loss of appetite, thirst, sore eyes, itchy skin, vomiting, diarrhea, urinary urgency, and abnormal calcium levels in the urine (indicative of kidney stones). High doses of vitamin D cause a build-up of calcium in soft tissues and in organs such as the liver, lungs, heart, kidneys, and muscles.

Vitamin E

Vitamin E represents a group of lipids with an isoprene structure and it is the only one possessing eight homologous forms. Vitamin E absorption requires the presence of fat; however, limited information exists on the influence of fat quantity on optimal absorption. In a study, the absorption of stable-isotope-labeled vitamin E following meals of varying fat content and source (Jeanes et al. 2004). These results demonstrate that both the amount of fat and the food matrix influence vitamin E absorption.

When cholesterol levels are in their normal state, they are balanced, typical and healthy. However, when cholesterol oxidizes, it becomes unsafe. Some isomers of vitamin E are protective antioxidants that battle with cholesterol oxidation. This happens as they can scrap free radical damage in the body, leading to cholesterol oxidation. Tocotrienol isomers of vitamin E have three double bonds that absolutely influence cardiovascular health due to their capability to diminish the activity of an enzyme (called HMG-CoA reductase) that regulates cholesterol synthesis. The isomers can also stop cell adhesion and hence slow down the process of atherosclerosis (thickening of the arteries). It is also important to note that synthetic vitamin E does not appear to have the same benefits as natural forms. Excessive alpha-tocopherol can, in fact, hinder with the cholesterol-lowering act of delta and gamma-tocotrienols, which are the two most bioactive tocotrienols and the types linked to cardioprotective activities.

Vitamin K

Vitamin K plays a vital role in the manufacture of proteins that mediate blood clotting. Vitamin K also supports healthy bones because it promotes the absorption and use of calcium, a mineral the body uses to build bone tissue. Vitamin K exists in two forms: K_1 is produced by plants and is the most common form in dietary sources, and K_2 is produced by beneficial bacteria in your intestines. In healthy people, the bacteria in the intestines produce enough vitamin K to support our needs, but many foods, such as vegetable oils and green leafy vegetables, contain vitamin K and contribute to the total level.

Vitamins A and D also cooperate to regulate the production of certain vitamin K-dependent proteins. Once vitamin K_2 activates these proteins, they help mineralize bones and teeth, protect arteries and other soft tissues from abnormal calcification, and protect against cell death. Magnesium, calcium, vitamin D_3, or vitamin K_2 also work in tandem. A water-soluble form of vitamin K can be converted to the active derivative, hydroquinone (Engelking 2010). The biologic half-life of vitamin K is shorter than the other fat-soluble vitamins. The liver is the main repository of vitamin K. Vitamin K is required for the hepatic post-synthetic transformation of several protein clotting factors. Vitamin K helps to facilitate the γ-carboxylation of glutamate residues, which in turn chelate Ca++. A vitamin K cycle exists in the endoplasmic reticulum of liver cells. An important therapeutic use of vitamin K is as an antidote in poisoning by dicoumarol or warfarin. Vitamin K plays a role in bone metabolism, as well as in the renal reabsorption of Ca++. Tissue distributions show that the vitamin actions are diversified.

Adults rarely suffer from a vitamin-K deficiency because internal bacteria produce vitamin K and many vegetables provide enough of it. However, a vitamin- K deficiency could impair blood clotting leading to nose bleeds, bleeding gums, blood in the urine and the formation of bruises (Malfait and De Paepe 2009). Doctors encourage patients on Coumadin, which is a blood-thinning medication, to maintain a consistent level of vitamin K (Innis 2008). Drastic changes in diet that significantly change the level of vitamin K could affect the medication's effectiveness.

Hypocholesterolemic Effects

Cholesterol and triglycerides are important fats in the blood. It is a vital constituent of cell membranes, brain and nerve cells, and bile, that assist the body to absorb fats and fat-soluble vitamins (Singh et al. 2015). Cholesterol is used by the body to make vitamin D and a range of hormones, like estrogen, testosterone, and cortisol. The body can make the entire cholesterol that it requires, however it also acquires cholesterol from food. Triglycerides contained in fat cells can be broken down, then used to provide energy for the body's metabolic processes, including growth. They are created in the intestine and liver from smaller fats called fatty acids. While some types of fatty acids are prepared by the body, others must be synthesized from food. Fats, such as cholesterol and triglycerides, cannot flow without restraint in the blood, because blood is mostly water. For free circulation in blood, cholesterol and triglycerides are put together with proteins and other materials to form particles called lipoproteins.

There are diverse types of lipoproteins. Each type has a purpose and is broken down and excreted in a faintly different way. Lipoproteins include chylomicrons, very low-density lipoproteins (VLDL), low-density lipoproteins (LDL), and high-density lipoproteins (HDL). Cholesterol elated by LDL is called LDL cholesterol, and cholesterol circulated by HDL is called HDL cholesterol. The body can control lipoprotein levels by adjusting the production rate of lipoproteins. The body can also normalize how swiftly lipoproteins penetrate and leave the bloodstream. From one measurement to the next, cholesterol levels can vary by about 10%, and triglyceride levels can vary by up to 25%. Lipid levels may be

- Too high (Dyslipidemia)
- Too low (Hypolipidemia)

Abnormal lipid levels may be because of the alterations that occur with aging, diverse disorders (including inherited ones), use of certain drugs, or lifestyle (following a diet high in saturated fat, being physically inactive, or being overweight). Abnormal levels of cholesterol can direct to long-term problems, such as atherosclerosis. In general, a high total cholesterol level (which includes LDL, HDL, and VLDL cholesterol), particularly a high level of LDL (the "bad") cholesterol, increases the risk of atherosclerosis and thus the risk of heart attack or stroke. However, not all types of cholesterol add to this risk. High levels of HDL (the "good") cholesterol may reduce the threat, and on the other hand, a low level of HDL cholesterol may add to risk. Hypocholesterolemia is the occurrence of unusually low levels of cholesterol in the blood. A deficiency in the body's manufacture of cholesterol can direct to unfavorable consequences. Cholesterol is a necessary constituent of mammalian cell membranes and is necessary to set up proper membrane permeability and fluidity. It is not clear if a lower than average cholesterol level is directly detrimental; it is frequently encountered in particular illnesses.

References

Alabdulkarim, B., Bakeet, Z. A. N., & Arzoo, S. (2012). Role of some functional lipids in preventing diseases and promoting health. *Journal of King Saud University-Science, 24*, 319–329.

Amir, H., Karas, M., Giat, J., Danilenko, M., Levy, R., & Yermiahu, T. (1998). Lycopene and 1,25-dihydroxyvitamin D3 cooperate in the inhibition of cell cycle progression and induction of differentiation in HL-60 leukemic cells. *Nutrition and Cancer, 33*, 105–112.

Awad, A. B., Fink, C. S., Williams, H., & Kim, U. (2002). In vitro and in vivo (SCID mice) effects of phytosterols on the growth and dissemination of human prostate cancer PC-3 cells. *European Journal of Cancer Prevention, 10*(6), 507–513.

Benavente-García, O., & Castillo, J. (2008). Dietary flavonoids as cancer prevention agents. Update on uses and properties of citrus flavonoids: New findings in anticancer, cardiovascular, and anti-inflammatory activity. *Journal of Agricultural and Food Chemistry, 56*(15), 6185–6205.

Boukhris, M., Bouaziz, M., Feki, I., Jemai, H., Feki, A. E., & Sami, S. (2012). Hypoglycemic and antioxidant effects of leaf essential oil of Pelargonium graveolens L'Hér. In alloxan induced diabetic rats. *Lipids in Health and Disease, 11*, 81.

De Sousa, E., Zanatta, L., Seifriz, I., Creczynski-Pasa, T. B., Pizzolatti, M. G., Szpoganicz, B., & Silva, F. R. M. B. (2004). Hypoglycemic effect and antioxidant potential of kaempferol-3,7-O-(α)-dirhamnoside from Bauhinia forficata leaves. *Journal of Natural Products, 67*(5), 829–832.

Dillard, C. J., & German, J. B. (2000). Phytochemicals: Nutraceuticals and human health. *Journal of the Science of Food and Agriculture, 80*, 1744–1756.
Engelking, L. R. (2010). *Textbook of veterinary physiological chemistry (updated)* (2nd ed.). Amsterdam, Netherlands: Academic Press.
Gunther, S., Patterson, R. E., Kristal, A. R., Stratton, K. L., & White, E. (2004). Demographic and health-related correlates of herbal and specialty supplement use. *Journal of the American Dietetic Association, 104*, 27–34.
Innis, M. D. (2008). Vitamin K deficiency disease. *Journal of Orthomolecular Medicine, 23*(1), 15–20.
Institute of Medicine. (2002). Dietary reference intakes for energy, carbohydrate, fibre, fat, fatty acids, cholesterol, protein and amino acids. *Journal of the American Dietetic Association, 102*(11), 1621–1630.
Jeanes, Y. M., Hall, W. L., Ellard, S., Lee, E., & Lodge, J. K. (2004). The absorption of vitamin E is influenced by the amount of fat in a meal and the food matrix. *British Journal of Nutrition, 92*(4), 575–579.
Kangralkar, V. A., Gavimath, C. C., Jadhav, N. A., & Burli, S. C. (2010). Potential hypoglycemic effect of essential oil of citrus reticulata in wistar rats. *International Journal of Applied Pharmaceutics, 1*, 6–9.
Kaur, G., Mukundan, S., Wani, V., & Kumar, M. S. (2015). Nutraceuticals in the management and prevention of metabolic syndrome. *Austin Journal of Pharmacology and Therapeutics, 3*, 1–6.
Li, P., Zhang, L., Tian, X., & Xing, J. (2014). Docosahexaenoic acid has an anti-diabetic effect in streptozotocin-induced diabetic mice. *International Journal of Clinical and Experimental Medicine, 7*(9), 3021–3029.
Madden, A., Trivedi, D., Smeeton, N., & Culkin, A. (2017). Modified dietary fat intake for treatment of gallstone disease. *Cochrane Database of Systematic Reviews, 2017*(3).
Malfait, F., & De Paepe, A. (2009). Bleeding in the heritable connective tissue disorders: Mechanisms, diagnosis and treatment. *Blood Reviews, 23*(5), 191–197.
Morishita, M., Kajita, M., Suzuki, A., Takayama, K., Chiba, Y., Tokiwa, S., & Nagai, T. (2000). The dose-related hypoglycemic effects of insulin emulsions incorporating highly purified EPA and DHA. *International Journal of Pharmaceutics, 201*, 175–185.
Nishikawa, Y., Wang, Z., Kerns, J., Wilcox, C. S., & Carr, B. I. (1999). Inhibition of hepatoma cell growth in vitro by arylating and non-arylating K vitamin analogs. Significance of protein tyrosine phosphatase inhibition. *Journal of Biological Chemistry, 274*, 34803–34810.
Noto, V., Taper, H. S., Jiang, Y. H., Janssens. J., Bonte. J., & De Loecker, W. (1989). Effects of sodium ascorbate (vitamin C) and 2-methyl-1,4-naphthoquinone (vitamin K3) treatment on human tumor cell growth in vitro. I. Synergism of combined vitamin C and K3 action. *Cancer, 63*, 901–906.
Palakurthi, S. S., Flückiger, R., Aktas, H., Changolkar, A. K., Shahsafaei, A., Harneit, S., … Halperin, J. A. (2000). Inhibition of translation initiation mediates the anticancer effect of the n-3 polyunsaturated fatty acid eicosapentaenoic acid. *Cancer Research, 60*, 2919–2925.
Pandalai, P. K., Pilat, M. J., Yamazaki, K., Naik, H., & Pienta, K. J. (1996). The effects of omega-3 and omega-6 fatty acids on in vitro prostate cancer growth. *Anticancer Research, 16*(2), 815–820.
Pirronen, V., Lindsay, D. G., Miettinen, T. A., Tovio, J., & Lampi, A. M. (2000). Review plant sterols: Biosynthesis, biological function and their importance to human nutrition. *Journal of the Science of Food and Agriculture, 80*, 939–966.
Sharma, B., Balomajumder, C., & Roy, P. (2008). Hypoglycemic and hypolipidemic effects of flavonoid rich extract from Eugenia jambolana seeds on streptozotocin induced diabetic rats. *Food Chemistry, 46*, 2376–2383.
Shimura, T., Miura, T., Usami, M., Ishihara, E., Tanigawa, K., Ishida, H., & Seino, Y. (1997). Docosahexanoic acid (DHA) improved glucose and lipid metabolism in KK-Ay mice with genetic non-insulin-dependent diabetes mellitus (NIDDM). *Biological & Pharmaceutical Bulletin, 20*(5), 507–510.

Simopoulos, P. A. (2008). The importance of the omega-6/omega-3 fatty acid ratio in cardiovascular disease and other chronic diseases. *Experimental Biology and Medicine, 233*(6), 674–688.

Singh, C. V., Verma, R. C. (2016). Effect of oral administration of vitamin a on blood glucose level in rabbits and its possible interactions with commonly used oral antidiabetic agents. *International Journal of Pharmaceutical science and Research, 7*, 2684–2691.

Singh, T. P., Malik, R. K., Katkamwar, S. G., & Kaur, G. (2015). Hypocholesterolemic effects of Lactobacillus reuteri LR6 in rats fed on high-cholesterol diet. *International Journal of Food Sciences and Nutrition, 66*(1), 71–75.

Stokes, C. S., Gluud, L. L., Casper, M., & Lammert, F. (2014). Ursodeoxycholic acid and diets higher in fat prevent gallbladder stones during weight loss: A meta-analysis of randomized controlled trials. *Clinical Gastroenterology and Hepatology, 12*(7), 1090–1100. e1092.

Suksomboon, N., Poolsup, N., & Sinprasert, S. (2011). Effects of vitamin E supplementation on glycaemic control in type 2 diabetes: Systematic review of randomized controlled trials. *Journal of Clinical Pharmacy and Therapeutics, 36*(1), 53–63.

Taghizadeh, M., Jamilian, M., Mazloomi, M., Sanami, M., Asemi, Z. (2016). A randomized-controlled clinical trial investigating the effect of omega-3 fatty acids and vitamin E co-supplementation on markers of insulin metabolism and lipid profiles in gestational diabetes. *Journal of Clinical Lipidology, 10*, 386–393.

Tahir, H. U., Sarfraz, R. A., Ashraf, A., & Adil, S. (2016). Chemical composition and antidiabetic activity of essential oils obtained from two spices (Syzygium aromaticum and Cuminum cyminum). *International Journal of Food Properties, 19*, 2156–2164.

Tanaka, M., Misawa, E., Ito, Y., Habara, N., Nomaguchi, K., Yamada, M., … Higuchi, R. (2006). Identification of five phytosterols from Aloe vera gel as anti-diabetic compounds. *Biological & Pharmaceutical Bulletin, 29*(7), 1418–1422.

Udupa, A., Nahar, P., Shah, S., Kshirsagar, M., & Ghongane, B. (2013). A comparative study of effects of omega-3 fatty acids, alpha lipoic acid and vitamin E in type 2 diabetes mellitus. *Annals of Medical and Health Sciences Research, 3*(3), 442–446.

Wang, X., Hang, Y., Liu, J., Hou, Y., Wang, N., & Wang, M. (2017). Anticancer effect of curcumin inhibits cell growth through mir-21/pten/akt pathway in breast cancer cell. *Oncology Letters, 13*, 4825–4831.

Winstein, S. J., Wright, M. E., Lawson, K. A., Snyder, K., Männisto, S., Taylor, P. R., Virtamo, J., Albanes, D. (2007) Serum and dietary vitamin E in relation to prostate cancer risk. *Cancer Epidemiology Biomarkers and Prevention, 16*, 1253–1259.

Wu, F. Y., Liao, W. C., & Chang, H. M. (1993). Comparison of antitumor activity of vitamins K1, K2 and K3 on human tumor cells by two (MTT and SRB) cell viability assays. *Life Sciences, 52*, 1797–1804.

Yu, H., & Huang, G. (2010). Enhanced in vitro anti-cancer activity of curcumin encapsulated in Hydrophobically modified starch. *Food Chemistry, 119*, 669–674.

Yu, N. H., Yin, J. J., & Shen, S. R. (2004). Growth inhibition of prostate cancer cells by epigallocatechin in the presence of Cu2. *Journal of Agricultural and Food Chemistry, 52*, 462–466.

Zhang, M., Binns, C. V., & Lee, A. H. (2002). Tea consumption and ovarian cancer risk: A case-control study in China. *Cancer Epidemiology, Biomarkers & Prevention, 11*, 713–718.

Index

A. Gani, B. A. Ashwar (eds.), *Food biopolymers: Structural, functional and nutraceutical properties*, https://doi.org/10.1007/978-3-030-27061-2